Regional Assessment of Global Change Impacts

Wolfram Mauser • Monika Prasch
Editors

Regional Assessment of Global Change Impacts

The Project GLOWA-Danube

Editors
Wolfram Mauser
Department of Geography
Ludwig-Maximilians-Universität München
(LMU Munich)
Munich, Germany

Monika Prasch
Department of Geography
Ludwig-Maximilians-Universität München
(LMU Munich)
Munich, Germany

ISBN 978-3-319-36640-1 ISBN 978-3-319-16751-0 (eBook)
DOI 10.1007/978-3-319-16751-0

Springer Cham Heidelberg New York Dordrecht London
Translated from German
Softcover re-print of the Hardcover 1st edition 2016
The Work was first published in 2006 by GLOWA-Danube-Projekt (Ludwig-Maximilians-Universität München, Department für Geographie, Lehrstuhl für Geographie und geographische Fernerkundung, Luisenstraße 37, 80333 München, Germany) with the following title: 'Global Change Atlas, Einzugsgebiet Obere Donau', ISBN: 978-3-00-026548-8. The Work was also published in electronic version, available at: http://www.glowa-danube.de/atlas/index.php.

Cover illustration: © ASTER GDEM is a product of METI and NASA. ERSDAC (EARTH REMOTE SENSING DATA ANALYSIS CENTER) (2009): The Ministry of Economy, Trade, and Industry (METI) of Japan and the United States National Aeronautics and Space Administration (NASA): Aster Global Digital Elevation Model (GDEM), available from http://gdem.ersdac.jspacesystems.or.jp/ (22.9.2014).

Printed on acid-free paper

Springer International Publishing AG Switzerland is part of Springer Science+Business Media (www.springer.com)

Preface

The onset of Future Earth marks a new era in global change research and acknowledges that only integrative and transdisciplinary research will allow science to contribute adequately to solving the challenges of global change that humanity is facing in the coming decades (Mauser et al. 2013). Future Earth will heavily rely on codesign through science and society of research and knowledge production, which will allow societies to both form a sustainable future environment and at the same time to adapt to changing global conditions, be it in the field of climate, economy, or environment. Adaptation to these changes will mostly take place on a regional level. Different adaptation options as well as alternative development paths will have to be formulated as scenarios. Their consequences and trade-offs will have to be carefully analyzed and explored in order to identify the most effective, efficient, and appropriate adaptation decisions. This requires a structured dialogue between scientists and regional stakeholders. This FutureEarth mode of operation marks a considerable paradigm shift towards a maturing global change science, which understands itself as science for society.

The new transdisciplinary role of global change science evolved from a number of regional global change projects, which have been carried out during the last decade in order to develop and test a suit of methodologies to integrate the scientific disciplines from natural sciences and humanities into a new whole. The new whole becomes more than its parts and for the first time allows to explore and understand the full range of processes and interactions, which led to rapidly changing environments and societies and which will have to be instrumentalized by societies to develop towards sustainability. At the same time these projects for the first time in the history of global change research motivated their participants to leave their scientific comfort zones and to strongly engage in exploring new, participatory ways to communicate with society and its administrative, economic and political governance structures.

GLOWA-Danube, which started in 2001 and finished in 2011, was among the first of these new projects. It was launched by the German Ministry for Education and Research (BMBF) as part of the Global Change of the Water Cycle (GLOWA) initiative. GLOWA for the first time aimed at systematically exploring integrative and transdisciplinary scientific approaches to identify decision alternations for regional adaptation to global change. GLOWA purposefully chose water issues for this new kind of projects because water is among the most important, most universal and at the same time most vulnerable natural resources. There is hardly a sector of society which is not affected by change in the water cycle, and water availability and water use will strongly be affected regionally by climate change as well as by changing water demands by society, industry and agriculture.

GLOWA-Danube followed the ambitious goal to for the first time explore in full the interactions between nature and society in the context of global change (climate, demography, economy , etc.) by looking at regional water resources and their management. We chose the Upper Danube basin and its inhabitants as the natural laboratory for the GLOWA-Danube project. It covers the full Danube and its tributaries until it leaves Germany in Passau, has an area of 80,000 km^2 and is the home of 12 Mio. inhabitants, which enjoy one of the largest per capita GDP on the globe. The Upper Danube challenges any simulation of present and future interaction of nature and society because it is spatially, politically, and administratively heterogeneous and complex including the Alps and their forelands as well as five countries and six states. Its water resources are sensitive to climate change because slight temperature changes have large impacts mainly in the Alps.

Integrative, transdisciplinary research in GLOWA-Danube was based on two approaches:

- Firstly to develop and use a new simulation tool, which integrates the relevant natural as well as societal and technical components and their interaction to represent and simulate a man-made water cycle and the related water uses as they are typically found in rich, densely populated watersheds. For this purpose, it couples the latest dynamic process models in hydrology, plant science, snow science, and glaciology with spatially distributed actor models, which simulate the behavior and decisions of human actors like farmers, households, water suppliers, and tourist facilities in an open parallel simulation model. This enables to study in detail the interaction between nature and humans and allows, on the basis of an understanding of actors, choices to better simulate the consequences of today's decisions on future environmental and societal conditions.
- Secondly to explore the transdisciplinary codesign of research and co-creation of knowledge through a close communication process of project scientists and regional stakeholders. It consists of a spiral of steps which include the formulation of scenarios of desirable futures based on the existing knowledge of likely climate change and other influencing factors, simulation of the transient evolution of environment and society in the Upper Danube watershed according to these scenarios, and translation of the simulation results into relevant information, which can be communicated with the stakeholders and which can be used to prepare decisions to adapt to global change. For this latter purpose, the Global Change Atlas of the Upper Danube was developed as a living document, which evolved during the course of GLOWA-Danube.

The GLOWA-Danube Atlas was purposefully published in German, the language of the stakeholders, and is the common, easy to understand by practitioners and stakeholders scientific knowledge base for the dialogue among stakeholders and scientists. It documents purpose, philosophy, architecture, methodologies, scenarios, and results of the project. After 2 years of consolidation, refinement, and further communication with the stakeholders, we decided to translate the Global Change Atlas of the Upper Danube into English and offer it to the growing Future Earth research community as one possible blueprint for successful future global change science for society in their respective regions.

Munich, Germany

Wolfram Mauser
Monika Prasch

Reference

Mauser W, Klepper G, Rice M, Schmalzbauer BS, Hackmann H, Leemans R, Moore H (2013) Transdisciplinary global change research: the co-creation of knowledge for sustainability. Curr Opin Environ Sustain 5:420–431, doi:10.1016/j.cosust.2013.07.001

Acknowledgements

The editors thank the following ministries and the LMU for funding the project GLOWA-Danube: German Federal Ministry of Education and Research, Heinemannstraße 2, 53175 Bonn, Germany, www.bmbf.de; German Federal Ministry for the Environment, Nature Conservation, Building and Nuclear Safety, Robert-Schumann-Platz 3, 53175 Bonn, Germany, www.bmu.de; Bavarian State Ministry of Education, Science and the Arts, Salvatorstraße 2, 80333 Munich, Germany, www.km.bayern.de; Ludwig-Maximilians-Universität München (LMU), Geschwister-Scholl-Platz 1, 80539 Munich, Germany, www.uni-muenchen.de; Ministry of Science, Research and the Art of Baden-Württemberg, Königstraße 46, 70173 Stuttgart, Germany, www.mwk-baden-wuerttemberg.de

The provision of data and photos by the referenced institutions and persons is gratefully acknowledged.

Additionally, we want to thank the authors of this book for their cooperation after the project GLOWA-Danube was finished. The graphic design by Vera Falck and the support of Andrea Reiter, Ruth Weidinger, and the student research assistants Leonie Keil, Stephanie Lumnitz, and Magdalena Mittermeier, all from the Department of Geography of the Ludwig-Maximilians-Universität München (LMU), are gratefully acknowledged.

Contents

Contributors

Josef Apfelbeck Production Theory and Resource Economics, Universität Hohenheim, Stuttgart, Germany

Heike Bach Vista Geowissenschaftliche Fernerkundung GmbH, Weßling, Germany

Roland Barthel Department of Earth Sciences, University of Gothenburg, Göteborg, Sweden

Jörg Bendix Department of Geography, University of Marburg, Marburg, Germany

Anja Berghammer Department of Geography, Ludwig-Maximilians-Universität München (LMU Munich), Munich, Germany

Jürgen Braun Institute for Modelling Hydraulic and Environmental Systems/ VEGAS, University of Stuttgart, Stuttgart, Germany

Hannah Büttner IFOK GmbH, Bensheim, Germany

Jan Cermak Department of Geography, University of Bochum (RUB), Bochum, Germany

Anja Colgan GB Dr. Schönwolf GmbH & Co. KG, Augsburg, Germany

Stephan Dabbert Production Theory and Resource Economics, Universität Hohenheim, Stuttgart, Germany

Alexander Dingeldey BWL Reiseverkehrsmanagement, Duale Hochschule Baden-Württemberg, Ravensburg, Germany

Matthias Egerer Bavarian Ministry of Economic Affairs and Media, Energy and Technology, Munich, Germany

Josef Egger Meteorological Institute Munich, Ludwig-Maximilians Universtiät München (LMU), Munich, Germany

Michael Elbers Micromata GmbH, Kassel, Germany

Andreas Ernst Center for Environmental Systems Research (CESR), University of Kassel, Kassel, Germany

Peter Fiener Institut für Geographie, Universität Augsburg, Augsburg, Germany

Barbara Früh Deutscher Wetterdienst DWD, Offenbach, Germany

Wilfried Hagg Department of Earth and Environmental Sciences, Università degli Studi di Milano-Bicocca, Milan, Italy

Tobias Hank Department of Geography, Ludwig-Maximilians-Universität München (LMU Munich), Munich, Germany

Christoph Heinzeller Department of Geography, Ludwig-Maximilians-Universität München (LMU Munich), Munich, Germany

Rolf Hennicker Institute for Informatics, Ludwig-Maximilians-Universität München (LMU Munich), Munich, Germany

Thomas Hörhan Austrian Federal Ministry of Agriculture, Forestry, Environment and Water Management, Vienna, Austria

Marco Huigen Production Theory and Resource Economics, Universität Hohenheim, Stuttgart, Germany

Daniela Jacob Climate Service Center, Hamburg, Germany

Stefan Janisch Institute for Informatics, Ludwig-Maximilians-Universität München (LMU Munich), Munich, Germany

Christoph Jeßberger Economics and Transportation, Sustainability and Renewable Energies, Bauhaus Luftfahrt e.V, Ottobrunn, Germany

Georg Kasper Hydrology, Applied Water Resources Management and Geoinformatics IAWG, Ottobrunn, Germany

Christian W. Klar Forschungszentrum Jülich, Jülich, Germany

Daniel Klemm wusoma GmbH, Munich, Germany

Peter Klotz Hydrology, Applied Water Resources Management and Geoinformatics IAWG, Ottobrunn, Germany

Franziska Koch Department of Geography, Ludwig-Maximilians-Universität München (LMU Munich), Munich, Germany

Andreas Kraus Institute for Informatics, Ludwig-Maximilians-Universität München (LMU Munich), Munich, Germany

Tatjana Krimly Production Theory and Resource Economics, Universität Hohenheim, Stuttgart, Germany

Michael Kuhn Institute of Meteorology and Geophysics, University of Innsbruck, Innsbruck, Austria

Silke Kuhn Center for Environmental Systems Research (CESR), University of Kassel, Kassel, Germany

Astrid Lambrecht Commission for Geodesy and Glaciology of the Bavarian Academy of Sciences and Humanities, Munich, Germany

Erich Langmantel Center for Energy, Climate and Exhaustible Resources, Ifo Institute – Leibniz Institute for Economic Research at the University of Munich, Munich, Germany

Anita Leinberger Albert-Ludwigs-Universtity Freiburg, Freiburg, Germany

Victoria I.S. Lenz-Wiedemann Institute of Geography, University of Cologne, Cologne, Germany

Matthias Ludwig Institute for Informatics, Ludwig-Maximilians-Universität München (LMU Munich), Munich, Germany

Ralf Ludwig Department of Geography, Ludwig-Maximilians-Universität München (LMU Munich), Munich, Germany

Thomas Marke Institute of Geography, University of Innsbruck, Innsbruck, Austria

Wolfram Mauser Department of Geography, Ludwig-Maximilians-Universität München (LMU Munich), Munich, Germany

Alejandro Meleg Consultant, Bogotá, Colombia

Markus Muerth Department of Geography, Ludwig-Maximilians-Universität München (LMU Munich), Munich, Germany

Thomas Nauss Department of Geography, University of Marburg, Marburg, Germany

Darla Nickel Deutsches Institut für Urbanistik (Difu), Berlin, Germany

Andreas Pfeiffer Institute of Atmospheric Physics, DLR, Oberpfaffenhofen, Germany

Monika Prasch Department of Geography, Ludwig-Maximilians-Universität München (LMU Munich), Munich, Germany

Swantje Preuschmann Climate Service Center, Hamburg, Germany

Markus Probeck GAF AG, Munich, Germany

Tim G. Reichenau Institute of Geography, University of Cologne, Cologne, Germany

Andrea Reiter Bavarian Research Alliance GmbH, Munich, Germany

Christoph Reudenbach Department of Geography, University of Marburg, Marburg, Germany

Fredy Alexander Peña Reyes Department of Earth and Environmental Sciences, University of Milano-Bicocca, Milan, Italy

Vlad Rojanschi Golder Associates Ltd., Calgary, AB, Canada

Thorben Römer Formerly, Institute of Hydraulic Engineering, University of Stuttgart, Germany

Mario Sax Wilhelm-Hauff-Str. 51, 84036 Landshut, Germany

Janus W. Schipper Süddeutsches Klimabüro, Karlsruher Institut für Technologie, Karlsruhe, Germany

Jürgen Schmude Department of Geography, Ludwig-Maximilians-Universität München (LMU Munich), Munich, Germany

Karl Schneider Institute of Geography, University of Cologne, Cologne, Germany

Carsten Schulz Center for Environmental Systems Research (CESR), University of Kassel, Kassel, Germany

Nina Schwarz Department of Computational Landscape Ecology, Helmholtz-Centre for Environmental Research – UFZ, Leipzig, Germany

Roman Seidl Institute for Environmental Decisions, ETH Zürich, Zürich, Switzerland

Judith Stagl Potsdam Institute for Climate Impact Research, Potsdam, Germany

Sara Stöber Concentris Research Management GmbH, Fürstenfeldbruck, Germany

Marta Stoll DeUkRus Translations, Marta Stoll, Ulm

Konstantin Stricker Hydrology, Applied Water Resources Management and Geoinformatics IAWG, Ottobrunn, Germany

Boris Thies Department of Geography, University of Marburg, Marburg, Germany

Alexandar Trifkovic Fichtner GmbH & Co, Stuttgart, Germany

Jan van Heyden Formerly, Institute of Hydraulic Engineering, University of Stuttgart, Germany

Markus Weber Commission for Geodesy and Glaciology of the Bavarian Academy of Sciences and Humanities, Munich, Germany

Ruth Weidinger Department of Geography, Ludwig-Maximilians-Universität München (LMU Munich), Munich, Germany

Winfried Willems Hydrology, Applied Water Resources Management and Geoinformatics IAWG, Ottobrunn, Germany

Alexander Wirsig MBW Marketing- und Absatzförderungsgesellschaft für Agrar- und Forstprodukte aus Baden-Württemberg mbH, Stuttgart, Germany

Volkmar Wirth Institute for Atmospheric Physics, Johannes Gutenberg University Mainz, Mainz, Germany

Jens Wolf Repository Safety Research Division, Gesellschaft für Anlagen- und Reaktorsicherheit (GRS) gGmbH , Braunschweig, Germany

Florian Zabel Department of Geography, Ludwig-Maximilians-Universität München (LMU Munich), Munich, Germany

Günther Zängl Deutscher Wetterdienst (DWD), Offenbach, Germany

Marcelo Zárate Production Theory and Resource Economics, Universität Hohenheim, Stuttgart, Germany

Ralf Ziller Institute of Hydraulic Engineering, University of Stuttgart, Stuttgart, Germany

Markus Zimmer Center for Energy, Climate and Exhaustible Resources, Ifo Institute – Leibniz Institute for Economic Research, University of Munich, Munich, Germany

Astrid Zimmermann Hydrology, Applied Water Resources Management and Geoinformatics IAWG, Ottobrunn, Germany

Part I
Introduction

Chapter 1
GLOWA-Danube

Wolfram Mauser, Monika Prasch, Ruth Weidinger, and Sara Stöber

Abstract The research project GLOWA-Danube, which was one of the first transdisciplinary research networks in the field global change research in Germany, is introduced. GLOWA-Danube between 2001 and 2010 developed integrative tools and scenarios for the sustainable use of water in the Upper Danube basin under conditions of global change. Research was conducted by a large university-based network of scientists from natural and socio-economic fields. The paper introduces the project overall philosophy, its aims, design, new methodological approaches to developing simulation tools, scenarios and stakeholder dialogue and the scope and architecture of the Global Change Atlas – the Upper Danube, the primary communication tool between science and stakeholders.

Keywords GLOWA-Danube • Global change • Hydrology • Transdisciplinary science • Stakeholder dialogue

1.1 The GLOWA-Danube Project

Global change is the term used to describe the sum of all changes in the living conditions for people across the world that result from increasingly intense impacts on the natural environment; these changes visibly alter the conditions under which people will live in the future. Key factors in this context include climate change, increasing internationalisation of the economy and globalisation of trade, changes in land use, demographic shifts, population growth, mobility, water pollution and the increasing use of renewable and non-renewable natural resources. Global change is accelerating, and its consequences are becoming more and more apparent; hence, it is also increasingly important to initiate and manage timely measures both to

W. Mauser (✉) • M. Prasch • R. Weidinger
Department of Geography, Ludwig-Maximilians-Universität München (LMU Munich), Munich, Germany
e-mail: w.mauser@lmu.de; m.prasch@lmu.de; R.Weidinger@lmu.de

S. Stöber
Concentris Research Management GmbH, Fürstenfeldbruck, Germany
e-mail: sara.stoeber@concentris.de

W. Mauser, M. Prasch (eds.), *Regional Assessment of Global Change Impacts*, DOI 10.1007/978-3-319-16751-0_1

avoid and mitigate the effects and to adapt to the unavoidable consequences along the way (Map 1.1).

The Earth's surface is more than 70 % covered by water. Freshwater amounts to only 2.53 % of the total water resources of the globe. People and ecosystems can use only a portion of this freshwater, since almost two-thirds of it is bound up in glaciers and permanent snow cover. Water is a subsistence product for both people and for nature. It is an element of all economic, cultural, social and ecological aspects of life. The hydrological cycle over the continents of the Earth, together with the biosphere, fills key roles within ecosystems, including provisioning of water through evaporation for food production, the purification of water and ecosystems, transport of natural and manmade waste and the processing of water for industrial and drinking purposes. It is thus the basis for the efficient cycling of materials and for a stable environment.

As a result of the expected impact of climate change and the looming change to the structure of population and industry, the availability, quality, demand for and allocation of water will experience massive changes both globally and regionally over the coming decades. It is expected that existing conflicts over the use of water will intensify, and new, as yet unseen, conflicts will arise.

Global change also has regional effects. For this reason, the regional conditions must be considered from natural, cultural, economic and political perspectives, especially in view of the measures to be taken to adapt to global change.

Complex administrative and political framework conditions (e.g. the EU Water Framework Directive or the EU Agricultural Reform with the Agenda 2007) should ensure that in the future the use of water by the various sectors is sustainably managed. These conditions must be implemented on a regional basis in the affected river basins, such that the overarching goal is a sustainable overall development under changing natural and socio-economic global conditions. This goal requires a concerted effort towards environmental management that encompasses all disciplines and uses. Conflicts between contradictory uses of water and water claims must be resolved in order to ensure the viability of natural ecosystems. Modern environmental management that focuses on sustainability can only meet the challenges of these tasks at the level of a river basin if the measures to be implemented are backed by scientific evidence from comprehensive analyses of the consequences and trade-offs of the different development scenarios and if there is wide-scale acceptance and support by the parties involved. This requires transparent tools embraced by the stakeholders within their decision support system.

1.2 GLOWA: Early Transdisciplinary Research for Global Change

The project consortium GLOWA (Globaler Wandel des Wasserkreislaufs = Global Change and the Hydrological Cycle, www.glowa.org) was initiated by the German Federal Ministry of Education and Research (Bundesministerium für Bildung und

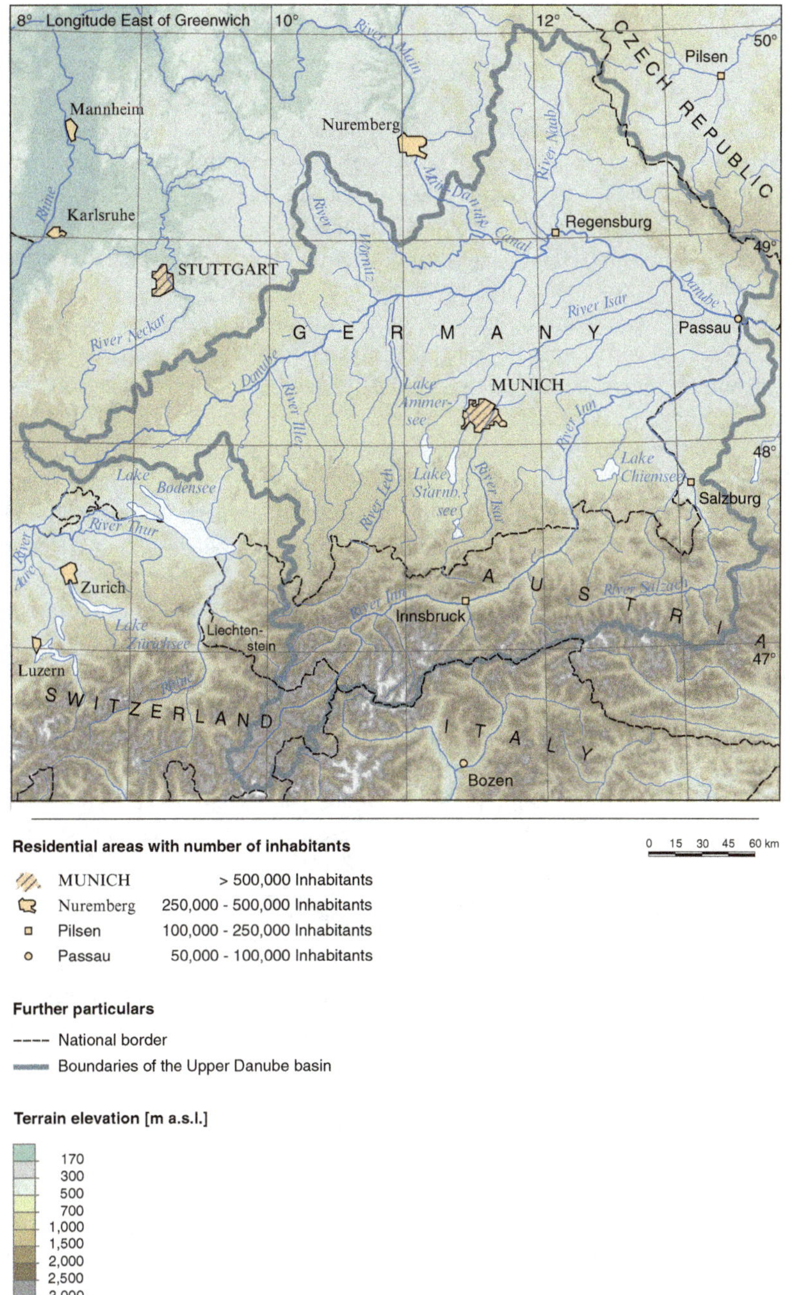

Map 1.1 Overview map (Data sources: Jarvis et al. 2008, Bayerisches Landesamt für Wasserwirtschaft 1983; ADAC 2003; DANUBIA - River network)

Forschung = BMBF) in 2001 with the goal of investigating the regional effects of global change on water resources using selected sample river basins, developing the tools that were lacking for a sustainable environmental management at the regional level of the average river basin as well as exploring options and trade-offs of a future sustainable and integrated management of drainage basins under conditions of global change. At the same time GLOWA for the first time in German global change research was designed to explore ways to ensure the societal relevance of the research. This marked the beginning in Germany of the exploration of efficient and effective ways of transdisciplinary communication between science and society, an approach which was now adopted by the Future Earth, the predecessor of the large global change research programmes of the twentieth century (Mauser et al. 2013).

GLOWA emerged as a research agenda in the late 1990s. At that time the central role of human emissions as causes of the observed climate change was proven through sophisticated fingerprinting. As an immediate consequence the exploration of ways to mitigate climate change was complemented by the understanding that societies need to adapt to the unavoidable consequences of climate change. Adaptation to climate change became the first regional environmental activity which was perceived as embedded in the context of a globe which was experienced to be changing in most of its aspects with ever-increasing speed.

The complexity and essential relevance of this issue and its enormous potential impact on the further development of societies strongly suggested that science, although endowed with the privilege to create knowledge for society but also strictly dissected into disciplines, in itself may not be able to identify the right questions to approach the issue and thereby may not find the answers society needs as orientation on its way into the future. A transdisciplinary process, beyond disciplines, even beyond science is needed, which allows science and society to codesign research relevant for society and to cocreate applicable knowledge on how to best adapt to upcoming future changes.

During the whole project duration from January 2001 to September 2010, GLOWA-Danube (www.glowa-danube.de) has been using the Danube as a natural and societal laboratory to develop and apply a new and integrated global change decision training system we call DANUBIA. It was used to explore successful ways to communicate with a network of regional stakeholders. At the start of the project, experiences were missing, though, on how to best cocreate the necessary knowledge to adapt to (climate) change. At the beginning of the project, transdisciplinarity was an exploratory experiment for both science and society. How to identify the right issues to be discussed between science and society, the right people (who then became stakeholders) and the most suitable way to communicate was the issue of theoretical and empirical analysis.

The basic idea behind DANUBIA is to simulate coupled human-environment systems with the best possible realism, to explore consequences of today's decisions suggested by stakeholders and to confront the same stakeholders with these simulated consequences. GLOWA-Danube thereby intended to create a cyber-environment for the stakeholders similar to that of a professional flight simulator, which trains all commercial pilots in their ability to make the right decisions in situations of danger and which is one of the essential reasons for the ever-improving statistics of casualties in commercial flying.

DANUBIA is a coupled system of model components that describes the principle processes and interactions in and between the natural and human system from the view of natural and social sciences that are associated with the use of water resources within a river basin. The developers of DANUBIA have been guided by the understanding that complex questions about the effects of global changes cannot have simple answers. The complex and mostly non-linear relationships and interactions of the natural and social processes involved in water use and the exploration of trade-offs inherent in different futures, which we as societies can choose, must be treated in an integrated manner (i.e. their interdependencies are considered beyond disciplinary and scientific boundaries). For this purpose, DANUBIA was developed from scratch with emphasis on maximising the ability to explore futures; hence, possible calibration of the models to the present conditions in the basins, which limits their applicability and analytic power for future conditions, was avoided as much as possible. In its ultimate configuration, DANUBIA is capable of simulating environmental topics related to water in terms of ecological, economic and cultural factors and of evaluating the sustainability of the proposed solution scenarios. Thus, DANUBIA contributes to finding the best possible solutions for sustainable environmental management in large, heterogeneous river basins.

New methodological approaches were developed, tested and applied within the context of GLOWA-Danube in order to bring together the traditionally distant fields of the natural, engineering and socio-economic sciences to identify long-term strategies for water use. Human actors entered DANUBIA as early as 2001 in the form of agents, which observe their natural and social environment, individually evaluate their options and make decisions which affect the environment. In this way the natural and social system becomes almost as closely linked as in reality. At the same time we learn about human decisions, which are at the heart of global change.

1.3 GLOWA-Danube and Stakeholders

GLOWA-Danube consisted of an intra-university research team, which in the third project phase was made up of 12 research groups at ten universities and research institutes from Bavaria, Baden-Württemberg, Hesse, North Rhine-Westphalia, Hamburg and Tyrol and which was generously sponsored by the Federal Ministry of Education and Research (BMBF) and the Free State of Bavaria and the State of Baden-Württemberg. The research team consists of more than 40 scientists from the following disciplines:

- Hydrology and Remote Sensing
  (Ludwig-Maximilians-Universität München (LMU Munich), Department of Geography)

- Stakeholder
(IFOK GmbH, Bensheim, Berlin, München; Ifo Institute – Leibniz Institute for Economic Research at the University of Munich, Center for Energy, Climate and Exhaustible Resources, München)
- Groundwater Management and Water Supply
(University of Stuttgart, Institute for Modelling Hydraulic and Environmental Systems)
- Water Resources Management
(Bavarian State Office for the Environment, Hof)
- Surface Waters
(Hydrology, Applied Water Resources Management and Geoinformatics, IAWG, Ottobrunn, Vista Geowissenschaftliche Fernerkundung GmbH, Weßling)
- Glaciology
(University of Innsbruck, Institute of Meteorology and Geophysics; Bavarian Academy of Sciences and Humanities, Commission for Geodesy and Glaciology)
- Informatics
(Ludwig-Maximilians-Universität München (LMU Munich), Institute for Informatics)
- Meteorology
(Ludwig-Maximilians-Universität München (LMU Munich), Institute for Meteorology; Institute for Atmospheric Physics, Johannes Gutenberg University Mainz)
- Regional Climate Modelling
(MPI Hamburg, Max Planck Institute for Meteorology)
- Ecosystems – Plant Ecology
(University of Cologne, Institute of Geography)
- Agricultural Economics
(Universität Hohenheim, Institute of Farm Management)
- Environmental Psychology
(University of Kassel, Center for Environmental Systems Research)
- Environmental Economics
(Ifo Institute – Leibniz Institute for Economic Research at the University of Munich, Center for Energy, Climate and Exhaustible Resources)
- Tourism Research
(Ludwig-Maximilians-Universität München (LMU Munich), Department of Geography)

The moderation and organisation of the transdisciplinary dialogue did not lie in the hands of this interdisciplinary research team. It was moreover organised by the IFOK (Institut für Organisationskommunikation, Bensheim), an SME in communication and moderation. This organisational form was chosen based on the confidence that only science can and should not moderate a transdisciplinary communication process with society, in which it is fully involved as one partner.

IFOK, together with the science steering group of GLOWA-Danube, approached key decision makers from the public and private water, energy and environment sectors in the Upper Danube watershed together with NGOs. Over 150 stakeholders ranging from ministries, public administrations and agencies, public and private water suppliers, energy enterprises, medium to small enterprises in the tourism sector, farming associations and large and local NGOs finally participated in the transdisciplinary communication process. It consisted of three steps of which the first was characterised by information exchange and mutual build-up of trust; the second was to organise technical communication on issues of interest to the stakeholders and the common design of future scenarios, which should be analysed with DANUBIA. The third step was the discussion of the results of the scenario simulations and the refinement of further scenarios, which again were analysed in a circle. The guiding principles of the dialogue were transparency, openness and inclusion with the consequence that all scenarios analysed were open to all stakeholders (no special scenarios for specific groups) and that uncertainties and lack of knowledge on both sides were clearly addressed, discussed and where possible considered.

1.4 The Upper Danube as a Pilot River Basin

GLOWA-Danube investigates the effects of global changes on the hydrological cycle in the Upper Danube basin down to gauge Achleiten, near Passau (see Fig. 1.1). With an area of 77,000 km^2, this represents a basin of regional scale, in which a variety of natural and anthropogenic factors influence the water balance. From the hydrological perspective it can be considered a large watershed. The natural hydrological and climatological conditions in the Upper Danube basin are very complex. They are determined to a large extent by processes in the Alps, which are influenced by steep elevation gradients, high precipitation totals and the dynamics of snow and ice.

In terms of its overall infrastructure and the moderate climate conditions, the Upper Danube basin offers an attractive living environment with a high standard of living and hygiene for approximately 12 million people. One requirement for this environment is the adequate supply of high-quality drinking water. In addition, the discharge generated in the Upper Danube represents an essential physical and economic base for downstream countries including Austria, Slovakia, Hungary, Croatia, Serbia, Bulgaria, Romania, Moldova and Ukraine. The water supply in the basin currently offers good conditions for all users.

Water is used in a variety of ways and serves as a vital economic factor: by households, in agriculture, in the industrial sector, for the generation of power and in the tourism industry.

In Bavaria, 95 % of drinking water comes from groundwater and spring sources, of which two-thirds can be delivered in its natural state and without chemical treatment to the supply network. Only 2.5 % is derived from abstraction from surface water. In Austria too, only very little surface water is used for drinking water reserves. Overall in Austria, the water demand is supplied 50 % from spring water, 49 % from

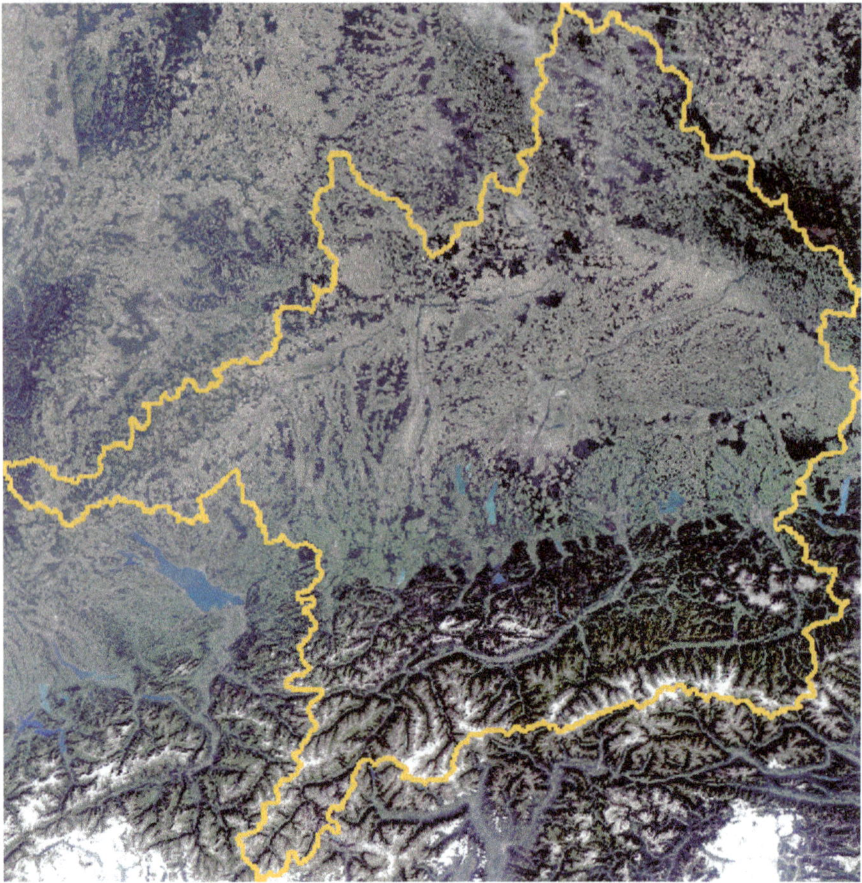

Fig. 1.1 The Upper Danube drainage basin (Data source: ESA ENVISAT-MERIS, 10.09.2004)

groundwater and only less than 1 % from surface water. In Baden-Württemberg, 45 % of the drinking water is sourced from groundwater, 34 % from river water and 17 % from spring water, of which the majority must be softened, purified, filtered and at times disinfected. The national average drinking water consumption is 135 l/person*day in Bavaria, 123 l/person*day in Baden-Württemberg and 150 l/person*day in Austria (the data are from the up-to-date websites: www.lfu.bayern.de (Bavarian State Office for the Environment), www.statistik-bw.de (Baden-Württemberg State Statistical Office), www.wassernet.at (Austrian Federal Ministry for Agriculture, Forestry, Environment and Water Management)). Consumption varies a great deal by region – very high consumptions are found in the cities and communities and in regions with high tourism flow. For the public water utilities, the long-term trends in water supply and demand are a key criterion for ensuring a secure supply of drinking water and regulated wastewater disposal even into the future. Monitoring the basin, processing of the untreated water and safeguarding the quality of the drinking water are prominent factors in this mandate.

Because of the relatively low quantities of water involved, ensuring the supply of drinking water for people in the future is more of a technological and economic task than a question of availability. In contrast, the burden of the available water resources for other sectors is far higher. By far, the most intensive water users in the Upper Danube basin are agriculture and the energy industry. For example, the production of a kilogram of flour requires around 550 l of green water, which is transpired by the wheat plants during their development. Because of the huge elevation gradients in the basin, large amounts of the discharge from the Upper Danube and its tributaries are already being utilised and hence regulated for the generation of power from this renewable natural resource. In addition, there is high and increasing use of water to support the winter tourism industry through the production of artificial snow.

Furthermore, the following factors support the choice of the Upper Danube basin as the pilot region for the development of DANUBIA:

- The vast extent of the basin, at almost 80,000 km^2, as well as its diverse administrative structure in terms of water management, which provides complexity and relevance for questions of management in terms of sustainable water use.
- The very steep spatial gradients of all of the parameters involved (e.g. climate, water, nitrogen, capital) provide distinct spatial interconnections and interactions within the basin in the form of lateral flows. This is reflected not only in the water supply but also in tourism and migration, for example.
- It is expected that climate changes will be quite marked as a result of the steep elevation gradients in the Upper Danube basin. This has ramifications for natural vegetation, agricultural structure and tourism, for example. Prevailing conflicts between agriculture, water management and tourism will thus be magnified.
- Both the excellent database and the requirement for harmonisation of the different databases from the various countries make the Upper Danube a suitable pilot basin for the mandatory water management of river basins specified within the EU Water Framework Directive (BMU 2005).

Thus, the Upper Danube is a highly suitable, natural and societal field laboratory and representative of the alpine and foreland regions of the temperate middle latitudes.

1.5 Climate Change in the Upper Danube Basin

There is a plethora of material that documents that the Upper Danube basin has undergone massive changes in climate over the past 800,000 years. These changes are the consequence of a regular sequence of glacial periods, which have led to glacial imprints of varying extents within the Alps, the alpine foothills and parts of the Bavarian Forest. As a result, there have been widespread elimination and resettlement in the vegetation, the formation of loess deposits in the Tertiary Hills and the sedimentation of huge quantities of moraine deposits in the alpine foothills and gravel in the Munich gravel plains. Causes for these massive glacial period changes in climate have been identified as the cyclical changes in the Earth's orbit around the Sun, which occurs within a temporal dynamic in the range of thousands of years.

1.6 Global Climate Change and the Effects that Can Already Be Perceived

Research over the past two decades has provided a wealth of specific information about the many dynamic changes in the natural world that have been caused by humans. These changes are predominantly related to global climate change. This can be seen in the measured global temperature, which has already increased by around 0.6 °C over the past century.

The regional consequences of global climate change can already be seen in the data time series from the network of meteorological stations within the Upper Danube. Analyses of the long-term measurements from the German Weather Service (DWD) confirm the worldwide trend of a temperature increase (KLIWA 2005a). In the Upper Danube basin, the ground-level atmospheric temperature has increased over the past 100 years. There has been a decline in precipitation in spring and summer, whereas winter precipitation has increased over large portions of the basin (BayFORKLIM 1999; KLIWA 2005b; Reiter et al. 2012). Beyond the basic increase in temperature, in recent years there have been other regional consequences of global warming.

Extreme meteorological conditions with intense rainfall, flooding and heat waves are a part of the story, but long-term trends and their consequences are equally striking. These include:

- The spectacular retreat of the alpine glaciers and the associated loss of both natural freshwater reserves for periods of low precipitation and the diminishment of the storage capacity during long-lasting precipitation events and the resulting loss of flood control. If warming proceeds at the same rate as it has over the past 20 years, at least the Eastern Alps will be completely ice-free within 70–100 years – with a long-lasting melt, the Schneeferner glacier (Zugspitze) could even disappear within 15–20 years (Weber 2008).
- The decrease in summer precipitation and associated change in the discharge regime with effects on the water supply in the rivers and the storage capacities of the lakes, which in turn has an effect on navigation, power generation, cooling water supplies and, last but not least, aquatic ecology.
- The changes in the duration of snow cover and the associated effects on ecology and the economy.
- The change in the seasonal mean temperatures and precipitation and the related spatial and temporal effects on the suitability and cultivation of agricultural crops.

- The change in the phenology of the natural vegetation and crops with an advance in flowering dates and a delaying of maturity that can already be observed (Menzel and Fabian 1999).

Global change is also evident and measurable in the Upper Danube basin and mostly relates to the past 100 years. A few examples include (see also Fig. 1.2):

- Structural changes in agriculture (there has been a significant decline in the number and size of farm units since 1950)

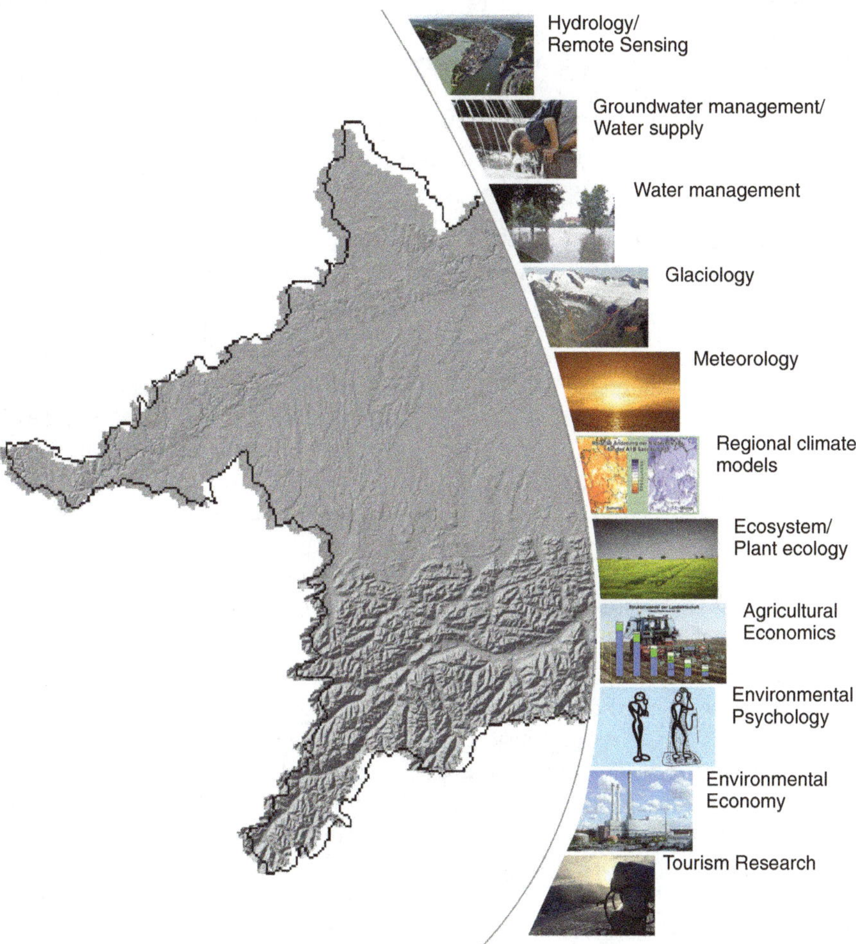

Fig. 1.2 Nature and social science subprojects within GLOWA-Danube (Photo credits: ©Passau Tourismus e.V.; schemmi/pixelio.de; Andreas/pixelio.de; Commission for Geodesy and Glaciology, Bavarian Academy of Sciences and Humanities; Christian Michelbach; MPI-M Hamburg; Christian Michelbach; Institute of Farm Management, University of Hohenheim; Centre for Environmental Systems Research, University of Kassel by Anna von Lilienfeld-Toal; Christian Michelbach; Alexander Dingeldey)

- An increase in the mean annual temperature (e.g. the measured time series of annual temperature from the Hohenpeissenberg Station (Bavaria) for the years 1880–2000 compared to the long-term average for 1971–1990)
- Changes in water consumption (e.g. as a consequence of temperature increase, people shower more frequently)
- Altered precipitation and discharge regimes with notable flooding events (e.g. numerous extreme flooding on the Danube during the past two decades)
- Change in the patterns of precipitation with more frequent intense rainfall (>25 mm/h) and thunderstorms involving downpours, torrents, hail, lightening and winds gusting more than 120 km/h (e.g. relative change in precipitation in Germany)
- Change in the duration of snow cover and snow depth in connection with changes to withdrawal of water for snowmaking machines (e.g. comparison of the size of the glacier in 1898 and 2005 at the Vernagt glacier in the Ötztal Alps)

1.7 Perspectives on Global Change in the Upper Danube Basin

The climate scenarios from the fourth and fifth status report of the IPCC indicate an increase in the mean global temperature for the twenty-first century (IPCC 2007): the best estimate for a low scenario is 1.8 °C (range 1.1–2.9 °C), and the best estimate in a high scenario is 4.0 °C (range 2.4–6.4 °C). In general, a temperature increase will lead to an intensification of the hydrological cycle, which will be expressed as elevated rates of evaporation and precipitation. It is expected that climate changes will be associated with significant effects on the supply of water. This will not only concern the altered average availability of water for nature and for people but also the changes in the frequency and intensity of the high- and low-water events.

Parallel to the change in climate and its effects on the hydrological cycle in the Upper Danube basin, there are other elements to the story; these aspects similarly involve visible changes that must be considered in an evaluation of the regional perspectives of global change in the Upper Danube basin. Over the next 30–40 years, there will be an increase in the population of the Upper Danube region. This is due to the excellent educational, scientific and economic infrastructure in the basin with Munich, Salzburg, Regensburg, Innsbruck, Augsburg and Passau as university cities and Munich as a dominating high-tech industry centre of Germany with a large airport hub. It can therefore be expected that by 2015 the water demand could increase by 1.3–9 % (BMU 2005). Afterwards, however, the change is anticipated to lead into the trend observed in general in Central Europe, which includes a population decline and sharp increase in the middle-aged population. The extent to which this will either reduce or exacerbate the competition between man and nature for water resources needs to be evaluated.

Winter tourism is one of the major sources of income and prosperity specifically in the remote valleys of the Alps. To maintain winter tourism, especially in the central alpine regions, more snow cannons will need to be used. In order to generate a snow depth of 30 cm, the snowmaking machines require one million litres of water per hectare. Sometimes, this means that waterways must be diverted or artificial reservoirs must be created to ensure the water supply (Lutz 2000).

During hot spells, a portion of the required water could until now be compensated for from meltwater from alpine snowpacks and glaciers. If these shrink and disappear, other technological solutions, which do not come for free, must be found to avoid potential shortages in drinking water and bottlenecks in the supply of irrigation water and energy in summer. This applies not only to the operation of hydroelectric power stations, but thermal power stations also require considerable quantities of water to generate steam and primarily to remove thermal discharge. That this problem is real was first experienced in the summer of 2003 (Weber 2007; Braun and Weber 2007) and again in 2005, as power stations needed to restrict their output due to high river water temperatures.

A change in the availability of evaporated water in agricultural areas in the Upper Danube can lead to vital changes in agriculture when critical thresholds are exceeded; these might include the introduction and spread of irrigation or new crops. This process is superimposed on a general change in the structure of agriculture and a progressive opening of the European agricultural market.

There must be increased awareness of the changes in precipitation and discharge, since flooding situations can arise in the winter in the alpine foothills, while in summer regions that are already relatively dry should expect low flows and a decrease in precipitation and groundwater recharge (BayFORKLIM 1999).

Under today's climate conditions, the Upper Danube is a significant exporter of water and can meet the majority of the water demands of the downstream users along the Danube down to the Black Sea with the discharges arising from the Alps, especially in summer. Most Danube countries have become members of EU during the last 25 years. They are substantially contributing to Europe's diversity and richness of cultures and to its future economic perspectives. It can be foreseen that also the non-EU Balkan countries will become members of the European Union over the course of the next 50 years and, as a result, the entire basin area of the Danube may become a part of the European Union. In this context questions of a new nature arise on efficient, equitable and – last but not least – sustainable transnational use of natural resources like water. These questions, for example, evaluate whether it is preferable to use Upper Danube water for an irrigation system in the Upper Danube as a consequence of climate change or to transport the same water through the Danube River and use it as irrigation water in an intensified agricultural structure in the tailwater regions in Hungary, Romania and Bulgaria, even if this inevitably leads to an extensification of agriculture in the headwater region. Reliable answers to these complex questions need new integrated tools to evaluate trade-offs under changing global boundary conditions.

1.8 Scenarios

The goal of GLOWA-Danube is to simulate as realistically as possible the consequences and trade-offs of decision alternatives leading to alternative futures of water in the Upper Danube river basin. The development of integrative and interdisciplinary methods and models in the context of DANUBIA, the formulation of scenarios for possible and probable future realities and a simulation of the consequences of the assumptions about future trends inherent within the scenarios are the core of GLOWA-Danube. They should be as close to reality as possible to provide, in a transdisciplinary communication process between science and society, policymakers and stakeholders with the opportunity to act out various courses of action and to evaluate their sustainability. The goal is to assess potential effects both on the ecosystem processes within the drainage basin and on the living conditions of the inhabitants.

The global change decision support system DANUBIA, developed in the context of the GLOWA-Danube project, contributes to this transdisciplinary evaluation process. This contribution consists of a model simulation of different futures combining aspects of society and nature. DANUBIA is thus confronted with various regulatory decision alternatives and global development trends (e.g. climate, the economy), which were developed by the participating stakeholders in the course of a structured communication between science and society and subsequently incorporated into the model.

From nature's point of view, only climate change scenarios are assumed to influence the futures of the Upper Danube basin. All other changes are either a natural consequence of climate change (e.g. glacier melt) or are initiated by external or internal developments, which affect human actors in the basin and therefore can only exert an impact within the Upper Danube via the actors that are represented within DANUBIA. Thus, for example, a change in the agricultural policy by the EU may represent a change in the external conditions for the farmers as actors in DANUBIA. The farmer-actors in DANUBIA respond (in various ways according to their lifestyles and attitudes) to the changes, by integrating these changes into their microeconomic calculations and by making further (and as a consequence different) decisions on land use, farming practices, resource use, etc., based on their evaluations and their experiences with the past. In the same way, the behaviour of households in terms of drinking water consumption can be treated in DANUBIA as the response of the household as actors to the price of water and their reaction on, for example, rising temperatures and potential shortages.

1.9 The Upper Danube Basin Global Change Atlas

It is of outstanding importance for the transdisciplinary dialogue to document and present the results of the simulations in a "descienced" form, such that they can be understood, evaluated and used by stakeholders. In the context of GLOWA-Danube, in 2004 the conceptualisation and development of the "Global Change Atlas – Upper Danube Basin" were initiated as both paper and online versions (www.glowadanube.de/atlas). The primary purpose of the atlas is to visualise the sectoral and integrative results of DANUBIA, to document them in a single source and to make them available as common knowledge base for the transdisciplinary discussion process with and among stakeholders and policymakers. To serve its purpose it was originally published in German, the mother language of the stakeholders. It was and after the completion of the GLOWA-Danube project in parts still is a living document. New findings can replace older versions and can easily be included in the Atlas. You hold in your hand the English version of the atlas, which has been translated and carefully edited.

In Part II, the participating research groups describe the natural and social scope of the pilot region of the Upper Danube basin. The data documented there form the basis for the extensive model simulation. Each contribution consists of one or more maps that illustrate the respective database and text that explains the source and significance of this dataset.

In Part III each specific discipline describes their sub-models and their context within DANUBIA, the data required for the sub-models and their validation with measured target variables. Selected datasets are calculated from each model and are depicted as maps. In this chapter too, each contribution comprises both a section of text and a map page.

Part V documents the results of the simulations of various scenarios on potential future changes and their consequences in the Upper Danube over the coming five decades. Naturally, this section is the most living part of the atlas. It displays the current state of the ongoing transdisciplinary discussion process among science, policymakers and stakeholders, through which common scenarios about potential future changes are commonly developed and their consequences and trade-offs evaluated. The open architecture of the atlas should allow this section to grow dynamically over time.

There is an appropriate logo located in the lower left corner of each map in order to clarify whether the map involves an input dataset for DANUBIA and/or is the result from the model simulation with DANUBIA:

Map pages with topics related to water consumption and abstraction have three units to help with comparability. Thus, the values commonly used in each respective discipline were converted into two other units. This results in discontinuous intervals in values for the corresponding map legends.

References

ADAC (2003) Der große illustrierte ADAC Weltatlas. ADAC Verlag, München

Bayerisches Landesamt für Wasserwirtschaft (1983) Deutsches Gewässerkundliches Jahrbuch. Donaugebiet, Abflussjahr 1980. ÜK 500 "Donaugebiet", München

BayFORKLIM (1999) Klimaänderungen in Bayern und ihre Auswirkungen. Abschlussbericht des Bayerischen Klimaforschungsverbundes

BMU Bundesministerium für Umwelt, Naturschutz und Reaktorsicherheit (2005) Die Wasserrahmenrichtlinie. Ergebnisse der Bestandsaufnahme 2004 in Deutschland, 2nd ed. Bonifatius, Paderborn

Braun L, Weber M (2007) Gletscher. Wasserkreislauf und Wasserspende. In: Bundesministerium für Umwelt, Naturschutz und Reaktorsicherheit (BMU) (ed) Klimawandel in den Alpen, Bonn

IPCC (2007) Climate change 2007: The physical science basis. In: Solomon S, Qin D, Manning M, Chen Z, Marquis M, Averyt KB, Tignor M, Miller HL (eds) Contribution of Working Group I to the Fourth Assessment Report of the Intergovernmental Panel on Climate Change. Cambridge University Press, Cambridge/New York

Jarvis A, Reuter HI, Nelson A, Guevara E (2008) Hole-filled seamless SRTM data V4. International Centre for Tropical Agriculture (CIAT). Available via DIALOG. http://srtm.csi.cgiar.org. Accessed 19 Sept 2014

KLIWA, Arbeitskreis Klimaveränderung und Konsequenzen für die Wasserwirtschaft (2005a) Langzeitverhalten der Lufttemperatur in Baden-Württemberg und Bayern. In: KLIWA-Berichte, H 5, 2005

KLIWA, Arbeitskreis Klimaveränderung und Konsequenzen für die Wasserwirtschaft (2005b) Langzeitverhalten des Gebietsniederschlags in Baden-Württemberg und Bayern. In: KLIWA-Berichte, H 7, 2005

Lutz G (2000) Beschneiungsanlagen in Bayern – Stand der Beschneiung, potentielle ökologische Risiken. In: Bayerisches Landesamt für Umweltschutz (ed) Technische Beschneiung und Umwelt Tagungsband zur Fachtagung am 15. November 2000, Augsburg

Mauser W, Klepper G, Rice M, Schmalzbauer BS, Hackmann H, Leemans R, Moore H (2013) Transdisciplinary global change research: the co-creation of knowledge for sustainability. Curr Opin Environ Sustain 5:420–431. doi:10.1016/j.cosust.2013.07.001

Menzel A, Fabian P (1999) Growing season extended in Europe. Nature 397:659

Reiter A, Weidinger R, Mauser W (2012) Recent climate change at the Upper Danube - a temporal and spatial analysis of temperature and precipitation time series. Clim Change 111(3–4):665–696. doi:10.1007/s10584-011-0173-y

Weber M (2007) Informationen zum Gletscherschwund. Gletscherschwund und Klimawandel an der Zugspitze und am Vernagtferner (Ötztaler Alpen), Kommission für Glaziologie der Bayerischen Akademie der Wissenschaften, November 2003

Weber M (2008) Mit den Gletschern geht's bergab. Das alpine Wasserreservoir schrumpft weiter. Bayern forscht 1/2008: 8–10

Chapter 2
DANUBIA: A Web-Based Modelling and Decision Support System to Investigate Global Change and the Hydrological Cycle in the Upper Danube Basin

Rolf Hennicker, Stefan Janisch, Andreas Kraus, and Matthias Ludwig

Abstract We describe architecture and design principles of the integrative modelling and simulation system DANUBIA. The system integrates the distributed simulation models of the socio-economic and natural science disciplines of the GLOWA-Danube project. During an integrative simulation the simulation models run in parallel and exchange iteratively data. The validity of the exchanged data with respect to model time is guaranteed by a central time controller which coordinates the single simulation models. The consistency of spatial data is ensured through the common proxel concept used by all models. The development of DANUBIA is based on object-oriented software engineering methods. A generic framework architecture has been constructed which implements general rules for the behaviour of all simulation models. It can be specialised and thus instantiated by concrete simulation models. We also have developed a particular framework for the socio-economic deep actor models.

Keywords GLOWA-Danube • Distributed system • Coupled simulation • Time controller • Proxel concept • Framework architecture

2.1 Overview

DANUBIA (Barth et al. 2004) is an integrative simulation and decision support system that was developed in the context of GLOWA-Danube. DANUBIA facilitates the study of scenarios related to the hydrological cycle from ecological and economic points of view in order to support scientists and policymakers in designing strategies for sustainable environmental management.

R. Hennicker (✉) • S. Janisch • A. Kraus • M. Ludwig
Institute for Informatics, Ludwig-Maximilians-Universität München (LMU Munich), Munich, Germany
e-mail: hennicker@ifi.lmu.de; stephanjanisch@gmail.com; AndiKraus@web.de; Matthias.Ludwig@ifi.lmu.de

W. Mauser, M. Prasch (eds.), *Regional Assessment of Global Change Impacts*, DOI 10.1007/978-3-319-16751-0_2

Sixteen simulation models from all of the participating research groups are integrated within DANUBIA. This permits the study of both sectoral and interdisciplinary issues while taking into account the interactions of interdependent processes.

The development of DANUBIA is based on methods from object-oriented software engineering and web engineering. At all stages of the development process, the *Unified Modeling Language* (*UML*, Booch et al. 1999) played a key role as the common notation that was used by all project partners to describe the integrative aspects of the systems.

The architecture of the DANUBIA system is illustrated in Fig. 2.1. The simulation models from the different specialist groups of GLOWA-Danube are realised within the *Components* component of DANUBIA. The *Core System* component of DANUBIA consists of a developer framework and a runtime environment (see Sect. 2.3). DANUBIA contains the DeepActor framework for integration of actors for the implementation of the socio-economic models.

At present, DANUBIA runs on a computing cluster with 56 processors. However, the system can be installed on individual computers or small networks for testing purposes.

2.2 Concept

2.2.1 *Main Components and Interfaces*

The sixteen simulation models integrated within DANUBIA are grouped by theme into five main components: *Atmosphere*, *Actor*, *Landsurface*, *Groundwater* and *Rivernetwork* (see Fig. 2.1). The exchange of data both among the main

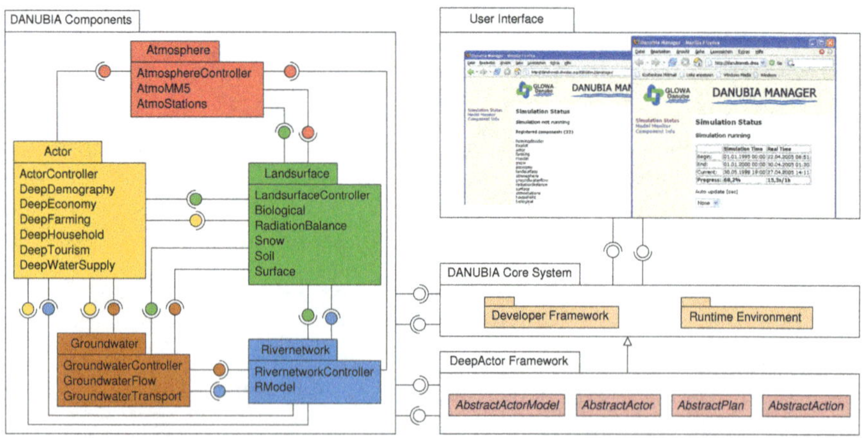

Fig. 2.1 Architecture of the DANUBIA system

components and between the individual simulation models takes place via interfaces. The interfaces contain the identifier and the data type for each parameter that is to be exchanged. The validity period of the exchanged data is ensured separately by an internal temporal coordination concept (see Sect. 2.2.3).

2.2.2 *Spatial Concept*

One central element of simulation of the environment involves the treatment of the simulation area. In DANUBIA, the simulation area is represented using a two-dimensional grid (see Fig. 2.2). The rows and columns of this grid are arranged using Lambert conformal conic projections of geographical coordinates. The coordinate points correspond to those in the Hydrological Atlas of Germany.

The length and width of the grid cells are each 1,000 m, resulting in a grid area of 1 km^2. The use of object-oriented paradigms gives the simulation area a structure that is both static and dynamic, the latter consisting of the processes that take place on the corresponding site in the grid. This led to the term proxel (an acronym for process pixel). A proxel object is identified by an identification number that is unique within the simulation area being considered; this identifier is known as the ProxelID (see PID in Fig. 2.2). All proxel objects are administered in a table object, called a ProxelTable. A universal proxel object stores the relevant characteristics of the specified grid point (coordinates, terrain elevation, land use, etc.) for all simulation models. Additionally, subject-specific characteristics of the proxels are added to the individual DANUBIA components via specialisation.

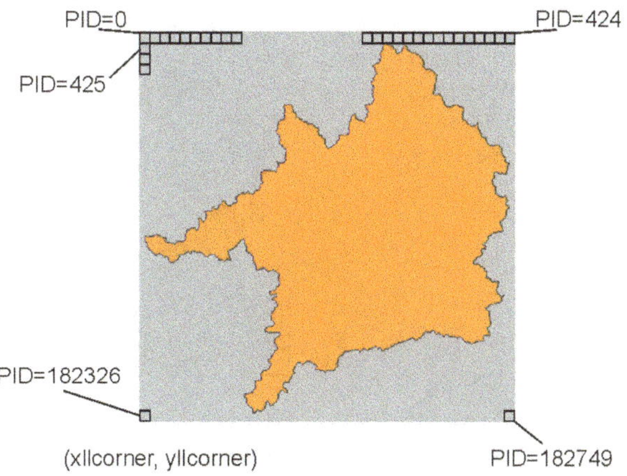

Fig. 2.2 The Upper Danube drainage basin

2.2.3 Temporal Concept

Over the entire simulation period, a simulation model calculates data at discrete time points; these data describe the current valid state of the modelled system. The interval between two such points in time is constant for each model and is known as the model time interval. The lengths of the model time interval in DANUBIA range from 1 h (e.g. in the *Atmosphere* and *Landsurface* models) to 1 month (in most of the actor models). Several models with differing time intervals are coupled within an integrated simulation, and periodic calculations are carried out, and data is exchanged during the runtime. In order to obtain reliable results from these simulations, the exchange of data must comply with the following conditions:

- The exchanged data must be stable; i.e. there must not be simultaneous access to the data for writing and reading.
- Each model must receive data upon a data request; this data must be valid with respect to its own local model time.

In order to comply with these requirements, DANUBIA has a component for temporal coordination among the individual models involved, known as a *Timecontroller* (Hennicker and Ludwig 2005, 2006). For the models themselves, there is thus the following life cycle (see Fig. 2.3):

waitForGetData: wait for the release to read data from other models using the *Timecontroller*.
getData: read data from models.
compute: calculate the relevant data at the next simulation time point.
waitForProvide: wait for the release to provide the new calculated data for other models using the *Timecontroller*.
provide: provide the new calculated data for other models.

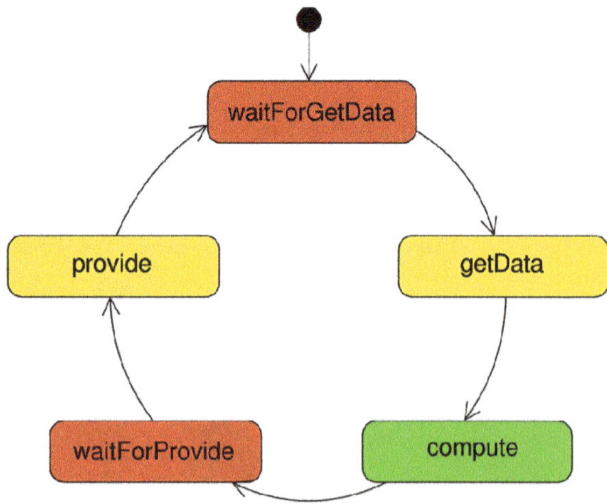

Fig. 2.3 Life cycle of a coupled simulation model

2.3 DANUBIA Core System

The DANUBIA core system consists of a developer framework and a runtime environment (see Fig. 2.4).

2.3.1 Developer Framework

The developer framework supports the model developer by providing the classes and interfaces that are used by the model developers to integrate their model implementations in DANUBIA.

The first aspect involves something known as base classes, which must be specialised according to the object-oriented inheritance principle. The most important base class is the *AbstractModel*, from which the specific model implementations are derived in DANUBIA. This base class already contains the implementation of the life cycle of a simulation model (see Sect. 2.2.3) and forms the basis for the connection of the model to the runtime environment (see Sect. 2.3.2). The *AbstractProxel* class is also in the base class category; this class realises the spatial concept described in Sect. 2.2.2.

Additional classes that are used by the model developers include the implementations of the jointly used data types in DANUBIA and tools for pre- and post-processing of input and output data. These tools allow the conversion of

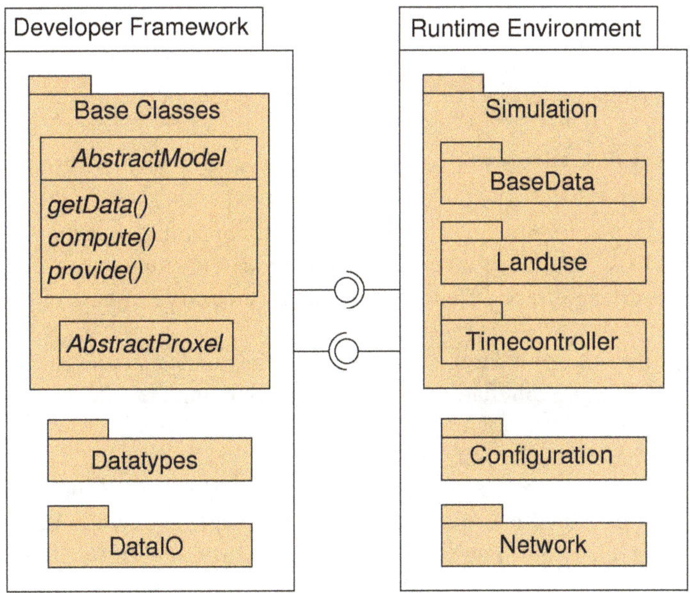

Fig. 2.4 DANUBIA core system

data between typical GIS format and the internal binary data format of DANUBIA, for example; they also allow temporal and spatial aggregation or averaging of data.

2.3.2 Runtime Environment

The runtime environment can handle either integrated simulation runs of the DANUBIA system on a single computing cluster or test runs of smaller model configurations on individual computers or smaller local networks. The runtime environment essentially consists of the components described below.

The main component of the runtime environment is the *Simulation* component with its subcomponents, *BaseData*, *Landuse* and *Timecontroller*. The *Simulation* component is responsible for the administration of the simulation models during a simulation, the subcomponents are assigned the following tasks: the *BaseData* component initialises the relevant universal characteristics of the proxel (see Sect. 2.2.2); the *Landuse* component administers the changing land use of the simulation regions during a simulation run and provides it to the simulation models that are involved. Finally, the *Timecontroller* component implements the temporal coordination of the individual simulation models described in Sect. 2.2.3 during an integrated simulation run.

The *Configuration* component generates the simulation configurations. In doing so, it interacts with the user interface (see Sect. 2.4). The distribution of the individual parts of the system over a network is carried out by the *Network* component. In particular, it is responsible for connecting the corresponding data import and data export interfaces of the models involved in an integrated simulation.

2.4 The User Interface

Distributed simulation runs can be configured, controlled and monitored with the DANUBIA web interface (see the *UserInterface* component in Fig. 2.1). The user interface offers different views depending on the status of the simulation system.

Prior to the start of a simulation run, the names of the simulation models are indicated. Once the simulation configuration is complete, the simulation can begin.

Once a simulation has commenced, the time progress of both the individual models and the integrated model group can be seen. The progress is indicated in real time and in simulation time. In addition, the user interface provides detailed information on the individual simulation models. For example, this information includes performance information such as processor and storage space usage, as well as metadata such as author and version of the simulation model.

2.5 The DeepActor Framework

The DeepActor framework enhances the developer framework provided by DANUBIA (see Sect. 2.3.1) with additional base classes and interfaces. It implements a joint conceptual basis of the DeepActor approach developed in the context of GLOWA-Danube and enables the socio-economic simulation models from GLOWA-Danube (the *Actor* main component in Fig. 2.1) to explicitly model the decision processes of the so-called deep actors. The DeepActor approach substantiates the method for agent-based simulation from the social sciences (Gilbert and Troitzsch 2005), which in turn is based on the agent concepts from the field of (distributed) artificial intelligence (Norvig and Russell 2003; Weiss 1999). The term actor is used to avoid confusion with the term software agent that is related only conceptually. A software agent is a programme that has the properties to operate autonomously, responsively, proactively and integratively. These properties are interpreted in technical terms for software agents; for example, the autonomous property is interpreted as the internal assignment of threads or processes to each agent. Here, in contrast, these kinds of properties are interpreted in conceptual terms for actors and hence always depend on the modelling within a specific simulation model. An actor represents an entity acting within the simulation area, which may be the private household, farmer or tourism infra- and suprastructure facility, for example. Simulation models that use the DeepActor framework are called DeepActor models.

The fundamental modelling elements of a DeepActor model are shown in Fig. 2.5. The *AbstractModel* base class pre-specified for DANUBIA is refined by a special base class, *AbstractActorModel*. In addition, there are base classes for actors, plans and actions. The static structure from Fig. 2.5, and hence the structural concept of the DeepActor approach in GLOWA-Danube, is explained in more detail below. The dynamic aspects of the method are explained in Sects. 2.5.1 and 2.5.2.

It should be noted that in general, the abstract elements of the framework are optionally specified in a DeepActor model. Depending on the type and complexity of the simulation model to be implemented, there are simple, reactive actors and complex actors that can learn. Both are implemented using the same basis of the DeepActor framework.

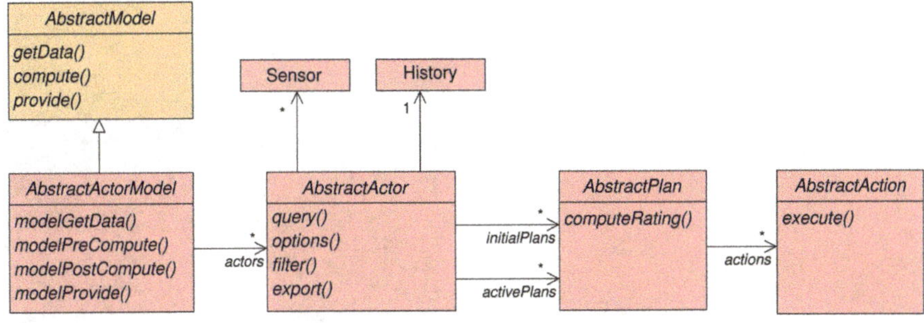

Fig. 2.5 Static structure of the DeepActor framework

In the simplest manifestation, the specified derivation of the *AbstractActorModel* base class is used for the administration and initialisation of the specific actors in a DeepActor model. The actors have sensors with which they perceive their surroundings, especially for the import of proxel data and for the import of data from other actors within the same model. Actors also have a history or memory, in which they store decision of previous simulation steps. Plans represent the behaviour options of an actor, and each plan contains a number of actions, which explicitly model the effects of a plan implementation. The number of initial plans (*initialPlans*) is the total number which is available to an actor over the simulation run. The number of active plans (*activePlans*) is the number of the initial plans that an actor has selected for implementation in a given time step.

2.5.1 Life Cycle of a DeepActor Model

The life cycle of a DeepActor model illustrated in Fig. 2.6 is a refined version of the cycle of a DANUBIA model (see Fig. 2.3).

The key difference is a distinction between the model and the actor calculations. The former represents the macro-level of the simulation, which is required at least for the coordinated exchange of data (*modelGetData*, *modelProvide*) with coupled simulation models, and the latter simulates the micro-level of individual decisions (*filter*, *options*), the effects of which are accounted for by the model calculations on the macro-level. The activities *getData*, *compute* and *provide* in Fig. 2.3 are implemented by the DeepActor framework using the abstract operations from Fig. 2.5. A specific DeepActor model supplies specific implementations for this as follows:

- *modelGetData* and *query:* the model imports data from coupled simulation models, and the actors sample the sensors.
- *modelPreCompute:* the model can initialise actor calculations.
- *options:* the actors activate the plans according to external conditions.
- *computeRating:* all active plans compute an evaluation attribute.
- *filter:* the number of active plans is reduced based on criteria specific to each actor.

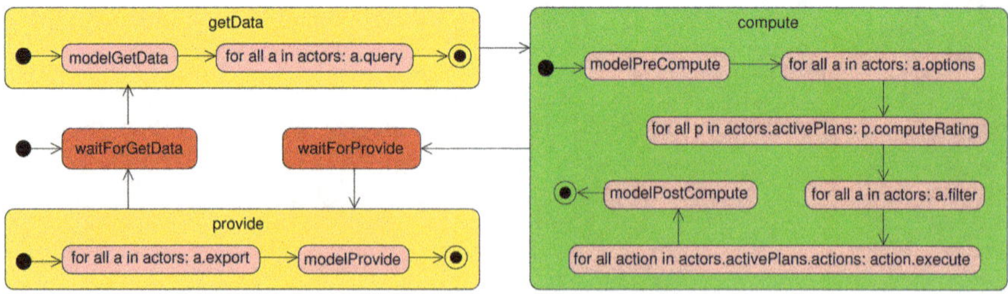

Fig. 2.6 Life cycle of a DeepActor model as refinement of Fig. 2.3

- *execute:* the actions of the remaining active plans are executed.
- *modelPostCompute:* the model can initialise actor calculations.
- *export* and *modelProvide:* the actors and the model export data for other actors and/or for coupled simulation models.

2.5.2 Decision Process of a Deep Actor

The decision by an actor comprises two steps: *options*, to determine the active number of plans in a given simulation step, and *filter*, to reduce the number of active plans to only those plans that actually should be executed. In the first step, plans are deactivated if, at the current simulation time point, they cannot be executed because of external conditions upon which the actor has no influence. In the second step, an actor considers his implicitly available goals and preferences and thereby determines the set of active plans to be executed in this time step. The calculation of an optional evaluation attribute (*computeRating*) happens in between. The resulting value can be used as the basis of the plan selection in the *filter* step. For example, in this way a selection algorithm based on a multiattribute utility theory (Norvig and Russell 2003) can be implemented.

Goals that motivate a decision by an actor are not an explicit element of the DeepActor method. Instead, the specific implementation of the *filter* step must account for the goals of each actor.

References

Barth M, Hennicker R, Kraus A, Ludwig M (2004) DANUBIA: an integrative simulation system for global change research in the Upper Danube Basin. Cybernet Syst 35:639–666

Booch G, Rumbaugh J, Jacobson I (1999) The unified modeling language user guide. Object technology series. Addison Wesley, Reading

Gilbert N, Troitzsch KG (2005) Simulation for the social scientist, 2nd edn. Open University Press, Berkshire

Hennicker R, Ludwig M (2005) Property-driven development of a coordination model for distributed simulations. In: Proceedings of the 7th IFIP international conference on Formal Methods for Open Object-Based Distributed Systems (FMOODS 2005), LNCS 3535. Springer, Berlin, pp 290–305

Hennicker R, Ludwig M (2006) Design and implementation of a coordination model for distributed simulations. In: Mayr HC, Breu R (eds) Proc. Modellierung 2006 (MOD'06). Lecture notes in informatics, vol P-82. Gesellschaft für Informatik, Bonn

Norvig P, Russell SJ (2003) Artificial intelligence: a modern approach. Prentice Hall, Englewood Cliffs

Weiss G (ed) (1999) Multiagent systems: a modern approach to distributed artificial intelligence. The MIT Press, Cambridge

Chapter 3
DeepActor Models in DANUBIA

Andreas Ernst, Silke Kuhn, Roland Barthel, Stefan Janisch, Tatjana Krimly, Mario Sax, and Markus Zimmer

Abstract This chapter describes the representation of decision processes of socio-economic actors by means of actor models, which is one distinctive feature of DANUBIA. An actor model (also called an agent model) describes socio-economic processes as the sum of the individual actions taken by a range of different actors. The DeepActor framework provides a basis for modelling and implementing the socio-economic DANUBIA models. Decision makers, such as individuals, organisations or businesses, are modelled as "actors". Each actor is localised within his physical and social environment (a proxel or a social network) and takes decisions as responses to his observations about which action to execute from a range of possible action options. Actors have various preferences and action options,

A. Ernst (✉) • S. Kuhn
Center for Environmental Systems Research (CESR), University of Kassel, Kassel, Germany
e-mail: ernst@usf.uni-kassel.de

R. Barthel
Department of Earth Sciences, University of Gothenburg, Göteborg, Sweden
e-mail: roland.barthel@gu.se

S. Janisch
Institute for Informatics, Ludwig-Maximilians-Universität München (LMU Munich), Munich, Germany
e-mail: stephanjanisch@gmail.com

T. Krimly
Production Theory and Resource Economics, Universität Hohenheim, Stuttgart, Germany
e-mail: T.Krimly@uni-hohenheim.de

M. Sax
Wilhelm-Hauff-Str. 51, 84036 Landshut, Germany
e-mail: sax.mario@googlemail.com

M. Zimmer
Center for Energy, Climate and Exhaustible Resources, Ifo Institute – Leibniz Institute for Economic Research, University of Munich, Munich, Germany
e-mail: zimmer@ifo.de

W. Mauser, M. Prasch (eds.), *Regional Assessment of Global Change Impacts*, DOI 10.1007/978-3-319-16751-0_3

represented by customised plans and decision procedures that are specific to each actor type. Further, actors have a "memory" (history) for recalling previous decisions. One special feature of DANUBIA is the coupling of physically based scientific models with socio-economic components. It is described how the transformation from quantitative states in nature to qualitative notifications for the actor model is realised using the flag concept.

Keywords DeepActor framework • Modelling framework • Agent-based modelling • Decision process • Socio-economic actor • Preferences

3.1 Introduction

One distinctive feature of DANUBIA is the representation of decision processes by socio-economic actors by means of actor models. An actor model (also called an agent model) describes socio-economic processes as the sum of the individual actions taken by a range of different actors. An actor is thus an entity that is capable of responding to changes in the system conditions in specific ways; hence, the actor makes "decisions". In this way, the individuality of the action is generated because each actor is characterised by customised characteristics and preferences.

3.2 Features of Multi-actor Modelling

Agent modelling is a concept from informatics. An agent, or actor, as it is designated hereinafter, is ultimately an independent component of a programme. The actions executed by an actor are oriented to the environment around him. A multi-agent or multi-actor model is the term used if several actors interact among each other (Gilbert and Troitzsch 2005).

The multi-actor modelling procedure was developed in the context of research on distributed artificial intelligence. Essentially, the method depicts emerging social phenomena from the individual actions. Depending on the field of application, any level of detail can be selected. Moss and Edmonds (2005) emphasise the importance of multi-actor models in the validation of theories in the social sciences.

According to Gilbert and Troitzsch (2005), actors have the following features: they essentially act independently (autonomy), they respond to changes in their environment (reactivity) and they are social to the extent that they can communicate with other actors (social interaction). In addition, they have memory and hence are capable of learning and can become active on their own and pursue a goal (proactivity).

Multi-actor modelling is utilised in the most diverse fields of application, but is especially applicable for simulating social systems. In contrast to social simulation models that are based on regression equations, multi-actor models consider individual decision makers.

Depending on the field of application, the participating actors possess a set of rules of conduct of varying complexities. Explicit representation of the actors allows their decisions to be traced back to the set of rules that were implemented and to the situational factors that influenced them.

The behaviour of this type of simulation model is generally more easily comprehended than that of a regression model, since the underlying processes are depicted. The phenomena observed result from the individual actions of the system.

For this reason, in those cases when there is explicit need for a decision between two or more action alternatives, the multi-actor concept is invoked in DANUBIA.

Models that utilise this concept are known as "deep" actor models as a contrast to the "shallow" models that are based on simpler algorithms. Accordingly, the designations DeepActor model and DeepActor method were selected, as well as the term DeepActor framework for the underlying framework.

3.3 The DeepActor Framework

The DeepActor framework (see Chap. 2) provides a basis for modelling and implementing the socio-economic DANUBIA models. The design of the object-oriented framework draws from the specific requirements of the individual models and implements a unified conceptual basis in this way.

Decision makers, such as individuals, organisations or businesses, are modelled as "actors". An actor is localised within his physical and social environment (a proxel or a social network) and makes decisions as responses to his observations about which action to execute out of a range of possible action options. The execution of the action in turn has an effect on his environment, which can form the basis of subsequent actions.

Actors have various preferences and action options, represented by customised plans and decision procedures that are specific to each actor type. Further, actors have a "memory" (history) for recalling previous decisions. A specific implementation is described by way of example in Chap. 42 for the *DeepHousehold* model.

The framework provides an abstract implementation of these concepts that can be optionally specified in a DeepActor model. For example, although the decision process is fixed in its sequence (perception of the environment, plan selection, execution of the chosen plan), it does not have a preset algorithm for plan selection. Depending on the type and complexity of the model, both simple actors that respond and complex actors that are capable of learning can be implemented in a DeepActor model.

Table 3.1 Lists the currently agreed upon flags, as well as the flag name and its significance, and specifies the models that calculate the flags and the models that use them

Flag description	Calculated by model	Signification	Used by model
GroundwaterQuantityFlag	Water supply	Quantitative state of groundwater resources	Water supply, farming, economy, tourism
GroundwaterQualityFlag	Water supply	Qualitative state of groundwater resources	Water supply, tourism
DrinkingWaterQuantityFlag	Water supply	Quantitative state of drinking water supply	Household, farming, tourism, economy
DrinkingWaterQualityFlag	Water supply	Qualitative state of drinking water supply	Household, tourism
RiverWaterQuantityFlag	River network	Quantitative state of surface water resources	Water supply, economy, tourism
RiverWaterQualityFlag	River network	Qualitative state of surface water resources	Water supply, tourism
RiverFloodFlag	River network	Flood event	Household
SoilSaturationFlagCollection	Land surface	Level of soil saturation as a measure of trafficability	Farming

The DeepActor framework enhances the developer framework that is prepared in DANUBIA. The key concepts of a simulation area, the exchange of data and a simulation period are adopted unchanged for the DeepActor models, so that the integration of DeepActor models into the model system of DANUBIA can take place transparently for the runtime environment of DANUBIA and other DANUBIA models (Table 3.1).

3.4 The Concept of Flags in the DeepActor Models

One special feature of DANUBIA is the coupling of physically based scientific models with socio-economic components. The output data from the science models serve as input data for the multi-actor models. The data supplied serve as the basis for decisions (plan selection) by the actors. For example, an actor in the *DeepHousehold* model makes a decision about the frequency of showering based on the temperature.

The basic idea of the flag concept is that the output from the science-based models often involves variables that can only be interpreted with the aid of expert knowledge and in a specific context. For example, the groundwater level in proxel at a certain time step says little about the supply of groundwater, unless the hydrogeological situation and the background history are known and understood.

Table 3.2 Flags (indices), used in DANUBIA for the transmission of natural or technical states to the actor models

DrinkingwaterQuantityFlag	Correspondent in reality	Reaction in the DeepHousehold model
1	No reports	No influence on actors' decisions, habits
2	First unspecific reports of possibly critical states of water supply	Influence on plan selection of especially sensitive actors
3	Calls for water saving	Influence on plan selection of all actor types
4	Acute water shortage	Strong influence on plan selection of all actor types, some plans are unfeasible

A method was developed to resolve this discrepancy, in which one or more scientific variables are converted into an index that summarily describes specific conditions (e.g. groundwater supply or flood risk).

These indices are known as flags in DANUBIA and can have values between 1 ("very good") and 4 or 5 ("catastrophic") (see Table 3.2). In this way, a flag represents a classified notification about the quantitative and qualitative status of groundwater and surface water resources, for example, or a potential risk and its extent.

In the context of the multi-actor modelling, the flag scheme, which initially appears to have very little differentiated classification, becomes the basis for individually adapted decision-making within various models with different actor types. Hence, flags can both be interpreted differently on an individual basis and can have predefined, differing significances. They can involve a restraint or express a new warning.

Details of the flag concept are presented in Barthel (2011).

A detailed description of the DeepActor concept and its implementation in DANUBIA is presented in Barthel et al. (2008). Soboll et al. (2011) and Barthel et al. (2012) illustrate the application and results of the approach using examples from various actor models and scenarios. Figure 3.1 gives a more detailed overview of the DeepActor models in DANUBIA.

model description and respective sub-project	***DeepHousehold*** Environmental Psychology (water consumption of private households)	***DeepWaterSupply*** Groundwater balance and management, Water supply	***DeepTourism*** Tourism Research	***DeepFarming*** Agricultural Economics	***DeepEconomy*** Environmental Economy	***DeepDemography*** Environmental Economy
1. Who is the actor?	Every actor represents all households of a type (one of five milieu groups) on a proxel.	The water supply company (WVU): each existing WVU represents one actor (except Switzerland).	Every actor represents facilites of a type of touristic infrastructure or suprastructure.	Every actor represents an agricultural holding of a specific type on a proxel.	Every actor represents a water intensive industrial unit on a proxel occupied by industry.	Every actor represents all households of a type (defined by household size, number of children, milieu group) on an inhabited proxel.
2. How many and actor types are there? How many actors are there altogether?	Five actor types per inhabited proxel • Post-materialists • Leading milieus • Traditional milieus • Mainstream milieus • Hedonistic milieus Sum: 9115 inhabited proxels * 5 actor types = 45575 actors	Three actor types: • Community water supply = small, one community, one WVU • Group water supply- = fusion of several community suppliers • Remote water supply = major, supraregional water supplier Sum: 1717	Nine actor types: • Skiing areas (with and without artificial snow) (253) • Golf courses (144) • Outdoor swimming pools (351) • Indoor swimming pools (202) • Outdoor and indoor swimming pools (39) • Water parks (35) • Thermal baths (15) • Restaurants (2013) • Tourist accomodations (2013) Sum: 5065	28 actor types: • 4 cash crops farm types • 3 finishing farm types • 13 forage crops farm types • 1 permanent crop farm type with hop cultivation • 7 mixed farm types In the Upper Danube basin there are altogether 58984 proxel with agricultural use; there is one actor typ per proxel	• 1 representative actor per industry proxel Sum: 1354 (equals the number of proxels occupied by industry)	10 types per milieu group (c.f. *DeepHousehold*): • 1 Person (P) / HH • 2 P / HH no child (K) • 2 P / HH with 1 K • 3 P / HH no K • 3 P / HH with 1 K • 3 P / HH with 2 K • 4 or more P/ HH no K • 4 or more P / HH with 1 K • 4 or more P / HH with 2 K • 4 or more P/ HH with 3 or more K Sum: 9115*50=455750
3. How do the actors differ?	The actors differ in their everyday, habitual water consumption behaviour, their innovation behaviour, their risk perception and in specific sociodemographic attributes, attitudes and values.	The actors differ in their size, that means in the number of supplied consumers, in the number of extraction points, in the size of the supplied area and the size of the extraction area.	Basically the actors represents different types of touristic infrastructure.	The actors differ in their land use management: arable land or grassland, crop rotation, animal husbandry and livestock.	The actors differ in their adopted production technology (production processes).	The actors differ in their migration preferences. Thus they evaluate the attractiveness of potential migration destinations (coming from socio-economic and natural factors) differently. The preferences are determined empirically for the average migrant and calibrated for the HH types on the basis of their distribution.
4. Empirical basis of distinction	The distinction relies on the so-called Sinus Milieus® from commercial market research (Sinus Sociovision, 2007). Following the so-called living-environment segmentation according to Sinus, five milieu groups were build and integrated into the model.	The reactions of water supply companies (known from the past and assumed) on changing demand and supply conditions serve as a basis. Mainly, the decision options are derived from the size of the extraction and supplied area.	The distinction relies on different functionalities und challenges of the real existing examples. They differ in size (e.g. number of allies of golfcourses or artificial snow covered area of skiing regions), in opening hours and water consumption. These data were established by means of surveys, investigations and literature reviews by the authors.	The distinction relies on statistical on the district level about the type of production, the land use and the number of farms in the respective production type in the Upper Danube basin dating from 1995. There are five different production types: cash crops farms, forage crops farms, finishing farm, permanent crops farms and mixed farms.	Companies of different sectors and location use different production processes for the production of their goods (stat. data). Because of the poor data records, only one representative company with a representative production process is simulated per proxel, which could be taken as the average production technology in the proxel.	The distinction relies on the required distinction of the Household model according to milieu groups (c.f. *DeepHouse hold*) and and in addition on the differentiation according to household size and number of children (stat. data), which are important for the water consumption behaviour.

5. What is decided by the actors?	The frequency of specific water use types (e.g. frequency of showering), the purchase of different water use concerning innovations (e.g. water-saving shower head).	If necessary the measures for the covering of deficits in the water supply infrastructure, e.g. the opening of a new extraction point.	They decide, if a facility of tourism infrastructure can be put into operation and if water is required or not for that purpose.	The actors decide on a daily basis on the cultivation management of the respective crops (sowing, appliance of fertilizer, harvest) and besides on the irrigation of their land areas.	The actors decide on the produced quantity of goods, on the required input of production factors and on changes of the established production technology.	They decide on the migration from a proxel within or outside of the basin area into a proxel within or outside of the basin area. The net changes of the migration balance are reproduced.
6. When do the actors decide? Which situations activate a decision and how is it calculated?	Hierarchic structure of decisions: An actor acts out of habits and decides consciously not till then when a critical threshold is exceeded. The decision is caculated by means of Multi-Attribute-Utility. A decision on the purchase of an innovation is taken of a specific percentage of all households in dependence of the product life cycle per time step. The calculation is carried out with a deliberative rule of decision, based on the Theory of Planned Behavior and the so-called Take-the-Best-heuristic. When no clear decision is taken, the Household actors imitate the behaviour of their social network.	An actor takes a decision, when the water demand of the actor models surpasses its initially defined extraction capacities. There are four different options for action: 1. standard behaviour: no modification of extraction and supply structures 2. increase capacities of existing extraction points, if possible. Elsewise: 3. open new extraction points, if possible. Elsewise: 4. crisis management: not otherwise specified measures, such as the delivery of water tankers. The decisions are based on the evaluation of the resources (groundwaterQuantityFlag). If the evaluation indicates a shortage of supply (flag 3, 4 or 5), the supply structure can not be expanded (option 2 not selectable, continue with option 3). The flag level, from which on no further expansion or exploitation is possible, is defined in scenarios.	A decision, which influences the water consumption is taken on a daily basis. Temperature, thickness of snow cover, data on precipitation, the date and flags on the water availability can be accounted for in the selection of appropriate options, this depends on the actor type. The decision-making process relies on the Multi-Attribute-Decision Making (MADM) approach. Depending on the actor type, the decision-making relevant environmental conditions are requested and interpreted with a type specific decision-related system of rules. The selection of a convenient plan for a skiing area actor is (in pseudocode): IF date >= season opening AND date <= end of season AND snow cover thickness >= 30 cm THEN open skiing area ELSEWISE close skiing area	The actors' land use management (crop rotation, livestock) is influenced indirectly by changes in agricultural policy (subsidies) or agricultural markets (prices) and by changes in the harvest crops. They are altogether calculatd in a coupled regional model. The actors' decision are directly activated by changes in plant growth stages, temperature, precipitation, soil traffi-cability and groundwater availability. Heuristic system of rules and algorithms, which include the above- mentioned information concerning the cultivation requirements for each crop type, are used for these decisions. Besides, the actor includes his experience and knowledge of weather and climate in recent years and thus he adjusts his behaviour (individual learning).	A company is weighing up, with which combination of produced volume of goods and the therefore required production factors, it is maximising its profit. Driving forces for the actor's decision are listed below: The current perceived conditions (e.g. prices and water supply), the actors' expectations (based on past experiences) of the development of disposal prices, costs of production factors, technological progress and limitations by the legal framework or restrictions concerning the availability of natural resources.	An actor is measuring the attractiveness of potential migration aims in every time step (monthly). A decision on migration is taken by a certain percentage of all actors of a type in a proxel per time step. The percentage results from the attractiveness of potential migration aims in comparison to the attractiveness of the area of origin.

Fig. 3.1 Overview of DeepActor models in DANUBIA

References

Barthel R (2011) An indicator approach to assessing and predicting the quantitative state of groundwater bodies on the regional scale with a special focus on the impacts of climate change. Hydrogeol J 19(3):525–546

Barthel R, Janisch S, Schwarz N, Trifkovic A, Nickel D, Schulz C, Mauser W (2008) An integrated modelling framework for simulating regional-scale actor responses to global change in the water domain. Environ Model Software 23(9):1095–1121

Barthel R, Reichenau TG, Krimly T, Dabbert S, Schneider K, Mauser W (2012) Integrated modeling of global change impacts on agriculture and groundwater resources. Water Resour Manage 26(7):1929–1951

Gilbert N, Troitzsch K (2005) Simulation for the social scientist. Open University Press, Berkshire

Moss S, Edmonds B (2005) Towards good social science. JASSS 8(4). http://jasss.soc.surrey.ac.uk

SinusSociovision (2007) Milieulandschaft 2007. http://www.sinus-sociovision.de/

Soboll A, Elbers M, Barthel R, Schmude J, Ernst A, Ziller R (2011) Integrated regional modelling and scenario development to evaluate future water demand under global change conditions. Mitig Adapt Strat Glob Chang 16(4):477–498

Chapter 4
Validation of the Hydrological Modelling in DANUBIA

Wolfram Mauser

Abstract The physically based, spatially distributed hydrological and land surface model PROMET is used in GLOWA-Danube to simulate the land surface and hydrologic processed as well as the impact of hydraulic structures on the natural water flows. The unique features, which make PROMET suitable for global change impact analysis in the field of hydrology within GLOWA-Danube, are explained, and PROMET was extensively validated for an extended climatic period from 1971 to 2003. The validation consisted of three steps which include the annual variability of the water balance over a climatic period, the daily variation of run-off and the annual peak discharges and low flows. The validation was carried out for the whole basin and selected subbasins differing in size and conditions. The validation shows good to very good results. Weak results can be attributed to human interventions as well as failures in hydraulic structures, which are not covered by PROMET.

Keywords GLOWA-Danube • PROMET • Ungauged watersheds • Hydrologic modelling

4.1 Introduction

The most accurate modelling of the hydrological cycle and the associated flows of water between the various hydrological components of the Upper Danube basin is a crucial feature for the exploration of future changes in the water cycle and the diverse consequences that result therefrom. The hydrological model that is used must be capable of providing accurate results under a broad range of environmental conditions, i.e. it can correctly simulate the flows of water in the basin, even under conditions in the future that may be remarkably different from today's situation.

W. Mauser (✉)
Department of Geography, Ludwig-Maximilians-Universität München (LMU Munich), Munich, Germany
e-mail: w.mauser@lmu.de

W. Mauser, M. Prasch (eds.), *Regional Assessment of Global Change Impacts*, DOI 10.1007/978-3-319-16751-0_4

The changes in the conditions relate to a wide range of factors, including, among others:

1. Climate change
2. Changes in land use and land cover (this includes the response of the vegetation to increased water stress and changes in CO_2 concentrations, as well as possible intensification of agriculture)
3. Changes in the water management and supply infrastructure as a result of the construction of new hydraulic structures (reservoirs, diversions, hydropower plants, flood protection measures, etc.)
4. Changes in the use of water, e.g. the introduction of irrigation

4.2 The Hydrological Model PROMET as a Supplement to DANUBIA

In the development of the DANUBIA components that describe the hydrological processes occurring at the land surface, considerable effort was made to realise the mutual interactions via appropriate interfaces, through which data is exchanged. The data represents the mutual interactions between the components. This effort generated the *Surface*, *Soil*, *Biological* and *Snow and Ice* components in DANUBIA. With a total of over 100 interfaces over which data exchange takes place, these components constitute a very comprehensive model system that reproduces well the complexity of the processes at the land surface. This structural complexity has corresponding effects on the performance of the *Landsurface* component. For this reason, it was decided to modify and make equivalent in its results the hydrological model PROMET (Mauser and Schädlich 1998; Mauser and Bach 2009) for operational purposes of simulating long and computationally intensive DANUBIA scenarios. PROMET was not developed in the context of the project and essentially is a non-object-oriented hydrological land surface model, which was developed in FORTRAN. It was integrated with DANUBIA to replace the DANUBIA's *Landsurface* component in order to increase the performance in the practical implementation of DANUBIA. The physical basis of PROMET is virtually identical with the original *Landsurface* component in terms of its description of the hydrological processes at the land surface, but avoids the high complexity of the interfaces because the components are internally coupled. The following principles are implemented in both DANUBIA and PROMET:

- Comprehensive physical and physiological description of the processes in the context of the flows of water in a mesoscale alpine drainage basin. This comprises the following processes:
  - Ingestion of meteorological drivers, either from regionally downscaled and bias-corrected climate models (Chap. 51) or from spatially and temporally interpolated station data (Chaps. 49 and 50).

- Energy and mass exchange between the land surface and the atmosphere, including the physiological control of gas exchange (interception, evapotranspiration, sensible heat flow, carbon exchange, radiation balance, momentum exchange).
- Dynamic, vegetation canopy development, which describes mechanistically the physiological processes related to photosynthesis, respiration, phenological development, canopy development and yield.
- The dynamics of snow and ice.
- Vertical and lateral, as well as saturated and unsaturated, flows of water (infiltration, interflow, surface discharge, groundwater flow).
- Channel run-off and flows through natural lakes as a self-organised process dependent on relief.
- Flows through hydraulic structures such as run-of-the-river hydropower stations, reservoirs and diversions, sewage plants, etc.

- Rigorous adherence to the laws of conservation of mass and energy.
- Spatial distribution in modelling the processes based on an isotropic spatial grid.
- Ensuring physical consistency and the predictive ability by largely avoiding calibration. This means that basin-wide the same rules apply to determine the spatially distributed values for the model parameters from first-order physical and/or physiological principles. This implies that parameter optimisation in subcatchments of the basin using measured discharges at singular locations of gauges is not applied. Instead, values from literature, measurements (on the ground or via remote sensing) and detailed relief analyses are used to initialise the spatially distributed model parameters. All methods used for determining parameter values are implemented in the same consistent way over the entire basin and that no separate rules and methods are applied to subbasins. In this way, it can be ensured that the method used for calculating the spatial distribution of a parameter also covers a broad range of values.

The interactions of the different elements of the hydrological cycle in PROMET are described in detail in its current form in Mauser and Bach (2009) and the cooperation between it and DANUBIA is shown in Fig. 4.1.

4.3 Validation of the Model

PROMET was set up as described above. The water flows for the meteorological period from 1970 to 2003 were modelled for the entire Upper Danube basin. This study period is longer than the typical climate period of 1971-2000. The extra initial year was included to spin up the groundwater reserves; the period was also intentionally extended to include the exceptionally warm year 2003 in the analysis. Simulation of the entire period was conducted continuously and with a 1-h model time interval. Analysis of the results was performed using measured discharges at sample gauges in the Upper Danube basin. These comprise both the outlet gauge at

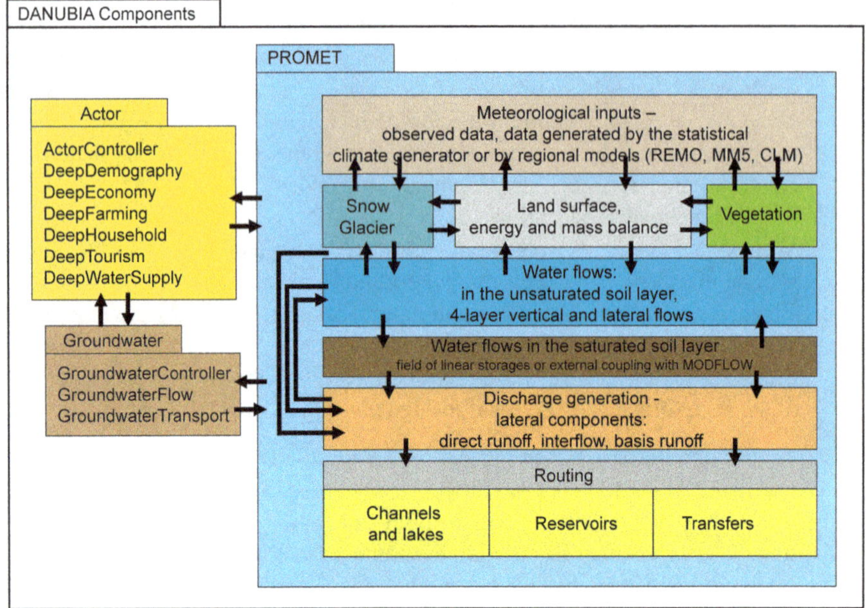

Fig. 4.1 Integration of subcomponents of the hydrological PROMET model and DANUBIA

Table 4.1 Selected gauges for the validation of PROMET 1971–2003

Gauge/river	Area [km^2]	Elevation gradient	MQ [m^3/s]	Discharge coefficient
Achleiten/Danube	76,673	2.96	1,590	0.55
Hofkirchen/Danube	46,496	2.09	640	0.57
Dillingen/Danube	11,350	1.67	162	0.57
Oberaudorf/Danube	9,715	4.07	307	0.78
Plattling/Isar	8,435	2.34	175	0.58
Laufen/Salzach	6,112	3.46	239	0.80
Heitzenhofen/Naab	5,431	1.48	49.8	0.40
Weilheim/Ammer	607	1.63	15.4	0.74

Reprinted from Mauser and Bach (2009, p. 370), with permission from Elsevier

Achleiten near Passau and gauges in the subbasins, selected such that a wide array of basin sizes and hydrological regimes were covered. The gauges selected and their basins are listed in Table 4.1.

Table 4.1 indicates the large variation in the size of the basins that were selected. They sometimes differ considerably in elevation gradient and in discharge coefficients from those at gauge Achleiten at the outlet of the Upper Danube. The modelled region of the Central Alps is primarily drained via the Inn and Salzach rivers. The corresponding gauges at Oberaudorf and Laufen have high discharge coefficients and elevation gradients. Parts of the northern alpine foothills drain via the Isar and Ammer/Amper rivers and their gauges (Plattling and Weilheim). They exhibit average discharge coefficients and elevation gradients. The lowest values come from the Naab subbasin in the northern part of the Upper Danube.

Validation of the hydrological model took place in three steps: first, the annual water balance was calculated for the whole basin and the subbasins. The results of this analysis give an indication of the validity of the surface model component to simulate water balance. Because of the high degree of temporal aggregation, they are not able to provide information about the validity of simulation of soil processes and lateral flows.

4.3.1 Annual Water Balance

Table 4.2 presents the results of a regression analysis (slope S, coefficient of determination R^2) as a complement to these results. The regression line was chosen such that it intersects the zero point. This approach results in the strictest possible criterion for a comparison between measurement and simulated values, since it is assumed that when a discharge is measured as 0 m^3/s, the modelling likewise yields 0 m^3/s and hence there is no offset. If the slope S has a value of 1.0, then there is no systematic under- or overestimation of the discharge by the model. If the coefficient of determination R^2 were also 1.0, then the measured and modelled datasets would be identical.

In general, the slopes approach a value of 1.0, with the highest positive deviation of 14 % originates from an overestimate at gauge Dillingen and the greatest underestimate of 7 % occurs at the gauge Salzach. The respective coefficients of determination are consistently high to very high, a result which suggests that the overall annual variability of the discharges over the 33-year period is well modelled both in the entire basin and in the partial subbasin areas.

Table 4.2 Gradient S and coefficient of determination R^2 of the linear regression of modelled and observed annual discharge at selected gauges in the Upper Danube basin, 1971–2003

Gauge	Gradient S	Coefficient of determination R^2
Achleiten	1.05	0.93
Hofkirchen	1.12	0.93
Dillingen	1.14	0.93
Oberaudorf	0.99	0.80
Plattling	1.03	0.88
Laufen	0.93	0.85
Heitzenhofen	1.01	0.86
Weilheim	1.09	0.88

Reprinted from Mauser and Bach (2009, p. 371), with permission from Elsevier

4.3.2 Daily Run-Off

In the second step, the hourly discharges were aggregated to form daily values. These were compared with the data measured at the gauges. The aggregation is required since artefacts would otherwise create larger discrepancies between the measured and modelled discharges. These discrepancies are caused by the fact that the temporal allocation of the hourly precipitation totals within the "Mannheim hours" (time intervals at 7:30, 14:30 and 21:30) (see Chap. 49 spatial interpolation) is not unequivocal (in contrast to hourly discharge measurements) and hence there are undefined temporal shifts in the modelled hourly discharges of approximately 5 h on average.

Figure 4.2 shows the pattern of measured and modelled daily discharges at gauge Achleiten for the years 1971–2003. It is notable that the measured and modelled courses are generally quite similar in terms of low flows and peak discharges. Relatively wet periods around 1980 are contrasted with dry periods, for example, in the years 1971 or 2003. To more closely examine the relationship between the two discharge curves in Fig. 4.2, they were subjected to a regression analysis using the same method that was used for the annual discharges above (see Table 4.2). The result is shown in Fig. 4.3. The modelled results indicate a slight trend towards overestimation of discharges (3 %) and a very high coefficient of determination, $R^2=0.87$. However, extreme discharges above 4,000 m³/s depict a fairly high degree of variation, suggesting that flood events are not modelled with the same precision as the more moderate discharges.

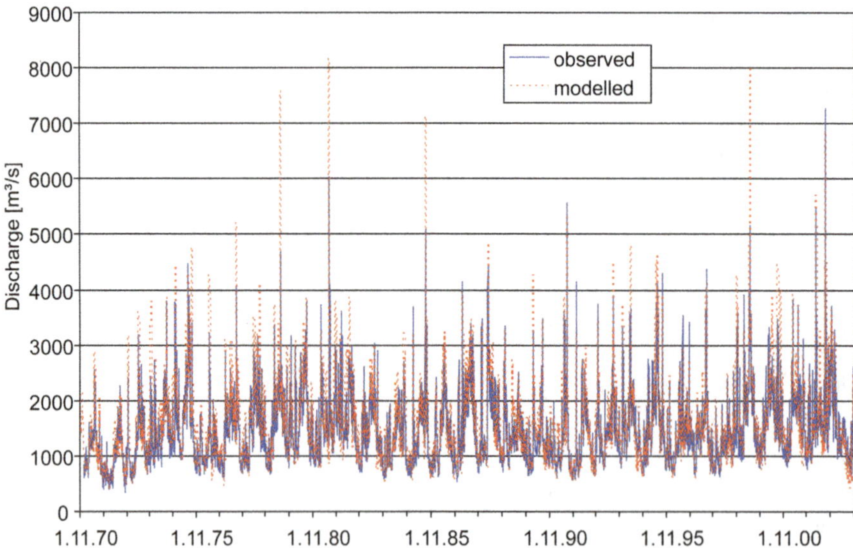

Fig. 4.2 Observed and modelled daily discharges at gauge Achleiten for the time period 1971–2003 (Adapted from Mauser and Bach 2009, p. 372)

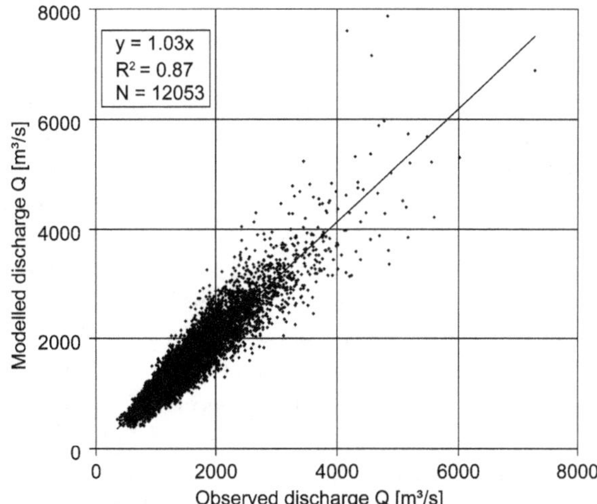

Fig. 4.3 Comparison of observed and modelled daily discharges at gauge Achleiten for the time period 1971–2003 (Reprinted from Mauser and Bach 2009, p. 372, with permission from Elsevier)

Table 4.3 Gradient *S* and coefficient of determination R^2 of linear regression and Nash-Sutcliffe coefficient of modelled and observed daily discharges at the selected gauges in the Upper Danube drainage basin, time period 1971–2003

Gauge	Gradient *S*	R^2	Nash-Sutcliffe coefficient
Achleiten	1.03	0.87	0.84
Hofkirchen	1.11	0.87	0.81
Dillingen	1.13	0.84	0.72
Oberaudorf	0.94	0.81	0.80
Plattling	1.08	0.75	0.47
Laufen	0.86	0.85	0.80
Heitzenhofen	0.99	0.78	0.79
Weilheim	0.98	0.73	0.69

Reprinted from Mauser and Bach (2009, p. 371), with permission from Elsevier

The method outlined in Fig. 4.3 was applied to all selected gauges and yielded the values shown in Table 4.3.

Compared to Table 4.2, in which the annual discharge volumes were contrasted for the various gauges, the differences in the slopes for the daily discharges are more marked. They range from a moderate overestimate by 13 % at Dillingen to a moderate underestimate of discharges by 14 % at gauge Laufen/Salzach. The coefficients of determination for the daily discharges are consistently high, but lower than those for the annual discharges (see Table 4.2); this result is the consequence of the greater variability of the daily discharges. Gauge Plattling is the most striking in terms of the coefficient of determination and the Nash-Sutcliffe coefficient; in this case, both values are relatively low and the modelling does not render a good result. The relatively poor correspondence between the measured and modelled daily discharges at this gauge is primarily the result of the influence of the Sylvenstein reservoir and the various diversions, for example, in the Inn region. Although the Sylvenstein reservoir is implemented within the model, its operation follows simple, normalised monthly rules that produce discharges based on the fill level (see also Chap. 29). These simple rules sometimes differ significantly from the actual operation of the Sylvenstein dam, which from an outside perspective is hard to comprehend. Data on the real operation of the dam was not available to the project. This highlights the limitations of physically based hydrological modelling that are imposed by human interventions, especially if the modelling should provide results about future changes in discharges based on climate change.

Therefore, if the catchment of the Isar up to gauge Plattling is excluded from further analyses, the coefficient of determination between the modelled and measured daily discharges in relation to the size of the subbasins can be examined. This analysis reveals a linear relationship between the coefficient of determination and the log of the subbasin area (see Fig. 4.4). The decrease in the coefficient of determination with decreasing subbasin area is not unexpected, since the number of proxels and hence the number of spatial sampling points in the model decrease at the same time. However, because of the fact that the indicated slope already explains 80 % of the variation in the data that form the basis of Fig. 4.4, it also appears that the coefficient of determination apparently only marginally depends on the location or the regime of the selected subbasin.

4.3.3 *Return Periods of Extremes*

For the third step in the validation, return periods were calculated from the measured and modelled annual discharge maxima and minima at gauge Achleiten and were then compared. The measured and modelled highest daily discharges for the period 1971–2003 are shown in Fig. 4.5.

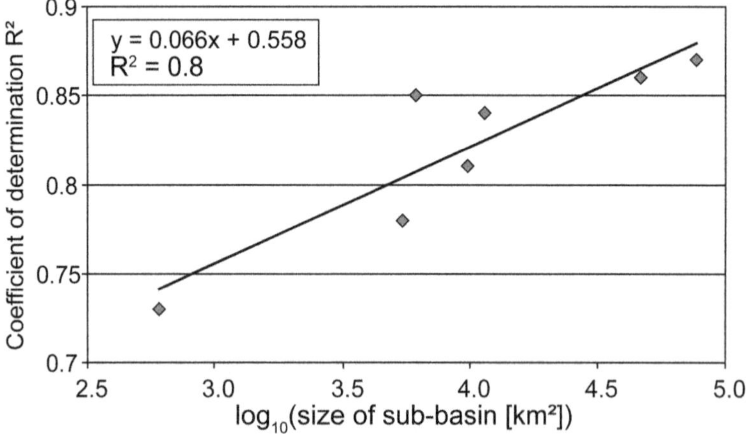

Fig. 4.4 Dependence of coefficient of determination R^2 on the area of the selected subbasin gauge Plattling wasn't included, due to the strong anthropogenic influences

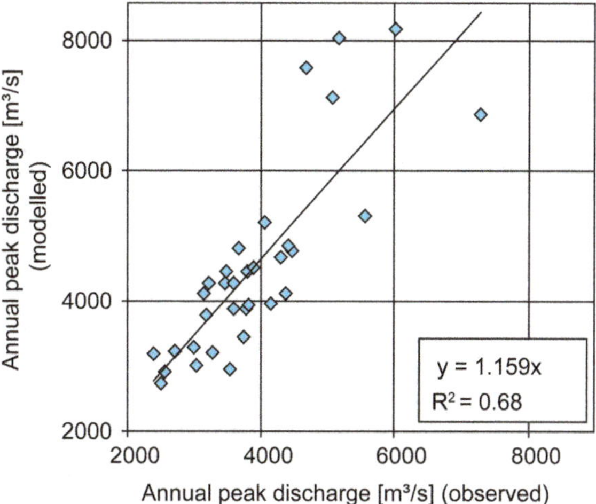

Fig. 4.5 Comparison of modelled and observed annual peak discharges at gauge Achleiten for the time period 1971–2003 (Adapted from Mauser and Bach 2009, p. 373)

As was indicated already in Fig. 4.2, in this figure there is a systematic overestimation of the peak discharges by an average of 16 % in the modelling results. The reason is the omission of inundation water diversions, which act as water storages in the case of larger flood events. The result is a reduction of the peak discharges. These cases of dam breaches are not accounted for in the modelling of the channel discharges. Also not included in the simulations are the specific operational control measures for the reservoirs present in the basin to partially retain the flood waves (e.g. by drawdown of the reservoir prior to the event). As shown by the example of the flood in August 2005 (LfU 2006), these strategies can contribute substantially to reducing the peak discharges. They add a totally new dimension to flood control and determination of design floods and need consideration of human decisions based on incomplete information. These decisions are currently not implemented within the model. Figure 4.6 illustrates the annual low flow discharges at gauge Achleiten. In this case, the mean 7-day discharge (NMQ7) is considered (see also Sect. 53.2).

There is no clear systematic deviation between measurements and simulations in case of low flows. Only 2 years stand out from the otherwise quite stable trend and show much higher measured values compared to the modelled low water discharges. The reasons for this deviation of simulated low flows could not be resolved.

Annualities of the return periods were determined from the annual maxima and minima by fitting probability distribution functions. The guidelines from the DVWK (1999) were used in this approach for the case of flood peaks. In the case of the low flows, it was assumed that a log-normal distribution best approximates the natural variability of low flow in the basin. The calculated return periods for high and low water are compared in Fig. 4.7. They correspond well with the values determined from measured data.

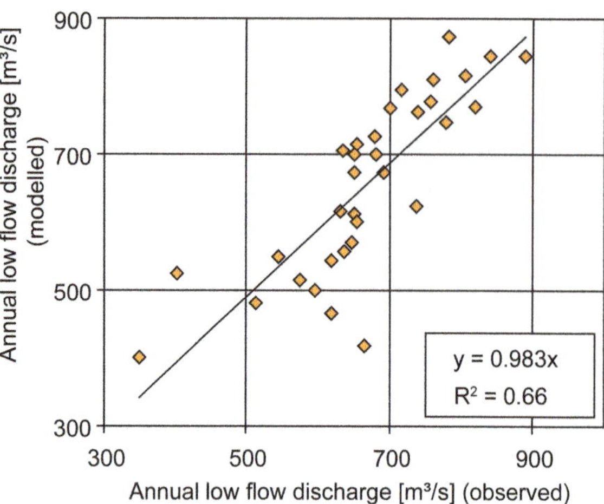

Fig. 4.6 Comparison of modelled and observed annual low flow discharges (NM7Q) at gauge Achleiten for the time period 1971–2003 (Adapted from Mauser and Bach 2009, p. 372)

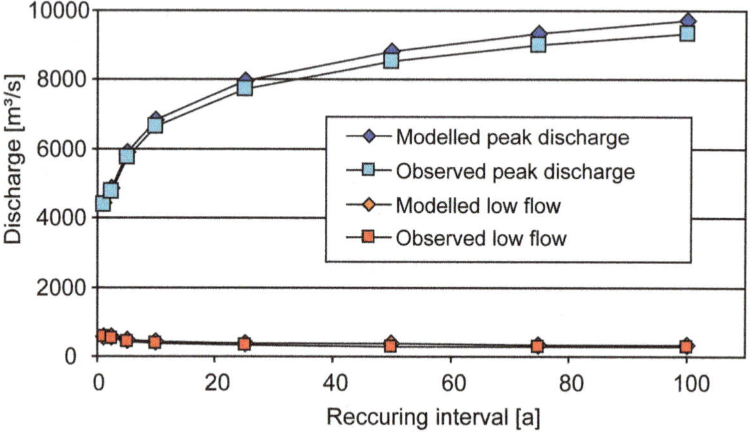

Fig. 4.7 Comparison of modelled and observed recurring intervals for flood and low flow discharges in Achleiten, 1971–2003 (Adapted from Mauser and Bach 2009, p. 374)

4.4 Summary

The validation process has shown that the method used in PROMET can model the spatial and temporal variability of the water flows in the Upper Danube basin with high accuracy. This is true both on seasonal and daily bases and for subbasins down to an area of around 1,000 km^2. The dynamics of the extremes are well modelled, with a slight tendency to overestimate the flood peaks. Since these sub-catchments include a large variety of natural conditions from high alpine watersheds to temperate lowland watershed, PROMET has proven to be able to cover a broad range of hydrological situations without specific calibration. We therefore conclude that it is capable of covering the changes in the hydrological situation induced by future climate change, which may happen in the analysis period from 2011 to 2060.

References

Bayerisches Landesamt für Umwelt (eds) (2006) Endbericht Hochwasser August 2005. Deutscher Verband für Wasserwirtschaft und Kulturbau, Augsburg

e.V. (DVWK) (1999) DVWK-Merkblatt 251/100. In: Statistische Analyse von Hochwasserabflüssen, Bonn

Mauser W, Bach H (2009) PROMET – large scale distributed hydrological modelling to study the impact of climate change on the water flows of mountain watersheds. J Hydrol 376:362–377

Mauser W, Schädlich S (1998) Modelling the spatial distribution of evapotranspiration on different scales using remote sensing data. J Hydrol 212–213(6):250–267

Chapter 5
The Stakeholder Dialogue in the Third Project Phase of GLOWA-Danube

Hannah Büttner

Abstract The article describes the process and results of the stakeholder dialogue in the third phase of the project. In this phase the objective of stakeholder engagement was to systematically orient the dialogue toward the development and evaluation of the scenarios that were designed, as well as to focus on the results of the simulations and their implications for relevant stakeholder groups. Another aim was to obtain the approval of the potential users of DANUBIA and its model results.

The dialogue process was developed and implemented by IFOK, whose role in the process was a neutral moderator and facilitator, acting as the mediator to promote the dialogue equally on both sides – scientists and stakeholders – for collectively developing scenarios and options for action. IFOK took care of the temporal coordination of the work and if necessary heed the responses of the stakeholder questions at GLOWA-Danube.

As a result of the stakeholder dialogue, vital groundwork was laid for implementation as the stakeholders were made aware of adaptive courses of action under uncertainty; the stakeholders accepted the methods of GLOWA-Danube and tabled relevant issues for which a response was provided. In this way, there was support for building trust in and enhancing the relevance of the results of GLOWA-Danube. Furthermore, it was evident that the dialogue and the workshops generated an open, multi-sectoral network on the subject of climate change and the supply of water in the region. Together, a mutual learning process and development of expertise were undertaken.

Keywords GLOWA-Danube • DANUBIA • Stakeholder dialogue

H. Büttner (✉)
IFOK GmbH, Bensheim, Germany
e-mail: hannah.buettner@ifok.de

W. Mauser, M. Prasch (eds.), *Regional Assessment of Global Change Impacts*, DOI 10.1007/978-3-319-16751-0_5

5.1 Implementing DANUBIA Globally – The Stakeholder Dialogue in the Third Phase

GLOWA-Danube focuses on the development of the integrated decision support system DANUBIA for long-term and sustainable water (use) management in the Upper Danube basin. To facilitate this process, the scientists that make up GLOWA-Danube have created a variety of coupled sub-models that can contribute diverse pioneering studies in the field of research on global change. The scientific work of GLOWA-Danube was accompanied by systematic dialogue with policymakers, affected groups and potential users of DANUBIA to ensure its relevance in practical terms. The scientists maintained an intensive exchange with these stakeholders throughout the entire project and could therefore draw valuable knowledge from practical experience.

In the third phase of the GLOWA-Danube project, the stakeholder dialogue was maintained through IFOK GmbH. The objective of this decision was to systematically orient the dialogue toward the development and evaluation of the scenarios that were designed, as well as to focus on the results of the simulations and their implications; another aim was to obtain the approval of the potential users of DANUBIA and its model results.

With the involvement of a key stakeholder, the Bavarian State Office for the Environment (LfU), as a project partner, an excellent condition for stakeholder dialogue was created for GLOWA-Danube in Phase III.

IFOK oriented and designed the project toward the following goals and responsibilities:

- *Establishing DANUBIA as a support system for decision-making.* The stakeholder dialogue should ensure that decision makers are informed about DANUBIA in a timely manner and that feedback and suggestions can be directed to GLOWA-Danube.
- *Enhancing the quality and relevance of the results.* The stakeholder dialogue has the task to organize the formulation and evaluation of the scenarios jointly with the stakeholders.
- *Supporting the development of adjustment strategies.* It is the mandate of the stakeholder dialogue process to facilitate a multi-sectoral discussion about project results in anticipation of conflicts and to discuss solution options.

Thus, key issues for the discussion process include: What should GLOWA-Danube accomplish to become even more relevant for the stakeholders? How can DANUBIA aid in solving specific issues, for example, for decisions regarding long-term investments in the context of river basin planning under the EU Water Framework Directive or in the development of strategies for adjustment to climate change?

As a result, from the outset, the question was addressed as to how the results of GLOWA-Danube can be established and implemented in the region, once the project has been completed and in view of the sustainability and practical relevance of the project.

5.2 Structuring of the Stakeholder Processes

In terms of the transdisciplinary research, the stakeholder dialogue for GLOWA-Danube was conceived by the IFOK as an ongoing and iterative process between the stakeholders and the scientists. This process facilitates open communication in both directions – from the scientists to the stakeholders and vice versa.

The following criteria were included to be able to identify the relevant stakeholders for GLOWA-Danube (see Fig. 5.1):

- Region: Upper Danube drainage basin (i.e. Bavaria, Baden-Württemberg and Austria)
- Field of action: water management, agriculture and forestry, power generation, shipping, water suppliers, tourism
- Social realms: politics, administration, economy, civil society/organisations

Ultimately, stakeholders were considered in terms of their potential roles in the project. In one respect, those involved should be approached as potential users of DANUBIA and its results.

In another sense, there should be active involvement by those individuals or groups who can contribute as decision makers, as users of water or as being potentially affected by the content of the scenarios and who can design options for action.

Overall, approximately 275 stakeholders were contacted; of this total, 90 were individuals from 40 institutions which actively participated in the workshops of the project, and another 30–40 were involved in the form of bilateral discussions and interviews.

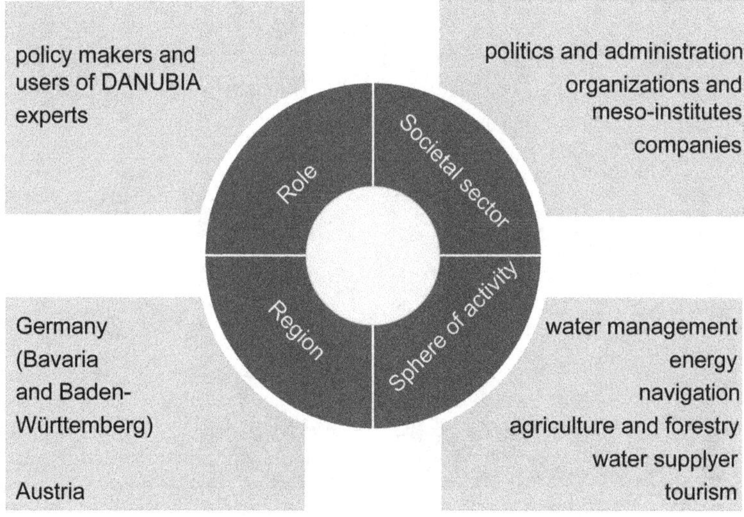

Fig. 5.1 Stakeholder of GLOWA-Danube

In the organisation of the dialogue process, there was close cooperation of the stakeholder participation in terms of content and time both with the individual tasks of the scientists and with the ongoing activities of the stakeholders. This close collaboration with the stakeholders was also a challenge for the scientists involved in GLOWA-Danube. They were constantly required to accommodate new proposals and suggestions from the stakeholders; however, these factors also provided valuable stimulus for their work.

The role of IFOK in this process was as a neutral moderator and facilitator, acting as the mediator to promote the dialogue equally on both sides – scientists and stakeholders – and to take care of the temporal coordination of the work and if necessary to heed the responses of the stakeholder questions at GLOWA-Danube (the "process motor").

5.3 Realisation of the Stakeholder Dialogue

In initial preparation, an agreement was required between the subprojects of GLOWA-Danube and IFOK. On the basis of a jointly developed understanding, the central working steps of the stakeholder dialogue were then drafted and implemented. As a result, the dialogue process consists of several systematically interconnected steps (see Fig. 5.2). This involves an alignment of the modelling and scenario development processes of GLOWA-Danube before and after every step.

First, the needs and expectations of the stakeholders from various fields of action and social realms were sounded out using guided interviews. This survey provided both content-related knowledge for the scientists of GLOWA-Danube and process-oriented information for the moderators to assist in further drafting of the dialogue process.

Next, select decision makers were informed at an early stage about the project and stakeholder dialogue (in a "Road Show"). This involved theme-based teams of

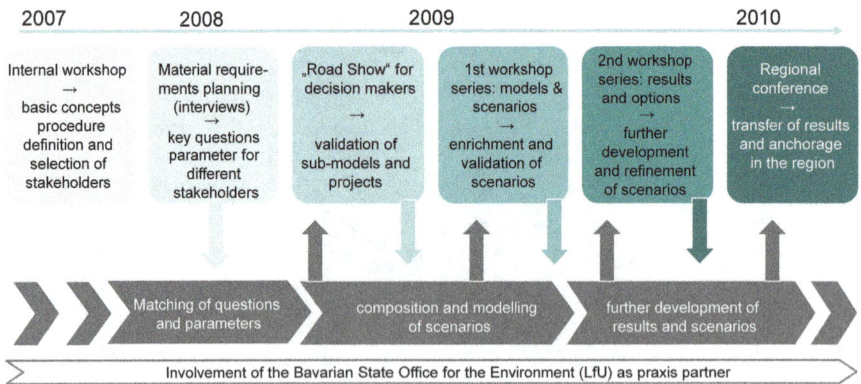

Fig. 5.2 Process scheme of the GLOWA-Danube stakeholder dialogue

scientists from GLOWA-Danube, who were brought together to hold mostly half-day facilitated round-table discussions on site, together with the moderators.

Building upon these two first steps, the central aspect of the dialogue process began in the form of two theme-based series of workshops separated by a year. Each series consisted of three full-day facilitated sessions on the topics of (1) energy and infrastructure, (2) agriculture and forestry and (3) water supply and tourism.

The first workshop series served in the securing of the sub-models, the joint development of the scenarios and the solicitation of the stakeholder requests and demands for GLOWA-Danube. The requests of the stakeholders were modelled in collaboration with the scientists and were prepared for the second workshop series. In the second series, the further developed results could be put forth. In addition, additional courses of action and needs, as well as the implications of the results for policy and communication, were discussed within the individual target groups.

In all sessions, the expedient exchange between the stakeholders and the scientists was essentially improved and facilitated by implementing the structured techniques of discussion moderation.

5.4 Results and Conclusion

The results of the stakeholder dialogue are distinguished in terms of content and the goals of the dialogue. The content-specific results can be found in several contributions, wherein the stakeholder wishes are addressed and responded to from the point of view of GLOWA-Danube (see Part V). In terms of the goals of the dialogue (primarily enhancing acceptance and relevance of the results), the following assertions can be made:

- Securing of the models and development of the scenarios took place in dialogue with the stakeholders.
- The stakeholders accepted the methods of GLOWA-Danube and tabled relevant issues for which a response was provided. In this way, there was support for building trust in and enhancing the relevance of the results of GLOWA-Danube.
- The stakeholders were made aware of adaptive courses of action under uncertainty. Thus, the vital groundwork was laid for implementation.

Furthermore, it was evident that the dialogue and the workshops generated an open, multi-sectoral network on the subject of climate change and the supply of water in the region. Together, a mutual learning process and development of expertise were undertaken. This network should be sustained, for example, through an intermediary multi-sectoral platform in the region. It is expected that this network is the foundation for further intermediary multi-sectoral cooperation within the region.

Chapter 6
GLOWA-Danube Results and Key Messages

Monika Prasch and Wolfram Mauser

Abstract The research project GLOWA-Danube used a transdisciplinary approach to explore the regional impacts of global change (climate, demography, economy) on the availability and use of the water resources in the Upper Danube basin in the period from 2011 to 2060. An intensive dialogue with stakeholders in the region resulted in consolidated scenarios for the future development, which were simulated with the integrated simulation tool DANUBIA, specifically developed for this purpose within the project. The key messages and results are complied in this paper. They show that water resources will become scarcer in the future but not scarce. The range of uncertainty depending on the selected scenario is documented. The significant impacts on a large variety of natural processes and societal activities are compiled.

Keywords GLOWA-Danube • DANUBIA • Upper Danube • Climate Change • Key messages

6.1 Introduction

In order to investigate the regional effects of climate change in the Upper Danube catchment, the integrative decision support system DANUBIA was developed within the frame of the research project GLOWA-Danube. It consists of various natural and social sciences-based components (see Chaps 2, 3, 4 and 24, 25, 26, 27, 28, 29, 30, 31, 32, 33, 34). DANUBIA was successfully validated with an extensive data set from the past, which proved the value of the data-rich Upper Danube watershed as a field laboratory for global change research. Several versions of combinations of impacts of global change factors on the regional water resources were simulated using an ensemble of scenarios for the future development of climate and society, which spanned a feasible range of possible developments and spanned the scenario funnel schematically described in Fig. 6.1.

M. Prasch (✉) • W. Mauser
Department of Geography, Ludwig-Maximilians-Universität München (LMU Munich), Munich, Germany
e-mail: m.prasch@lmu.de; w.mauser@lmu.de

W. Mauser, M. Prasch (eds.), *Regional Assessment of Global Change Impacts*, DOI 10.1007/978-3-319-16751-0_6

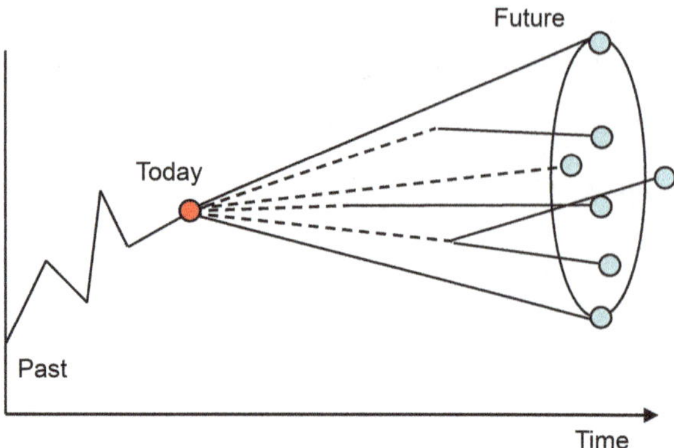

Fig. 6.1 Schematic funnel of a range of scenarios exploring the future

The main focus was placed on examining the natural resource water and its usage. The climate scenarios developed cover the time period 2011–2060. They are based on the moderate A1B emission scenario from the Fourth IPCC Assessment Report. The following statements summarise the results achieved by the GLOWA-Danube team of scientist.

6.2 The Methodological Approach of GLOWA-Danube

Both the regional impacts of climate change and the potential adaptation strategies are complex due to the multiple linkages and interactions between climatic, geographic and social factors. These complex linkages often make the analysis of direct cause-effect relations difficult or even impossible. DANUBIA was therefore developed as a simulation tool from scratch to take these linkages into consideration and to represent complex interactions and feedbacks. DANUBIA makes use of the latest software tools, such as Unified Modeling Language (UML) for design and parallel distributed computing. The development of DANUBIA was completed successfully. It has proven itself as a flexible framework to interactively couple the various components of the different disciplines involved in GLOWA-Danube and to realistically represent their interactions. Meanwhile, the framework has proved its operational capability in a variety of different applications.

DANUBIA is an expandable, scale-independent model that can be regionalised. It is available as open-source for further simulations of future human-environment interactions, applicable to a wide range of questions.

6.3 The Regional Development of Climate

We were able to show that a significant rise of surface air temperature already took place in the past in the Upper Danube watershed (see Chap. 23). The measured temperature increase in the Upper Danube from 1960 to 2006 was 1.6 °C, which is more than twice as high as the global average. Beyond this general trend of increasing temperatures, the available scientific knowledge on the possible future temperature development opens up a wide range of scenarios. Our discussions with the stakeholders regarding our findings and the available knowledge on climate scenarios led to the common denominator that the impacts of a likely temperature increase ranging between 3.3 and 5.2 °C over the period 1990–2100 should be analysed further (see Chaps. 47, 48, 49, 50, and 51).

On the basis of measured precipitation data from the past, a tendency for decreasing precipitation in the summer months and increasing precipitation in the winter months was observed (see Chap. 13). This trend has been taken into account in the development of the regional climate scenarios for the Upper Danube. We can expect more precipitation in winter (between +8 and +47 %) and less precipitation in summer (between −14 and −69 %) in the Upper Danube watershed. Overall, the annual precipitation will decrease slightly in the future (see Chaps. 47, 48, 49, 50, and 51).

Although moderate assumptions for the future climate were made based on the consensus reached among and with the stakeholders and none of the "worst-case" scenarios of the global emissions or climate development were considered, the change in the climate of the Upper Danube watershed is clearly more significant than the global average – especially when considering the strong climate change signal already observed in the past.

6.4 The Development of Society

The socio-economic scenarios in GLOWA-Danube were based on the societal megatrends from SinusSociovision of Sinus Institut, Heidelberg (http://www.sinus-institut.de/), which refer to the conditions in society as a whole and their likely change in the future. Thus, the societal scenarios hold a correspondingly high level of abstraction. To implement these in GLOWA-Danube, specifications of the more general megatrends were developed and adapted to each participating subproject. For the development of the social-based orientation towards the future, three scenarios were developed which take into account inter alia new technologies, the globalisation and demographic, economic and political development (see Chap. 52). The scenario baseline represents the current status quo and assumes this status quo valid in the future. The specific realisation depends on the respective discipline and its component in DANUBIA. The scenario public welfare describes a society, which is characterised by a return to the responsibility of the whole society and by placing a high value on general welfare and sustainable development. The scenario *Performance* describes the opposite trend to the scenario public welfare. In this scenario more emphasis is placed on economic efficiency and the performance of the individual. Figure 6.2 shows the menu of scenarios, which allows to formulate a total of 60 scenarios as combinations of selections 1–3. They can be used to investigate the effectiveness and efficiency of adaptation actions in selection 4, which are proposed by the stakeholders.

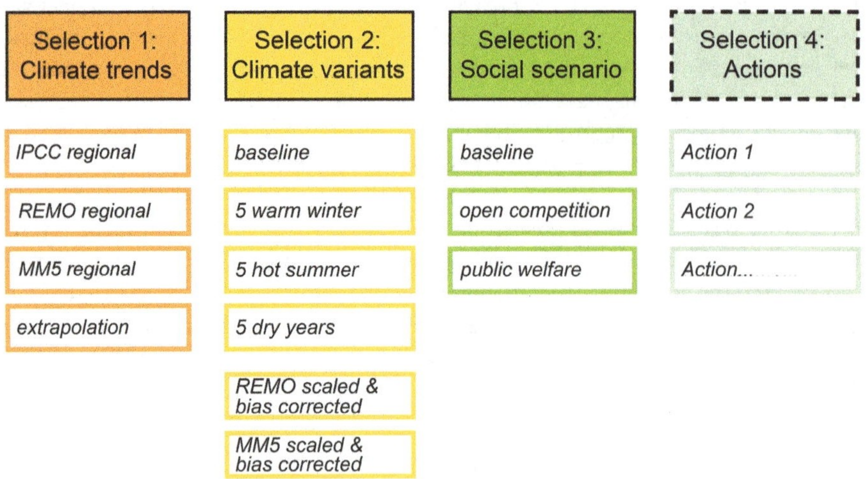

Fig. 6.2 Menu of climate scenarios and social scenarios in GLOWA-Danube, which can be combined

6.5 Water Balance

The simulated results show that the water availability in the Upper Danube will decrease during the period 2011–2060, but water will not become scarce (see Chap. 53). Comparing the period 2036–2060 with 1971–2000, a decrease in water availability is found which varies between 5 and 25 % depending on the climate scenario.

Apart from the slight decrease in precipitation, a complex network of interactions, primarily the increasing air temperature coupled with a strongly increasing evapotranspiration, is responsible for this development. As such, the evapotranspiration will increase on average by about 10–25 %, depending on the applied climate scenario (see Chap. 58). As a result of the decreasing precipitation, the rising temperature and the higher evapotranspiration, the river discharge in the Upper Danube catchment will be reduced in the future as well. The reduction varies between 5 and 35 % until the year 2060 depending on the applied climate scenario. At a regional level, the reduction will be least in the Alps, whereas along the Danube River, it will be the strongest. Thus, the annual water delivery of the Upper Danube from the gauge in Achleiten to the downstream users will be reduced by 9–31 %, by 2060, based on the different scenarios (see Chap. 53).

The groundwater recharge in the total catchment area will be reduced by 5–21 %, when comparing the time period 2036–2060 with 1971–2000, due to the increase of evapotranspiration and the slight decrease in precipitation (see Chap. 59). Increasing temperatures lead to a strong reduction in the depth of the snow cover and to a reduction of snow cover duration by 30–60 days at all altitudes, until the year 2060 (see Chap. 57). Snow conditions, which in 2006 prevail at altitudes of about 1,000 m asl, are likely to be found at altitudes of 2,000 m asl around 2060. In the summer months, precipitation on high peaks will increasingly fall in the form of rain instead of snow. As a consequence, there will be less snow storage available in the mountains; therefore, a decrease in the portion of snowmelt water contribution on the total water discharge is expected (see Chap. 60). The intense reduction of the snow storage and the earlier snow melting in the Alps lead to a pronounced forward displacement of the annual peak of river discharge from summer to spring, as well as to a strong to very strong reduction in the low-flow discharges of the main rivers in the Upper Danube catchment (Fig. 6.3).

The low flow at the gauge in Achleiten near Passau will be reduced by 25–53 % by the year 2060. In combination with increasing water temperatures, this may result in a reduction of the availability of water for cooling thermal power stations and to restrictions posed on navigation during summer (see Chaps. 54 and 61). The strong reduction in the low-flow discharges along the Danube will be opposite to an expected increase in the low-flow discharges in the Alpine valleys. Reasons for this are the complex interaction between a higher proportion of precipitation expected to fall as rain in winter and an increased evapotranspiration with reduced rainfall in the summer. The increased proportion of rain and snow melt contribute to an increase of low-flow discharges in the Alpine valleys, while the higher evapotranspiration and decrease in precipitation in the summer are conducive to accentuated low-water situations in the Alpine foothills and in the northern part of the Upper Danube catchment.

Fig. 6.3 Glacier skiing area “Schneeferner” at the Zugspitze (Photo Markus Weber)

The reduction in the mean annual run-off and the shift of the peak in spring in turn influence the energy production from hydropower. Thus, under a constant, business-as-usual reservoir management, a change in the seasonal cycle of the inflows and outflows can be expected, as well as a more even filling of the reservoirs towards a more balanced seasonal cycle (see Chap. 68) (Fig. 6.4).

The electricity generated from the installed hydropower plants is a main source of renewable energy in the Upper Danube catchment. The amount of electricity generated will be significantly reduced by 3–16 %, due to the decrease in run-off in the future, the extent of which varies depending on the chosen future climate scenario and the subcatchment considered. The decrease is particularly strong after the occurrence of several consecutive dry years. In the southern area of the watershed, the drop in hydropower energy production is slightly attenuated during the first-scenario years because of an increase in run-off from the melting glaciers (Fig. 6.5).

Indeed, the glaciers in the catchment of the Upper Danube will have disappeared almost completely between 2035 and 2045. The simulations show that the water stored in glaciers at present cannot provide an essential additional water contribution necessary to ensure and maintain today's water balance in the Upper Danube catchment. However, the melting of glaciers will lead to a slight increase in discharge in Passau of about 2 % during the period 2011–2035. In the headwater regions, the glaciers therefore make a non-negligible contribution to the increase in the low flows during this period. After the year 2035, the meltwater flow from the glaciers will go down due to disappearing glaciers. This will further amplify the general decrease in run-off (see Chap. 56).

Regarding the development of the natural flood discharge in the Upper Danube basin, no clear trend emerged from our research. The results suggest that at the gauge in Achleiten, there will be no major changes to the flood peaks and the frequency of their occurrence. However, the results do show a clear increase in the flood peaks in the Alpine valleys and in the head watersheds. In the head watersheds, the flood peaks increase until 2060 in part by a factor of 3 (see Chap. 55).

Fig. 6.4 The "Finstertal" reservoir in the Kühtai/Tyrol (Photo Markus Weber)

Fig. 6.5 Glacial ice in the Ötztal (Photo Markus Weber)

Fig. 6.6 Flooding in a mountain stream in the Ötztal in August 2006 (Photo Markus Weber)

Both the increasing flood peaks and the aforementioned changes in the low-flow discharge can be attributed mostly to the change in the precipitation type in alpine regions from snow to rain and the resulting reduction in water storage contribution from snow (Fig. 6.6).

The results presented show the partly severe consequences of climate change on the water resources of the Upper Danube. The water supply to the downstream users, who depend on the Danube river water and who use it intensively, will be moderately to significantly reduced in the future. The synopsis of the results shows that the role of the Upper Danube as a "surge chamber" for the Danube downstream users should be re-evaluated for the future.

6.6 Water Consumption and Water Supply

The simulations of the water consumption behaviour of households show that the private per capita consumption of water in the Upper Danube basin during the period under consideration, from 2011 to 2060, will be significantly reduced (Fig. 6.7).

In the second half of the simulation period, a significant slowdown of this reduction is observed. The decrease can mainly be attributed to a widespread implementation of water-saving technologies in households and the changing consumer behaviour. The adoption of water-saving technologies varies depending on innovation, environment and social scenario. In some cases even, a universal spread of technologies is achieved. The decline in the per capita water consumption is partly compensated by the rising population numbers at the beginning of the simulation period. However, overall, the simulations show a decline in private drinking water consumption by about 20–25 % until 2060 (see Chaps. 65 and 67).

The reduction in the groundwater recharge, which is observed in all climate scenarios, leads to only occasional isolated, local, temporary shortages in drinking water supply, and this is only in the second simulation period (2036–2060) under

Fig. 6.7 Different types of water consumption (Artwork: Center for Environmental Systems Research, University of Kassel by Anna von Lilienfeld-Toal)

declining withdrawals. The intensity of shortages naturally increases with decreasing rates of groundwater recharge. The areal extent of the decline in abstractions plays only a minor role thereby. Areas that are particularly affected are those with very small-scale supply structures, which withdraw the groundwater from shallow, spatially limited aquifers (e.g. in north-eastern Bavaria). At the same time, there is a tendency towards a strong precipitation decline at the northern edge of the Alps and in parts of the alpine foothills which affects groundwater recharge. Under an assumed declining water demand, the water availability in the catchment area will be sufficient secure public drinking water supply, even under extreme climatic conditions. However, this will require appropriate adaption measures to buffer the local and temporary shortages, for example, tapping into alternative areas, or enabling water delivery from water companies in the proximity, or from remote water supply systems. Currently, it is not possible to estimate the impacts of the potential need for irrigation water for agriculture. The impacts of climate change on deep water aquifers during the simulation period cannot be significantly determined as these react slowly to changes, and the uncertainties in the water-bearing stratum of the models are too high.

6.7 Winter and Summer Tourism

The changes in winter tourism are characterised by dwindling snow cover durations, depending on regional and elevation factors. This may span 30–60 days, depending on the chosen climate scenario. The diminishing guaranteed snow cover at lower elevations intensifies the concentration of winter tourism at higher-located and well-developed ski areas with adequate infrastructural facilities. In these skiing areas, due to the increase in precipitation expected in winter, the snow conditions will not deteriorate despite higher temperatures. In fact, snow conditions may even improve in some regions. However, a decrease in the number of optimum skiing days is expected in the whole study area. Optimum skiing days are characterised by different factors, such as a lack of rainfall, adequate snow cover, sunshine, little wind speed and pleasant temperatures (Fig. 6.8).

Due to the high investment costs for producing artificial snow and due to a lower guarantee of snow cover, an economically viable operation will not be sustained in some low-lying ski areas, especially since the advent of higher temperatures will often cause the use of snow cannons to be impossible. In the second half of the simulation period, depending on the chosen scenario, between 20 and 50 % of today's ski areas will no longer be able to secure their existence through ski tourism (see Chap. 62). As a result of higher temperatures in the summer, locations with a high percentage of holiday travellers may reckon with a growth in the number of visitors, which may compensate for the losses experienced during the winter season, to some extent. Climate change therefore also affects summer tourism, although to a lesser extent than the winter tourism.

Fig. 6.8 Snow cannon in the Stubai (Photo Markus Weber)

6.8 Agriculture and Forestry

All of the investigated climate change scenarios show that the increasing atmospheric CO_2 concentrations and the higher temperatures will lead to an increase in crop yields (see Chap. 70). The water-use efficiency of the vegetation (ratio of biomass production to water transpired) will improve significantly for C3 plants. Thus, transpiration amounts do not increase proportionally to the amount of biomass produced. Occasionally, there may be reductions in the yields of crop grown on light soils due to water stress during dry years as a consequence of soil drying, especially along the Danube River (Fig. 6.9).

The mineralization of organic matter in the soil will increase. Thus, the soil nitrogen availability will improve, provided that the soil contains sufficient organic matter. These effects, which are influenced inter alia by the temperature increase of the upper soil layer (see Chap. 71), vary to different degrees, depending on the local climate factors at a small scale.

The harvest dates of grain cereals will be advanced by about 3 weeks. In addition, the harvest dates of summer grain and winter grain will be closer in the season. However, the ratio of different crop types will hardly change during the period under consideration. The number of days with rain at the time of the cereal harvest will be reduced. However, since the interannual variability will increase, it can be generally assumed that the planning reliability for farmers will decline in the future (see Chap. 76).

Fig. 6.9 Mature spring wheat (Photo Markus Weber)

The concentration of nitrate in the percolation water increases slightly during the scenario period; however, the climate impacts on nitrate leaching can be controlled from a perspective of water pollution management through appropriate adaptation strategies such as fertiliser application management (see Chap. 72). In the future, merely locally, there will be a menace to the quality of groundwater through nitrate leaching. This will occur in regions where already high background concentrations of nitrate loads are present and a continuing need for action will exist (see Chap. 73). None of the investigated societal scenarios demonstrated a climate-related deterioration of the income situation of farmers (see Chap. 70).

In all climate scenarios, an increase in the wildfire risk is expected due to the higher potential evaporation and to less rainfall occurring in the spring and summer months (see Chaps. 74 and 75).

6.9 Industrial Water Usage

The vulnerability of the industrial sector to climate change can be assessed as being low in the Upper Danube catchment. There are solely regionally limited losses of growth, of up to 0.4 per mill per year. In some regions, economic growth even benefits from climate change. The industry responds to water shortages foremost by optimising their processes and subsequently with circuit or multiple uses, and thus they can avoid a production constraint as a consequence of a resource shortage (see Chap. 66).

Part II
Data

Chapter 7
Digital Terrain Model

Anja Colgan and Ralf Ludwig

Abstract The digital terrain model (DTM) is one of the basic data sets for the characterisation of the process pixel of the DANUBIA model. It is fundamental for describing vertical and lateral processes in the study area. Its accuracy is of utmost importance for the modelling and the exchange of water flow between the DANUBIA model objects, Soil, Rivernetwork and Groundwater. To assemble a hydrologically consistent DTM for the catchment area of the Upper Danube, difficulties due to differences in the resolution and the data quality of the input data sets from the various national sources had to be resolved. The initially derived 50 m resolution DTM was aggregated step by step to the model's 1,000 m resolution. It was found that retaining the minimum elevation of river network pixels at each step reduced the undesired smoothing effects during aggregation. The river network was subsequently derived using the software TOPAZ. The resultant 1,000 m resolution DTM and river network had a good accuracy. The generated river network correlated well with the actual river network of the Upper Danube.

Keywords GLOWA-Danube • Upper Danube • DTM

7.1 Introduction

The Upper Danube catchment area is characterised by an exceptionally diverse landscape. This landscape includes the glacial high mountain regions (Piz Bernina is the highest point at 4,052 m a.s.l.) and forested low mountain ranges as well as hilly landscapes, deeply carved river ways and vast plains. The drainage outlet at gauge Achleiten, near Passau, is situated at 288 m a.s.l. The relief energy is

A. Colgan
GB Dr. Schönwolf GmbH & Co. KG, Augsburg, Germany
e-mail: anjacolgan@aol.com

R. Ludwig (✉)
Department of Geography, Ludwig-Maximilians-Universität München (LMU Munich), Munich, Germany
e-mail: r.ludwig@lmu.de

W. Mauser, M. Prasch (eds.), *Regional Assessment of Global Change Impacts*, DOI 10.1007/978-3-319-16751-0_7

therefore 3,764 m. The resulting gradients are reflected in the climatic conditions and the patterns of land use (Map 7.1).

In all spatial model approaches, the depiction of topography as a digital terrain model (DTM) is fundamental for describing vertical and lateral processes. The topographic data of the study area forms an important empirical basis for the numerous modelling aspects of the project groups involved and is therefore established as one of the basic data sets in the characterisation of the DANUBIA proxel. Sensitivity studies have shown that the 1 km^2 resolution of the original digital terrain model (GTOPO30) used by the USGS (US Geological Survey) is not able to provide sufficient precision for the requirements specified by the project partners for a realistic representation of water runoff in the Upper Danube catchment area. Especially the interfaces between the model objects, Soil, Rivernetwork and Groundwater, require a highly accurate topography, because in this case the exchange of water runoff requires hydrological consistency in the data. Furthermore, high-resolution information about topography is an essential prerequisite for high-quality geometric and radiometric processing of remote sensing data and for the derivation and interpretation of other surface data sets, such as soil distribution.

The quality of digital terrain models is of particular importance at model scales of 1 km^2. This coarse resolution presents the risk of huge loss of information through generalisation. Methods must be in place to ensure a depiction of a hydrologically consistent river network that is true to location and which preserves the key aspects of the topographical variability.

7.2 Data Processing

Recording and processing of topographical data is usually carried out by national surveying agencies. Because there is not yet one international standard that is implemented by all national agencies, there are occasionally serious problems, if the study area crosses state or federal boundaries. The resulting data sets differ in data status, technical generation and processing, geometry, resolution, quality and availability. The topography of the international catchment area of the Upper Danube is thus only rendered by means of a mosaic of digital terrain models from the various countries. The need for a high-resolution digital terrain model requires access to a variety of data sources. Processing of the data must resolve the problems that arise from the variation in data quality. Figure 7.1 illustrates the geographical coverage by the individual data sets. For the white region (6 – Italy, Switzerland, Czech Republic), only the global digital elevation model (DEM) of the USGS, with a resolution of 1,000 m, was available. Table 7.1 lists the data sources and their resolutions.

To begin with, the individual data sets were transformed into a single projection (Lambert conformal conic projection, international ellipsoid after Hayford, Potsdam

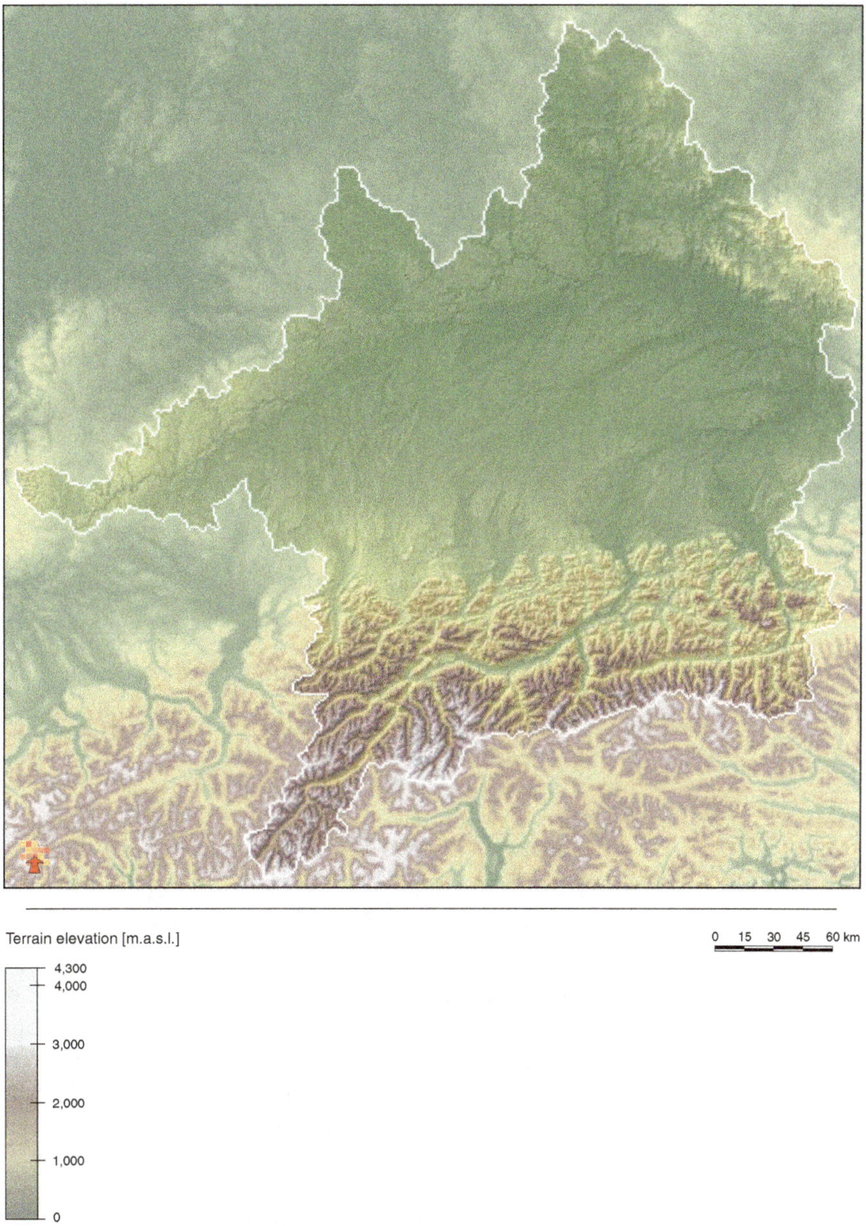

Map 7.1 Digital terrain model relief representation (twofold vertical exaggeration, Jarvis et al. 2008) (*Note*: Model calculations are based on DTMs shown in Table 7.1, but in maps and figures in this book hole-filled seamless SRTM data V4 (Jarvis et al. 2008) is shown)

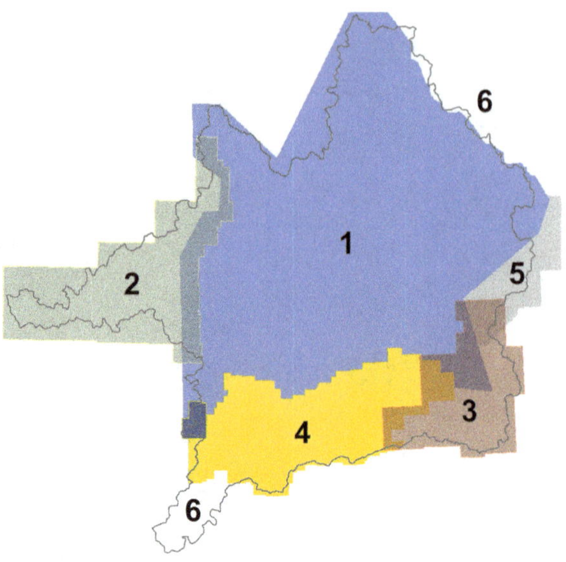

Fig. 7.1 Data sources for the DTM in the Upper Danube catchment area

Table 7.1 Data sources

No.	Resolution (m)	Data source
1	50	Bavarian Agency for Surveying and Geographic Information, Munich, Germany
2	50	State Agency for Spatial Information and Rural Development Baden-Württemberg, Stuttgart, Germany
3	25	Federal Office for Metrology and Surveying, Vienna, Austria
4	75	Regional Government of Tyrol, Innsbruck, Austria
5	50	SRTM (Shuttle Radar Topography Mission) X-Band
6	1,000	USGS (US Geological Survey) GTOPO30 (Global DEV), USA

Note: In all figures the hole-filled seamless SRTM data V4 (Jarvis et al. 2008) is shown

Datum). This corresponds to the projection used for the Hydrological Atlas of Germany. For some data sets, it was also necessary to adjust the resolution to 50 m. It was ensured, however, that the transformations for projection and resolution were performed in a single step. The cubic convolution was selected as the method for interpolation. This process utilises a high-pass filter in a 4×4 pixel window in the source data in order to calculate a new altitude value with a very high positional accuracy. The individual DTMs were then combined. In regions where data sets of similar quality overlap (Fig. 7.1, green and purple), only negligible differences in altitude were detected. In these cases, averaging provided minimal smoothing. In other overlap regions, the data from the Bavarian Agency for Surveying and Geoinformation and the Federal Office of Metrology and Surveying in Vienna was given preference over the less accurate SRTM, Tyrol and USGS data.

The primary goal of this data fusion was the generation of a hydrologically consistent reference data set with geographical resolution of 1 km^2. To minimise the loss of data quality during the transformation from 50 to 1,000 m resolution, the process was executed incrementally (50–100–250–500–1,000 m). Bilinear interpolation was chosen as the method of aggregation after comparative calculations. At each aggregation step, the minimum elevation was retained for the river network proxel, in order to ensure a significant reduction in the unwanted smoothing effects that otherwise occur during changes of resolution. The river network derived in each instance served as a reference for evaluating the quality of the terrain model. Digital terrain analysis was carried out using the software package TOPAZ (Garbrecht and Martz 1995). TOPAZ is based on the D8 algorithm that calculates the potential flow path from the local elevation values surrounding a pixel. The term D8 stands for deterministic eight. To calculate the flow path, the D8 algorithm compares the elevation values of neighbouring pixels with the value of the central pixel and selects the steepest downward slope as the direction of flow. With this principle, exactly one flow direction is determined for each pixel. The course of the surface drainage was successively routed through the pixels in accordance with the greatest downward slope, whilst determining the exact number of upstream pixels by addition of the all individual contributors for each grid cell. Hydrographic segmentation was implemented to identify the actual river network, the upstream or lateral sub-catchments and their inflow into each downstream river channel, as well as the catchment boundaries. By the spatially distributed allocation of threshold values, this technique can be used to determine the maximum size of a source area (critical source area) and the minimum source channel length.

7.3 Results

Figure 7.2 illustrates the resulting digital terrain model for the Upper Danube catchment and the derived river network at a spatial resolution of 1,000 m.

The overall outcome of scaling the various data inputs to the model's 1,000 m resolution can be rated as good. The derived river network shows a high degree of correlation to the actual river network in the Upper Danube catchment. The incorporation of subscale data by retaining the minimum elevation values during the upscaling has been shown to be an efficient method for preserving the topographic depth contours. The lateral drainage paths in the catchment area are plausible and have a reasonable degree of accuracy. Figure 7.2 illustrates the digital terrain model for the Upper Danube catchment as a coloured and shaded relief representation.

Fig. 7.2 Digital terrain model of the Upper Danube catchment area with the resulting river network at a spatial resolution of 1,000 m

References

Garbrecht J, Martz L (1995) TOPAZ – Version 1.1. National Agricultural Water Quality Laboratory, USDA, Agricultural Research Service, Durant, Oklahoma

Jarvis A, Reuter HI, Nelson A, Guevara E (2008) Hole-filled seamless SRTM data V4. International Centre for Tropical Agriculture (CIAT). Available via DIALOG. http://srtm.csi.cgiar.org. Accessed 19 Sept 2014

Chapter 8
Soil Textures

Markus Muerth and Ralf Ludwig

Abstract The spatial distribution of soil types in the Upper Danube basin represents an important basic data set for the simulation of distributed soil water contents and soil temperatures. The hydraulic and thermal processes in soil models are influenced by the physical parameters of the model layers, which in turn are determined by the physical properties of the respective soil horizon. The Upper Danube basin is characterized by a wide variety of soil types comprising different textures. As a result of the significant regional differences in bedrock, relief, climate, vegetation and the duration of soil genesis, the soil types found vary from the shallow raw soils of the high alpine regions to the fertile soils on loess deposits. The main soil textures found in the basin range from loamy clay to coarse sand and thus include almost every grain size. The distribution of soil types in the Upper Danube basin is presented in the 1:1,000,000 soil overview map (BÜK1000) of Germany. To reclassify the soils of the Upper Danube for DANUBIA, pedotransfer functions were used to calculate the hydraulic characteristics of the soil layers of each soil type. These in turn were used for a better representation and classification of the individual soils into textural classes based on their hydraulic properties. As a result, the 32 soilscapes of the BÜK1000 found in the drainage basin were summarized into 15 textural classes aggregating hydrologically similar soils.

Keywords GLOWA-Danube • DANUBIA • Soil • Soil texture • Upper Danube • Hydraulic properties

8.1 Introduction

Soils that are at a comparable developmental stage and have similar horizon profiles as a result of characteristic pedogenic processes can be grouped into soil types. In contrast, soil texture is the description of a soil or soil horizon in terms of the

M. Muerth (✉) • R. Ludwig
Department of Geography, Ludwig-Maximilians-Universität München (LMU Munich), Munich, Germany
e-mail: m.muerth@lmu.de; r.ludwig@lmu.de

W. Mauser, M. Prasch (eds.), *Regional Assessment of Global Change Impacts*, DOI 10.1007/978-3-319-16751-0_8

dominant grain size of its mineral components. The fine particles of soils are categorized on the basis of grain size diameter into the basic groups clay (<2 μm), silt (<63 μm) and sand (<2 mm). Coarser elements are characterized by the so-called skeletal content of a soil. The spatial distribution of soil textures in the Upper Danube basin (Map 8.1) represents an important basic data set for the simulation of distributed ground water recharge and soil temperatures (see Chaps. 24 and 71). The hydraulic process descriptions of the DANUBIA *Soil* component are influenced by the physical parameters of each soil layer; these in turn are determined by the soil physical properties of each soil horizon (Table 8.1).

The Upper Danube basin is characterized by a wide variety of soil types comprising different textures. As a result of the significant regional differences in bedrock, relief, climate, vegetation and duration of soil genesis, the soil types found vary from the shallow raw soils of the high alpine regions to the fertile Haplic Luvisols on loess deposits. In the young moraine landscapes and areas where ground- and backwater accumulate, moorland also occurs. The main soil textures in the basin range from loamy clay to coarse sand and thus include almost every grain size. In the Alpine region, the soil types found at certain altitudes are largely determined by the local climate and bedrock. Hence, below the nival zone with its very shallow Lithic Leptosols, mainly shallow Rendzic or Umbric Leptosols (German: Rendzina or Ranker) occur. Below, in the colline and mountain forest zones of both the Alps and other mountainous regions, either Eutric and Vertic Cambisols on calcareous bedrock or Dystric and Spodic Cambisols and also Podzols on crystalline bedrock can be found. Calcaric Regosols, Eutric Cambisols and Haplic Luvisols have developed on loose rock deposits as well as Gleysols, alluvial and fen soils in the river valleys (Kuntze et al. 1994). In the Alpine foothills, Haplic Luvisols have formed on virtually all the moraine deposits of the Würm glacial period. The soil texture of those varies only slightly from south to north between clayey and sandy loam. A notable exception is the Munich gravel plain with its Calcaric Regosols of loamy-sandy texture. Moraine deposits from older glacial periods are often covered by fertile loess layers (silty loam texture), where primarily Haplic Luvisols occur. The Tertiary Hills with to some extent thick loess layers are largely characterized by deeply decalcified Eutric Cambisols and Haplic Luvisols, which can have stagnic or gleyic characteristics, depending on bedrock substrate and groundwater level. Depending on the amount of loess deposits, the soil textures vary between silty loam and clayey silt (BGR 1998; Kuntze et al. 1994). The floodplains along rivers and streams can have fluvial sediments of varying grain sizes. Fens have formed in depressions with no drainage, in regions with high groundwater levels or in areas of frequent flooding. Transitional and upland moors are also common, primarily near the northern Alpine fringe. On the Swabian and Franconian Jura Mountains, Rendzic Leptosols and Chromic Cambisols of loamy-clayey texture on Malm limestone prevail as well as widespread Vertic Cambisols and Stagnic Gleysols from marlstone and claystone weathering of the Dogger and Lias region. The sandstones of the Keuper are typically accompanied by Cambic and Haplic Podzols that are always of sandy texture. Umbric Leptosols and Dystric Cambisols dominate the crystalline rocks of the Black Forest, the Bavarian Forest, the Bohemian Forest, the

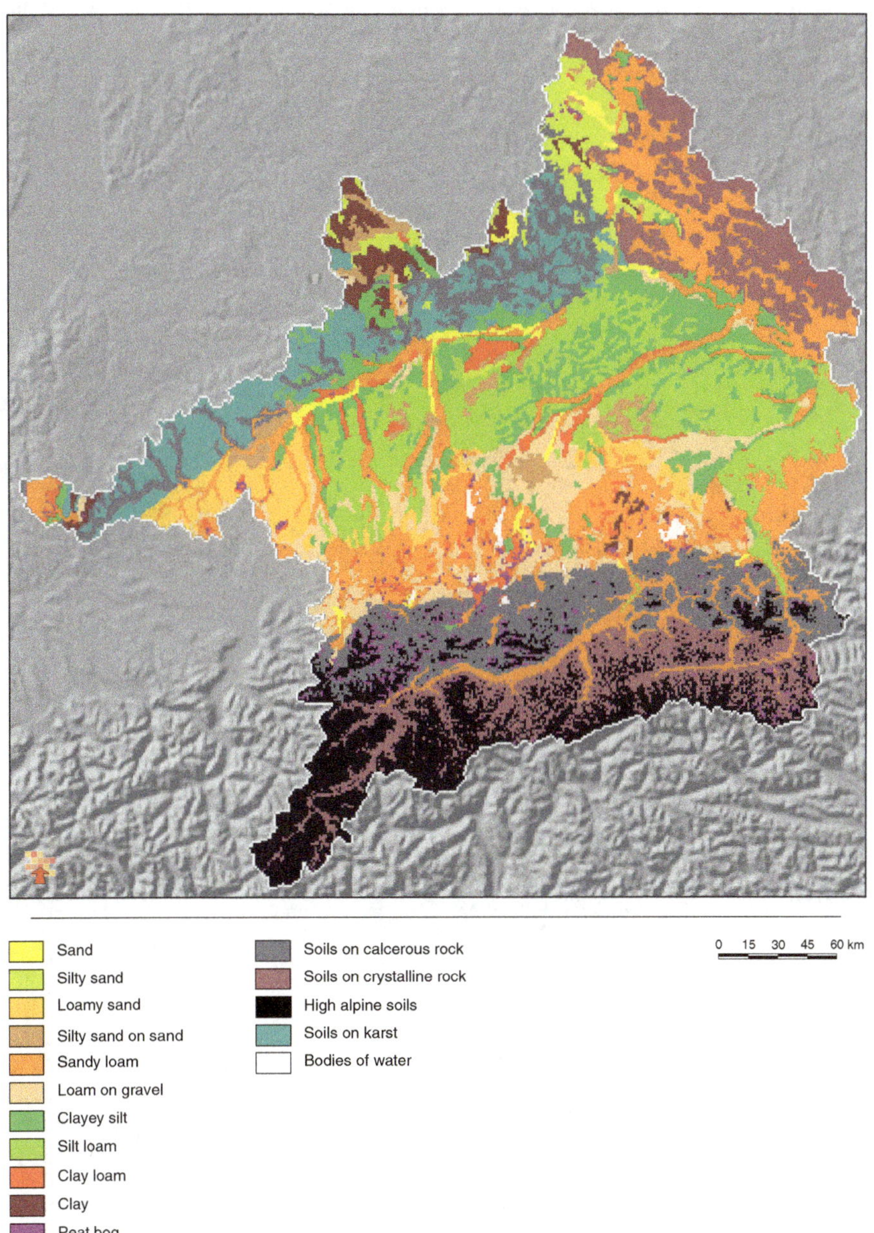

Map 8.1 Soil textures (Data source: Bundesanstalt für Geowissenschaften und Rohstoffe (BGR 1998), BÜK 1000 V 1.0)

Table 8.1 Soil textural classes integrating hydrological similar soil textural groups and the physical properties assigned to each soil horizon by the model component *Soil*

Soil type class	Depth (cm)	Soil type	Pore volume (vol%)	Skeleton (vol%)	Humus (wt%)	Clay (wt%)	Sand (wt%)
Sand	5	mS	52	0	10	1	95
	20	mS	46	0	4	1	95
	65	mS	46	0	4	1	95
	200	mS	34	0	0	1	95
Silty sand on sandstone	5	Su4	43	0	5	8	46
	20	Su4	43	0	2	8	46
	65	Su4	40	0	1	16	52
	200	Su4	36	0	0	22	67
Loamy sand	5	Sl4	48	0	3	15	60
	20	Sl3	42	0	1	10	65
	65	Ls4	28	20	0	20	60
	200	Ls4	28	20	0	20	60
Silty sand on sand	5	Su3	46	0	3	8	56
	20	Su3	46	0	3	8	56
	65	Uls	42	0	2	15	33
	200	sS	38	0	0	5	91
Sandy loam	5	Sl4	48	0	5	15	60
	20	Sl4	48	0	3	15	60
	65	Lt2	35	0	0	30	30
	200	Ls2	39	0	0	20	35
Sandy loam on gravel	5	Sl4	44	20	10	18	52
	20	Sl4	44	20	3	18	52
	65	Sl4	34	20	1	18	52
	200	gS	38	0	0	5	91
Clayey silt	5	Ut3	53	0	4	15	10
	20	Ut3	53	0	2	15	10
	65	Ut4	45	0	1	20	10
	200	Ut4	42	0	0	24	6
Silty loam	5	Lu	47	0	5	20	20
	20	Lu	47	0	1	20	20
	65	Tu3	40	0	0	35	10
	200	Lu	36	10	0	20	20
Clayey loam	5	Lt2	47	0	3	30	30
	20	Lt2	47	0	1	30	30
	65	Lu	46	0	1	30	15
	200	Tu3	42	0	1	35	10

(continued)

Table 8.1 (continued)

Soil type class	Depth (cm)	Soil type	Pore volume (vol%)	Skeleton (vol%)	Humus (wt%)	Clay (wt%)	Sand (wt%)
Silty loam on Lias clay	5	Lu	46	0	3	21	23
	20	Lu	46	0	3	21	23
	65	Lt3	43	0	2	45	10
	200	Tu2	43	0	0	51	5
Moors	5	H	77	0	100	0	0
	20	H	77	0	100	0	0
	65	H	77	0	100	0	0
	200	Lu	54	0	20	25	15
Soils on limestone	5	Lts	62	0	15	35	45
	20	Lts	46	0	7	40	40
	65	Tl*	18	20	0	50	30
	200	Lts*	7	85	0	45	35
Soils on crystalline rocks	5	Sl3	45	0	5	12	55
	20	Sl3	45	0	5	12	55
	65	Sl3	43	0	2	9	67
	200	Sl3	35	10	0	9	73
Extreme raw soils	5	Lt2	48	20	10	30	30
	20	Slu*	10	80	3	17	38
	65	Sl3*	7	85	0	10	65
	200	Sl3*	7	85	0	10	65
Karst soils	5	Ut4	46	0	3	24	6
	20	Ut4	46	0	3	24	6
	65	Tu2	48	10	1	58	3
	200	Tu2	40	20	0	58	3

Horizons with an * are notably affected by their content of coarse material

Upper Palatinate Forest and the Fichtelgebirge. These are relatively nutrient rich as a result of alkaline igneous rocks, but tend to be acidic and podzolic on granite and gneiss bedrock (BGR 1998; Kuntze et al. 1994).

8.2 Data Processing

The distribution of soil textures in the Upper Danube basin was derived from the 1:1,000,000 soil base map (BÜK1000) of the German Federal Institute for Geosciences and Natural Resources (*Bundesanstalt für Geowissenschaften und Rohstoffe*, BGR). Despite a poor spatial resolution, this map provides detailed information on the horizons of the characteristic soil types for each of the 72 soilscape classes, including the data on individual grain size fractions, porosity and skeletal content. Thirty-two of the seventy-two soilscapes have significant coverage in the German part of the Upper Danube basin. Yet, there was no comparable data available for the respective areas in the Upper Danube basin from the neighbouring countries. To obtain a homogeneous data set, the soil textural distribution of these regions was artificially generated based on the relationship between physiographic landscape features and the soil distribution in the German portion of the Upper Danube basin. In addition, the soils of the Alpine region were spatially differentiated as bog or forest using the available data on land use (see Chap. 9).

The calculation of the soil physical parameters for modelling was based on the detailed breakdown of the characteristic pedologic values for each horizon in the BÜK1000 (BGR 1998) and on information on the typical soils of Bavaria from the Bavarian State Research Centre for Agriculture (LfL Bayern 2005).

8.3 Results

The Rawls and Brakensiek (1985) pedotransfer functions for soil water retention parameters and the Wösten et al. (1999) pedotransfer function for soil water conductivity provide the physical soil parameters according to Brooks and Corey (1964), which allow the calculation of the hydraulic characteristics of the soil layers of each given soil type. These in turn are used for a better representation and classification of the individual soils into textural classes based on their hydraulic properties. The calculation requires data on total pore volume and the proportions of sand, clay and organic material on fine soil for each soil horizon. Furthermore, the skeletal content of a soil layer, expressing the fraction of material coarser than sand, is determinative for the evaluation of the highly variable pore volume parameter and is therefore required to determine the hydraulic properties of a characteristic soil texture. As a result, the 32 soilscapes of the BÜK1000 found in the drainage basin are summarized into 15 textural classes aggregating hydrologically similar soils (see Map 8.1).

In order to evaluate the calculated hydraulic parameters, the amount of water in the soil column that is available to plants (in mm) is calculated for each soil textural class and compared with the plant available water content values given in the BÜK1000 map legend (*PAW_Map*). The so-called extractable field capacity (plant available water capacity) is defined in Germany as the soil water that is stored in 0.2–50 μm pore diameters and can be determined from soil horizon texture by using Ad-hoc-AG Boden (2005) tables (*PAW_real*). As a comparison, the somewhat lower plant available water content according to the American definition (Dingman 2002) was calculated with the parameters of Brooks and Corey (*PAW_B&C*), as these parameters are the basis for the soil water model in DANUBIA. *PAW_B&C* includes only the quantity of water that is stored in 0.2 to approximately 20 micrometer pore diameters. Both calculated values ultimately span the actual range of values that are realistic for the actual plant available soil water contents (see Fig. 8.1).

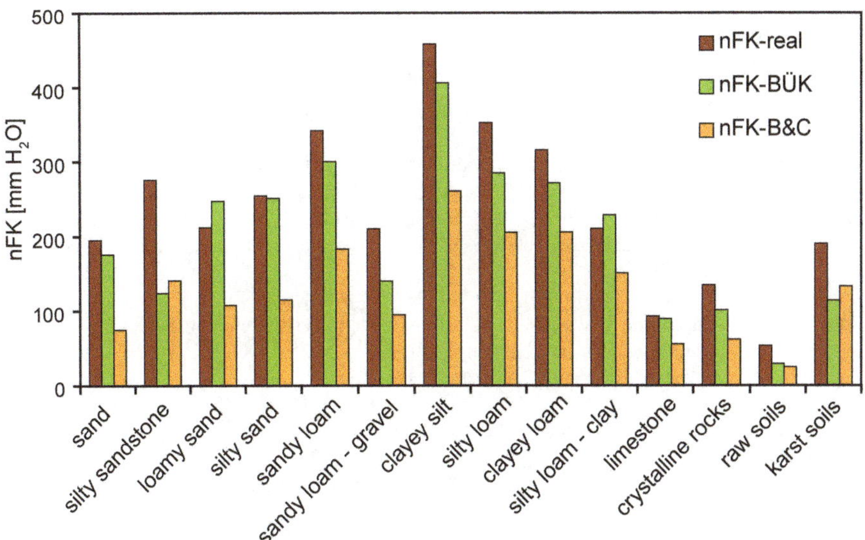

Fig. 8.1 Total plant available soil water (PAW_real) from Ad-hoc-AG Boden (2005) tables and plant available soil water calculated according to Brooks and Corey (1964) (PAW_B&C) compared to PAW_Map values taken from the BÜK1000 map description (BGR 1998), each in (mm) water column

Comparing *PAW_real* values for the dominant soil classes with the corresponding data from the BÜK1000 reveals a slight overestimate of *PAW_real* compared to *PAW_Map* for most soil textural classes. This difference is related to the simplification of the soil horizons in the model as well as to the use of empirical equations to determine hydraulic properties.

Since the BÜK1000 does not include any hydraulic conductivities for its soil horizons, these values cannot be directly compared to the values calculated according to Wösten et al. (1999). Instead, the calculated values were compared with typical conductivities for each soil type class, as tabulated in Ad-hoc-AG Boden (2005). However, since the saturated water conductivity of a soil layer varies significantly depending on soil density and composition, both calculated and reference values vary strongly with small differences in soil physical properties. For this reason, only the plausibility of the calculated hydraulic conductivities can be assessed based on reasonable simulated water flows and the meaningful range of its values. Overall, the validation of simulated runoff and soil moisture values at certain reference points showed that the chosen soil classes and parameters describe the hydrological characteristics of the soils in the Upper Danube region in a meaningful way.

References

Ad-hoc-AG Boden (2005) Bodenkundliche Kartieranleitung, 5th edn. Schweizerbart Verlag, Hannover, 438p

Brooks RH, Corey AT (1964) Properties of porous media affecting fluid flow. In: Hydrology paper no. 3. Civil Engineering Department, Colorado State University, Fort Collins

Bundesanstalt für Geowissenschaften und Rohstoffe (BGR, 1998) Bodenübersichtskarte der Bundesrepublik Deutschland 1:.000.000 (BÜK 1000 Vers. 1.0). Digitales Archiv FISBo BGR, Hannover und Berlin

Dingman S (2002) Physical hydrology. Prentice Hall, Upper Saddle River

Kuntze H, Roeschmann G, Schwerdtfeger G (1994) Bodenkunde. Ulmer, Stuttgart

LfL Bayern (2005) Böden und ihre Nutzung. Available via DIALOG. http://www.lfl.bayern.de/iab/bodenschutz/13755/index.php. Accessed 5 Jan 2015

Rawls WJ, Brakensiek DL (1985) Prediction of soil water properties for hydrologic modelling. In: Jones EB, Ward TJ (eds) Watershed management in the eighties symposium. ASCE, New York, pp 293–299

Wösten JHM, Lilly A, Nemes A, Le Bas C (1999) Development and use of hydraulic properties of European soils. Geoderma 90:169–185

Chapter 9
Land Use and Land Cover

Markus Probeck, Anja Colgan, Tatjana Krimly, Marcelo Zárate, and Karl Schneider

Abstract One of the key challenges in Global Change Research is the modelling of future changes in land use and land cover as a result of socio-economic and global climate change. Such changes are calculated dynamically in the Global Change decision support system for the Upper Danube (DANUBIA), thus helping to predict the effects of Global Change in this region. For any such spatially explicit modelling, land use/land cover is the key information layer which integrates all involved natural scientific and socio-economic process models. Land cover characterises the condition of the earth's surface and hence the properties that most directly influence water and energy fluxes, whereas land use describes the type of anthropogenic use, thus being a key parameter for modelling socio-economic processes. In order to meet all model requirements, a new tailored land use/land cover map was created with a representative overall area approximation and localisation of land use and land cover classes on basin scale, using a combination of CORINE Land Cover (CLC) data, official agricultural statistics and rule-based GIS operations. The resulting land use/land cover map can be used as initial state for modelling in DANUBIA. A total of 27 unique land use/land cover categories are distinguished in DANUBIA. The top hierarchical level is used consistently in all models, whereas the additional levels (e.g. detailed arable land classes) are individually used depending on the information needed by each model.

M. Probeck (✉)
GAF AG, Munich, Germany
e-mail: markus.probeck@gaf.de

A. Colgan
GB Dr. Schönwolf GmbH & Co. KG, Augsburg, Germany
e-mail: anjacolgan@aol.com

T. Krimly • M. Zárate
Production Theory and Resource Economics,
Universität Hohenheim, Stuttgart, Germany
e-mail: T.Krimly@uni-hohenheim.de

K. Schneider
Institute of Geography, University of Cologne, Cologne, Germany
e-mail: karl.schneider@uni-koeln.de

W. Mauser, M. Prasch (eds.), *Regional Assessment of Global Change Impacts*, DOI 10.1007/978-3-319-16751-0_9

Keywords Land use • Land cover • Global change • Upper Danube • DANUBIA

9.1 Introduction

For any spatially explicit modelling in DANUBIA, land use and land cover are the key information layer which integrates all involved natural scientific and socio-economic process models. Land cover characterises the condition of the earth's surface and hence the properties that most directly influence water and energy fluxes, whereas land use describes the type of anthropogenic use, thus being a key parameter for modelling socio-economic processes. Land use and land cover can be considered closely associated but not synonymous terms.

The particular importance of land use and land cover information necessitates meeting a range of diverse requirements for this data set, in particular:

- Differentiation of a significant number of land use/land cover classes in order to meet the various needs of the processes being modelled.
- High level of agreement with the areal statistics data of agricultural land use for realistic agricultural economic modelling in *Farming*.
- The data set must be verifiable and updatable with remote sensing data.

Available maps of land use and land cover do not meet these criteria. Thus, a new land use/land cover map was created which has been used as a basis for all further modelling in DANUBIA (Map 9.1).

The decision support system DANUBIA initially used only one land use/land cover class for modelling per 1 km^2 proxel. However, in reality, an area of 1 km^2 is composed of a variety of different categories. On basin scale, a strict assignment by majority class per proxel would inevitably lead to significant aggregation losses and hence an overestimation of classes with high area fraction at the expense of those with lower area fraction. Therefore, a dedicated scaling technique was developed which considers all land use and land cover classes according to their actual surface fractions and provides a land use/land cover data set with a most representative overall area approximation and spatial allocation of land use and land cover classes on basin scale.

In the current development stage of DANUBIA, land use and land cover can be represented as subscale area fractions. This means that every DANUBIA proxel may contain several land use/land cover classes with different area percentages. Such higher-resolution representation of the spatial distribution of land use and land cover results in additional improvement of the modelling results (Ludwig et al. 2003).

Modelling the expected future changes of land use and land cover resulting from socio-economic and global climate change is currently one of the key challenges in Global Change Research. Such changes are calculated dynamically in DANUBIA, thus helping to predict the effects of Global Change in the Upper Danube region.

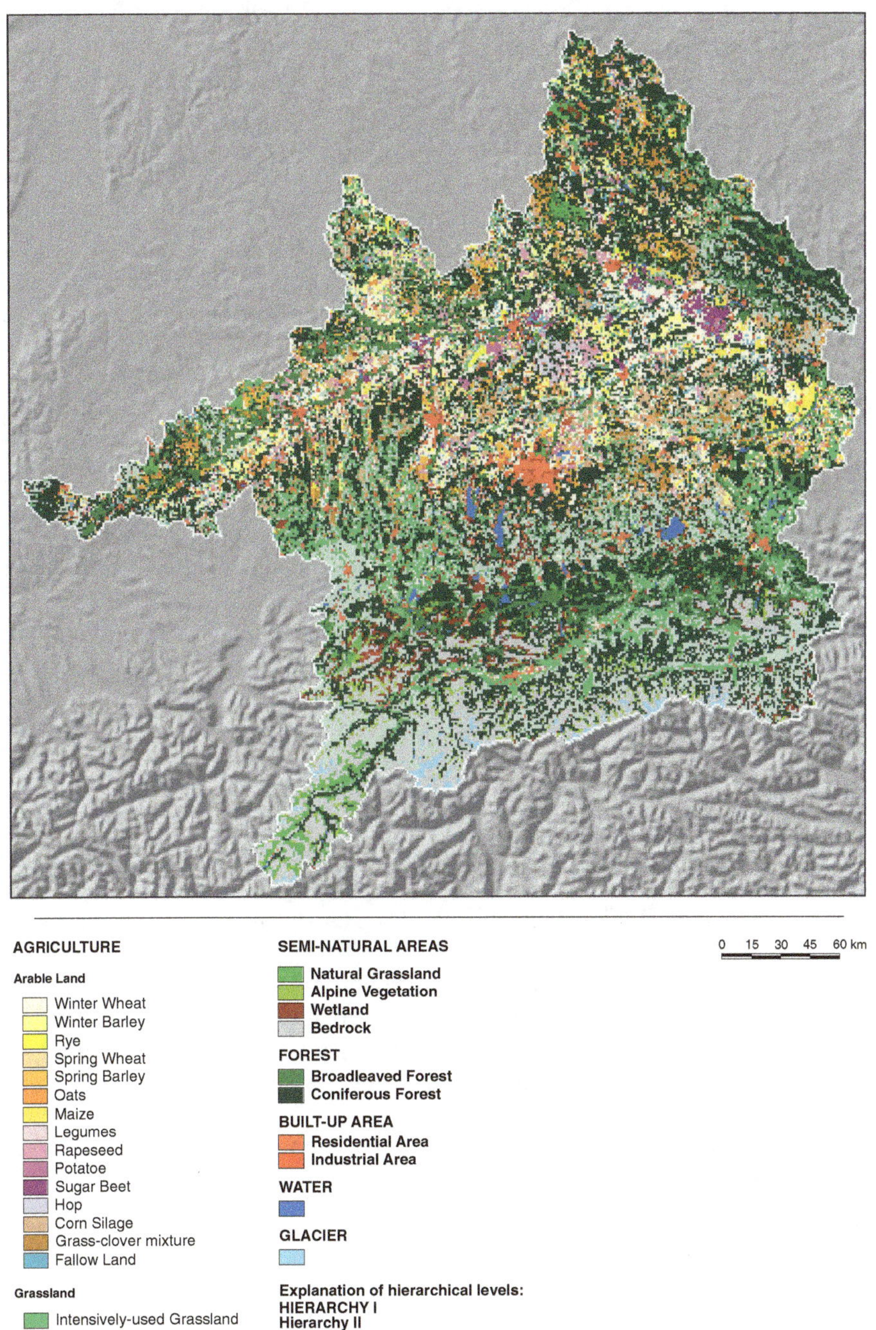

Map 9.1 Land use and land cover (Data sources: Bavarian State Office for Statistics and Data Processing 2001–2003; State Statistical Office Baden-Württemberg 1997; Statistics Austria 1996; Corine Land Cover (European Commission 2005))

9.2 Data Processing

Due to the considerable size of the Upper Danube region, neither terrestrial mapping nor analogue land use maps could provide a consistent classification of land use and land cover for the whole basin area; this can only be achieved by means of remote sensing techniques.

Therefore, the present land use and land cover map has been created as basis for modelling in DANUBIA, using a combination of CORINE Land Cover (CLC) data (European Commission 2005), official agricultural statistics figures and rule-based GIS operations. The underlying CLC 2000 data had originally been derived from high-resolution optical satellite imagery and consist of 44 possible classes for the German and Austrian part (CLC Level 3) and 15 classes for the Swiss part of the Upper Danube basin (CLC Level 2), respectively. In addition, agricultural land use and yield statistics on district (i.e. "Landkreis"/"Bezirk") level (Statistics Austria 2004; Statistics Austria 1996; Bavarian State Office for Statistics and Data Processing 2001–2003; State Statistical Office Baden-Württemberg 1997) were incorporated into the map creation process. These statistics comprise information on the spatial extent and average yield for arable and pasture land, on the areal proportions of up to 15 agricultural crop types and on animal husbandry. Such information is required for calibrating the agronomic model *Farming* (see Chap. 39), since crop growing and animal husbandry are intimately linked.

In a first preparatory step, the land use and land cover classes of the CLC data set were converted to the classes used in DANUBIA to establish class interoperability. Mixed land use/land cover classes from CLC, which cannot unambiguously be assigned to just one DANUBIA class, have been proportionately split and assigned to other classes, using additional information layers and local expertise. For example, the CORINE Land Cover class 3.3.3 "sparsely vegetated areas" has been divided to the DANUBIA classes natural grassland, alpine vegetation and bedrock as a function of slope and terrain elevation.

After this thematic transformation, the derived DANUBIA land use and land cover classes were spatially aggregated to the 1 km^2 proxel resolution. In this intermediate data set, the subscale areal percentages of the individual land use/land cover classes per proxel are preserved for further processing in the subsequent steps.

The following method has been applied to generate the presented final DANUBIA land use and land cover data set with only one discrete land use/land cover class per 1 km^2 proxel:

- In a first step, the classes water, glacier and built-up areas were directly assigned to those proxels where they represent the majority class.
- Then, the extent and spatial distribution of the agricultural areas in the intermediate data set (as derived from CLC) were adjusted to the figures provided by the

agricultural land use statistics. The resulting adjustments (decrease or increase) of agricultural land at district level have in turn been compensated by altering the area percentages of forest or seminatural areas.

- In a next step, the statistically given cumulative sum of cultivated area for all 15 crop types per district was spatially disaggregated over all "agricultural proxels", starting from those having the highest agricultural area fractions per proxel, until the statistically given totals per district and crop type were fully reflected in the map. This step ensured that the high degree of agricultural crop differentiation from the statistics was integrated into the map, with the spatial distribution of land use and land cover classes remaining largely consistent with the remote sensing-based observations (CLC).
- Finally, the spatial attribution of intensively and extensively used grassland to the "pasture" pixels from CLC has been realised with the help of altitude and slope information from a digital elevation model (DEM), such that extensive grassland has been preferentially assigned to steep or high-altitude proxels where farming is not easily possible.
- Forests and other (semi)natural land use/land cover classes were assigned to the remaining proxels, as far as possible preserving the overall land use/land cover distribution patterns as provided by CLC and the overall surface percentages of the classes on district and basin scale.

The resulting data set on land use and land cover is used as initial state for modelling in DANUBIA. It is structured in three hierarchical levels (see map legend):

- The highest hierarchical level comprises the categories agricultural areas, forest, seminatural areas, waterbodies, glaciers and built-up areas.
- In the second hierarchical level, agricultural areas are further distinguished into grassland and arable land, forest areas are differentiated into broadleaved and coniferous forest, and seminatural areas are split up into natural grasslands (e.g. alpine meadows), alpine vegetation, wetlands (e.g. swamps and peat bogs) and bedrock. Built-up areas are further subdivided into industrial and residential areas.
- In addition, a third hierarchical level is introduced for the 15 different crop types of arable land use (e.g. winter wheat, corn, hop) and to further distinguish grassland into intensively used and extensively used grassland.

Hence a total of 27 unique land use and land cover categories is distinguished in DANUBIA. The top hierarchical level is used consistently in all models. The additional levels are individually used depending on the information needed by each model.

Starting from this initial state, the actual land use for all arable land and grassland proxels is annually recalculated dynamically by *DeepFarming* (see Chap. 40), serving as input to further modelling in *Biological* and other models.

9.3 Results

As a result of various climatic, geologic and geomorphologic drivers as well as anthropogenic interventions, land use and land cover in the Upper Danube basin show a spatially quite heterogeneous distribution pattern. Whereas in the Alpine region, the vertical land cover zoning is primarily caused by increasingly harsher climatic conditions prevailing with higher altitudes, the overall land use in the drainage basin is dominated by arable and pasture land as well as forestry. The highest degree of built-up areas can be found in an area in and around the biggest cities, i.e. Munich, Augsburg, Salzburg, Regensburg, Ingolstadt and Ulm.

As a result of high annual precipitation rates, alpine meadows and grassland dominate in the Alpine region and the southern Alpine Foreland. Dairy farming plays an important role in this region. Arable land occurs only in a few climatically favoured spots where cereals and forage crops can be grown. Towards the northern Alpine Foreland, the proportion of arable land increases up to 60 % due to more favourable climatic conditions. Winter cereals, corn and root crops are the main crops grown there, with a considerable portion of corn silage used for fattening bulls. An arable land share of up to 80–90 % can be found in the Tertiary Hills north of Munich and in the vast Danube River basin lowlands. In addition to cereal crops, hop is cultivated in the Hallertau region in the Tertiary Hills and sugar beet in the Gäuboden region around Straubing. In contrast to that, the Swabian and Franconian Jura regions have unfavourable agronomic conditions due to thin soil layers and the ubiquitous karst formation. Therefore, the proportion of grassland is high (80–100 %), while grassland farming and livestock breeding are less specialised here than in the Alpine region (see also Chap. 19).

Historic agricultural cultivation practice led to forested regions in the Upper Danube basin being mostly limited to areas which are not suited for farming due to unfavourable climatic, topographic or soil conditions. Therefore, the largest forested areas are found in the low mountain ranges of the Swabian and Franconian Jura and the Black Forest, as well as in the Alpine regions of the basin. Especially in the Alpine Foreland, commercial forestry has been favouring coniferous forests (mostly spruce) until the recent past, whereas broadleaved and mixed forests are more widely spread in riparian areas, in the Swabian and Franconian Jura as well as (to a minor degree) in the Northern Alps.

References

Bavarian State Office for Statistics and Data Processing (2001–2003) Results of the farm structure survey 1995. Database query of the Bavarian State Office for Statistics and Data Processing, Munich

European Commission (2005) CORINE Land Cover updating for the year 2000: IMAGE2000 and CLC2000 – products and methods. Publication no. EUR 21757 EN, Ispra

Ludwig R, Probeck M, Mauser W (2003) Mesoscale water balance modelling in the Upper Danube watershed using sub-scale land cover information derived from NOAA-AVHRR imagery and GIS-techniques. Phys Chem Earth 28:1351–1364

Statistical Office of the Federal State of Baden-Wuerttemberg (1997) Agricultural reporting 1995. Results for agricultural comparable areas and counties, 519, Stuttgart

Statistics Austria (1996) Results of the agricultural statistics 1995. Beiträge zur österreichischen Statistik, 1205, Wien

Statistics Austria (2004) Results of the farm structure survey 1995. Provided Dataset, M. Dötzl, Statistics Austria, Vienna

Chapter 10
Climate Stations

Anja Colgan and Ruth Weidinger

Abstract Data from the climate stations was on the one hand used as an input source for the temporal and spatial interpolation of meteorological data by the model *AtmoStations* (see Chaps. 1.5, 1.6, and 1.7) and on the other hand to validate the results of the component *Rivernetwork* (Chap. 2.3.1) and the models *Snow* (Chap. 2.4.1) and *AtmoMM5* (Chap. 2.5.1). Data from all available climate stations (totalling 377) were obtained from the respective German and Austrian national meteorological agencies for the modelling period between 1960 and 2004. The analysis of the source data showed differences in the measurement procedures over time (change to automated data collection) and between the two national agencies (time of measurement and measurement units). However, these differences are taken into account in the model *AtmoStations* as much as possible, so that the degree of inhomogeneity in the data is kept as low as possible. The individual stations cover different time periods. Only a quarter of the stations (approx. 100) have a complete record for the modelling period. Therefore, the number of stations per time increment available for spatial interpolation varies considerably. The altitude of the climate stations varies from 96 m above sea level (Mannheim) to 3,105 m above sea level (Sonnblick). The analysis of the altitude distribution of the climate stations shows that the vast majority of stations are below 1,000 m above sea level. However, station density approximately corresponds to the area-altitude distribution of the catchment area.

Keywords GLOWA-Danube • Upper Danube • Meteorological observations • Station data • AtmoStations

A. Colgan
GB Dr. Schönwolf GmbH & Co. KG, Augsburg, Germany
e-mail: anjacolgan@aol.com

R. Weidinger (✉)
Department of Geography, Ludwig-Maximilians-Universität München (LMU Munich), Munich, Germany
e-mail: R.Weidinger@lmu.de

W. Mauser, M. Prasch (eds.), *Regional Assessment of Global Change Impacts*, DOI 10.1007/978-3-319-16751-0_10

10.1 Introduction

Almost all DANUBIA model components require values at ground level for meteorological parameters at high temporal and spatial resolution. Such data is already generated at 1 km^2 resolution and at hourly intervals either via the mesoscale atmosphere model *MM5* or by interpolation of observed time series. The measurement data from climate stations form the basis for the temporal and spatial interpolation of the meteorological parameters with the model *AtmoStations* and are used to validate the results of the DANUBIA component *Rivernetwork* (Chap. 29) and the models *Snow* (Chap. 30) and *AtmoMM5* (Chap. 32).

In order to rely on a network of measurement stations with good coverage and of guaranteed data quality, homogeneity and continuity, data from all available stations of the German Weather Service (Deutscher Wetterdienst=DWD), the national meteorological service of the Federal Republic of Germany and the Central Institution for Meteorology and Geodynamics (Zentralanstalt für Meteorologie und Geodynamik=ZAMG) and the national meteorological and geophysical service of the Federal Republic of Austria were obtained for the Upper Danube catchment area and its surroundings. For the modelling period from 1960 to 2004, there were a total of 377 weather stations available for the region in question, of which 280 were German and 97 were Austrian. For the region within Switzerland, there was discharge data at the boundary gauges and climatologies for various meteorological parameters. Thus, there was no need for a costly acquisition of data.

10.2 Data Processing

The DWD data come from climate stations with up to three daily observations and from the main synoptic-climatological network (hourly measurements, enhanced with climatological data). The data obtained from the ZAMG include ground-level observations of meteorological parameters that are captured at conventional climate stations with manual measurements and at semiautomated weather stations. Table 10.1 presents the parameters for which data was captured at these stations.

On April 1, 2001, the method for capturing climatological data at the DWD was converted. New automatic data acquisition meant that additional data for calculating daily values could be utilized. One main aspect of the change in data management at the DWD was the shift in scheduled weather measurements from 7:30, 14:30 and 21:30 CET to 7:00, 14:00 and 19:00 CET. The actual observation time is 10 min prior to the reference time. This shift took effect at first for the main synoptic-climatological automated network. The new times were also adopted with the subsequent automation of the secondary climate and precipitation network (www.dwd.de). This change entailed only a minor shift in the time points for the temporal interpolation of the parameter with the model *AtmoStations*. However, the change entailed a significant change in the mean values (Gattermayr 2001). The model *AtmoStations* did not use the mean values from the weather stations, but instead calculated these from the interpolated values.

Table 10.1 Parameters of weather stations with time-scheduled measurement or measurement duration

Parameter	DWD	ZAMG
Air temperature (2 m above ground)	(1)	(2)
Air humidity (2 m above ground)	(1)	(2)
Wind direction and force	(1)	(2)
Visibility	(1)	(2)
Soil condition	(1)	(2)
Cloud coverage	(1)	(2)
Precipitation amount	(3)	(4)
Type of precipitation	(3)	(4)
Air pressure	(1) (5)	(2)
Cloud density and weather events	(1)	(2)
Snow depth, type of snow cover	7:30 CET	7:00 LMT
Depth of fresh snow	(6)	(6)
Snow water equivalent	7:30 CET	
Precipitation total	(8)	(9)
Mean air temperature	(10)	(11)
Mean air humidity	(10)	(10)
Mean wind force	(10)	(10)
Mean cloud coverage	(10)	(10)
Mean air pressure	(10)	(10)
Information on fallen precipitation, condensation/sublimation and weather events	(12)	
Sunshine duration	(12)	(12)
Maximum gust of wind	(12)	(12) (15)
Maximum air temperature	(13)	(16)
Minimum air temperature	(13)	(16)
Minimum ground temperature	(14)	(17)
Global radiation		(12)

CET Central European Time, *LMT* local mean time

1. Three scheduled measurements: 7:30, 14:30 and 21:30 CET until March 2001 and 6:50, 12:50 and 18:50 CET since April 2001
2. Three scheduled measurements: 7:00, 14:00 and 19:00 LMT
3. Three measurements: 21:30 on the previous day to 7:30, 7:30 to 14:30 and 14:30 to 21:30 CET until March 2001 and 18:50 on the previous day to 6:50, 6:50 to 12:50 and 12:50 to 18:50 CET since April 2001
4. Two measurements: 19:00 on the previous day to 7:00 and 7:00 to 19:00 LMT
5. Only at synoptic-climatological network stations
6. 7:30 on the previous day to 7:30 CET (since April 2001: 6:50 on the previous day to 6:50)
7. 7:00 on the previous day to 7:00 LMT
8. 7:30 to 7:30 CET on the following day (since April 2001: 6:50 to 6:50 on the following day)
9. 7:00 to 7:00 LMT on the following day
10. Daily from three scheduled measurements
11. (Max + min)/2
12. Daily (between 00:00 and 24:00 legal time, at full-time stations UTC)
13. Until March 2001: 21:30 CET on the previous day to 21:30 CET (since April 2001; see 10)
14. Until March 2001: 21:30 CET on the previous day to 7:30 CET (since April 2001; see 10)
15. Additional information at the time of daily maximum
16. 19:00 on the previous day to 19:00 LMT
17. 19:00 on the previous day to 7:00 LMT

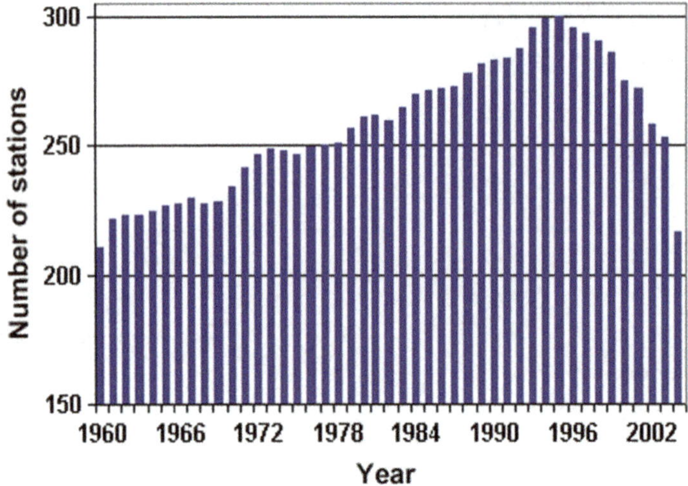

Fig. 10.1 Number of stations with daily measures per year

In addition to the change in measurement procedure during the study period, there is also the difference in the methods used by the DWD and ZAMG to be considered (e.g. different measurement times, timing according to CET or LMT or different units of the measurements). However, these differences are taken into account in the model *AtmoStations* as much as possible, so that the degree of inhomogeneity in the data is kept as low as possible.

For the period from 1960 to 2004, a total of over four million measurement days (data records) are available. The individual stations cover different time periods and also sometimes contain significant gaps in the data. Roughly a quarter of the stations (approx. 100) have a complete record for this period. Therefore, the number of stations per time increment available for spatial interpolation varies considerably. For the year 1960, data from 211 stations could be used for the model, whereas for the years 1994 and 1995, up to 300 stations were available (see Fig. 10.1).

10.3 Results

Map 10.1 shows the locations of the DWD and ZAMG climate stations within the study area. The station with the highest altitude is located in Austria on Sonnblick Mountain (3,105 m a.s.l.), and the second highest is in Germany on the Zugspitze (2,960 m a.s.l.); third highest is the device on the Pitztal Glacier at 2,850 m a.s.l. The two highest stations in the Bavarian Forest are on the Great Arber (1,437 m a.s.l.) and the Great Falkenstein (1,307 m a.s.l.). The lowest altitude measurements

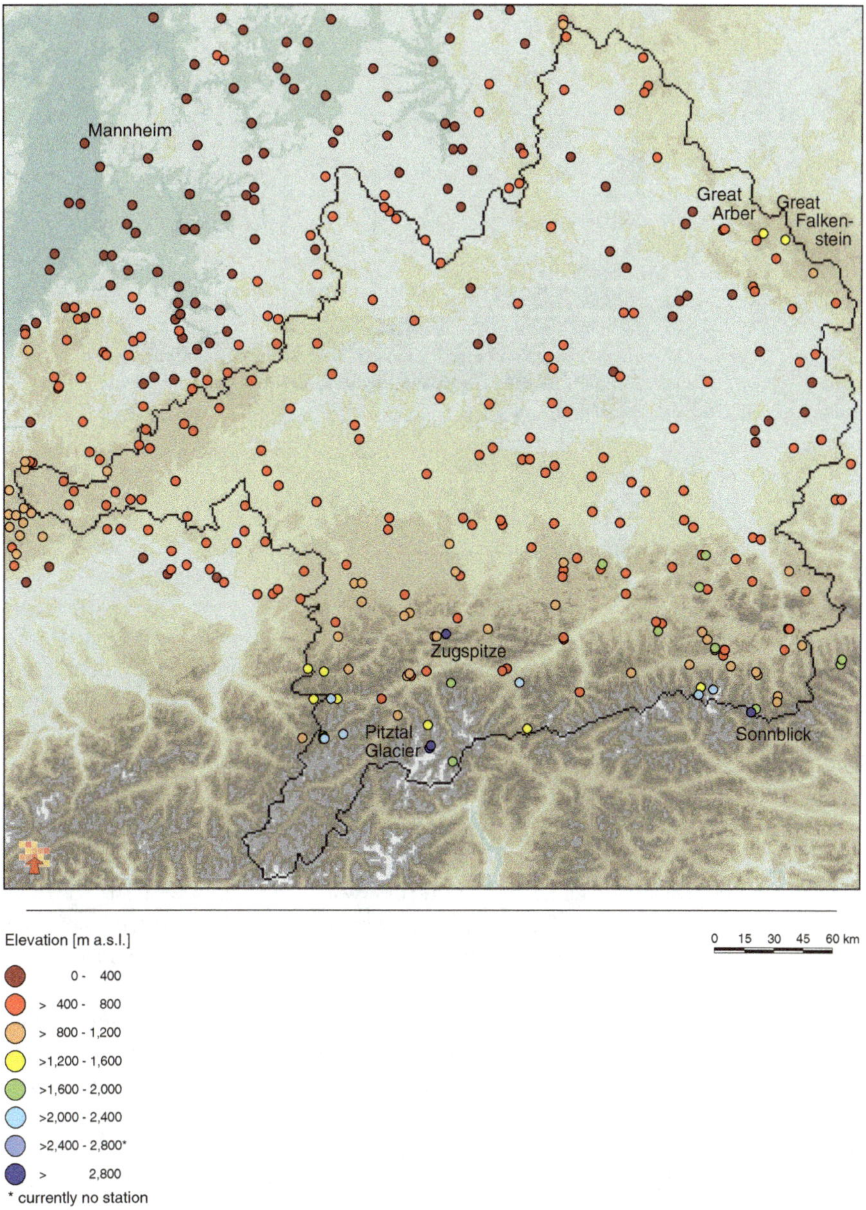

Map 10.1 Climate stations (Data source: DWD, Deutscher Wetterdienst, ZAMG, Zentralanstalt für Meteorologie)

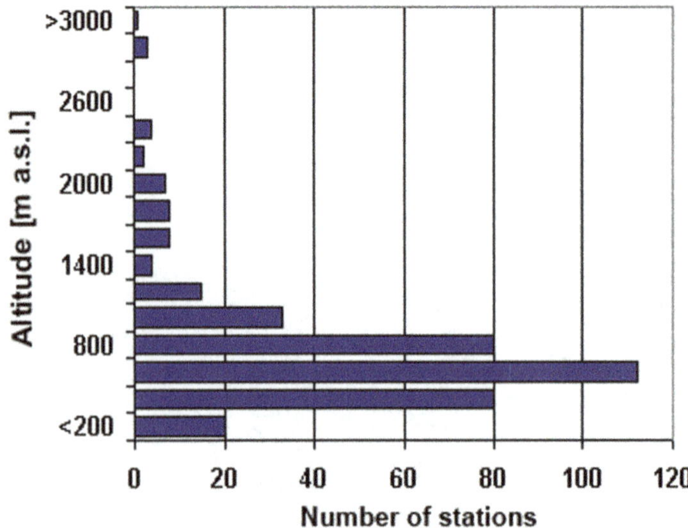

Fig. 10.2 Altitude distribution of the climate stations

are taken in Mannheim at 96 m a.s.l. It is noteworthy that at several Austrian stations, two symbols are directly adjacent or virtually on top of each other. In part, this situation derives from the fact that these stations are carried forward under a new designation if the type of station changes (e.g. as a result of automation). Another reason is that stations on private property must be relocated and supplied with a new number if the observer changes.

The altitude distribution of the climate stations indicates that above 1,000 m a.s.l. there are significantly fewer stations in operation than below 1,000 m a.s.l.. Half of the sites are below 533 m a.s.l. and 75 % lie below 785 m a.s.l. (see Fig. 10.2). Eight stations are located above 2,000 m a.s.l., of which seven are in Austria. However, station density approximately corresponds to the area-altitude distribution of the catchment area.

The high altitude and the strong small-scale variability in the Alpine region have an important impact on the climate and water supply in the Upper Danube catchment area. Relief enhances the thermal and dynamic processes in the atmosphere that can, for example, lead locally to more precipitation and hence can be very significant for the water supply. For the spatial interpretation and simulation of climatic state variables in regions with sharp relief patterns, there is therefore a need for a greater number of stations with regular high-quality data even at high sites and the most extreme locations of the study area. Unfortunately, at these sites logistical problems for the operation of the stations are often experienced. In addition to the measurements from the stations depicted on Map 10.1, data from climate stations at academic institutions were used to validate the model results. For example, the "Vernagtbach Climate Station" (at Vernagtferner) in Ötztal at an altitude of 2,640 m a.s.l., which belongs to the Commission for Geodesy and Glaciology of the Bavarian

Academy of Sciences and Humanities, supplied year-round hourly data for all parameters listed in Table 10.1 except for cloud cover and visibility. Another station is located also in Ötztal on the Schwarzkögele Glacier (3,074 m a.s.l.), and there are several measurement sites on the glacier itself that are at 3,000 m a.s.l.

Reference

Gattermayr W (2001) Hydrometeorologische Erhebungen am Mühleggerköpfl / Nordtiroler Kalkalpen. In: Herman F, Smidt S, Englisch M (eds) Stickstoffflüsse am Mühleggerköpfl in den Nordtiroler Kalkalpen, FBVA-Berichte, No 119. Bundesministerium für Land- und Forstwirtschaft, Wiien, pp 53–59

Chapter 11
Spatial and Temporal Interpolation of the Meteorological Data: Precipitation, Temperature and Radiation

Wolfram Mauser and Andrea Reiter

Abstract Simulations of the impacts of climate change on the water resources need accurate meteorological drivers. In GLOWA-Danube, hourly fields of air temperature, air humidity, wind, short- and longwave radiation and precipitation were determined from weather station data for the past period from 1970 to 2006 as well as for future climates (using the statistical climate generator (Chap. 49)) with a spatial resolution of 1 km. Interpolation is based on hourly regressions of elevation gradients and a DEM and inverse square distance weighting of the regression residuals. For precipitation, additional bias correction from high-spatial-resolution monthly rainfall data was applied. The maps of average precipitation, temperature and radiation in the Upper Danube watershed are displayed. They show the overall distribution of major meteorological inputs and demonstrate that small-scale effects like dry valleys in the Alps and albedo effects of cities on radiation are documented.

Keywords GLOWA-Danube • Meteorological drivers • Spatial interpolation • Bias correction

11.1 Introduction

Meteorological parameters are key factors in modelling hydrological processes. The data used for the study area were delivered by the German DWD and the Austrian ZAMG (see Chap. 10). They were measured three times daily at the "Mannheimer

W. Mauser (✉)
Department of Geography, Ludwig-Maximilians-Universität München (LMU Munich), Munich, Germany
e-mail: w.mauser@lmu.de

A. Reiter
Bavarian Research Alliance GmbH, Munich, Germany
e-mail: reiter@bayfor.org

W. Mauser, M. Prasch (eds.), *Regional Assessment of Global Change Impacts*, DOI 10.1007/978-3-319-16751-0_11

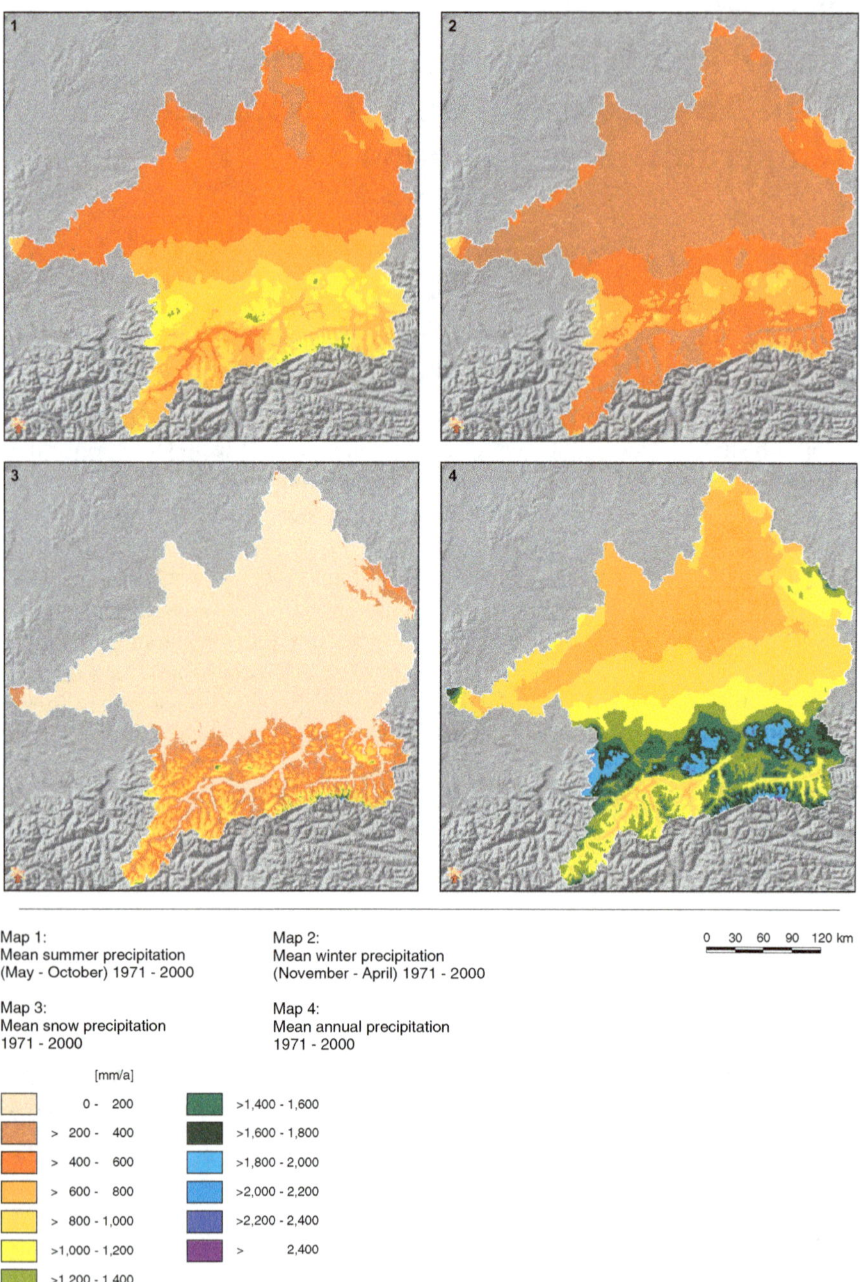

Map 11.1 Calculated precipitation (Data sources: DWD Deutscher Wetterdienst; ZAMG, Zentralanstalt für Meteorologie und Geodynamik)

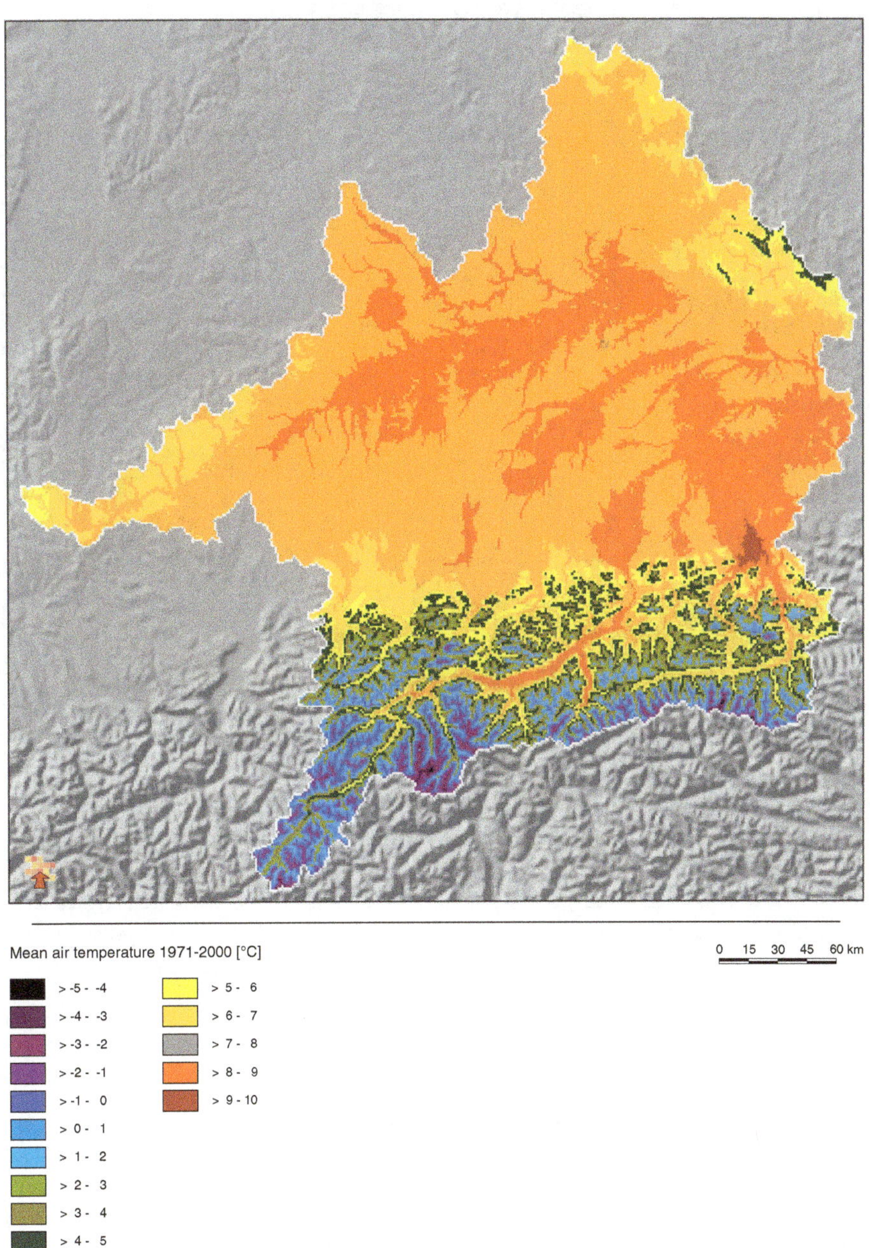

Map 11.2 Calculated air temperature (Data sources: DWD Deutscher Wetterdienst; ZAMG, Zentralanstalt für Meteorologie und Geodynamik)

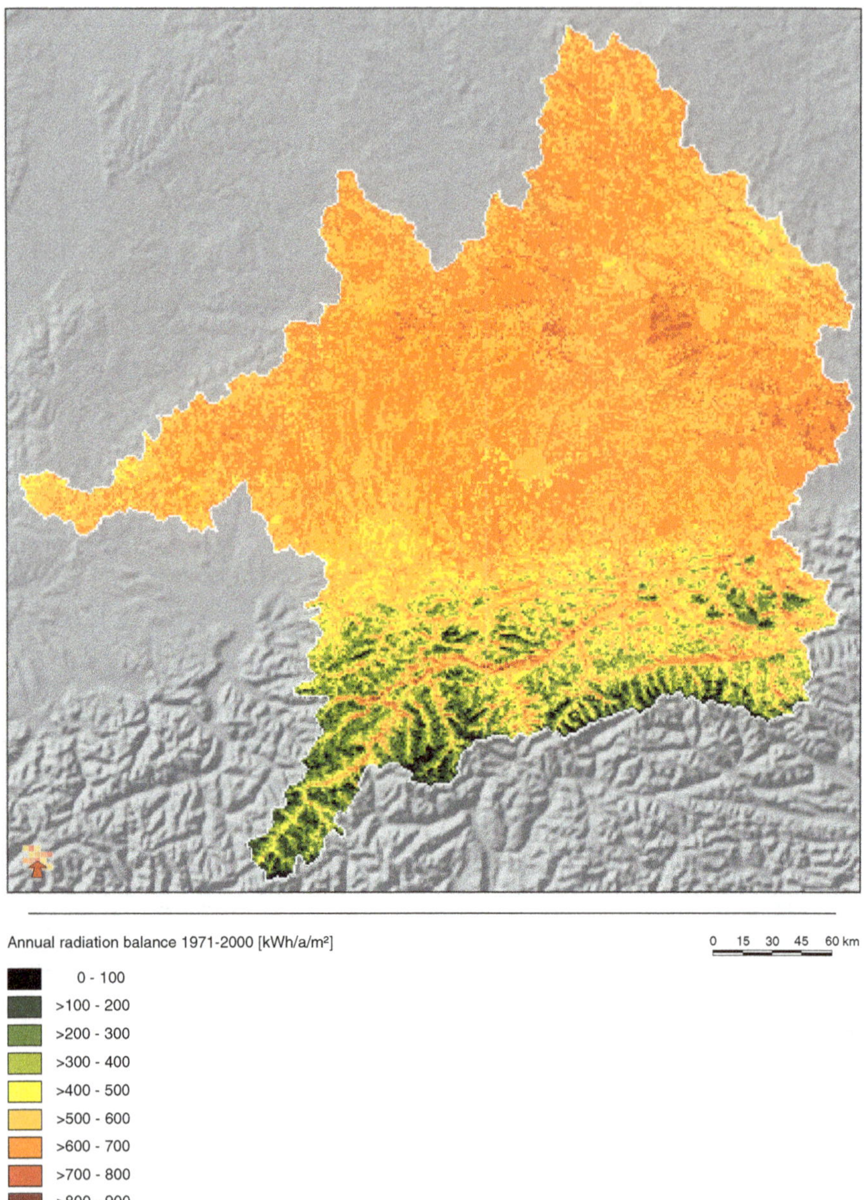

Map 11.3 Calculated radiation balance (Data sources: DWD Deutscher Wetterdienst; ZAMG, Zentralanstalt für Meteorologie und Geodynamik)

Stunde" at 7:30, 14:30 and 21:30 local time. They exist only as station point data and moreover have the advantage of being available at a large number of stations at the expense of too low a temporal resolution for a detailed process modelling. Therefore, temporal and spatial interpolation must be applied in order to obtain discrete values for each time step and every proxel (process pixel: a geo-located pixel, in which processes take place, see Chap. 2); the interpolated values can then serve as input data for the model.

The land surface processes that exhibit the greatest temporal variability and spatial heterogeneity determine the requirements for temporal and spatial resolution of the meteorological drivers of DANUBIA. Evapotranspiration, sensible heat flux, infiltration, snow accumulation and ablation as well as the fast components of the interflows from the run-off formation process are of particular temporal variability. These parameters have pronounced diurnal variation that is significantly affected by the daily energy availability pattern (net radiation), air temperature and precipitation.

In contrast to many other hydrological model approaches that describe the land surface processes using daily time steps (e.g. The Soil and Water Assessment Tool SWAT 2000, Neitsch et al. 2002) and therefore must use highly parameterized approaches with low deterministic process representation, DANUBIA uses nonlinear modelling that is based on detailed physical descriptions of processes. In order to consider the high temporal variability in the meteorological data and the nonlinearity of the processes under consideration, a 1-h time interval was selected in DANUBIA. With a total of 288 available precipitation stations in the study area and a spatial resolution of $1 \times 1 km^2$, one precipitation station covers an average area of $16 \times 16 km^2$. At a mean precipitation intensity of 2.6 mm/h for individual events, the mean maximum spatial precipitation gradient is thus 0.16 mm/km between a station with precipitation and a neighbouring station without precipitation. The actual spatial precipitation gradients are significantly higher during convective precipitation events in the Alpine foothills. However, at present, this fact cannot be sufficiently accounted for as a result of the low density of stations.

The challenge in developing the interpolation described below is thus to devise a computationally efficient, stable and precise method that is capable of processing the list of meteorological data listed below and that can take into account the prevalent and sizeable topographical and climatic gradients of the Upper Danube basin.

11.2 Data Processing

The following meteorological parameters were spatially modelled on an hourly basis and in fields of 1×1 km^2 resolution as input data for the various simulation model components in DANUBIA:

- Atmospheric temperature [°C], relative humidity [%] and wind velocity [m/s], in each case 2 m above the ground surface
- Precipitation intensity [mm/h]

- Direct and diffuse shortwave radiation (0.3–3 μm) and longwave radiation (3–100 μm) [W/m^2]
- Standard air pressure and daily CO_2 partial pressure based on selectable IPCC SRES scenarios [hPa]

11.2.1 Temporal Interpolation

Interpolation values were generated from the three available measurements per day of the meteorological parameter's atmospheric temperature, relative humidity, wind velocity, cloud cover and precipitation for every desired 1-h time step for all stations. Temporal interpolation between the measured values was performed using a cubic spline interpolation function (third-order polynomial) for all parameters except precipitation; the polynomial function was uniquely determined through four consecutive measurements. The hourly values between measurements could then be determined from the course of the polynomial function.

However, precipitation is always tied to a precipitation event and is, therefore, contrary to the above-mentioned meteorological parameters, not representing a spatially continuous field. In addition, precipitation values are aggregated values, in that they refer to a value that includes the total precipitation falling within a specific time period. For this reason, the precipitation totals measured at 7:00, 14:00 and 21:00 need to be temporally disaggregated. This is accomplished in two ways: whenever there was no precipitation before and after a measured precipitation, a convective event is assumed. The intensity of the measured precipitation is then distributed using a Gaussian distribution across the 7 h prior to the time of measurement. Similarly, whenever there was a precipitation before and after a measured precipitation, an advective event is assumed. The measured total precipitation at the time of measurement is distributed evenly to the hours before the measurement. If two precipitation measurements follow a measurement point with no precipitation, the precipitation intensity increases linearly. The same likewise applies to the opposite situation, when a measurement with no precipitation followed two precipitation measurements.

11.2.2 Spatial Interpolation

For each hour, the temporally interpolated values at the measurement stations are transferred to the area of the Upper Danube basin. The method used is based on the interpolation of residuals based on the spatial support parameter topography and is applied to all temporally interpolated parameters.

It is generally assumed that all parameters are related to terrain elevation and that this relationship allows a regression line to be determined between the temporally interpolated values for each meteorological parameter at the meteo-station and the altitude of the stations. This regression line then allows the calculation of an altitude-dependent parameter value for all locations of meteo-stations. It cannot be expected that with this regression, the parameter values at the location of the station will exactly reproduce the measured value. Instead, the residuals of the regression analysis will yield the differences between the measured values and the regression values determined using the digital elevation model. The residual field of the regression is thus corrected for the topographical trend of the considered parameter and represents the local specificity of each measurement at the location of the considered station and at the considered time. These site-specific deviations from the mean topographical property of a considered parameter are then spatially interpolated. The residual values at the six neighbouring stations of the considered proxel are used to spatially interpolate the residuals. To do this, first, the closest station is determined in each quadrant. Then, the two next closest of the remaining stations are determined. The relative weight for these stations for calculation of the relevant residual value at the position of the proxel in question is determined using the inverse of the square of its distance to the selected station. The final step then involves the summation, for each proxel, of the mean topographical field determined from regression and the interpolated residual of the considered meteorological parameter.

With this method of interpolation, computationally efficient spatial fields of the meteorological parameters can be determined. For atmospheric temperature, relative humidity, percentage clear sky and wind velocity, this approach has proven suitable. For the presentation of the spatial distribution of precipitation, however, it was shown that the use of the terrain model alone did not sufficiently reproduce the spatial pattern of precipitation with satisfactory precision. Many spatial precipitation processes related to topography cannot be accounted for by the altitude gradients and the sparse measurement network. Therefore, the climatological analysis of B. Früh (see Chap. 32) was used. In her study, the measurements from over 2,000 precipitation collectors in the drainage basin over 10 years were analysed and interpolated for each month. These monthly precipitation values with high spatial resolution served to determine local correction factors for the hourly interpolated precipitation. A monthly correction factor was determined for every proxel using the quotient of the total monthly precipitation determined there and the mean total precipitation for the entire area. As an example, the outcome from this correction led to a significant reduction in precipitation in the dry valleys of the Alpine region, as can be clearly seen in Map 11.1. This small-scale specific feature that is not dependent on local topography but rather on the Alpine ensemble of mountains protecting the inner valleys from inflowing precipitation cannot be accounted for by conventional interpolation methods.

A radiation model was developed for the determination of direct and diffuse shortwave radiation and longwave radiation, which uses the topography of the proxel in question (geographic coordinates, terrain elevation, inclination, exposure) as well as the actual percentage of visible clear sky derived from the surrounding mountain topography. The astronomical parameters solar zenith and azimuth angle that were calculated from the geographical location of the proxel, together with the solar constant, yielded the top-of-atmosphere shortwave radiation input (Brutsaert 1982). The attenuation of direct shortwave radiation from absorption and dispersion and the diffuse shortwave radiation were calculated according to McClatchey et al. (1972), and the shortwave radiative flux directed towards the earth's surface was calculated taking into account the cloud cover according to Möser and Raschke (1983). Longwave radiation was estimated from atmospheric temperature and percentage cloud cover using the approaches of Czeplak and Kasten (1987) and Swinbank (1964). Radiation balance resulted from the determination of reflection and emission at the land surface, taking into account short- and longwave albedo and atmospheric temperature.

11.3 Results

Maps 11.1, 11.2 and 11.3 present the average annual precipitation [mm/a], atmospheric temperature [°C] and radiation balance [kWh/a/m^2] as important meteorological variables for hydrological modelling for the Upper Danube basin from 1971 to 2000.

Map 11.1 shows that precipitation in the Alpine region can reach peak values of up to 1,400 mm during the summer season, whereas the northern part of the basin receives up to a maximum of 800 mm summer precipitation. Precipitation during the winter season is significantly lower and even in the Alpine region only infrequently amounts to more than 800 mm. Snowfall above 200 mm only takes place in the Alpine mountain regions, the low mountain ranges and in the western part of the basin (the Swabian Jura).

Easily discernible in Map 11.2 is the dependency of atmospheric temperature on terrain elevation. Thus, the mean temperature in the higher sites of the Central Alps, the Alpine foothills and the low mountain ranges are noticeably lower than in the lower lying regions. Even the extremely small-scale temperature differences caused by the strong relief between valley and mountain sites stand out.

Larger built-up areas in which the net radiation is between 400 and 500 kWh/a/m^2 can be clearly seen in Map 11.3 of radiation balance due to their difference in albedo. The heterogeneity of net radiation in the drainage basin can be explained as the result of the diverse land use and the varying terrain elevation.

References

Brutsaert W (1982) Evaporation into the atmosphere – theory, history and application. Springer, Netherlands/Heidelberg

Czeplak G, Kasten F (1987) Parametrisierung der atmosphärischen Wärmestrahlung bei bewölktem Himmel. Meteorol Rundsch 40:184–187

McClatchey RA, Fenn RW, Selby EA, Volz FE, Garing JS (1972) Optical properties of the atmosphere. Air-Force Cambridge Research Laboratories, AFCRL 72 0497, Environmental research paper no. 411

Moser W, Raschke E (1983) Mapping of global radiation and cloudiness from METEOSAT image data. Meterol Rundsch 36(2):33–37

Neitsch SL, Arnold JG, Kiniry JR, Williams JR, King KW (2002) Soil and water assessment tool theoretical documentation version 2000. TWRI report TR-191, Texas Water Resources Institute, College Station, Texas

Texas Water Resources Institute, College Station, TX, Swinbank WC (1964) Long-wave radiation from clear skies. Q J R Meteorol Soc 89:339–348

Chapter 12
Ice Reservoir

Markus Weber, Monika Prasch, Michael Kuhn, Astrid Lambrecht, and Wilfried Hagg

Abstract In the alpine regions of the Danube drainage basin, glaciers play a key role in the water balance of the headwater regions. They not only contain an important reservoir of freshwater, but they also have a regulating effect on run-off in alpine rivers. To calculate ice melt in a numerical model, information about surface topography and ice thickness must be provided at the highest possible resolution. Ninety-two percent of the glaciers within the Upper Danube have surface areas that are smaller than the 1×1 km^2-grid (proxel) used for DANUBIA, and these glacial areas' basin can be subdivided into 556 subareas, which are distributed over 1,196 proxels within the limits of the DANUBIA model.

For each proxel, an area-altitude distribution was prepared using the available digital glacial boundaries and a high-resolution digital elevation model from the glacial inventory of Austria, the Swiss glacier inventory and the Bavarian Glacier project. Additional for each partial area value for the thickness of the ice in the year 2000 was determined either by measurements or by estimation.

In Map 12.1, the water equivalent of the entire ice mass is calculated for the area of a proxel and distributed evenly across the area. The values are thus directly comparable to other hydrological variables such as the total annual precipitation or discharge rate. The value for the potential meltwater contribution distributed across the entire drainage basin of 213 mm is approximately comparable to the precipitation in the basin for the two summer months.

M. Weber (✉) • A. Lambrecht
Commission for Geodesy and Glaciology of the Bavarian Academy of Sciences and Humanities, Munich, Germany
e-mail: Wasti.Weber@kfg.badw.de; Astrid.Lambrecht@keg.badw.de

M. Prasch
Department of Geography, Ludwig-Maximilians-Universität München (LMU Munich), Munich, Germany
e-mail: m.prasch@lmu.de

M. Kuhn
Institute of Meteorology and Geophysics, University of Innsbruck, Innsbruck, Austria
e-mail: Michael.Kuhn@uibk.ac.at

W. Hagg
Department of Earth and Environmental Sciences, Università degli Studi di Milano-Bicocca, Milan, Italy
e-mail: wilfriedhagg@gmail.com

W. Mauser, M. Prasch (eds.), *Regional Assessment of Global Change Impacts*, DOI 10.1007/978-3-319-16751-0_12

Keywords GLOWA-Danube • Danube basin • Glacier • Ice thickness • Glacier inventory

12.1 Introduction

In the alpine regions of the Danube drainage basin, glaciers play a key role in the water balance of the headwater regions. They not only contain an important reservoir of freshwater, but they also have a regulating effect on run-off in alpine rivers, since during warm, dry periods, the meltwater is almost instantly fed into the channels. This fact has particular significance in dry periods of summer, because extremely low water levels are prevented as a result of the glacial meltwater.

Melting in glaciers takes place primarily at the surface. The quantity of meltwater is therefore proportional to the area of the glacier. However, changes in size over time do not simply depend on snow accumulation in winter and melt in the summer; change is largely the result of thickness distribution of the ice mass and the rearrangement of ice through ice flow. This sort of information must be gathered at the highest possible resolution for a model and be available at model initialization.

When described in terms of the 1×1 km^2-grid size used for DANUBIA, 92 % of the glaciers have surface areas that are smaller than one proxel (see Fig. 12.1). Over one-third cover only 10 % of the area of a proxel. At the same time, the glaciers within a proxel extend to up to 1,000 m altitudes and are therefore exposed to very different climatic conditions across short distances. Glaciers experience the most obvious notable changes in their geometry as a result of climate change in the lowest elevations and at their edges, where the ice is thinnest and almost no ice movement is observed.

The detailed data for glaciers form the basis for the simulation of the contributions of meltwater and the future status of the glaciers. The location and extent of the glacial ice reservoir in the year 2000 within the Upper Danube basin are depicted on Map 12.1 at the scale of the computational grid used by DANUBIA; also illustrated is the varying hydrological significance of the ice for the respective Alpine portions of the basin.

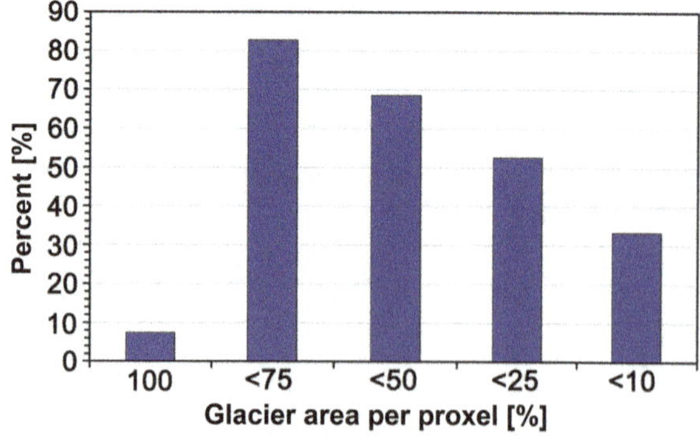

Fig. 12.1 Percentage of proxels where the percentage of glacier area is below the given threshold in %

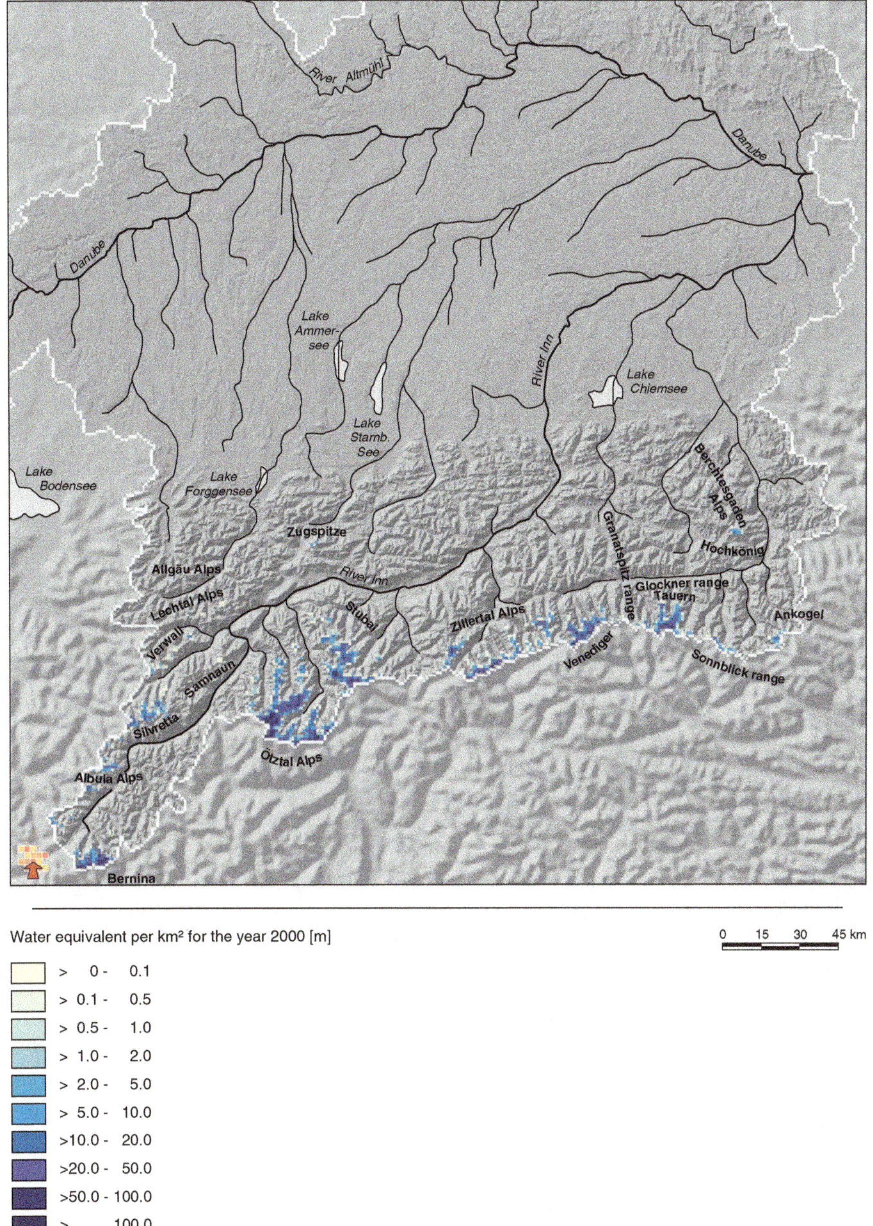

Map 12.1 Ice reservoir (Data sources: Austrian Glacier Inventory 1998 of the Institute of Meteorology and Geophysics, University of Innsbruck, Hagg (2006), Swiss Glacier Inventory of the University of Zurich, Switzerland, Jarvis et al. 2008, DANUBIA river network)

12.2 Data Processing

Depending on nomenclature, the current glacial areas within the Upper Danube basin can be subdivided into 556 subareas, which can be assigned to the Eastern Alps according to Table 12.1. Within the limits of the DANUBIA model, these are distributed over 1,196 proxels. For each proxel, an area-altitude distribution into classes up to a maximum of 50 m elevation was prepared using the available digital glacial boundaries and a high-resolution digital elevation model (DEM). For the Austrian glacier that comprises approximately 90 % of the glaciated area in the basin, the data from the new glacial inventory of the Institute of Meteorology and Geophysics at the University of Innsbruck (Institut für Meteorologie und Geophysik Innsbruck, IMGI) (Lambrecht and Kuhn 2007) was used; this inventory uses elevation models with a grid width of 10 m and glacial masks prepared using aerial photographs that were taken in the period from 1996 to 2002.

The boundary line polygons for the areas of the glaciers in Switzerland could be generated using the remote sensing data kindly provided from Swiss glacier inventory from the year 2000 (Paul et al. 2002). With this information and the 90 m DEM from the SRT mission flown by NASA in February 2000 (Jarvis et al. 2008; Rabus et al. 2003), the Swiss glaciers were distributed into 50 m elevation levels.

Table 12.1 Distribution of glaciers in the drainage basin on different mountain ranges

Mountain range	NG	*A* [km^2]	*V* [km^3]	PS [mm]
Abula Alps	23	8.72	0.260	3.40
Allgäu Alps	1	0.09	0.002	0.02
Ankogel range	9	2.51	0.061	0.79
Berchtesgaden Alps	3	1.99	0.038	0.50
Bernina Alps	15	30.35	1.629	21.26
Glockner range	42	26.75	1.358	17.71
Goldberg range	17	5.07	0.135	1.76
Granatspitz range	6	2.24	0.115	1.50
Lechtal Alps	5	0.29	0.004	0.05
Ötztal Alps	152	129.39	6.581	85.85
Samnaun	3	0.08	0.001	0.01
Silvretta	36	17.65	0.584	7.63
Stubai Alps	101	52.20	2.234	29.14
Venediger range	32	31.97	1.507	19.65
Verwall	23	2.02	0.045	0.58
Wetterstein	3	0.73	0.012	0.15
Zillertal Alps	85	45.87	1.838	23.98
Total	556	357.92	16.402	213.98

NG number of glaciers, *A* glacier area, *V* stored water volume, *PS* potential meltwater discharge, scaled to the size of the drainage basin

For completeness, the five German glaciers in the Bavarian mountains were also included in the dataset. The terrain data required to determine the areas at 20 m intervals of elevation were taken from the Bavarian Glacier project of the German Research Foundation (Deutsche Forschungsgemeinschaft, DFG) (Hagg 2006).

Each partial area needed to be assigned a realistic value for the thickness of the ice in the year 2000. Although the surface of glaciers can all be measured very precisely, in general, the knowledge about the shape of the glacier is still rather incomplete but yet essential for a realistic model of the glacier. For this reason, many glaciology research groups have recently made an effort to eliminate these knowledge gaps using measuring techniques applicable in the area, such as radio-echo sounding. Measurements of ice thickness were already made for approximately 50 glaciers of varying size and location within the study area (Span et al. 2005; Fischer and Kuhn 2013), and the list is steadily growing. For example, in 2007, detailed distributions of ice thickness were determined for the Vernagtferner, a large glacier in the Ötztal Alps. But even smaller areas such as the Schneeferner on the Zugspitze or the Schwarzmilzferner in the Allgäu Alps now have precise measurements (Hagg et al. 2012). These allow ice thicknesses to be determined directly for the partial areas (see Fig. 12.2).

However, measurements of ice thickness are still lacking for the majority of glaciers. Estimated or historical empirical values were adopted for these glaciers; these values are further supported by (1) data from neighbouring glaciers with known ice volumes or known ice thickness patterns, (2) formulae for estimating the maximum possible ice thickness with a given gradient (Paterson 1981) and (3) a relationship between the volume of a glacier V [km^3] and its area A [km^2] according to Bahr et al. (1997) that uses an exponent determined for alpine regions:

$$V = 0.02\ A^{1.36}$$

As a consequence, it is likely that data on ice masses for the glaciers with no thickness measurements are containing errors, which in some cases can amount to as much as 100 %, but in most cases, the 10–20 % tolerance range of the process

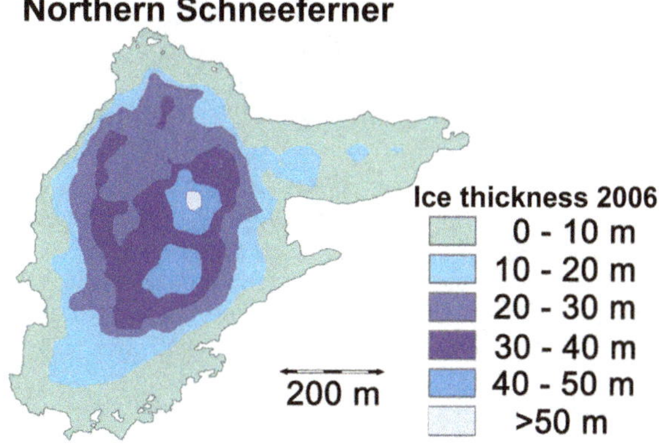

Fig. 12.2 Measured ice thicknesses at the Northern Schneeferner

model is not exceeded. The total integrated volume of water equivalent from all glaciers across the simulation area of 16.4 km^3 listed in Table 12.1 does not significantly deviate from previous integral estimates for the Eastern Alps; similarly, the value for the potential meltwater contribution distributed across the entire drainage basin of 213 mm is approximately comparable to the precipitation in the basin for the two summer months. Still, as a result of the scope and the quality of the base data in the existing inventory, this is assumed to be the most precise and detailed representation of the ice reservoir in the Eastern Alps.

12.3 Results

In Map 12.1, the water equivalent of the entire ice mass is calculated for the area of a proxel at 1 × 1 km^2 and distributed evenly across the area. The values are thus directly comparable to other hydrological variables such as the total annual precipitation or discharge rate.

The glacial annual storage of ice portrayed in the map can be considered as local sources for freshwater in hot dry periods. The extent of their usefulness as a source increases with the amount of area occupied by the ice mass. Higher values for water equivalent generally mean a thick ice cover and are thus associated with a larger area of the glacier, as described by the relationship given above between the volume and the area of glaciers. Hence, high values for water equivalent do not only represent a large reserve of glacial ice but also a productive source of meltwater.

Map 12.1 makes clear that the truly glaciated areas in the Danube basin are found in the main Alpine ridge of the region, in which the majority of peaks reach well over 3,000 m asl. More than half of the ice mass in the Upper Danube basin is situated within the Ötztal and Stubai Alps mountain ranges. In addition, the Zillertal Alps, Venedig and Glockner ranges as well as the Bernina range are all highly glaciated; the latter, at 4,049 m asl, is the highest point of the study area. Ice storage up to 50 times the annual precipitation is still stored in a few places there.

References

Bahr DB, Meier MF, Peckham SD (1997) The physical basis of glacier volume-area scaling. J Geophys Res 102:20,355–20,362. doi:10.1029/97 JBO1696

Fischer A, Kuhn M (2013) Ground-penetrating radar measurements of 64 Austrian glaciers between 1995 and 2010. Ann Glaciol 54(64):179–188. doi:10.3189/2013AoG64A108

Hagg W (2006) Digitale Aufbereitung historischer Gletscherkarten in Bayern. Mitteilungen der Geographischen Gesellschaft München 88:67–88

Hagg W, Mayer C, Mayr E, Heilig A (2012) Climate and glacier fluctuations in the Bavarian Alps in the past 120 years. Erdkunde 66:121–142

Jarvis A, Reuter HI, Nelson A, Guevara E (2008) Hole-filled seamless SRTM data V4. International Centre for Tropical Agriculture (CIAT). http://srtm.csi.cgiar.org. Accessed 19 Sept 2014

Lambrecht A, Kuhn M (2007) Glacier changes in the Austrian Alps during the last three decades, derived from the new Austrian Glacier inventory. Ann Glaciol 46:177–184. doi:10.3189/172756407782871341

Paterson WSB (1981) The physics of glaciers. Pergamom Press, Oxfrod/New York

Paul F, Kääb A, Maisch M, Kellenberger T, Haeberli W (2002) The new remote-sensing-derived Swiss glacier inventory. I. Methods. Ann Glaciol 34:355–361

Rabus B, Eineder M, Roth A, Bamler R (2003) The shuttle radar topography mission- a new class of digital elevation models acquired by spaceborne radar. ISPRS J Photogramm 57:241–262

Span N, Fischer A, Kuhn M, Massimo M, Butschek M (2005) Radarmessungen der Eisdicke Österreichischer Gletscher. Band 1: Messungen 1995 bis 1998. Österreichische Beiträge zu Meteorologie und Geophysik, 33, Vienna, Austria

Chapter 13
Trends in Temperature and Precipitation

Andrea Reiter and Ruth Weidinger

Abstract Climate change can already be detected in the behaviour of temperature and precipitation time series, especially over the recent decades. A data analysis of meteorological stations in the Upper Danube region revealed 83 sufficient out of 377 available time series covering the years 1960–2006 to calculate temperature and precipitation time series for the summer months (June, July, August) and the winter months (December, January, February). For a spatial analysis of the time series, the Upper Danube region was divided into five subregions with homogenous climates. The analysis comprises a representation analysis, a linear trend analysis and the Mann-Kendall test for determining the level of statistical significance of the calculated trends. The results showed significant increase of summer temperatures and slight increase of winter temperatures, both with high significance levels. The precipitation trends showed less significant or even negligible values and revealed both decreases and increases in both investigated seasons. Furthermore, the summer precipitation showed mostly decreasing values in the western part of the investigated area while the winter precipitation showed decreasing values in the Alps, especially in the eastern part.

Keywords Upper Danube • GLOWA-Danube • Regional climate change • Time series analysis • Temperature • Precipitation • Trend analysis

13.1 Introduction

The impact of climate change on water resources in the Upper Danube basin is a key focus of the GLOWA-Danube project. Although copious statements about global change have been made over the past decades, changes that can already be detected

A. Reiter (✉)
Bavarian Research Alliance GmbH, Munich, Germany
e-mail: reiter@bayfor.org

R. Weidinger
Department of Geography, Ludwig-Maximilians-Universität München (LMU Munich), Munich, Germany
e-mail: R.Weidinger@lmu.de

W. Mauser, M. Prasch (eds.), *Regional Assessment of Global Change Impacts*, DOI 10.1007/978-3-319-16751-0_13

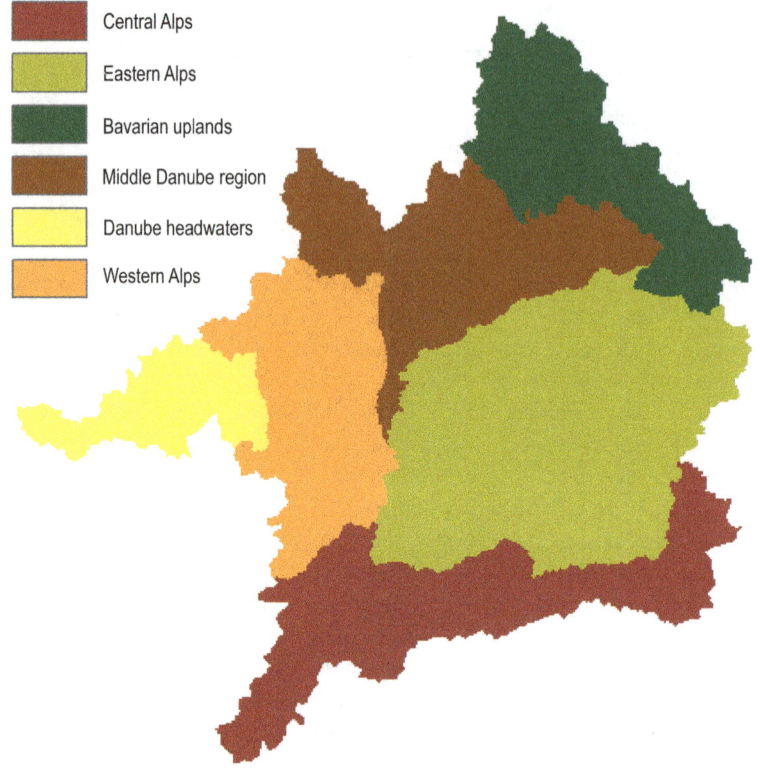

Fig. 13.1 Subregions in the Upper Danube basin based on the KLIWA study areas (Based on Kliwa 2005a, p. 15)

in variables describing the most important elements of climate – atmospheric temperature and precipitation – at the regional level of the basin are of high interest. The data available for the project allow a pointwise trend analysis of the temporal patterns of the climate variables precipitation and temperature in the GLOWA study area (Reiter et al. 2012). These changes are obvious over the past 10 years, particularly in the summers but also in winters (see Fig. 13.1). In order to determine the changes that have occurred over the recent decades, a statistical trend analysis was carried out for both summer and winter time series.

13.2 Data Processing

As basic data source, the station data from the DWD and ZAMG (see Chap. 10) were used. For the analysis method, the data had to comply with the following requirements (Fliri 1972): the area and altitude range of the basin had to be sufficiently represented, and the time series had to cover as long a period as possible but no less than 30 years. The goal of this analysis was to use time series that included up to 2006. Out of 377 available stations, 83 stations met these requirements for the trend analysis of summer and winter temperatures and precipitation amounts (in each case 83 stations – but sometimes different ones), covering the years from 1960 to 2006 (47 years). Extending the start of the time period to before 1960 was not an option, since then there were an insufficient number of stations to adequately cover the area of the basin.

For each year in the study period, the mean temperature and precipitation total was calculated from daily values for the three summer months (June, July and August) and for the three winter months (December, January and February).

The basin area was further divided into subregions for a regional study of temperature and precipitation trends; these subregions are based on the division of the study area used for the KLIWA project (see KLIWA 2005a, b and Fig. 13.2). These subregions comprise areas that have homogenous climates with respect to precipitation, which is important for the spatial analysis (the names of the subregions were retained from KLIWA). (KLIWA is a co-operation project on climate change and consequences for water management in Southern Germany).

In order to make reliable statements about the temporal progression of temperature and precipitation while largely avoiding influences that distort trends, statistical techniques were applied to the time series that are outlined below.

First, a representation analysis was performed to obtain information about the extent to which climate variable's temporal fluctuations (thus also its trend pattern) are representative for a given region (Rapp and Schönwiese 1996). Therefore, all time series for a subregion were compared to each other using Pearson's correlation

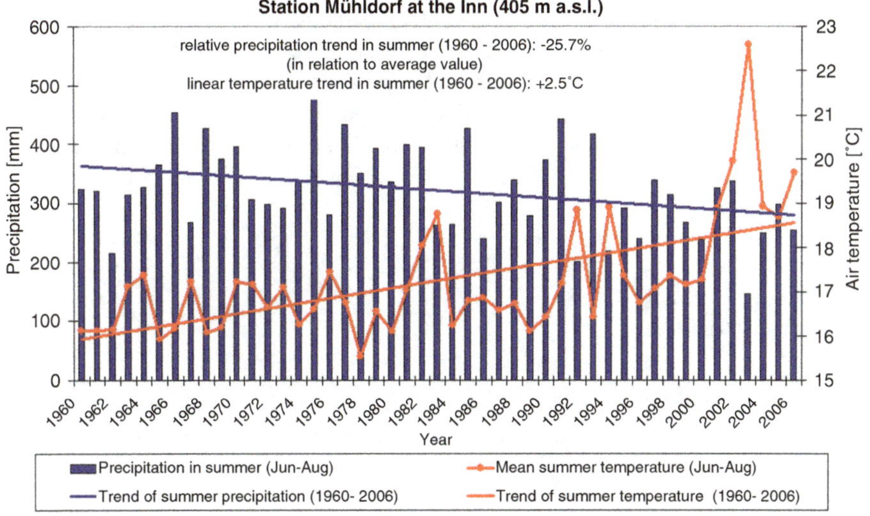

Fig. 13.2 Example for a temporal progression of temperature and precipitation for the summertime in the Upper Danube basin (Data sources: DWD Deutscher Wetterdienst)

coefficient ($-1 \leq r \leq +1$). If two time series have similar fluctuation patterns, the coefficient is very high, for example, $r \approx 0.9$. The results of this analysis are crucial for subsequent averaging of a period trend across a subregion.

The correlations for temperature time series both in the summer and winter months are mostly quite high, and the coefficients are almost always higher than 0.9. Both in the Danube headwaters and the East and Central Alps subregions (see Fig. 13.1), the individual station periods are somewhat less correlated ($r \approx 0.83$). In the East and Central Alps subregions, there appears to be lower correlation among some stations, but this result is not surprising based on the small-scale nature of the relief and the commonness of orographic effects.

The correlations for precipitation time series are significantly weaker in summer than in winter, since summertime convective precipitation events are geographically limited and therefore vary within each subregion more than advective precipitation in winter. A coefficient of 0.7 was stipulated as the minimum for a generally adequate value for precipitation (Rapp and Schönwiese 1995).

For summertime precipitation values, the correlation coefficients are between 0.31 and 0.86, such that only approximately 30 % of the periods are sufficiently correlated. Significantly, higher correlations are seen for wintertime precipitation values. Values are between 0.65 and 0.98 for almost all regions of the analysis. In the Central Alps region, coefficients decrease with increasing altitude, reaching values of 0.57–0.84 (between 1,200 and 1,600 m a.s.l.) and 0.41–0.71 (above 1,600 m a.s.l.).

In order to determine the temporal patterns of the climate variables, the linear trend in average temperatures and precipitation totals were calculated for each station for each of the 47 summer and winter seasons; this took place in form of regression equations using the least squares method.

For a correct interpretation of the individual trends, a certain level of statistical significance also needs to be demonstrated; this level indicates how 'probable' a calculated trend is. For this purpose, the Mann-Kendall test was used to estimate the significance of a temporal series trend (Salmi et al. 2002; Schönwiese 2000). This test is especially suited to the analysis of temperature and precipitation trends since it does not make any assumptions with respect to the frequency distribution of the data and is very robust. This test provides a significance level (expressed as a %) for each time series that is set to at least 80 % for temperature and precipitation trends (KLIWA 2005a).

13.3 Results

Because of the highly variable results for the representation analysis for precipitation, only for temperature a mean trend was calculated for the subregions and depicted on Map 13.1.

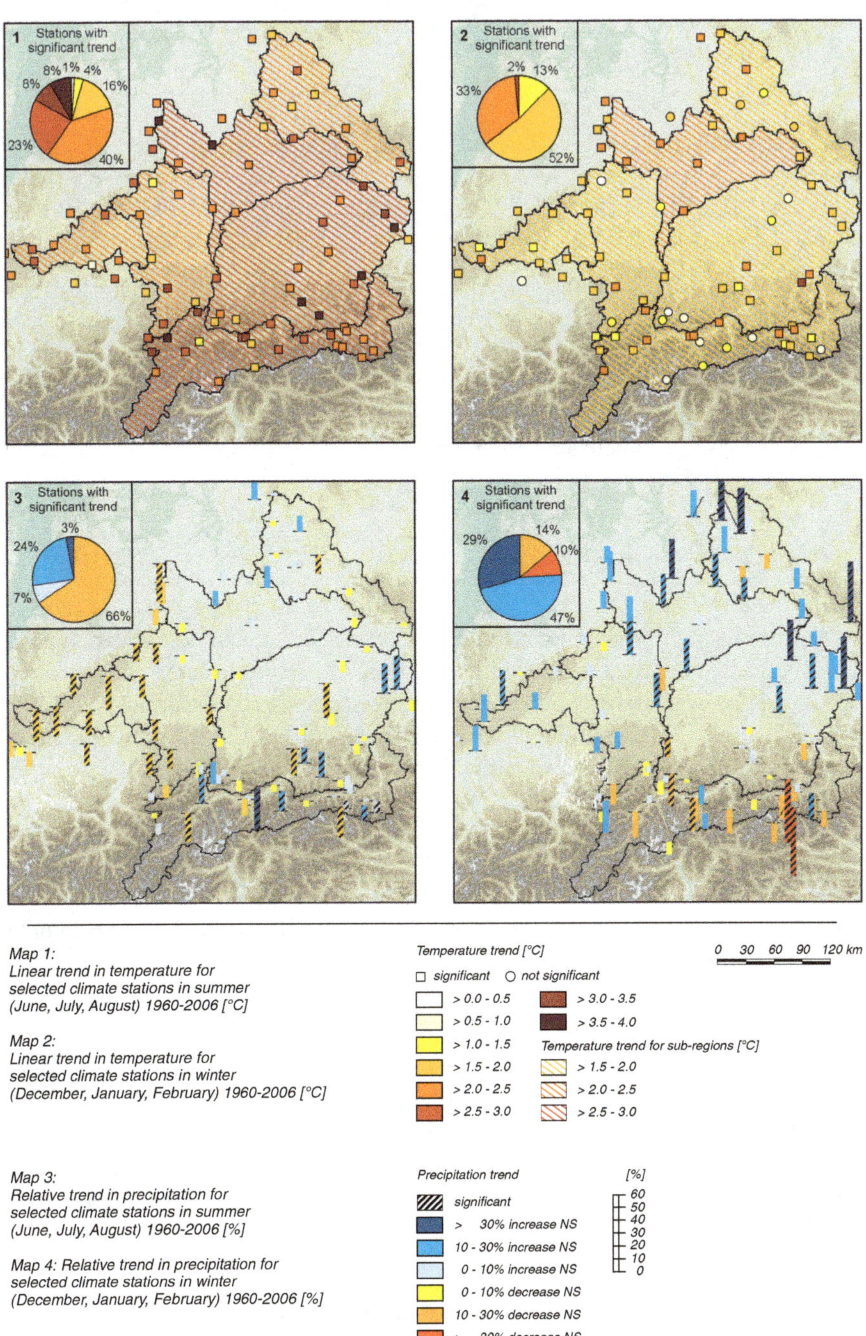

Map 13.1 Trends in temperature and precipitation (Data sources: DWD Deutscher Wetterdienst; ZAMG, Zentralanstalt für Meteorologie und Geodynamik; With kind permission from Springer Science+Business Media: Reiter et al. 2012, Figs. 3, 7, 14 and 18)

13.3.1 Temperature

Summer (Map 13.1 – 1) The Mann-Kendall test showed significant increase in summer temperatures at all 83 stations included from the basin during the period from 1960 to 2006. At almost 80 % of the stations, a temperature increase of more than 2 °C was detected, and at only one station, this increase was less than 1 °C.

Winter (Map 13.1 – 2) In winter, temperature trends were generally less pronounced than in summer. However, there was still an increase in temperature at all 83 time series studied, with 63 of the stations having a significant trend. Although the temperature increase at 85 % of the stations with significant trends was between 1.5 and 2.5 °C, a temperature increase of more than 2.5 °C was only evident at a single station (Bad Reichenhall, Germany).

13.3.2 Precipitation

In contrast to the trends in temperature, the trends for precipitation were notably less significant, since precipitation events vary to a much greater extent in both space and time. The trend in precipitation is specified as a relative trend; this means, the calculated linear trend (the difference between the final value x_n and the initial value x_1 of the trend line) is divided by the mean x of each time series studied.

$$\text{relative trend} = \frac{x_n - x_1}{\overline{x}} \cdot 100\%$$

This yields the change in % of the mean compared to the situation at the beginning of the period (Rapp and Schönwiese 1996).

Summer (Map 13.1 – 3) In summer, only 29 of the 83 time series analysed were significant with trends that were both negative and positive (−30 % to +30 %). Even the nonsignificant time series showed both decreases and increases in precipitation; thus, there has been no definitive trend in summer precipitation patterns over the past 47 years. Half of the 83 time series studied revealed no distinct trend (−10 % to +10 %), 26 series had a clear negative trend (10–30 % less precipitation), and only 13 showed a noticeable positive trend. There was a clustering of the negative (and in this case, also significant) trend in the western part of the basin and a clustering of positive trends in the eastern alpine regions and in the north.

Winter (Map 13.1 – 4) In the analysis of the 83 time series with winter precipitation events, 28 stations revealed only negligible changes. There were 21 time series (25 %) with significant trends; the majority of them (16) had trends with increasing precipitation (the relative trend was 28 % more precipitation on average); only in the alpine region there was a tendency towards decreasing precipitation. The changes were more striking towards the east. Even the nonsignificant time series confirmed this trend, where there was a notable tendency towards drier winters in the heights of the Alps, particularly in the east (Kitzbühel Alps).

Finally, it should be pointed out that trend analyses can only ever presume to make statements about the period of time being studied. Even minor changes or shifts in the period under investigation or extreme values at the beginning or end of a period can significantly alter the value of a trend or even reverse it. Forecasts for the future that are derived from the trend should therefore only be made with extreme caution.

References

Fliri F (1972) Statistik und Diagramm. Westermann, Braunschweig

KLIWA (ed) (2005a) Langzeitverhalten der Lufttemperatur in Baden-Württemberg und Bayern. KLIWA-Heft 5. Bayerisches Landesamt für Wasserwirtschaft, München, p 76

KLIWA (ed) (2005b) Langzeitverhalten des Gebietsniederschlags in Baden-Württemberg und Bayern. KLIWA-Heft 6. Bayerisches Landesamt für Wasserwirtschaft, München, p 160

Rapp J, Schönwiese C-D (1995) Atlas der Niederschlags- und Temperaturtrends in Deutschland 1891–1990. Frankfurter Geowissenschaftliche Arbeiten Serie B – Meteorologie und Geophysik, Bd 5, Frankfurt

Rapp J, Schönwiese C-D (1996) Niederschlag- und Temperaturtrends in Baden-Württemberg 1955–1994 und 1895–1994. In: Lehn H et al (eds) Wasser – Die elementare Ressource. Materialienband. Akademie für Technikfolgenabschätzung in Baden-Württemberg, Stuttgart, Arbeitsbericht Nr. 52, 113–170

Reiter A, Weidinger R, Mauser W (2012) Recent climate change at the Upper Danube – a temporal and spatial analysis of temperature and precipitation time series. Clim Change 111(3–4):665–696. doi:10.1007/s10584-011-0173-y

Salmi R, Määttä A, Anttila P, Ruoho-Airola T, Amnell T (2002) Detecting trends of annual values of atmospheric pollutants by the Mann-Kendall test and Sen's Slope estimates – the excel template application MAKESENS. Publications on air quality, No. 31, Helsinki

Schönwiese C-D (2000) Praktische Statistik für Meteorologen und Geowissenschaftler. Gebrüder Borntraeger, Berlin

Chapter 14
Hydrogeology – A Consistent Basin-Wide Representation of the Major Aquifers in the Upper Danube Basin

Roland Barthel, Jürgen Braun, Vlad Rojanschi, and Jens Wolf

Abstract Modelling the changes of groundwater resources under conditions of global change is a crucial task in the UDC, where groundwater forms the main source of drinking water. The first step in the development of a numerical modelling of groundwater flow and storage changes is to develop a conceptual hydrogeological model that adequately represents the relevant geological and hydrological features of the model area. The large area of the UDC and the coarse resolution of the common DANUBIA model thereby demand strict simplification of the actual hydrogeological conditions. A central decision in this conceptualisation process is the definition of the major regional hydrogeological units to be considered by the model. This chapter describes the process of data collection, data homogenisation and aggregation into ten main hydrogeological units, called "base classes". The properties and the relevance of these base classes for the numerical model are explained. The most important simplifications made are briefly discussed.

R. Barthel (✉)
Department of Earth Sciences, University of Gothenburg,
Göteborg, Sweden
e-mail: roland.barthel@gu.se

J. Braun
Institute for Modelling Hydraulic and Environmental Systems/VEGAS,
University of Stuttgart, Stuttgart, Germany
e-mail: juergen.braun@iws.uni-stuttgart.de

V. Rojanschi
Golder Associates Ltd., Calgary, AB, Canada
e-mail: Vlad_Rojanschi@golder.com

J. Wolf
Repository Safety Research Division, Gesellschaft für Anlagen- und
Reaktorsicherheit (GRS) gGmbH, Braunschweig, Germany
e-mail: jens.wolf@grs.de

W. Mauser, M. Prasch (eds.), *Regional Assessment of Global Change Impacts*, DOI 10.1007/978-3-319-16751-0_14

Keywords GLOWA-Danube • Upper Danube • Hydrogeology • Groundwater • Regional aquifers

14.1 Introduction

Modelling of groundwater flow which is a key element of the hydrological cycle requires an adequate representation of the hydrogeological conditions of the drainage basin. In the case of a large regional basin, vast simplification of the hydrogeology is required to match the common spatial resolution of the DANUBIA modelling approach and to achieve reasonable computation times (see also Barthel et al. (2005, 2011) and Wolf et al. (2008). Considering the complexity of the hydrogeological conditions of the Upper Danube basin, this is a rather difficult undertaking (Map 14.1).

14.2 Data Processing

Comprehensive hydrogeological maps covering the entire UDC are not available. Instead, a large variety of sources of information had to be analysed and homogenised and subsequently merged into one consistent representation of the basin's main aquifers. For that purpose, geological formations with comparable hydraulic properties were summarised into "hydrogeological units". A large part of the data base used for this task was extracted from the maps provided by the Geological Offices in Baden-Württemberg, Bavaria, Austria and Switzerland (LGRB 1998; BayGLA 1996; Weber et al. 1997; BuWaG 2002). Detailed geological maps are available for nearly the entire drainage basin, but they exist at different scales and have variable coordinate systems. More problematic than the differences between data sources that relate simply to presentation (scale, coordinate system) is the fact that different maps and charts use varying classification schemes for the established geological formations. Thus, varying attention to detail (vertical resolution), the use of different bio-, chrono- and lithostratigraphic classifications and diverse approaches to the drawing of boundaries between geological units are all factors that sometimes lead to significant inconsistencies that create problems when pooling the data into a single hydrogeological map. It is furthermore problematic that the hydrostratigraphic classification only rarely matches the traditional geological map units.

Once a consistent, simplified hydrogeological map of the UDC was created, a vector-raster conversion for the implementation of the derived conceptualisation into the numerical model had to be performed. This step resulted further loss of detail as the coarse 1 km raster does not allow an accurate representation of the complex geological boundaries, which is in particular problematic for the small, yet

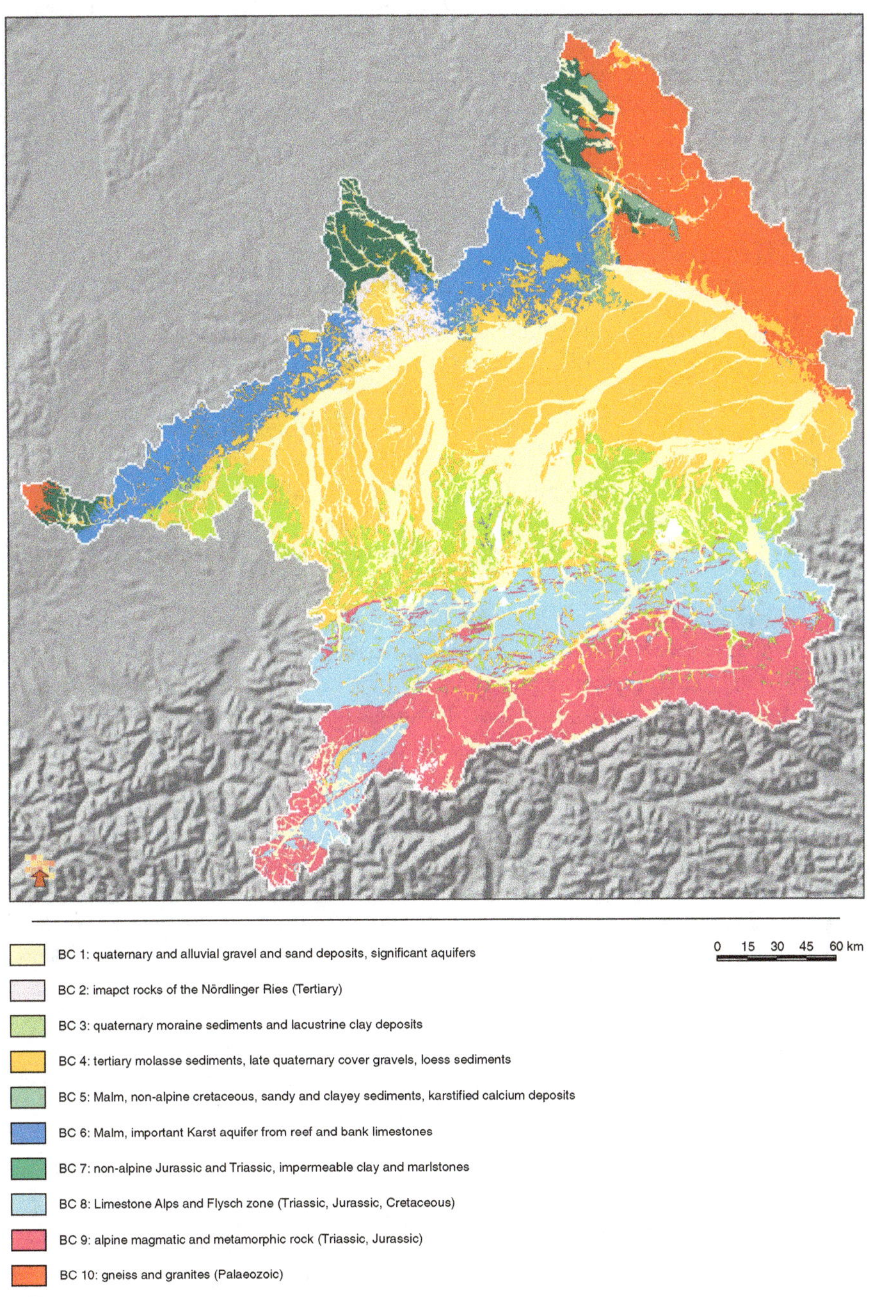

Map 14.1 Hydrogeology (Data sources: Geological map of Bavaria 1:500,000 @ Bayerisches Landesamt für Umwelt, www.lfu.bayern.de, Geowissenschaftliche Übersichtskarten 1:350,000 (Landesamt für Geologie, Rohstoffe und Bergbau, Baden-Württemberg, Freiburg, 1998) Metallogenetic Map of the Republic of Austria 1:500,000 (Weber 1997); @ Geological Map of Switzerland 1:500,000, Federal Office for Topography swisstopo 2002)

hydrogeologically very important alluvial deposits of the basin. The gridded hydrogeological map for the Danube basin is thus the product of a long sequence of steps to adapt and simplify the data in order to provide an adequate basis for the numerical interpretation of the groundwater system.

14.3 Results

Ten base classes (BC) with similar hydraulic and hydrogeological characteristics were derived from the geological units; these ten classes represent the hydrogeological conditions.

Figure 14.1 explains the base classes and their horizontal and vertical extent. The geology of the model area is characterised by the alpine belt in the south, the uplifting of which over the past 50 million years has basically filled the large valleys of the foothills (the Molasse basin) with debris. Naturally, the greatest thickness of the Molasse sediments, in the range of 5,000 m, is found at the edges of the alpine zones. The pre-alpine sediments (mostly Malm – Upper Jurassic) are located below the Molasse sediments, in great depth at the alpine fringe, and outcropping to the surface in the Swabian and Franconian Jura in the northern part of the basin. Through the Quaternary period until present time, the gradient from the alpine body to the Danube provides for a distinct south-north and southwest-northeast water network that currently transports and accumulates the alpine debris into the river valleys and gravel plains.

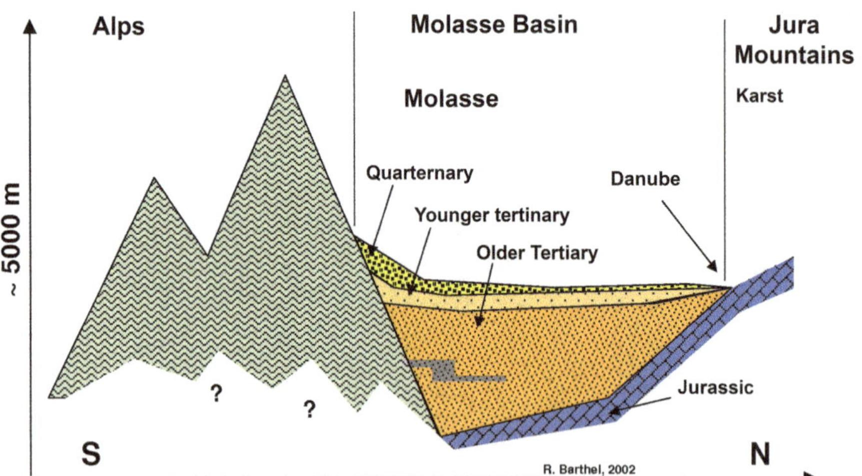

Fig. 14.1 Schematic geological cross section through the study area

14.3.1 Base Class 1

The most significant aquifers in the Danube basin are the *quaternary valley aquifers*. These aquifers are alluvial gravel and sand deposits with very high permeabilities that were deposited during or since the most recent glacial period. The majority of the short-term and intermediate-term turnover of water flows over these aquifers, (see Jie et al. 2011) which sometimes have extremely low dimensions. The Malm aquifer (base class 6) assumes this function only in the Swabian and Franconian Jura.

14.3.2 Base Class 2

The impact rocks of the Nördlinger Ries are special within the drainage basin because they are the consequence of a meteorite impact. Hydrogeologically, they are largely poor conductors of water. They have only negligible impact on the subterranean water balance of the basin because they are limited in extent.

14.3.3 Base Class 3

The *quaternary moraine deposits* are mostly quite impermeable because of their high content of fines. In some places, between moraine and molasse, there are thin but very rich sand-gravel active aquifers that are not modelled, however, since they only have local relevance. Lacustrine clay deposits are also among the moraines in this base class because they too are very impervious.

14.3.4 Base Class 4

The tertiary consists of mostly *molasse sediments* from the Lower Marine Molasse as far as the Upper Freshwater Molasse (Oberen Süßwassermolasse, OSM). As a result of the interbedding of sandy (fluvial) and clayey (marine) loose sediments, these sediments are only very slightly permeable and thus form the most significant groundwater aquitards within the aquifer system. The upper layers of the OSM are also regionally significant aquifers and must be separated from the foremost marine deposits that lie at a lower level. With hydraulic contact, early quaternary cover gravels are also considered in the OSM (these are sometimes very fertile but not regionally contiguous and therefore do not form their own aquifer).

14.3.5 Base Class 5

The *non-alpine cretaceous* (Cenoman to Campan) is predominantly composed of sand and clay deposits. However, there are also karstified limestone deposits that are generally found in hydraulic contact with Malm but which have lower permeabilities than those in base class 6.

14.3.6 Base Class 6

As an important Karst aquifer composed of reef and bank limestones with sometimes very high permeabilities, the *Malm* forms the highest formation in the Swabian and Franconian Jura and plays a direct role in the hydrological cycle. The almost complete lack of surface waters in the outcropping area of Malm is a result of the intensive karstification and is characteristic of this area. Towards the south, the layers disappear under the molasse layers (BC4). At the transition zone, from Karst that is not covered to Karst that is covered, there is a discharge of the majority of the alpine waters into the Danube valley aquifer (BC1).

14.3.7 Base Class 7

In the *non-alpine Jurassic (Lias and Dogger) and the Triassic*, the impermeable clay and marlstones of the Mesozoic form the basis of the Malm aquifers (BC6) and hence the entire aquifer system in the Upper Danube basin. The smaller aquifers Triassic to Cretaceous sandstone and limestone in the northern parts of the Nördlinger Ries have no significance for water balance in the region.

14.3.8 Base Class 8

The Northern *Limestone Alps* contain alpine deposits from the Triassic, Jurassic and Cretaceous, chiefly limestone, dolomite and marl. They are in part strongly karstified with a mid to high permeability. Very little is known about the vertical extent of the karstification.

This class also includes the Lower Engadine Window which is also notable for its high proportion of karstified elements, as well as the *Flysch zone* that lies directly to the north.

14.3.9 Base Class 9

The Alps are composed of *alpine magmatic and metamorphic rocks*, largely from volcanic, granitoids, gneisses and shales from the mid-eastern Alps of the Triassic and Jurassic. They are characterised by very low permeabilities and can thus be termed aquicludes.

14.3.10 Base Class 10

In the Palaeozoic of the Bavarian and Upper Palatinate Forests in the east and the Black Forest in the west, there is a large quantity of *gneiss and granites* with low permeability. Especially the granites can be deeply eroded and serve as local aquifers where this occurs. They thus form important local groundwater resources that mostly are manifested at sources – however, they may not be depicted at the regional scale. With the use of such groundwater resources, the security of the supply can be at risk during long dry periods.

References

Barthel R, Rojanschi V, Wolf J, Braun J (2005) Large-scale water resources management within the framework of GLOWA-Danube. Part A: The groundwater model. Phys Chem Earth 30(6–7):372–382

Barthel R, Reichenau TG, Muerth M, Heinzeller C, Schneider K, Hennicker R, Mauser W (2011) Global change impacts on groundwater in Southern Germany-Part 1: Natural aspects [Folgen des Globalen Wandels für das Grundwasser in Süddeutschland – Teil 1: Naturräumliche Aspekte]. Grundwasser 16(4):247–257

BayGLA (1996) Geologische Karte von Bayern 1:500.000. Bayerisches Geologischen Landesamt, Munich

BuWaG (2002) Geologische Karte der Schweiz 1:500.000. Bundesamt für Wasser und Geologie, Berne

Jie Z, van Heyden J, Bendel D, Barthel R (2011) Combination of soil-water balance models and water-table fluctuation methods for evaluation and improvement of groundwater recharge calculations. J Hydrogeol 19(8):1487–1502

LGRB (1998) Geowissenschaftliche Übersichtskarte von Baden Württemberg 1:350.000. Landesamt für Geologie, Rohstoffe und Bergbau, Freiburg

Weber L (Hrsg.) (1997) Handbuch der Lagerstätten der Erze, Industrieminerale und Energierohstoffe Österreichs. Erläuterungen zur metallogenetischen Karte von Österreich 1:500.000 unter Einbeziehung der Industrieminerale und Energierohstoffe.- Arch. Lagerst.forsch. Geol. B.-A., 19, 607 S., 393 Abb., 37 Tab., 2 Ktn., 2 Lis-ten, Wien

Wolf J, Barthel R, Braun J (2008) Modeling ground water flow in alluvial mountainous catchments on a watershed scale. Ground Water 46(5):695–705

Chapter 15
Mean Daily Discharge and Discharge Variability

Winfried Willems, Georg Kasper, Peter Klotz, Konstantin Stricker, and Astrid Zimmermann

Abstract Long-term discharge characteristics and their variability provide basic information on the availability of surface water. Because discharges are measures at a limited number of gauges within a drainage network, a regionalisation procedure is needed in order to get a complete picture of the spatial distribution of discharge in the catchment. The following discharge characteristics are taken into account here: minimum discharge (NQ), 25th percentile (Q25), 50th percentile (Q50), mean discharge (MQ), 75th percentile (Q75) and maximum discharge (HQ). The regionalisation method developed is based on the close statistical relationship between the logarithmic cumulative stream length and the logarithmic discharge characteristic in the Upper Danube drainage basin. All measures are calculated for the reference period from January 1980 to December 1999. Map 15.1 presents the regionalised mean discharge MQ and its variability, determined from the quotient of regionalised values of NQ and HQ. Primarily, it is alpine tributaries such as the Iller, Lech, Isar and Inn Rivers that influence the drainage patterns of the Danube.

Keywords GLOWA-Danube • Long-term discharge characteristics • Regionalisation • Log-log regression

15.1 Introduction

The term discharge refers to the volume of water per unit time that flows through a defined cross-sectional area within a stream. Discharge rate is given by dividing by the appropriate area of the drainage basin. The regional patterns of distribution for the mean annual discharge provide the basic data on the availability of surface water. Information about the range of the discharges at a given site describes the

W. Willems (✉) • G. Kasper • P. Klotz • K. Stricker • A. Zimmermann
Hydrology, Applied Water Resources Management and Geoinformatics IAWG, Ottobrunn, Germany
e-mail: willems@iawg.de

W. Mauser, M. Prasch (eds.), *Regional Assessment of Global Change Impacts*, DOI 10.1007/978-3-319-16751-0_15

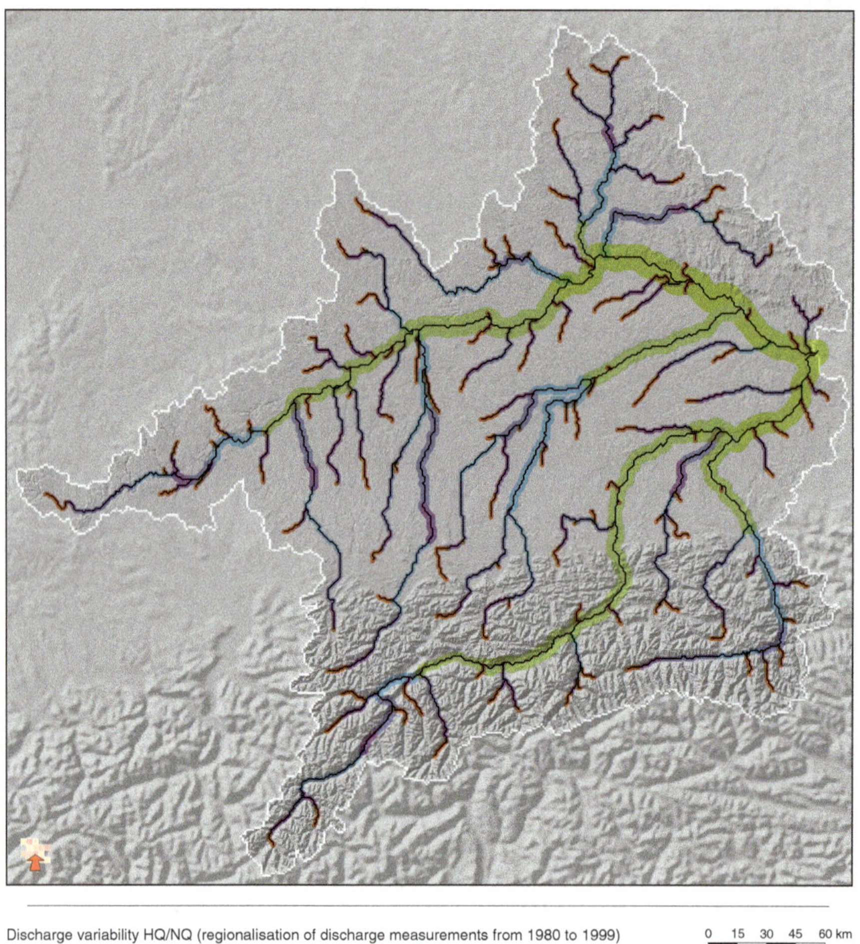

Map 15.1 Mean daily discharge and discharge variability (Data sources: DANUBIA – river network; Discharge Data © Bayerisches Landesamt für Umwelt, www.lfu.bayern.de; State Office for the Environment, Measurements and Nature Conservation Baden-Württemberg (LUBW), Karlsruhe, 1999; Austrian Federal Ministry for Agriculture, Forestry, Environment and Water Management, BMLFUW, Vienna, 1999, Jarvis et al. 2008)

discharge variability that is calculated here using the quotient of flood discharge (HQ) and low flow discharge (NQ).

Discharges are measured at a limited number of gauges within a drainage network. If discharge values are required for unmeasured cross-sectional areas, then an appropriate transfer of the measurements is needed (regionalisation). As a result, the discharge values (MQ, HQ, etc.) can be determined for every required point within the drainage network using the regionalisation method developed for the surface water subproject. An additional step yields the discharge variability for the entire Upper Danube drainage basin. This also provides the input data for the model parameterisation for the surface water subproject (see Chap. 29). The regionalisation method allows statements to be made about the runoff characteristics of unobserved sections in the drainage network. This evidence can be used in the planning of hydraulic engineering facilities, for industrial, public and private water consumption or for river navigation purposes. Significant changes in mean discharges and discharge variabilities that are induced by global climate change, for example, may lead to far-reaching consequences within the environment. There are already trends in the changes in discharge data that can be detected today (KLIWA 2000).

15.2 Data Processing

The establishment of the regionalisation method requires discharge data from as many gauges as possible within the drainage basin. Thus, series of mean daily discharge are used as data base that have been made available from the Bavarian State Office for the Environment (Bayerisches Landesamt für Umwelt = LfU); the Baden-Württemberg State Office for the Environment, Measurements and Nature Conservation (Landesanstalt für Umwelt, Messungen und Naturschutz Baden-Württemberg = LUBW); and the Austrian Federal Ministry for Agriculture, Forestry, Environment and Water Management (Bundesministerium für Land- und Forstwirtschaft, Umwelt- und Wasserwirtschaft Österreich = BMLFUW). These series of data were first used to calculate principal primary statistics for hydrological variables (e.g. minimum discharge NQ, mean discharge MQ, peak discharge HQ) for each gauge. Then regression equations were compiled in order to regionalise these aggregated values based on general and area-wide features. Figure 15.1 shows the values for NQ, MQ and HQ determined using the series of data on mean daily discharges for the period from January 1980 to December 1999.

The regionalisation method developed within the surface water subproject for determining discharge values is based on the close statistical relationship between the logarithmic cumulative flow length (LC) and the logarithmic discharge (Q) in the Upper Danube drainage basin.

$$\mathrm{Log}(Q) = c_1 + c_2 \cdot \mathrm{Log}(\mathrm{LC})$$

Figure 15.2 shows this relationship based on the mean discharges (MQ) in a log-log graphical representation. The coefficient of determination here is 0.95.

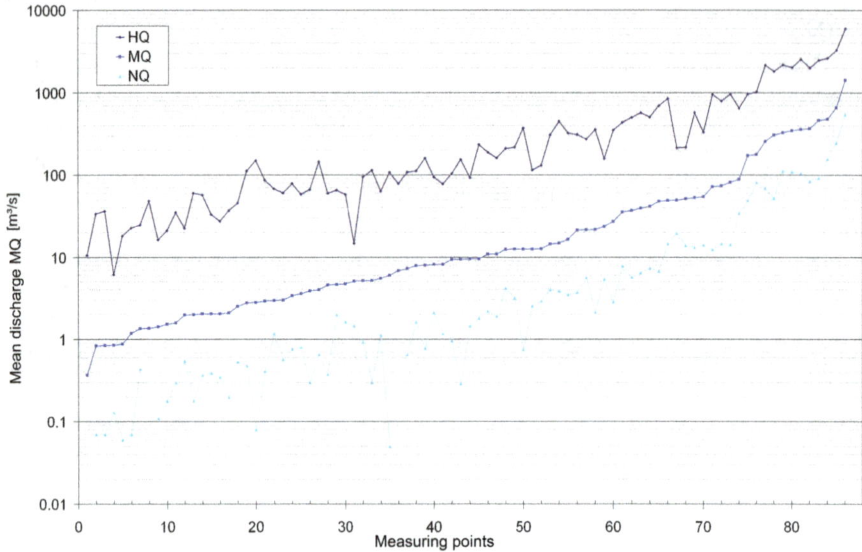

Fig. 15.1 Minimum (NQ), mean (MQ) and peak discharge at selected gauges

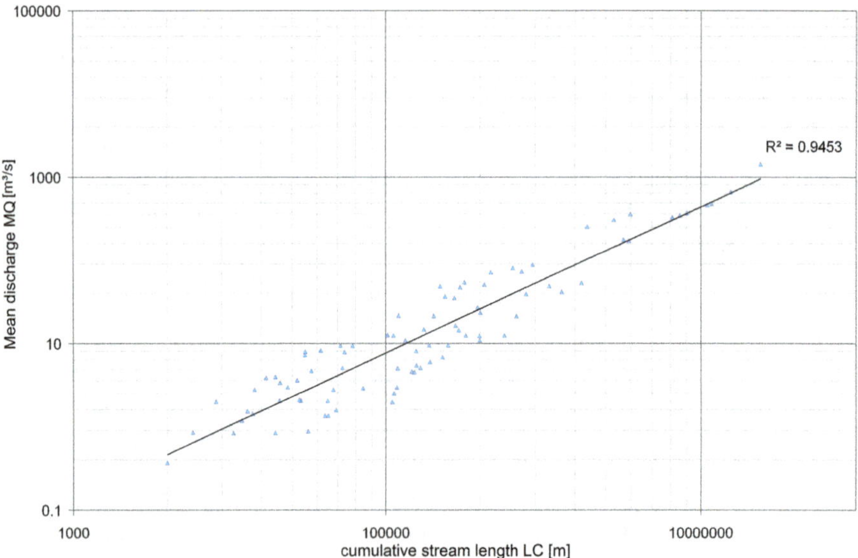

Fig. 15.2 Regression between the cumulative stream length and mean daily flow

Table 15.1 Parameter estimates c_1 and c_2 of the regression between logarithmic cumulative stream length (LC) and logarithmic discharge (Q)

Q	c_1	c_2
Ln NQ	−10.933793	0.9711137
Ln Q25	−9.029499	0.9024768
Ln Q50	−8.325971	0.8747306
Ln MQ	−7.562846	0.8350463
Ln Q75	−7.480112	0.8410268
Ln HQ	−2.717849	0.6481749

The parameters c1 and c2 for the following six hydrological variables are given in Table 15.1: minimum discharge (NQ), 25th percentile (Q25), 50th percentile (Q50 is equivalent to the median), mean discharge (MQ), 75th percentile (Q75) and peak discharge (HQ). The hydrological variables for each point in the Upper Danube drainage network can be calculated based on the regression parameters, since the cumulative flow lengths are known for each point.

15.3 Results

The regionalised mean discharge (MQ) and discharge variability, determined from the quotient of regionalised values of NQ and HQ, are presented. The mean discharge is classified into five categories that are represented by different line thicknesses. Discharge variability is divided into six classes. These variabilities are distinguished on Map 15.1 by different colours.

Mean discharge increases with increasing size of the drainage basin. Where tributaries merge with the main watercourse, there is a significant increase in discharge. Primarily, it is alpine tributaries such as the Iller, Lech, Isar and Inn rivers that influence the runoff characteristics of the Danube. Thus, in early summer, extreme discharges from alpine snowmelts occur that can lead to high water levels in the Danube. However, also the inflows from northern low mountain ranges and lowlands contribute to elevated discharges in the Danube during long-lasting rainfalls. In addition to spatial variability, the range of the discharges also varies. There are highly variable discharges in the smaller drainage basins and in the headwaters of the water bodies. Variability decreases with increasing size of the drainage basin, because extreme situations can be better compensated for.

References

Jarvis A, Reuter HI, Nelson A, Guevara E (2008) Hole-filled seamless SRTM data V4. International Centre for Tropical Agriculture (CIAT). Available from http://srtm.csi.cgiar.org

KLIWA (ed) (2000) Klimaveränderung und Konsequenzen für die Wasserwirtschaft. Fachvorträge beim KLIWA-Symposium am 29. und 30.11.2000 in Karlsruhe. KLIWA-Berichte, Heft 1, Karlsruhe

Chapter 16
Population

Matthias Egerer, Markus Zimmer, and Markus Probeck

Abstract This chapter presents the method developed for modelling absolute population trends and dividing the absolute population numbers into household size and income, parameters incorporated in the DANUBIA model *Demography*. The model provides information about the overall population and the summarized household incomes by district. The population trend was integrated in RIWU in an individually adjustable time trend separate from the model that determined the absolute population trend for the entire drainage basin. Unpooling of values for population and income determined by RIWU by district is done by means of remote sensing, and the disaggregated values are then assigned to proxels. Results show that, on average, 1,142.2 people live in a modelled, populated proxel. The results obtained with the aid of remote sensing represent a close approximation of reality.

Keywords GLOWA-Danube • Upper Danube • Economics

16.1 Introduction

In the context of modelling the global change scenarios, population is a key factor in determining household water demand. In addition to absolute population trends, key determinants for calculating specific water requirements include household size and income. Both parameters are incorporated in the DANUBIA model *Demography*.

M. Egerer (✉)
Bavarian Ministry of Economic Affairs and Media, Energy and Technology, Munich, Germany
e-mail: Matthias.Egerer@stmwi.bayern.de

M. Zimmer
Center for Energy, Climate and Exhaustible Resources, Ifo Institute – Leibniz Institute for Economic Research, University of Munich, Munich, Germany
e-mail: zimmer@ifo.de

M. Probeck
GAF AG, Munich, Germany
e-mail: markus.probeck@gaf.de

W. Mauser, M. Prasch (eds.), *Regional Assessment of Global Change Impacts*, DOI 10.1007/978-3-319-16751-0_16

Table 16.1 Mean percentage distribution of population in Germany

Households	Income						
	<1,100€ (%)	1,100–1,999€ (%)	2,000–2,999€ (%)	3,000–5,000€ (%)	>5,000€ (%)	Total Germany (%)	Total Austria (%)
1 person	14.86	14.60	3.71	1.03	0.18	34.37	33.53
2 persons	1.90	10.91	10.38	6.86	1.75	31.79	28.55
3 persons	0.18	3.22	5.65	5.68	1.43	16.16	16.30
4 persons	0.00	1.76	4.13	4.94	1.76	12.59	13.91
=5 persons	0.00	0.38	1.63	2.17	0.91	5.09	7.71
Total	16.93	30.86	25.49	20.68	6.02	100.00	100.00

STATISTIK AUSTRIA (2003) and Federal Statistical Office (1998)

The method developed for dividing absolute population numbers between these two determinants is introduced below.

The environmental economics subproject developed the regional economic simulation model RIWU (Regional Industrial Water Use) as a basis for the DANUBIA model *Economy*. In addition to modelling economic development and industrial water use (see Chaps. 17 and 45), this model provides information about the total population and the sum of the household incomes by district. The goal of the independent DANUBIA model *Demography* is to divide the population by proxel into one of five categories for household size and income onto a 5×5 matrix (see Table 16.1). The data calculated by the RIWU model for population and household income are read into each calculated time increment (month) in *Demography*. In order to prepare the 5×5 matrix, it is first necessary to disaggregate the values for population and income determined by RIWU at the district level onto the proxel and then to assign these to the 25 fields on the matrix. These two steps are described below.

16.2 Preparation of the Data

The data for the RIWU model are introduced in the caption for the "Gross Domestic Product" map (see Chap. 17). The statistics from the economic calculation of the Federal Statistics Office (Statistisches Bundesamt 1998) form the basis for calculating the 5×5 matrix. These statistics are based on a representative population survey conducted in Germany in 1993 (28,917 households); each household size category listed in Table 16.1 is assigned to 28 income categories. For simplification, these 28 income categories from the official statistics are summarized into five categories in the model presented here. Table 16.1 reveals the concise averaged population distribution according to the official statistics.

Algorithms were developed within the DANUBIA model *Demography* that adapt for each proxel this average distribution or the average income on which it is

Table 16.2 Percentage distribution for Switzerland

	Income					
Households	<1,985€ (%)	1,985–3,308€ (%)	3,309–4,632€ (%)	4,633–5,956€ (%)	>5,956€ (%)	Total (%)
1 person	5.52	7.21	5.83	3.40	3.18	25.15
2 persons	0.93	5.79	6.65	6.63	14.78	34.78
3 persons	0.07	0.35	2.40	4.75	7.08	14.65
4 persons	0.00	0.82	3.79	3.09	10.90	18.59
=5 persons	0.00	0.18	1.06	1.54	4.05	6.83
Total	6.52	14.34	19.74	19.42	39.99	100

based. That is, a 5×5 matrix is determined for every possible average income (combination of the variables household income and population) that can be calculated by RIWA (see Sect. 16.3).

A comparison of available values about household size and income distribution for Austria indicates good consistency with comparable values for Germany. Thus, the percentage of individual household size categories in the overall population is consistent with the distribution in Germany, differing by only a few percentage points (see the two right-hand columns of Table 16.1). Even the Gini coefficient as a proxy variable for the income distribution is almost identical in both countries (Germany: 30; Austria: 31; Central Intelligence Agency 2004). Therefore, the average matrix used for Germany and the algorithms on which it is based can be applied to the Austrian portion of the drainage basin.

This approach cannot be applied for Switzerland. Both the population distribution into household size categories and the absolute incomes deviate too much from the German and Austrian values. Thus, a separate method for determining the 5×5 matrix was developed for Switzerland. Because Switzerland constitutes only a very small portion of the drainage basin and even there only a few proxels are populated, the simplest method possible was selected. A 5×5 matrix was developed based on official statistics (Schweizerisches Bundesamt für Statistik 2003); in contrast to the German and Austrian portions of the basin, the percentage allocations in this matrix were independent of the modelled average incomes of each proxel. The percentage distribution shown in Table 16.2 is thus constant for all simulations for each populated proxel in the Swiss portion of the basin (Grisons).

16.3 Description of the Model

The population trend was integrated in RIWU in an individually adjustable time trend separate from the model. In addition to the economic development, this time trend also determines the absolute population trend for the entire drainage basin. The distribution of population by district is positively associated with the level of economic activity of a region. Thus, in the relatively socially homogenous region of the Upper Danube basin, it is assumed that the dominant effect on population trend is from population movements to take up job offers.

16.3.1 Calculation of Proxel Values

Since a uniform distribution of the district values calculated in RIWU over all proxels would yield meaningless results because all proxels would falsely appear populated at a uniform density, a tool for disaggregating the data was developed for DANUBIA. The tool is applied in its general form for all variables calculated by the environmental economy subproject. The important steps of this tool to calculate proxel values for the population variable are presented below:

Remote sensing is the key tool for unpooling the district values. Their images make it possible to estimate the distribution of administrative parameters by means of the land use classifications. For the distribution of the population values, a GIS was prepared based on the CORINE land use data from the hydrology/remote sensing subproject; within a municipality, this GIS matches linearly the population known from the official statistics to the CORINE land use category designated as "development with residential buildings". For example, this means that a proxel with 40 % of its area built-up has four times as large a population as one with only 10 % built-up.

However, this procedure is not sufficient, since in many municipalities that are populated according to statistics, there are no populated areas included in the land use category, since the community is too small or too thinly populated for the 250×250 m CORINE classification to identify. This applies predominantly to regions in the Central Alps. For this reason, a GIS and decision rules were used to seek an "optimum" population proxel for these communities. All inhabitants listed in the community statistics are assigned to this "artificial" population proxel. In locating all such proxels, it is first assumed that proxels with large areas of water, frozen or rocky areas should have no artificially placed populations. This assumption already excludes many proxels as regions for "artificial" population. Of the remaining proxels, the minimum of the so-called population index is sought for each community concerned:

$$\text{Population index} = \text{Altitude above sea level} + \left(\text{slope}\left[^\circ\right] * 50\right)$$

All inhabitants from this community are assigned to the appropriate proxel. The distribution that results serves as a relative, constant key to distributing the district values for population and household income calculated in RIWU to the proxels. The percentage of the population out of the total population of a district calculated in this way is deemed to be fixed. The population and its income therefore develop proportional to the proxels populated from the outset.

16.3.2 Configuration of the 5×5 Matrix

The goal of the DANUBIA *Demography* model is the allocation of the calculated proxel populations into each of five income and household size categories. To do this, the population is transferred to the *Demography* model from RIWU as the total number of inhabitants of a proxel and the household income as the sum of the available incomes in the entire population of the same proxel. The starting point is the consideration that the 5×5 matrix is calculated based on the average income of a proxel in relation to the distribution of the official statistic. It is therefore first necessary to determine the average income for the distribution of the national average presented in Table 16.1. For simplicity it is assumed that the incomes within individual categories in the official statistic are uniformly distributed. In the "under 1,100€" income category, a minimum income of 200€ is assumed, in order to match the depiction of social welfare benefits. In the income category with more than 5,000€, a maximum income of 10,000€ is assumed for simplifying the calculation. Under these assumptions, the average household income of the distribution given in Table 16.1 is 2,504.25€.

The distribution of population into the 25 fields is finally adjusted either up or down for each proxel according to a fixed algorithm depending on the difference between the proxel-specific average income and the statistical average income.

16.4 Presentation of the Results

Map 16.1 depicts the proxel population for January 1995 generated by RIWU. The RIWU district values were assigned to the proxels using the method described above. Large cities like Munich or Innsbruck are thus easy to pick out. On average, 1,142.2 people live in a modelled, populated proxel.

The RIWU model is calibrated for 1995. A model simulation for 2001 serves as a test of the predictive power. The comparison with statistical values yields an average forecasting error of −10 % with a variance of 18 % for the population. Precise validation of the proxel values is not possible because of missing statistical values. However, it can be assumed that the results obtained with the aid of remote sensing represent a close approximation of reality.

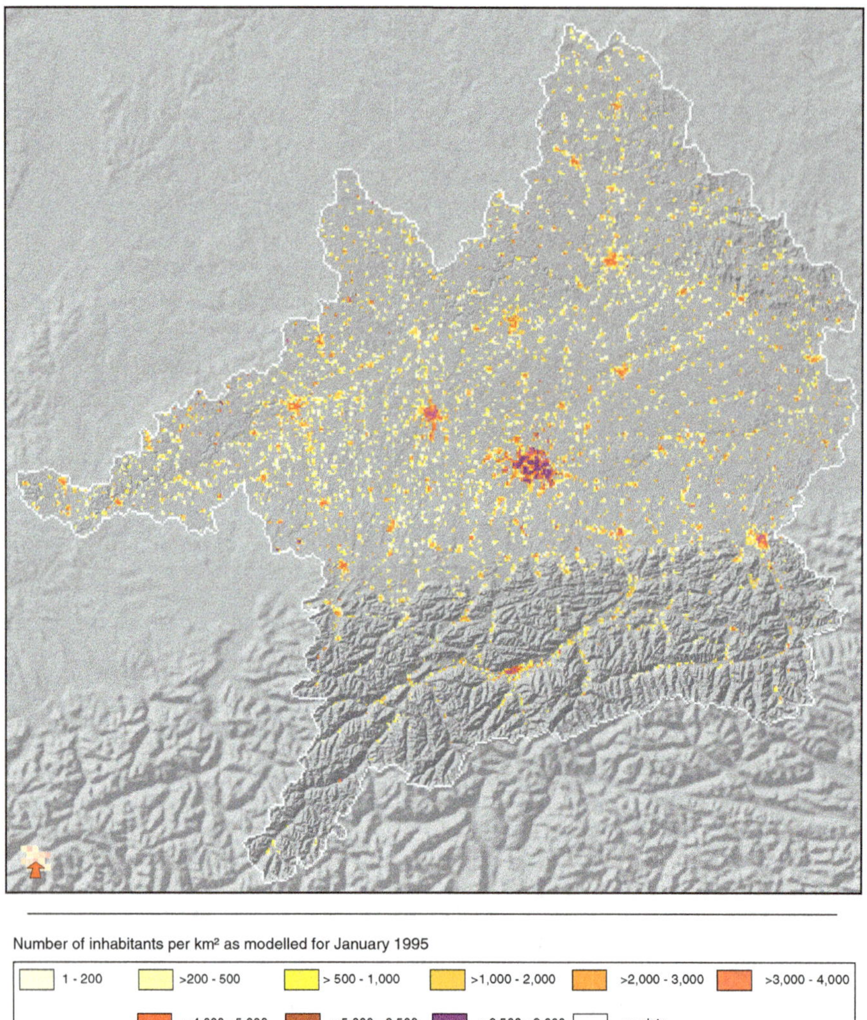

Map 16.1 Population (Data sources: © Bavarian State Office for Statistics and Data Processing, Munich, 2004; Federal Statistical Office, Wiesbaden, 2004; @ State Statistical Office Baden-Württemberg, Stuttgart, 2004; © STATISTIK AUSTRIA; Federal Statistical Office, Neuchâtel 2003; EEA, European Environment Agency, CORINE Land Cover, Copenhagen, 2005)

References

Central Intelligence Agency (2004) The World Factbook 2004. Availabe via DIALOG. http://www.cia.gov/cia/publications/factbook/fields/2172.html. Accessed 5 Jan 2005

Schweizerisches Bundesamt für Statistik (2003) Einkommens- und Verbrauchserhebung 2001 (EVE 2001), Neuchâtel

Statistik Austria (2003) Statistisches Jahrbuch Österreichs 2004

Statistisches Bundesamt (1998) Wirtschaftsrechnungen – Einkommens- und Verbrauchsstichproben 1993. Fachserie 15, Heft 6, Wiesbaden

Chapter 17
Gross Domestic Product

Matthias Egerer, Erich Langmantel, and Markus Zimmer

Abstract The regional economic simulation model RIWU (Regional Industrial Water Use) forms the basis for the DANUBIA model *Economy* presented in this chapter. RIWU allows for the modelling of economic development, expressed as the gross domestic product (GDP), and industrial water consumption at the level of the district. The gross domestic product is the decisive variable for measuring economic performance and serves as both the basis for estimating the volume of industrial water consumption and quantification of the effects of climate-induced changes in the usable water supply on economic development. A central goal of the model is to explain spatial structures of economic development for different water supplies. Industrial water abstraction (WAUF) is positively associated with the level of economic activity, represented by the value added (WI), and is negatively associated with the price of water (PEWA). The results for the gross domestic product by district as modelled by RIWU are mapped and reflect quite well the spatial distribution of gross domestic product.

Keywords GLOWA-Danube • Upper Danube • Gross domestic product

17.1 Introduction

The regional economic simulation model RIWU (Regional Industrial Water Use) forms the basis for the DANUBIA model *Economy*. RIWU models economic development on the district level. The key variable of interest is the gross domestic product (GDP), which is the main determinant of the industrial water consumption. In DANUBIA this variable serves as the basis for estimating the volume of industrial

M. Egerer (✉)
Bavarian Ministry of Economic Affairs and Media, Energy and Technology, Munich, Germany
e-mail: Matthias.Egerer@stmwi.bayern.de

E. Langmantel • M. Zimmer
Center for Energy, Climate and Exhaustible Resources, Ifo Institute – Leibniz Institute for Economic Research at the University of Munich, Munich, Germany
e-mail: zimmer@ifo.de

W. Mauser, M. Prasch (eds.), *Regional Assessment of Global Change Impacts*, DOI 10.1007/978-3-319-16751-0_17

water consumption, on the one hand, and at the same time permits the quantification of the effects of climate-induced changes in the usable water supply on economic development.

The model *Economy* is presented below. First, the used statistical data are presented and then the basics of the theoretical model and model equations are explained.

17.2 Data Preparation

Data from the official statistics from Germany, Austria and Switzerland from the years 1980, 1988 and 1995 form the basis for the model. For Bavaria, the data for all variables are available at the district level, but for Baden-Württemberg, Austria and Switzerland, this is not always the case. In Austria and Switzerland, some variables are available only at the broader level of an administrative region or state or their corresponding counterparts. For this reason, the data district values needed to be complemented with own calculations based on the available information.

The values for the following variables from the official statistics from the individual countries are the basis for the model and are used for its quantification:

- BIP Gross domestic product
- WI Industry value added
- WDL Service sector value added
- WS State value added
- BLAP Price for development land
- B Population
- FL Area
- SVFL Residential and traffic area
- YH Household income
- WAUF Industry water consumption
- π Industry labour productivity
- PEWA Water price (estimated costs for private water extraction by the industry)

The variable PEWA holds a special position in the dataset. Since the water used by industry largely comes from its own private sources (in Bavaria up to approximately 84 %, in Baden-Württemberg up to approximately 92 %), there is no official price equivalent to the price for drinking water for customers of water supply companies. In order to complete the dataset, these variables must be determined on the basis of theoretical considerations. Thus, it is assumed that water is an input factor for industrial production and supports production processes in the company for which the substitution of individual production factors (labour, capital and water) are more expensive and hence more unprofitable with increasing usage. In this way the (implicit) price for industrial water abstraction is equivalent to the so-called marginal product of water. This corresponds in the profit maximum to the relationship between industrial production and water use. Because these two variables are known from statistics, the price of water can be calculated.

In the context of DANUBIA modelling, the district values for BIP modelled by RIWU are disaggregated further to the level of the proxel. Remote sensing images and the information they contain about the distribution of industrial areas (see Chap. 9) are useful in this regard. The gross domestic product is distributed among the proxels in a manner that is proportional to the percentage of this land use category for each proxel in a district. However, the results obtained are only meaningful for mapping the exchange of data between the individual models at the level of the proxel. In reality, a large portion of the gross domestic product is also generated outside of industrial regions (e.g. in private service sectors and in the public sector). Therefore, the representation of proxel values would not yield meaningful results.

17.3 Description of the Model

Economic development normally takes place within a spatial structure that is characterised by agglomerations and the associated peripheries. According to the economic theory from the so-called New Economic Geography, these kinds of spatial concentrating processes for economic activities result from an interaction between centripetal and centrifugal forces. The centripetal forces that promote the formation of industrial agglomerations can be more or less assigned to one of two categories. First, companies favour locations where other firms are already settled, for example, to take advantage of a strong local market. Second, spatial agglomeration reduces transportation costs for suppliers that as a result often settle close to the companies they supply. The growth of an agglomeration obtains therefore a self-reinforcing tendency (Krugman 1998). The centrifugal effects result from the increasing costs of agglomeration that limit the growth. In particular, these effects are related to immobile production factors such as land or natural resources. A concentration of economic activities leads to increased demand for these factors, which in turn leads to higher prices and hence to negative incentives to further concentrate industrial activity in these areas. In the RIWU model the price for local development land is used as an indicator for the costs of agglomeration (Brakman et al. 2001). Because a goal of the model is to explain spatial structures of economic development for different water supplies, the price of industrial water abstraction (PEWA) is introduced as a second factor limiting spatial agglomeration. Just like land, water is treated as a scarce local production factor.

The model is quantified using the above data for the period from 1980 to 1995 by calculating for every variable a behavioural equation using regression analysis. The exceptions are the variables district area (FL) and residential and traffic area (SVFL) that are specified by political authorities as well as the price of water (PEWA). RIWU consists of a total of ten model equations with which the variables WAUF,

WI, WDL, BIP, WS, BLAP, B, and YH are estimated (Langmantel 2004a, b). The phase period of the model is a month.

The key variable for predicting the economic growth of a district is the productivity development (π). It can be expected that it will continue to grow in the future as a result of competition within the markets. At the moment, based on values observed in the past, a monthly increase of 0.33 % is assumed for the average across all districts. Since regional deviations from the average district productivity have hardly changed in the past, it is reasonable to assume that the given regional productivity differences will also remain the same for model predictions.

Increased productivity improves competitiveness and has a positive effect on the value added by industry (WI). The size of the local market also has a positive influence; this size is represented by the value added in the service sector (WDL). Higher land prices (BLAP) and higher water prices (PEWA) have affect negatively the level of industrial activity. Neighbourhood effects also play a role since industrial clusters in the proximity create positive spillover effects.

Industrial water abstraction (WAUF) is positively associated with the level of economic activity, represented by the value added (WI) and is negatively associated with the price of water (PEWA). The general technical progress that is reflected in the increase in productivity (π) lowers the specific water consumption over time.

The value added in the service sector (WDL) is highly dependent on the extent of the central function of the affected area. Many service offers are only economically viable for a certain minimum size of the market, so that these services concentrate at central locations from where they can serve the market potential of the surrounding area. The gross domestic product in the neighbouring regions and the size the industry and the state institutions at the location thus determine the level of service activities in a region. The service sector is also positively influenced by the long-term structural change, which is expressed in the increase of industrial productivity (π).

The prediction for regional population development is based on the consideration that in the relatively socially homogenous region of the Upper Danube drainage basin, population movements to take up job offers are far more important than the local birth and death rates. In the RIWU model, the population density of a district (B/FL) is positively influenced by the level of economic activity in this region (BIP/FL). The regional employment offer has a negative effect on immigration; an indicator for this is the regionally specific relationship between BIP and labour productivity.

It is presumed that household income (YH) develops in proportion to the BIP within each region and the neighbouring regions.

The cost of land (BLAP) is all the higher, the higher the degree of economic concentration in a region (BIP/FL). Higher cost of land in the proximity leads likewise to increased land prices in a region, while expanding the share of residential and traffic use area (SVFL/FL) has a decreasing effect.

Finally, for simplicity it is assumed that the gross domestic product (BIP) and the value added by state (WS) develop in a manner that is proportional to the level of industry and services.

17.4 Presentation of the Results

Map 17.1 depicts the results for the gross domestic product by district as modelled by RIWU for January 1995. The results reflect quite well the spatial distribution of gross domestic product.

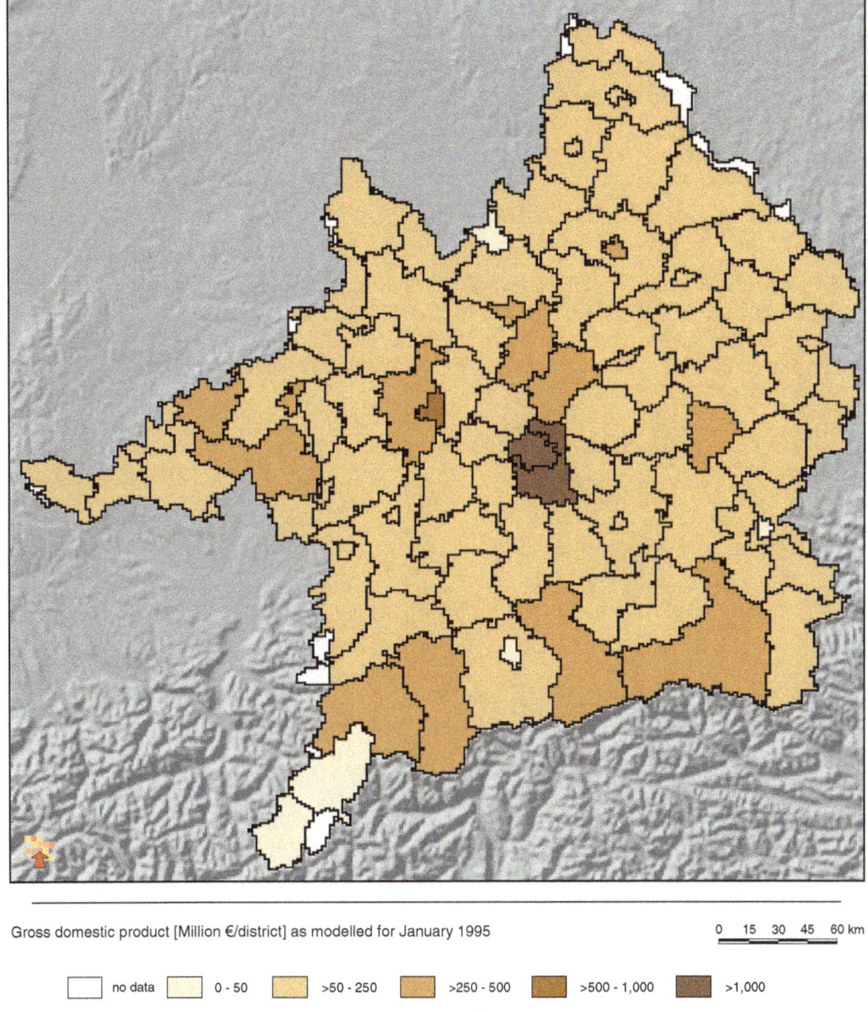

Map 17.1 Gross domestic product (©Bavarian State Office for Statistics and Data Processing, Munich, 2004; Federal Statistical Office, Wiesbaden, 2004; State Statistical Office Baden-Württemberg, Stuttgart, 2004; ©STATISTIK AUSTRIA 2003; Federal Statistical Office, Neuchâtel 2003; EEA, European Environment Agency, CORINE Land Cover, Copenhagen, 2005, borderlines: ESRI Deutschland GmbH, Kranzberg, Germany http://ESRI.de, ©BEV – Bundesamt für Eich- udn Vermessungswesen 2014, GG85 ©Swisstopo)

The districts which show no industrial use areas according to the remote sensing data are problematic. In these cases, no values were generated in the attempt to disaggregate the district values to the proxel level. These districts are indicated in the legend as having "no value".

The model is calibrated for 1995. A model simulation for 2001 serves as a test of the predictive power. The comparison with actual values yields an average forecast error for BIP of −1.6 % with a variance of 30 %. It was not unexpected that the realised values for 2001 are underestimated, since DANUBIA provisionally works with unchanged land use and hence generates an artificial scarcity, in contrast to the actual increasing residential and traffic areas. This fact leads to an overestimate of the price of land and an underestimate of economic growth.

References

Brakman S, Garretsen H et al (2001) New economic geography in Germany: testing the Helpman-Hanson Model. HWWA discussion paper 172, Hamburg

Krugman P (1998) What's new about the economic geography? Oxf Rev Econ Policy 14(2):7–17

Langmantel E (2004a) Regional industrial growth and water demand an empirical analysis for the upper Danube catchment. Jahrb Reg 24:161176

Langmantel E (2004b) Der Beitrag von Industrie und Dienstleistungen zum regionalen Wirtschaftswachstum in Bayern. ifo Schnelldienst 57(15):8–13

Chapter 18
Water Demand in Tourism Facilities

Mario Sax, Jürgen Schmude, and Alexander Dingeldey

Abstract Water represents in the Upper Danube drainage basin a resource that at present is available in sufficient quantity. With regard to climate change it is conceivable that water supply for tourism infrastructure facilities will no longer be possible at the present level. This can affect the operability of these tourism facilities, and, as a consequence of this, the affected regions could suffer serious economic losses. In the context of GLOWA-Danube, the objective of the Tourism research group is to analyse water use by the tourism industry on a small scale within the Upper Danube basin. A supply-oriented actor model was developed to help achieve this goal. To quantify water demand by the tourism industry, the quantities of water required by various infrastructure facilities (e.g. golf courses, ski regions with snowmaking machines, swimming pools) was determined using a study of literature and by consultation with experts. The relevant infrastructure facilities in the study area were compiled for a spatially differentiated representation in DANUBIA. The water demand of tourism infrastructure is concentrated in the Alpine Region and Bavarian Forest (ski areas) and in major cities (Munich, Regensburg, Augsburg) due to the hospitality sector. The water user of golf course and swimming pools is distributed throughout the study area, with some concentration of golf course in the environs of Munich.

Keywords Climate change • Tourism facilities • Water use • Tourism sector • Upper Danube basin • Tourism industry • Actor model

M. Sax (✉)
Wilhelm-Hauff-Str. 51, 84036 Landshut, Germany
e-mail: sax.mario@googlemail.com

J. Schmude
Department of Geography, Ludwig-Maximilians-Universität München (LMU Munich), Munich, Germany
e-mail: j.schmude@lmu.de

A. Dingeldey
BWL Reiseverkehrsmanagement, Duale Hochschule Baden-Württemberg, Ravensburg, Germany
e-mail: Dingeldey@dhbw-ravensburg.de

W. Mauser, M. Prasch (eds.), *Regional Assessment of Global Change Impacts*, DOI 10.1007/978-3-319-16751-0_18

18.1 Introduction

In the Upper Danube drainage basin, water represents a resource that at present is available in sufficient quantity. However, it is conceivable that with climate change in the coming years the water supply for tourism infrastructure facilities will no longer be possible at the usual level. This can affect the operability of these tourism facilities, and, as a consequence of this, the affected regions could suffer serious economic losses.

Water is consumed within the tourism industry as a commodity for the tourism infra- and suprastructure. Tourism infrastructure includes usable facilities such as golf courses, ski areas, swimming pools, cable cars, etc. The term suprastructure includes the gastronomy and hospitality sectors.

In the context of GLOWA-Danube, the objective of the Tourism research group is to analyse water use by the tourism industry on a small scale within the Upper Danube basin. A supply-oriented actor model was developed to help achieve this goal; this model is based on the tourism infra- and suprastructures.

The estimate of the potential change to the demand for water by the tourism industry that is tied to the change in water supply as a result of global change has important ramifications for policymakers in the tourism sector because of the associated economic consequences. For example, since the supply of natural snow during the winter sport season is not always adequate, but a guaranteed snow cover is absolutely essential for the ski industry, attempts are increasingly made to compensate using artificial snowmaking machines (Hahn 2004; Appel 2003; Görl 2004; FAZ, March 2 2004). Thus, Bavaria can expect further construction of snowmaking machines in the near future, since the Bavarian State Government tends to facilitate the authorisation of snowmaking machines in winter sport regions (SZ, 15.12.2003).

18.2 Data Processing

The preparation of the maps used compilations of data on water-intensive tourism infrastructure facilities. Water-intensive facilities include various types of swimming pools, golf courses, ski areas with artificial snow production and lodging establishments. The spatial distribution of these facilities and their water demand values form the base data for modelling water use by the tourism industry in DANUBIA.

To quantify water demand by the tourism industry, the quantities of water required by various infrastructure facilities was determined using a study of literature and by consultation with experts. The relevant infrastructure facilities in the study area were compiled for a spatially differentiated representation in DANUBIA. The investigations required for this compilation include the analysis of documents and websites, inquiries with authorities and telephone interviews. As a result, a database with the infrastructure of the communities in the study area could be established that also serves as the basis for surveys.

On this basis, it is necessary to categorise the infrastructure facilities at the level of the proxel in order to ensure the collected data is integrated in DANUBIA. For each infrastructure facility, the geographic coordinates were determined using topographical maps for Germany, Austria and Switzerland. The locations of the infrastructure sites were then transferred into a Geographic Information System. After projecting this data into the coordinate system of the GLOWA-Danube project, the located infrastructure facilities were intersected with the proxel grid. This procedure allowed an explicit assignment of a proxel identification number (PID) to each infrastructure facility.

Differing values were reported for the water demands for golf courses. Baartz (1994) assumed a water requirement of 50,000 m^3/a for an 18-hole course. According to their own information, the German Golf Association quotes the figure of approximately 24,000 m^3/a. If only greens, aprons and drives are watered, the annual requirement drops to 12,000 m^3 (Baartz 1994). Currently, a total water quantity of 30,000 m^3/a is assumed for modelling the necessary amounts of water in DANUBIA. Watering normally takes place during the months of May to September. Based on this value, the water demands of a golf course per season are determined in proportion to the number of fairways. The proportion of the water demand of a golf course that is drinking water is approximately 5 % of the total demand. The rest is supplied from other water sources. Ponds are a common way to store backup water sources.

The water demands of ski regions with snowmaking machines can be determined either using known consumption values or with the rule of thumb "one ha of snow-covered ski slope requires 1,000 m^3 of water per season" (Hahn 2004) while taking into account the snow-covered area. Since artificial snow production begins as early as November in preparation for the season opening (Lutz 2000) and can continue as late as into March, the water demand can be spread evenly across these 5 months. This demand is supplied by withdrawal from flowing and standing water sources, springs and the drinking water supply (Lutz 2000). In Bavaria, the proportion of the demand by snowmaking machine equipment that is supplied by drinking water is approximately 10 % of the total water demand (expert discussion, StmUG = Bavarian State Ministry of the Environment and Public Health July 2002).

Public swimming pools in the Upper Danube basin are divided into different subclasses, each with a specific water demand. Swimming pools are separated as either outdoor pools, indoor pools, indoor/outdoor pools, water parks or thermal spas. The water demand is between 2,000 and 4,400 m^3 of drinking water per month (BOEB e.V. 2003 and internal surveys). Depending on the type of pool, seasonal opening times are also considered in the determination of tourism-related water demand.

Additional information such as the water demand values presented in Map 18.1(1–3)was stored in order to prepare this information for modelling tourism-related water demand (see Chap. 46).

Map 18.1(4), which depicts the water demand for the hospitality sector, was prepared as follows: For economical reasons, it is not possible to directly map all

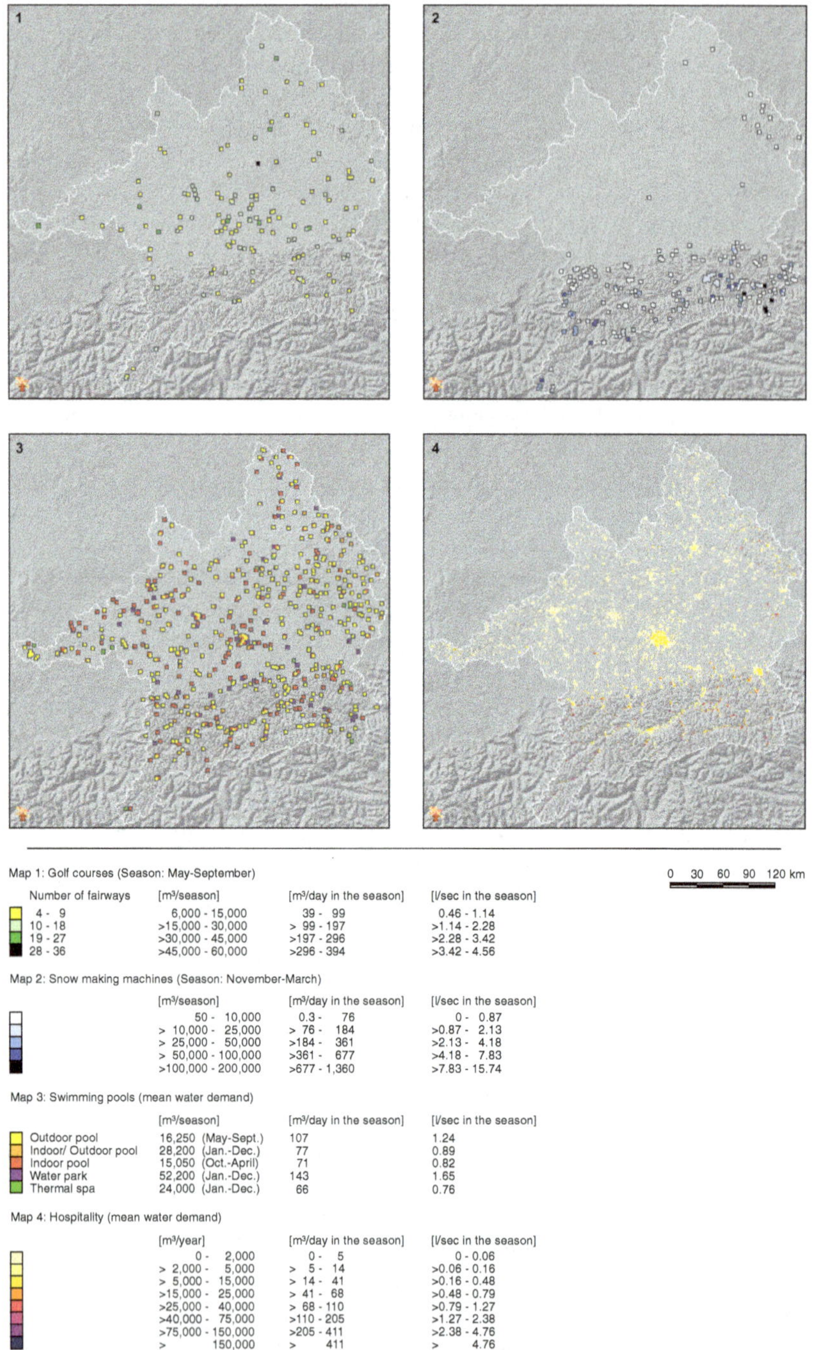

Map 18.1 Water demand in tourism facilities (Data sources: Investigations and calculations carried out by the authors; DANUBIA – Population, see Chap. 16)

hospitality establishments, which is why here a simplified method was used. The presentation of water demand by hospitality establishments is based on the assumption that this activity can occur only in populated areas (wherein the areas of a district are generally only partly populated). Thus, each populated proxel was attributed the ability to have a hospitality establishment. The water demand by the hospitality sector calculated from the number of bed nights per community (a bed night corresponds to 180 l of water (Sax 2008)) was converted to the individual populated proxels using the population proportion for the proxel with respect to the total population of a community. Using this procedure, a water demand value associated with the hospitality sector was assigned to each populated proxel in communities that have overnight guests:

$$WD_{Prox} = WD_{b} * BN * PopShare_{Prox}$$

WD_{Prox}: water demand per proxel
$WD_{ÜB}$: water demand per guest and overnight stay
BN: number of bed nights
$PopShare_{Prox}$: share of population (0–1) of the proxel to the total for the community

This approach allows a certain generalisation of reality. The inventory of hospitality establishments within a community is subject to more or less steady ongoing change that cannot be accounted for in this approach. Changes in the locations of hospitality establishments from closures or new constructions thus do not need to be considered (Sax 2008).

The disadvantage of this approach is that all populated proxels are assigned a water demand value related to the tourism industry, even if this is not actually true. This is of particular relevance in agglomerated areas such as Munich, where large areas are residential. However, the error that arises therefrom becomes smaller with decreasing density of a community. Therefore, this approach is best suited for communities with only one populated proxel. As a further example, hotels that are in single locations that are not represented as a populated area based on the proxel grid are not accounted for even if in reality there is a veritable water demand related to tourism there. However, the distribution of water demand from the hospitality sections projected onto the populated proxels can be considered an appropriate method for the use of the proxel concept (Sax 2008).

18.3 Results

The four maps illustrate the spatial distribution of water demand by the tourism industry within the study area and are to be viewed as independent of each other.

They provide an overview as to which areas characterised by tourism can expect a particularly significant change in the availability of water.

Golf courses are found throughout the study area, although there is a concentration of them in the environs of Munich (see Map 18.1(1)). The number of fairways varies for the golf courses found throughout the study area. This value is a good attribute to investigate because it represents the size of a course and this in turn is determinative of the water demand.

Ski areas with equipment for artificial snow production are generally located in the Alpine region and in the Bavarian Forest. Ski areas with high water demands are situated in the Alps (see Map 18.1(2)). Swimming pools are widespread throughout the GLOWA-Danube study area (see Map 18.1(3)).

Map 18.1(4) highlights the fact that on the one hand, the water demand from the hospitality sector is concentrated in the cities (Munich, Regensburg, Augsburg, etc.) and, on the other hand, is also high at certain sites within the Alps and in the Bavarian Forest.

The depiction of the water-intensive infrastructure facilities is a snapshot in time. In order to further support the data on water demand, additional primary research will be conducted on this subject throughout the course of the project.

References

Appel D (2003) Urlaubsorte kämpfen mit Schneekanonen gegen den Klimawandel. In: SZ Freising vom 04./05./06. January 2003, Munich, p R2

Baartz R (1994) Der Konflikt zwischen Sport und Umwelt, dargestellt am Beispiel der Entwicklung des Golfsports im Raum Brandenburg-Berlin. Franz Steiner Verlag, Stuttgart

BOEB e.V. (2003) Bundesfachverband Öffentliche Bäder e.V. Unveröffentlichte Vorauswertung für den überörtlichen Betriebsvergleich Bäderbetriebe 2003, Essen

Görl W (2004) Rettung aus allen Rohren. In: Süddeutsche Zeitung vom 27. February 2004, Munich, p 3

Hahn F (2004) Künstliche Beschneiung im Alpenraum. Ein Hintergrundbericht. Cipra International, Schaan

Lutz G (2000) Beschneiungsanlagen in Bayern Stand der Beschneiung, potentielle ökologische Risiken. In: Technische Beschneiung und Umwelt. Tagungsband zur Fachtagung am 15. November 2000. Bayerisches Landesamt für Umweltschutz, Augsburg, pp 7–14

o.A. (2003) Staatsregierung will mehr Kunstschnee. In: Süddeutsche Zeitung vom 15. Dezember 2003, Munich, p 45

o.A. (2004) Auf Frau Holle ist kein Verlass. In: Frankfurter Allgemeine Zeitung vom 02. März 2004, Frankfurt a.M, p T6

Sax M (2008) Entwicklung eines Konzepts zur computergestützten Modellierung der touristischen Wassernutzung im Einzugsgebiet der oberen Donau unter Berücksichtigung des Klimawandels. In: Schmude J (ed) Beiträge zur Wirtschaftsgeographie Regensburg, vol 11. Universität Regensburg

StmUGV (2002) Bayerisches Staatsministerium für Umwelt, Gesundheit und Verbraucherschutz; expert interview, July 2002

Chapter 19
Agriculture

Alexander Wirsig, Tatjana Krimly, Marta Stoll, and Stephan Dabbert

Abstract Analysis and evaluation of agricultural land use is a key factor in assessing agricultural water demand. Interactions between agriculture and water include groundwater recharge, water withdrawal and impact on water quality. In the Upper Danube drainage basin, agriculture plays a key role. Data available at district level served as basis for calculations of an agroeconomic model and was downscaled by using a specially developed tool to deliver results with a higher spatial resolution and to produce a more detailed picture of agricultural production. The results indicate that the study area involves a great diversity in agricultural landscapes. Grassland dominates in the Alpine region as well as in the Alpine foothills, where dairy farming predominates. In the Tertiary Hills region, arable land dominates and pig farming is widespread. Along the Isar River from Munich to Passau, there is very intensive arable cropping and pig farming. In the Eastern Bavarian Uplands along the border with the Czech Republic, farming with cattle production is less specialised than in the Alpine region.

Keywords Agricultural land use • Water demand • Global change • Regional optimisation model

19.1 Introduction

Agriculture plays a key role in the Upper Danube drainage basin, because 55 % of the area is used for agriculture. The basin thereby comprises 18 % of the utilised agricultural area (UAA) in Germany as well as 27 % of the cattle population and

A. Wirsig (✉)
MBW Marketing- und Absatzförderungsgesellschaft für Agrar- und Forstprodukte aus Baden-Württemberg mbH, Stuttgart, Germany
e-mail: wirsig@mbw-net.de

T. Krimly • S. Dabbert
Production Theory and Resource Economics, Universität Hohenheim, Stuttgart, Germany
e-mail: T.Krimly@uni-hohenheim.de; dabbert@uni-hohenheim.de

M. Stoll
DeUkRus Translations, Marta Stoll, Beimerstetter Str. 39, 89081, Ulm

W. Mauser, M. Prasch (eds.), *Regional Assessment of Global Change Impacts*, DOI 10.1007/978-3-319-16751-0_19

28 % of the agricultural holdings, of which 43 % are full time farms (Dabbert et al. 2002). Moreover, agriculture influences important environmental parameters because of numerous interactions (Verburg et al. 2000). The interactions between agriculture and water as a resource are multifold. In many places, agriculture has a key function in groundwater recharge. At the same time, agriculture may pollute ground- and surface waters and impacts the capacity for flood retention in a landscape (SRU 2004).

The objective of the agricultural economics subproject is to analyse and to evaluate the agricultural water demand and agricultural land use that influence water quality and quantity from a production-related (different animal and plant production processes) and socioeconomic viewpoint. On this basis, a spatially differentiated process analytical model of the agricultural sector was developed for the study area (Henseler et al. 2009; Winter 2005); this model serves in the analysis of various political, economic and meteorological scenarios within the interdisciplinary decision support system of DANUBIA.

The most important interactions between agriculture and water include groundwater recharge, water withdrawal and impact on water quality. These interactions are outlined below: the distribution of agricultural land use (arable land and grassland) and the cultivation of various types of crops both influence groundwater recharge and other hydrological factors such as the capacity for flood retention within a landscape. A crucial parameter for differing groundwater recharge is the water consumption of the crops through evaporation. Agricultural land use in turn is dependent on natural location factors such as precipitation and soil characteristics (Krimly et al. 2003). Crop rotation is not the only factor influencing water balance; soil tillage also has an effect. For example, mulch seeding or direct sowing fosters the infiltration of precipitation and results in lower surface water run-off (Winter 2005). Agriculture withdraws water from the public water system both for animal and plant production. In animal husbandry water is used for sanitary purposes and to water the animals. An adequate supply of water for the livestock is also necessary for ethical reasons. Table 19.1 provides an overview of the watering demands from livestock. In addition, drinking water is used for diluting pesticides. For most crops, between 200 and 400 L of water is needed per pesticide measure per hectare (Winter 2005).

Table 19.1 Guide values for the drinking water demand of farmed animals

Type of animal, age/ stage of production	Average water demand litres/(animal*day)
Dairy cow	50 (30–70)
Cattle >1 year	25 (15–35)
Cattle <1 year	20 (15–25)
Lactating sow	30 (20–40)
Fattening pig	8 (5–10)
Sheep	5 (2–8)
Riding and cart horse	35 (25–45)

Löffler (2002)

At the same time, agriculture may pollute ground- and surface waters. High livestock densities, primarily in off-land production, such as pig and poultry production, lead to increased pollution levels (SRU 2004). The extent of the water pollution from agriculture is therefore dependent on the site characteristics and the climatic conditions, which in turn have an impact on the type and intensity of agricultural land use (Winter 2005).

19.2 Data Processing

The maps are based on data from the respective statistical offices for the year 1995. In Bavaria and Baden-Württemberg, these data are available at district level and in Austria at the level of the political region. This data also serves as the basis for calculations of the agroeconomic model and are downscaled to the scale of the proxel collectively used in DANUBIA using a specially developed tool (see Chap. 39).

As a consequence of the different differentiations and allocation methods for agricultural holdings and the utilised agricultural area (UAA) in Bavaria, Baden-Württemberg and Austria, there are minor, but negligible inconsistencies among the datasets. The livestock densities are expressed in livestock units (LU) per hectare of utilised agricultural area. The calculation of LU includes the three livestock categories of cattle, pigs and poultry, based on the livestock unit key in the agricultural statistic (HLBS 1996). The Bavarian State Office for Statistics and Data Processing, the Baden-Württemberg State Statistical Office and the Austrian Statistical Office were the sources of the data. Mapping agriculture in the Swiss portion of the Danube basin was not undertaken because of its negligible significance.

19.3 Results

The UAA in the Upper Danube basin comprises 3.9 million hectares. Grassland makes up 52 % of the UAA, while arable land represents 44 and 2 % are attributed to production areas for permanent crops, horticultural crops and fallow land. Grassland dominates in the Alpine region as well as in the Alpine foothills, where dairy farming predominates. In the Austrian Alps the average farm size is partly larger than in the rest of the drainage basin (see Map 19.1(1)). Arable cropping plays a minor role in these regions, with only a maximum of 40 % of the UAA. Mostly cereals and forage for cattle are cultivated in these regions. Further north the proportion of grassland and dairy farming decreases, while arable land increases up to 60 %. Bull fattening replaces dairy farming (Winter 2005). On large areas of the arable land, corn silage is produced as fodder for the fattening bulls. In the Tertiary Hills region, the proportion of arable land is approximately 80–90 % and pig farming is widespread. Along the Isar River from Munich to Passau, there is very intensive arable cropping and pig farming. Map 19.1(3, 4) depicts the livestock densities

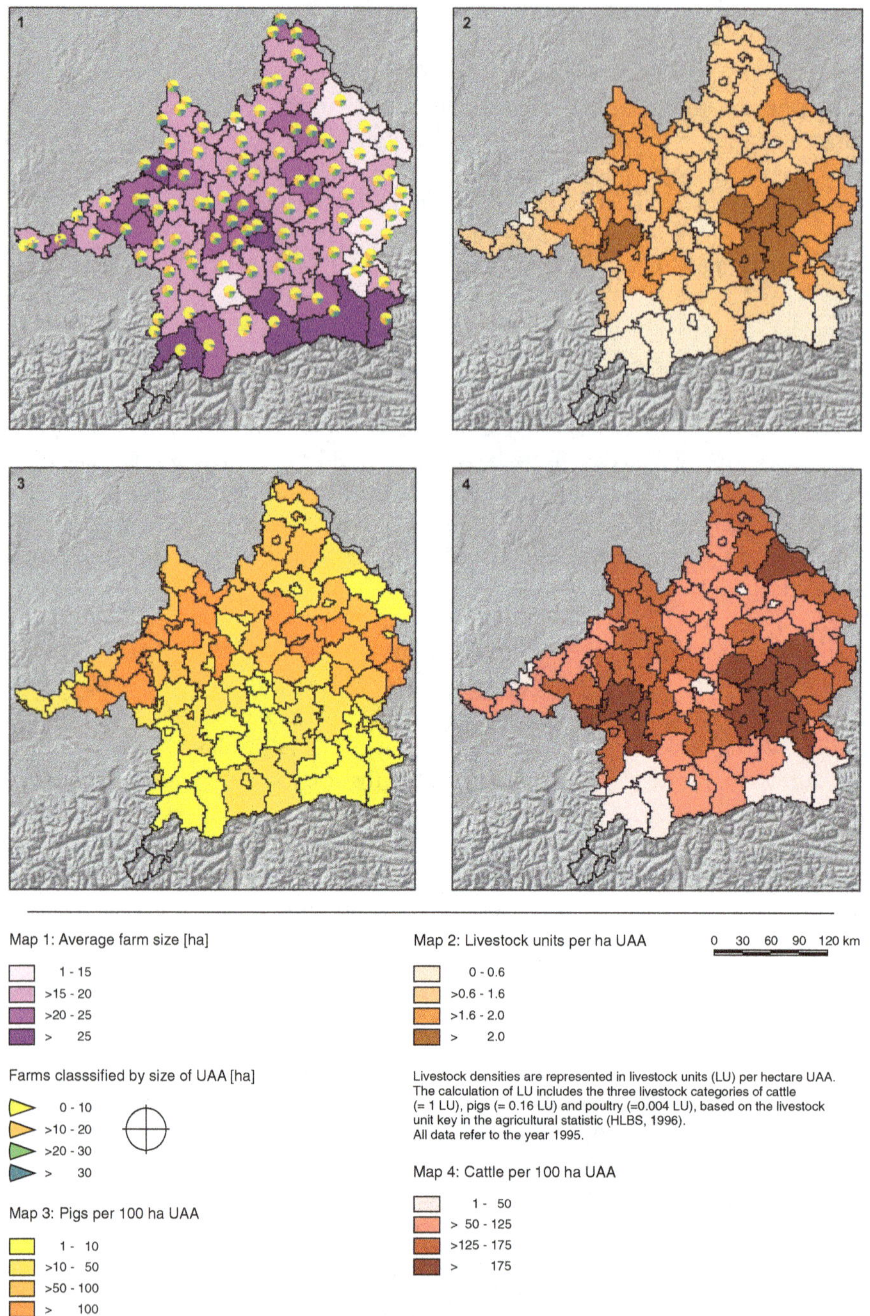

Map 19.1 Agriculture (Data sources: ©Bavarian State Office for Statistics and Data Processing, Munich, 2004; Federal Statistical Office, Wiesbaden, 2004; State Statistical Office Baden-Württemberg, Stuttgart, 2004; ©STATISTIK AUSTRIA 2003; Federal Statistical Office, Neuchâtel 2003; EEA, European Environment Agency, CORINE Land Cover, Copenhagen, 2005 borderlines: ESRI Deutschland GmbH, Kranzberg, Germany http://ESRI.de, ©BEV – Bundesamt für Eich- und Vermessungswesen 2014, GG85 ©Swisstopo)

of the regions for cattle and pigs. Map 19.1(2) provides an overview of the overall livestock densities. North of the Danube River, the local agronomic conditions are not favourable for arable cropping. In this region, the proportion of grassland is 80–100 %. In the Eastern Bavarian Uplands along the border with the Czech Republic, farming with cattle production is less specialised than in the Alpine region (Winter 2005). The production of plant products is influenced by the local climatic and natural conditions, animal husbandry as well as by socioeconomic factors and political framework conditions (Winter 2005). The local conditions have a direct effect on the production process. Whether a product is used as fodder or cash crop for human consumption depends on animal husbandry. Cereal crops are cultivated on more than 50 % of the arable land and are widespread across the study area. Wheat is the most important cereal, as it can be used for human consumption as well as fodder in animal production. Forage crops (crop silage, clover and grass-clover mixture) make up 27 % of the arable land and serve as fodder for cattle. Rapeseed is cultivated as a catch crop on 6 % of the arable land in sites with cereal-dominated crop rotations. These locations are typically those with unfavourable conditions for arable cropping and low cattle numbers. Potatoes and sugar beets are planted on 6 % of the arable land, mostly in locations with favourable conditions for arable cropping. In regions with mild climatic conditions and fertile soil, annual and perennial special crops are cultivated. Fruit, vegetable and hop are the most important lines of production in special crop cultivation and constitute 2 % of the arable land in the Upper Danube basin (Winter 2005).

References

Dabbert S, Herrmann S, Winter T, Vogel T (2002) Land use and agricultural actors in GlOWA-Danube. GLOWA status-conference, Munich, 6–8 May 2002

Henseler M, Wirsig A, Herrmann S, Krimly T, Dabbert S (2009) Modeling the impact of global change on regional agricultural land use through an activity-based non-linear programming approach. Agric Syst 100(1–3):31–42

HLBS (1996) Betriebswirtschaftliche Begriffe für die landwirtschaftliche Buchführung und Beratung. Schriftenreihe des Hauptverbandes der landwirtschaftlichen Buchstellen und Sachverständigen (HLBS), H.14, Pflug u. Feder, Sankt Augustin

Krimly T, Winter T, Dabbert S (2003) Agrar-ökonomische Modellierung der Landnutzung im Einzugsgebiet der oberen Donau zur Integration in das interdisziplinäre Entscheidungsunterstützungssystem DANUBIA. In: Dabbert S, Grosskopf W, Heidhues F, Zeddies J (eds) Perspektiven der Landnutzung – Regionen, Landschaften, Betriebe, Entscheidungsträger und Instrumente. Schriften der Gesellschaft für Wirtschafts- und Sozialwissenschaften des Landbaues e.V, vol 39. Landwirtschaftsverlag, Münster-Hiltrup, pp 191–199

Löffler K (2002) Anatomie und Physiologie der Haustiere. Ulmer Verlag, Stuttgart

SRU (2004) Umweltgutachten 2004. Umweltpolitische Handlungsfähigkeit sichern. Nomos Verlagsgesellschaft, BadenBaden

Verburg P-H, Chen Y, Soepboer W, Veldkamp T (2000) GIS-based modeling of human-environment interactions for natural resource management – applications in Asia. In: Proceedings of the 4th international conference on integrating GIS and environmental modeling (GIS/EM4) "Problems, Prospects and Research Needs", Banff, Alberta, 2–8 September 2000

Winter T (2005) Ein nichtlineares prozessanalytisches Agrarsektormodell für das Einzugsgebiet der oberen Donau. Ein Beitrag zum Decision-Support-System GLOWA-Danubia. Dissertation, Universität Hohenheim

Chapter 20
Extraction of Water for Public Drinking Water Supply

Roland Barthel, Alejandro Meleg, Darla Nickel, and Alexandar Trifkovic

Abstract The supply of drinking water in the Upper Danube drainage basin is generally a municipal responsibility and based almost exclusively on extraction from ground- and spring water sources. This section presents maps showing an overview of the quantities extracted and regional variations. Extraction depends both on the water yield, which may be constrained by local hydrogeological conditions, and on consumption, which is mainly a result of population density. The map shows the aggregated quantities of water extracted from ground-, spring and surface water sources in the communities and the points of drinking water withdrawal within the drainage basin for the year 1998. This information is used by the *WaterSupply* model, which provides the connection between water availability (*Groundwater* and *Rivernetwork* models) and water demand (actor models) within DANUBIA. This chapter describes the data sources and the processing steps taken in preparation of the presented map. Limitations and sources of uncertainty are discussed.

Keywords GLOWA-Danube • Upper Danube • Water supply • Drinking water • Groundwater

R. Barthel (✉)
Department of Earth Sciences, University of Gothenburg,
Göteborg, Sweden
e-mail: roland.barthel@gu.se

A. Meleg
Consultant, Bogotá, Colombia
e-mail: alemeleg@hotmail.com

D. Nickel
Deutsches Institut für Urbanistik (Difu), Berlin, Germany
e-mail: nickel@difu.de

A. Trifkovic
Fichtner GmbH & Co, Sarweystr. 3, 70191 Stuttgart, Germany
e-mail: atrifkovic@hotmail.com

W. Mauser, M. Prasch (eds.), *Regional Assessment of Global Change Impacts*, DOI 10.1007/978-3-319-16751-0_20

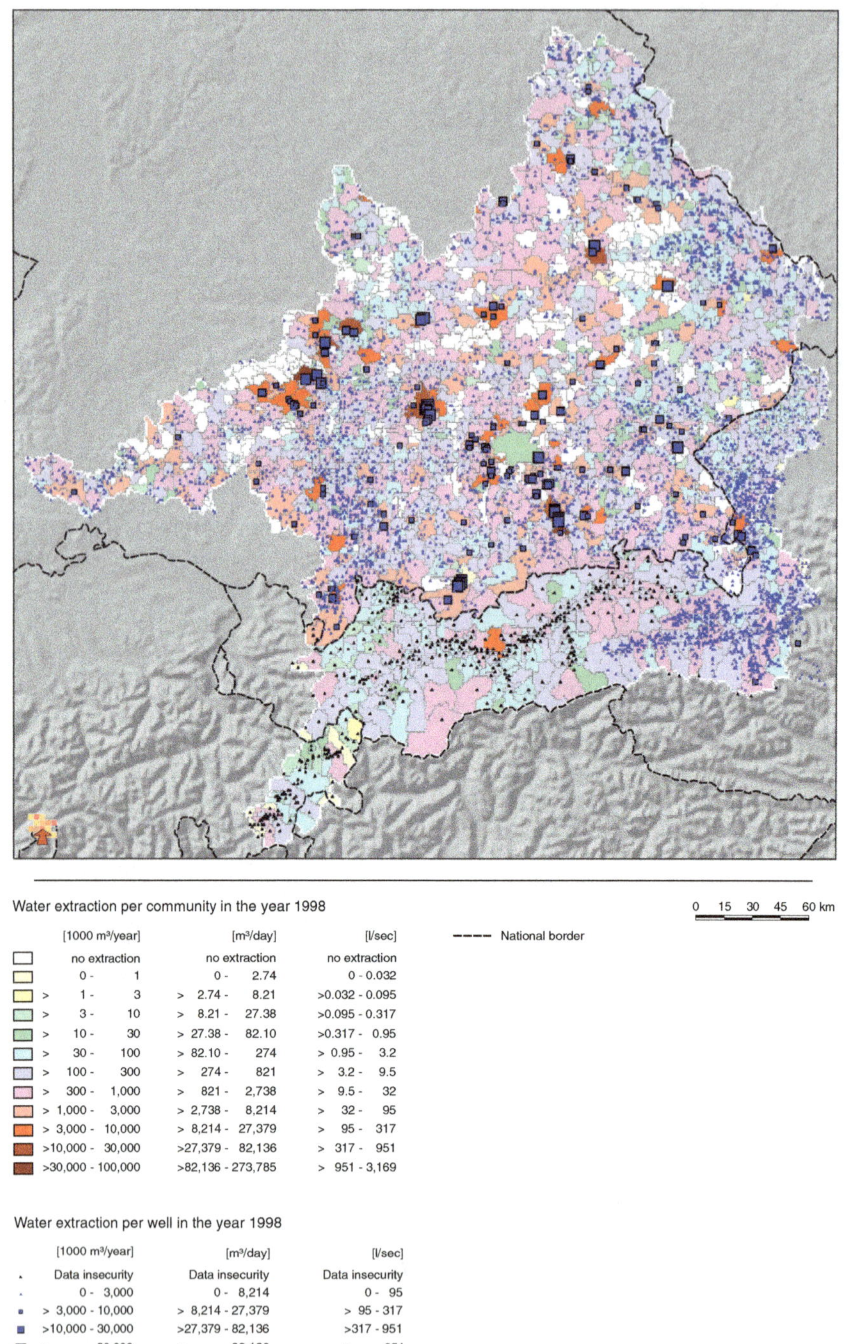

Map 20.1 Extraction of water for public drinking water supply (Data sources: Bavarian State Office for Statistics and Data Processing, Munich, 1998; Bavarian State Office for the Environment (LfU), 1998; Federal Association for German Gas and Water Industry e.V. (BGW) Bonn: 112.

20.1 Introduction

The supply of drinking water in the Upper Danube drainage basin is generally a municipal responsibility and based almost exclusively on extraction from ground- and spring water sources. Map 20.1 provides an overview of the quantities extracted and regional variations. It should be noted that extraction depends both on the water yield and on consumption. The map shows the aggregated quantities of water extracted from ground-, spring and surface water sources in the communities and the points of drinking water withdrawal within the drainage basin for the year 1998. This information is used by the *WaterSupply* model, which provides the connection between water availability (*Groundwater* and *Rivernetwork* models) and water demand (actor models) within DANUBIA.

20.2 Data Processing

The maps are primarily based on data from the statistical offices and water management authorities in Germany, Austria and Switzerland. The quantity, quality and availability of drinking water supply data provided by the individual countries exhibit extensive heterogeneity. Therefore, the data processing steps taken in preparation of the map are described. This is followed by a discussion of the underlying hydrogeological conditions and the organisation of the drinking water supply.

The aggregated withdrawals for drinking water purposes in Germany were derived directly from the water statistics collected by the State Statistical Offices in Bavaria and Baden-Württemberg. Comparable and public water statistics at communal level are not administrated by the Austrian and Swiss authorities Regierung des Kantons Graubünden (2000). In these cases, withdrawals were extrapolated from the population numbers in the communities in 1998 and the average per resident consumption values from the Austrian states of Upper Austria, Salzburg and Tyrol and the Swiss canton Grisons. The extraction estimated in this way therefore only indicates the order of magnitude.

Map 20.1 (continued) Statistik Wasser, 2001; State Statistical Office Baden-Württemberg, Stuttgart, 1998; District Offices: Alb-Donau, Biberach, Breisgau-Hochschwarzwald, Heidenheim, Ostalb, Ravensburg, Reutlingen, Schwarzwald-Baar, Sigmaringen, Tuttlingen, Zollemalb, city of Ulm, 1998–2003; Operation results of Austrian Waterworks 1999. Statistik DG 1. ÖVGW Austrian Association for Gas and Water, Vienna.; Data on withdrawal of water of the Austrian Federal Ministry for Agriculture, Forestry, Environment and Water Management, Strategy Paper 'Grundwasserentnahmen', 2004, and 'Lage und Abgrenzung von Grundwasserkörpern', 2003; Geographical Information System of Salzburg SAGIS, Office of the Provincial Government of Salzburg, 2004; EcoGIS, Web-based Map Browser for Environmental Data of Switzerland, BUWAL, 2001–2003, Swiss Agency for the Environment, Forests and Landscape, borderlines: ESRI Deutschland GmbH, Kranzberg, Germany http://ESRI.de, ©BEV – Bundesamt für Eich- udn Vermessungswesen 2014, GG85 ©Swisstopo)

Data on locations and total capacities for the individual points of withdrawal were made available in varying forms and extents by the former Bavarian State Office for Water Management, by the district offices within Baden-Württemberg and by the state offices of Upper Austria and Salzburg. The indicated capacities were derived from the water regulations, from the actual withdrawals, from the number of residents supplied, directly from company reports and from a comparison with the available community statistics.

The locations of the points of withdrawal in Switzerland were determined based on the locations of designated water protection areas. In contrast to Germany and Austria, the designation of water protection areas for drinking water is mandated. Thus, it can be assumed that the points of withdrawal are almost all included. The precise geographical locations and capacities of each facility are documented centrally by the Federal Office for the Environment, Forests and Landscape in the form of a water atlas, but for reasons of national security, this information is not made public nor is it available for research purposes.

Alone the Austrian state of Tyrol provided no information on locations and capacities for the points of withdrawal. Therefore, for each community and groundwater aquifer, a point of withdrawal was artificially defined as the basis for the *WaterSupply* model. Details on the data processing are also presented by Nickel et al. (2005).

20.3 Results

The characteristics of the drinking water supply in the Upper Danube basin include:

- Preferential use of ground- and spring water sources
- A three-tier supply system structure, consisting of municipal, regional and long-distance water suppliers (mostly in the hands of local authorities)
- A high percentage of population connected to the public water supply

In the Baden-Württemberg region of the Danube basin, a total of 172 million m^3 of water is abstracted for the supply of drinking water. The largest percentage, at 59 %, is extracted from groundwater and an additional 22 % is from springs. With the extraction of water from the Danube near Leipheim, the Landeswasserversorgung (a long-distance water utility) is the only company that withdraws surface water. The withdrawal of water from the Danube is equal to at least 19 % of the total withdrawal. In Bavaria, the total quantity withdrawn (678 million m^3) almost entirely comes from groundwater (78 % groundwater and 18 % spring water). Only 1 % is sourced by the inter-municipal utility ‘Wasserversorgung Bayerischer Wald’ from the Frauenau Dam in the district of Regen. The municipal utilities at Passau and Augsburg also operate bank filtration facilities that are counted as groundwater withdrawal for the purpose of this study. There is no use of surface water in the Austrian and Swiss parts of the drainage basin.

Groundwater availability within the Upper Danube basin is significantly higher in the south compared to the north as a result of meteorological and hydrogeological conditions. In the Alps, there are numerous springs and the coarse-grained porous aquifers in the alpine valleys are fed by abundant precipitation and relatively low evaporation rates. Similar conditions are also found in the gravel and sand valley aquifers and the gravel plains of the alpine foreland, which also feed the largest water suppliers. In the north and northwest, the karstified carbonate aquifers of the Upper Jurassic are productive aquifers with their numerous abundant springs. Unfavourable conditions for groundwater extraction exist primarily in the Eastern Bavarian crystalline regions as well as in the northern regions that are characterised by hard rock aquifers from the Mesozoic (see Chap. 14).

The hydrogeological conditions are mirrored by the structure of water supply. In regions with poor geological conditions (Eastern Bavaria), numerous very small water intake structures are common. In contrast, in the Karst and in the valleys of the alpine fringe, large water intake structures are in operation with relevance well beyond the region. An overview of the regional distribution of drinking water sources in Bavaria is provided, for example, in BayLfW (1982).

In the German part of the Danube basin, supply security in arid regions and regions with naturally or anthropogenically stressed water resources is further ensured by inter-municipal and long-distance water supply utilities. According to data from the district offices, 42 out of a total of 162 water suppliers in Baden-Württemberg are inter-municipal utilities that supply more than one community. The largest of these, the Landeswasserversorgung, delivers the majority of the water sourced from within the Danube basin to municipalities and cities outside of the drainage basin. According to the information from the former Bavarian State Office for Water Management, approximately 900 municipal companies, 150 inter-municipal companies and numerous smaller companies are responsible for supplying the population of Bavaria with drinking water. The largest companies are the municipal utilities in Munich, Augsburg and Regensburg, as well as some long-distance water suppliers in the northern part of the basin, which also supply areas outside of the Danube drainage basin where water is scarce. Connection rates to the public supply of drinking water are high in both the German states of Baden-Württemberg and Bavaria (99.5 and 98 %).

Compared to the German part, in both the Austrian and the Swiss parts of the river basin, inter-municipal and long-distance suppliers play a lesser role, while individual suppliers and small water cooperatives (i.e. non-public drinking water suppliers) assume greater significance. In the province of Salzburg, for example, over 12 % of the population is not connected to the public drinking water supply; similar, though somewhat lower values can be assumed for the states of Upper Austria and Tyrol and for the Swiss canton Grisons. This situation can be attributed to the alpine nature of the southern Danube basin and to the numerous alpine springs with high-quality drinking water. The use of locally available springs spares the expenses associated with treatment and transport: in the Austrian part of the basin, 50 % of the drinking water is derived from springs, and in Grisons, this value is 60 %.

Only 2 of the 3,022 springs for supplying drinking water in Tyrol can be associated with inter-municipal companies (Kutzschbach and Fleischhacker 1997).

The structure of the water supply system described above ensures an overall high degree of supply security. Long-term, supraregional shortages in drinking water supply have not been experienced in the Upper Danube basin in recent decades. Exceptionally dry years with high demand and simultaneous lower supplies were documented for the years 1976 and 2003 (e.g. BUWAL 2004). Since demand and infrastructure in 1976 were not the same as today, only 2003 can serve as a reference for potential drier future conditions. Analyses of the dry year have been presented for Switzerland (BUWAL 2004), Bavaria (BayLfW 2004), Baden-Württemberg (LfU 2004) and Austria (Eybl et al. 2004). According to these, no significant shortages in the supply of drinking water were recorded for the entire drainage basin. Calls for water conservation and the implementation of emergency supply measures were recorded only in regions characterised by very small water suppliers not connected to long-distance or group suppliers and mainly dependent upon small spring. For example, this was the case for a few municipalities in the outlying parts of the Bavarian Forest or in the alpine regions of Switzerland. The situation in Switzerland, for instance, was overall much less dramatic than in 1976 (BUWAL 2004). This can be attributed to the fact that in 2003 there was lower per-person demand and better distribution capabilities.

References

BayLfW (1982) Die mittel und langfristige Trink- und Brauchwasserversorgung in Bayern. Informationsberichte Bayerisches Landesamt für Wasserwirtschaft 6/82, München

BayLfW (2004) Wasserwirtschaftlicher Bericht Trockenperiode 2003 Kurzfassung. Bayerisches Landesamt für Wasserwirtschaft, München

BUWAL (2004) Auswirkungen des Hitzesommers 2003 auf die Gewässer. Dokumentation Gewässerschutz. Schriftenreieh Umwelt NR. 369. Bundesamt für Umwelt, Wald und Landschaft, Berne

Eybl J, Godina R, Lalk P, Lorenz P, Müller G, Weilguni V (2004) Trockenheit in Österreich im Jahr 2003. Ein hydrologischer Situationsbericht, Vienna

Kutzschbach W, Fleischhacker E (1997) Wasserwirtschaftskonzept Tirol. Amt der Tiroler Landesregierung, Innsbruck

LfU (2004) Das Niedrigwasserjahr 2003. Landesanstalt für Umweltschutz Baden-Württemberg, Karlsruhe

Nickel D, Barthel R, Braun J (2005) Large-scale water resources management within the framework of GLOWA-Danube – the water supply model. Phys Chem Earth 30(6–7):383–388

Regierung des Kantons Graubünden (2000) Kantonaler Richtplan Graubünden. Amt für Raumplanung Graubünden, Chur

Chapter 21
Topsoil Organic Carbon Content

Christian W. Klar, Peter Fiener, and Karl Schneider

Abstract Soil organic carbon (SOC) is a key soil component. SOC stores large amounts of carbon and it also affects water fluxes as well as the availability of nutrients. Soil water fluxes are impacted by SOC directly due to its effect upon soil hydraulic parameters. Also many indirect effects of SOC upon water fluxes have to be recognised, such as the impact upon plant growth and nitrogen turnover. SOC maps for all modelled soil layers in the DANUBIA simulation system for the Upper Danube watershed are therefore a necessary prerequisite to model plant growth, soil nitrogen turnover and soil water fluxes. For the majority of the Upper Danube watershed, these maps were easily derived from the 1:1,000,000 soil map (BÜK 1000, BGR, Bodenübersichtskarte 1:1 Million. Bundesanstalt für Geowissenschaften und Rohstoffe, Berlin, 1995). The BÜK 1000 distinguishes 33 different soil types within the Upper Danube catchment. Soil properties including SOC content are associated to soil-type-specific pedogenetic horizons. The SOC content and the C/N ratios for each of the soil layers were derived as a weighted mean of the SOC contents given for the BÜK 1000 soil layers. The BÜK 1000 does not cover areas outside of Germany. For these mainly alpine regions, the SOC contents were estimated using a three-step rule-based approach by (a) establishing a statistical relationship between soil-type unit and elevation, (b) estimating the SOC based on the assigned soil type and (c) selecting the predominate soil type for grid cells with multiple soil types (majority principle).

Keywords GLOWA-Danube • DANUBIA • Global change • Soil organic carbon • SOC

C.W. Klar (✉)
Forschungszentrum Jülich, Jülich, Germany
e-mail: c.klar@fz-juelich.de

P. Fiener
Institut für Geographie, Universität Augsburg, Augsburg, Germany
e-mail: peter.fiener@geo.uni-augsburg.de

K. Schneider
Institute of Geography, University of Cologne, Cologne, Germany
e-mail: karl.schneider@uni-koeln.de

W. Mauser, M. Prasch (eds.), *Regional Assessment of Global Change Impacts*, DOI 10.1007/978-3-319-16751-0_21

21.1 Introduction

Soil organic matter is one of the main components of soils as it stores large amounts of carbon and contains important amounts of macronutrients (nitrogen, N, phosphorus, P, and potassium, K). In many ecosystems the N and C availability control the overall soil turnover and functioning (Batlle-Aguilar et al. 2011). In this context the C/N ratio is of particular significance because it determines the availability of microbial nitrogen. Moreover, soil organic matter or soil organic carbon (SOC) content is an important proxy variable to estimate infiltration capacity as increasing SOC contents are associated with larger aggregate stability, larger unsaturated hydraulic conductivity as well as higher biological activity affecting macro-porosity (Fiener et al. 2013).

In general SOC and associated N are important parameters affecting the water balance in the Upper Danube basin via their potential effect on plant growth and hence transpiration as well as their effect on partitioning precipitation into surface runoff and infiltration. Therefore, any change in climate and land use affecting SOC contents may result in a substantial climate feedback. Besides indirect effects on water balance, the SOC and associated N contents are essential for N leaching from the vadose zone, the latter representing one of the most important hazards for groundwater quality in the basin.

21.1.1 Data Processing

To model N turnover and N leaching (SNT; Chap. 38), spatially distributed SOC amounts in all three modelled soil layers (0–20, 20–80, 80–200 cm) are needed. Given the spatial resolution of 1 km grid cells used in DANUBIA, the German 1:1,000,000 soil map (BÜK 1000, BGR 1995) is most suitable. The BÜK 1000 does not cover the study area outside of Germany (Fig. 21.1). Thus, the soil layer-specific SOC contents in the remaining, mostly mountainous areas outside of Germany were derived by extrapolating the BÜK 1000 data to the entire basin.

Overall a three-stage preprocessing of the BÜK 1000 was applied to assimilate the data into the DANUBIA modelling system:

(a) A rule-based approach was used employing statistical relations between the soil-type units in the southern, alpine regions in the BÜK 1000 and elevation taken from the project digital terrain model.
(b) Based on the derived rules, elevation-specific soil types with their associated SOC contents were estimated for the regions not covered by the map.
(c) Due to the spatial resolution of DANUBIA, it is not possible to include a smaller-scale soil variability as partly represented in the BÜK 1000. If soils are not homogeneous within a DANUBIA proxel, the SOC content is taken from the major soil type within the proxel (majority principle).

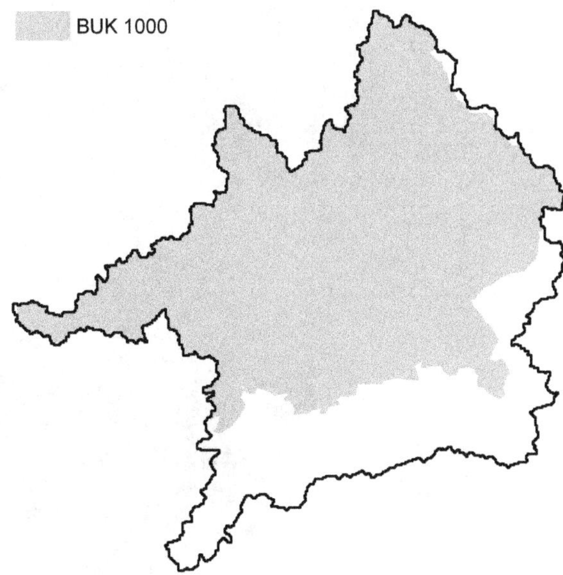

Fig. 21.1 Coverage of the Upper Danube basin by BÜK 1000 (BGR 1995)

The BÜK 1000 represents 33 different soil types within the Upper Danube basin. Soil properties including SOC content are associated to soil-type-specific pedogenetic horizons. The depths of these horizons differ spatially. Hence, to meet the DANUBIA requirement of consistent soil layers (0–20, 20–80, 80–200 cm), it was necessary to reprocess the soil depth profiles. Therefore, the given horizon-specific SOC contents were weighted by the proportion of the respective soil layer and the weighted average was used for parameterisation of the three soil layers. Analogously the C/N ratios were converted to the soil layer system.

21.2 Results

Map 21.1 presents the topsoil SOC amounts (0–20 cm) as used in DANUBIA for modelling N turnover and N leaching. In order to improve the readability of the map, the SOC contents were categorised in 11 classes.

In general, most topsoil SOC amounts are below 6.25 kg m^{-2} (Fig. 21.2) which results in a SOC content of 2.3 % (assuming a typical bulk density of 1,350 kg m^{-3}). A latitudinal change of topsoil SOC amounts can be identified from south to north. In the southern, mostly alpine part of the basin, topsoil SOC amounts are relatively high in poorly developed soil profiles on relatively steep slopes and climatic conditions favouring topsoil SOC accumulation. However, the values shown here for these areas may be associated with a high degree of uncertainty, as generally little soil information is available for alpine areas, and, more importantly, most of the SOC values presented here are based on a simple extrapolation approach. Since

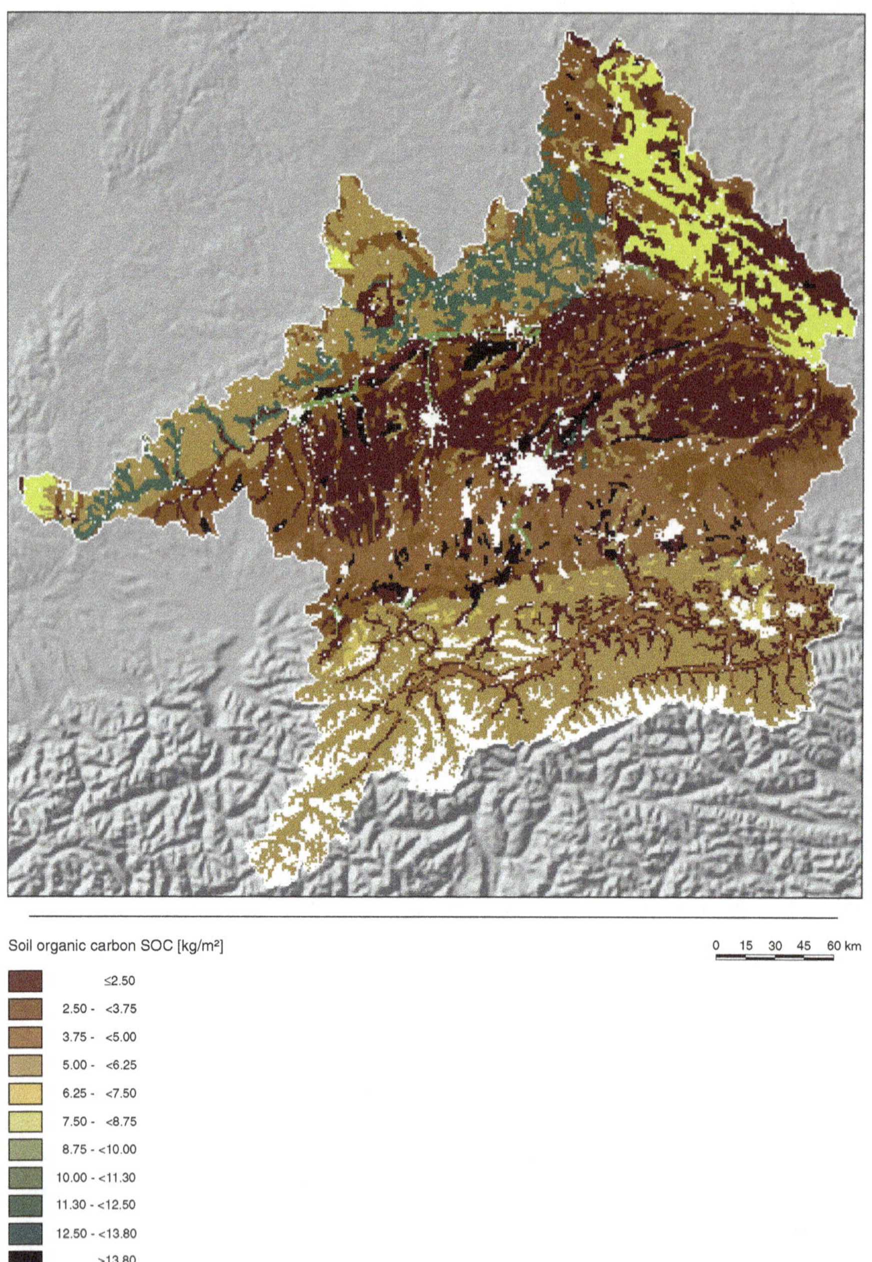

Map 21.1 Topsoil organic carbon content (Data source: BÜK 1000, BGR 1995)

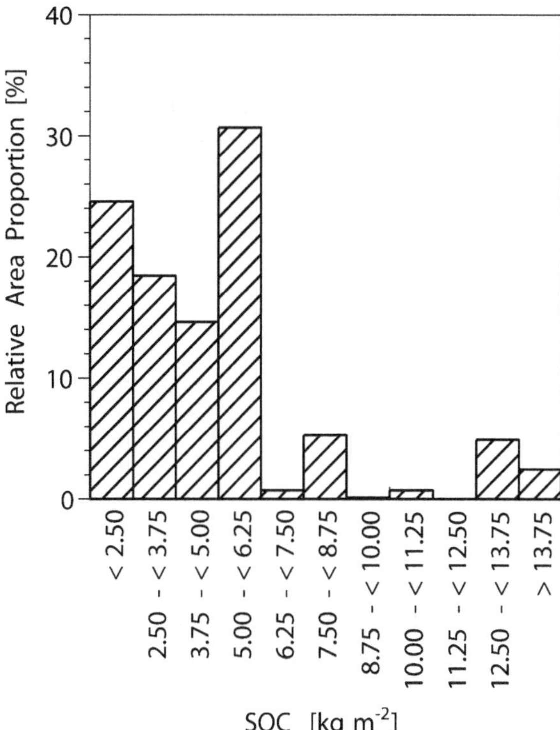

Fig. 21.2 Relative area proportion of different soil organic carbon (SOC) amount classes in the Upper Danube basin based on BÜK 1000 (BGR 1995)

the map is used to model N turnover and especially N leaching, this fact may pose a problem particularly along the inner-alpine valley floors. Here the map indicates alpine soils, while intensive agricultural use may occur with high fertiliser inputs (e.g. along the Inn valley). However, as proxels with agricultural use are relatively rare within the alpine region (see land use, Map 9.1), the chosen method should result in a reasonable model input for the vast majority of proxels in this area.

North of the Alps, in a zone where climatic conditions do not allow for intensive cropping, a grassland zone exists with relatively high topsoil SOC amounts. The lowest topsoil SOC amounts are associated with an area of intensive cultivation between this grassland zone and the Danube River. Partly soils with SOC contents below 1 % can be found. North of the Danube River, the SOC pattern becomes more patchy, with generally higher topsoil SOC amounts in the north-western to northern part (Swabian Jura and Franconian Jura) as compared to the north-eastern part (Bavarian Forest and Bohemian Forest). Apart from a mostly latitudinal sequence of topsoil SOC amounts, it is worth noting that areas with highest SOC values are located along the larger valley floors, where high groundwater levels reduce SOC mineralization. This is most pronounced in the (former) peat lands, e.g. the Donaumoos.

References

Batlle-Aguilar J, Brovelli A, Porporato A, Barry DA (2011) Modelling soil carbon and nitrogen cycles during land use change. A review. Agron Sustain Dev 31:251–274

Bundesanstalt für Geowissenschaften und Rohstoffe (ed) (1995) Bodenübersichtskarte 1:1 Million. BGR, Berlin

Fiener P, Auerswald K, Winter F, Disse M (2013) Statistical analysis and modelling of surface runoff from arable fields. Hydrol Earth Syst Sci Discuss 10:3665–3692

Chapter 22
Data on Quantity and Quality of Groundwater

Thorben Römer, Jan van Heyden, and Roland Barthel

Abstract Water supply in the Upper Danube catchment (UDC) is mainly based on groundwater. The development of models to adequately describe the quantity and quality of groundwater resources and to simulate the impact of global change requires a good understanding and conceptual representation of the regional hydrogeological conditions. Observation data is needed to evaluate the validity of the model results by comparing measured and simulated values for distinct locations. Thus, the comprehensive collection and analysis of groundwater monitoring data is an essential prerequisite for reaching the objectives of GLOWA-Danube. This chapter describes the data sources used to establish the conceptual hydrogeological model and gives an overview of the quantity and spatial and temporal distribution of groundwater monitoring data. It is shown that the focus of observations on shallow quaternary groundwater aquifers and the temporally and spatially sparse groundwater quality monitoring presents a great challenge to understanding the past and present behaviour of groundwater resources in the UDC and thus a challenge to simulate the future behaviour under conditions of global change.

Keywords GLOWA-Danube • Upper Danube • Water supply • Groundwater quality • Groundwater quantity • Time series • Observation wells • Monitoring

---

T. Römer • J. van Heyden
Formerly, Institute of Hydraulic Engineering, University of Stuttgart, Germany
e-mail: torben.roemer@gmail.com; javahey@gmail.com

R. Barthel (✉)
Department of Earth Sciences, University of Gothenburg, Göteborg, Sweden
e-mail: roland.barthel@gu.se

W. Mauser, M. Prasch (eds.), *Regional Assessment of Global Change Impacts*, DOI 10.1007/978-3-319-16751-0_22

22.1 Introduction

The majority of the water supply in the Upper Danube drainage basin is based on the use of groundwater resources. The ramifications of this use and of global change can be evaluated using information about interactions and can be simulated using models. However, the opportunity to deduce interactions and to validate the model results can only be realised by comparisons with observations. Measurements and monitoring networks on groundwater quantity and quality are therefore key elements in groundwater management and in the protection of this resource. Estimating the quantity of groundwater is carried out using the indirect measurement variables of spring discharge and the piezometric level of groundwater observation wells (GWOW). The evaluation of groundwater quality makes use of various chemical parameters that are calculated for water samples from springs, GWOW and extraction wells in the water supply system either on site or in the laboratory. In addition to the sample itself, the configuration of the monitoring network is a key element in later composing the data pool.

22.2 Data Processing

The data on the quantity and quality of groundwater are derived from several sources; there is no cohesive monitoring network that covers the entire drainage basin.

22.2.1 Groundwater Quantity

To quantitatively evaluate groundwater, data on spring water discharges and groundwater levels were obtained for Baden-Württemberg (BW) from the State Office for the Environment, Measurements and Nature Conservation (LUBW), for Bavaria (BY) from the State Office for the Environment (LfU) and for Austria (AT) from the Federal Ministry for Agriculture, Forestry, Environment and Water Management.

The public authorities produced a total of approximately 10.9 million groundwater measurements and approximately 230,000 spring discharge measurements. There are a large number of groundwater measurements here and there as far back as 1915, largely from the Bavarian region of the basin (see Map 22.1(1) and Fig. 22.1).

Map 22.1 (continued) Office for the Environment (LfU), 2002, 2007 and 2008; former Bavarian State Geology Office, 2003; Water quality survey in Austria conformable to the law of hydrography; BGBl. No. 252/90, i.d.g.F.; BMLFUW, Section VII water management planning; offices of the federal state governments Environment Agency Austria GmbH Vienna, at the request of 02.04.2001 (UBA-Zl.: 134-164/01) and at the request of 26.04.2007 (UBA-Zl.: 134-82/07); CCM River and Catchment Database, © European Commission – JRC, 2007)

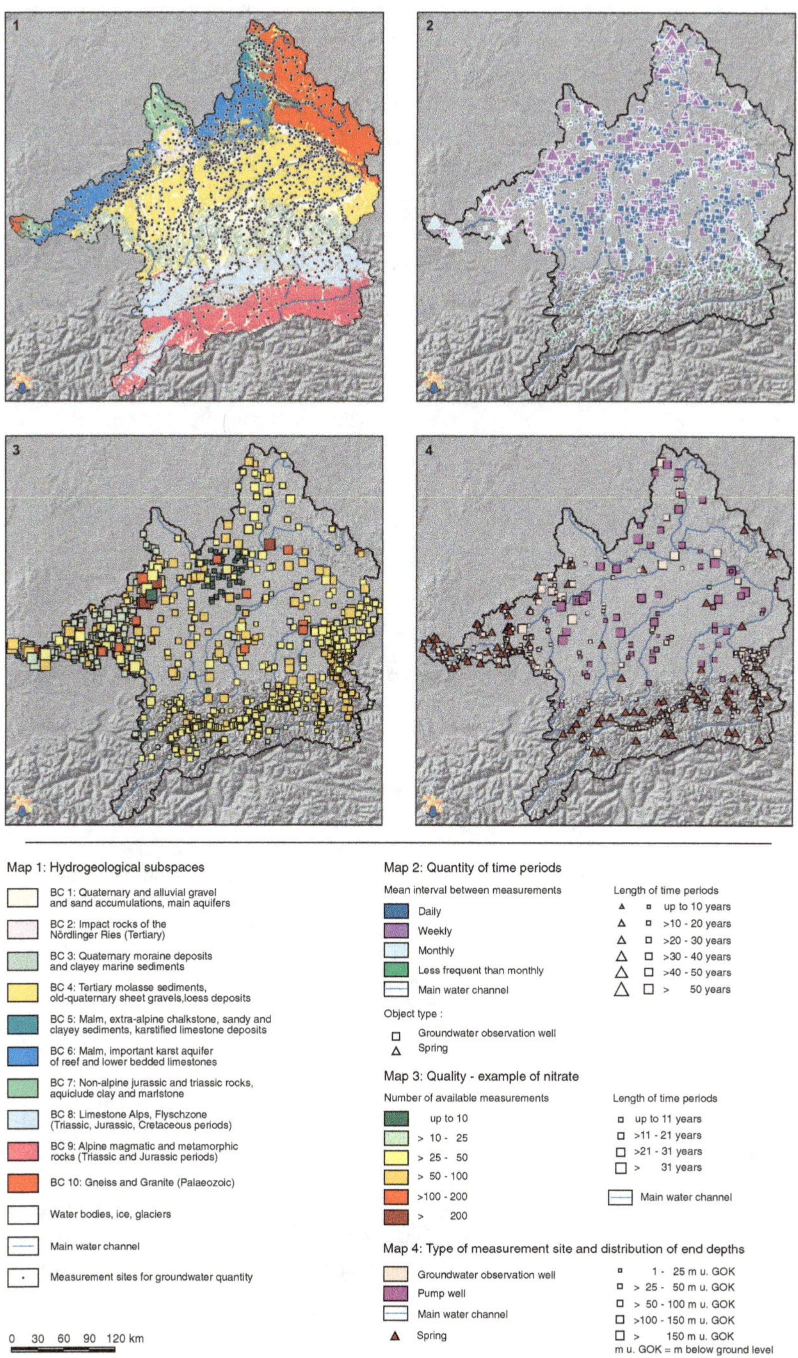

Map 22.1 Data on quantity and quality of groundwater (Data sources: State Office for the Environment, Measurements and Nature Conservation of Baden-Württemberg (LUBW), 2006 and 2007; former State Office for Environmental Protection Baden-Württemberg, 2005; Bavarian State

In addition, there are measurements after 1975 from AT and comparatively fewer measurements from BW. There is a clear steep increase in groundwater measurements beginning in the 1980s that led to a comprehensive dataset.

A different story is provided by the presentation of the number of spring discharges compiled (see Fig. 22.2). In this case, the total number of measurements in BY and BW since the end of the 1950s has been nearly constant, whereas the measurements in BW have increased slightly. Data on spring discharges in AT are only available since the 1990s. The spatial distribution of the measurement sites (see Map 22.1(1, 2)) is oriented along the rivers of the Upper Danube basin. Fluvioglacial valley sediments along the channels represent especially productive, easy-to-develop groundwater aquifers and must therefore be observed more closely (Bavarian State Office for the Environment, former Bavarian State Geology Office 2003). The

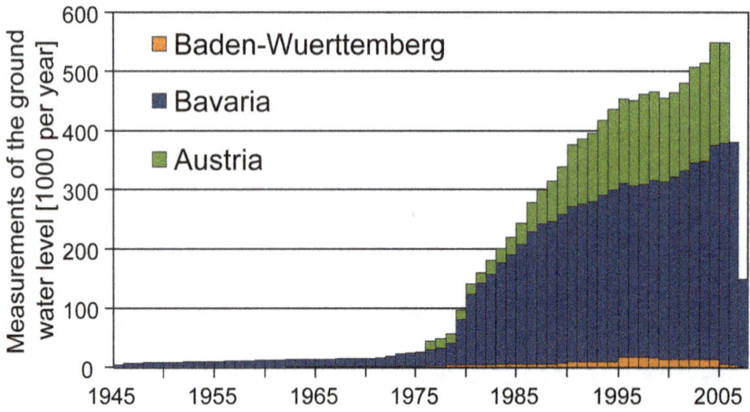

Fig. 22.1 Number of groundwater level measurements (stacked) in thousands per year

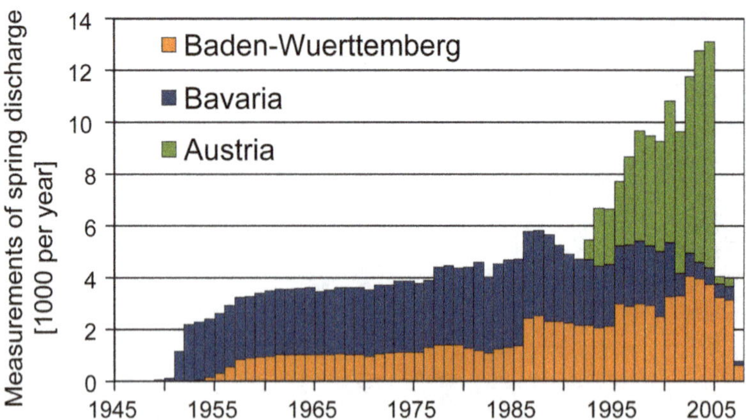

Fig. 22.2 Number of spring discharge measurements (stacked) in thousands per year

locations of the GWOW (black dots on Map 22.1(1)) can be assigned to the individual hydrogeological base classes (see also Chap. 14). More than half of the measurement sites, approximately 58.6 %, lie within the quaternary valley deposits (BC 1), although this makes up only approximately 18.8 % of the area. An additional number, approximately 18.3 %, are found in the molasse region (BC 4), which accounts for a greater area (23.5 %) than the quaternary valley deposit region. Another 5.8 % of the measurement sites are located in Malm (BC 5 and 6), which covers 8.2 % of the area. The remaining 17.3 % of the measurement sites are distributed among the other regions that make up 49.5 % of the area of the drainage basin. The majority of the data is gathered on a daily or weekly basis (see Map 22.1(2)).

22.2.2 Groundwater Quality

Like the data on groundwater quantities, the data on groundwater quality comes from several data pools that have varying compositions. The structure of the monitoring network or different legal regulations, particularly in AT (Austria 1993), has effects on the available data (see Fig. 22.3 and Map 22.1(3)). Figure 22.3 presents the change in the frequency of nitrate measurements. Nitrate is the longest and the most frequently calculated parameter. As a result, nitrate is used below as a representative parameter for all the variables calculated in order to present the nature of the measurements of quality. The input of nitrate into groundwater varies in both space and time (see Sect. 30.3). In addition to nitrate, full analyses are available from around 1950 for some of the measurement sites. With

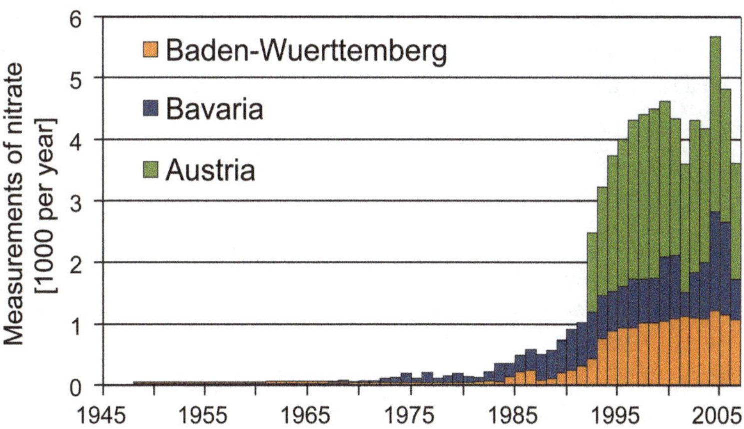

Fig. 22.3 Number of quality measurements (stacked) for the example of nitrate in thousands per year

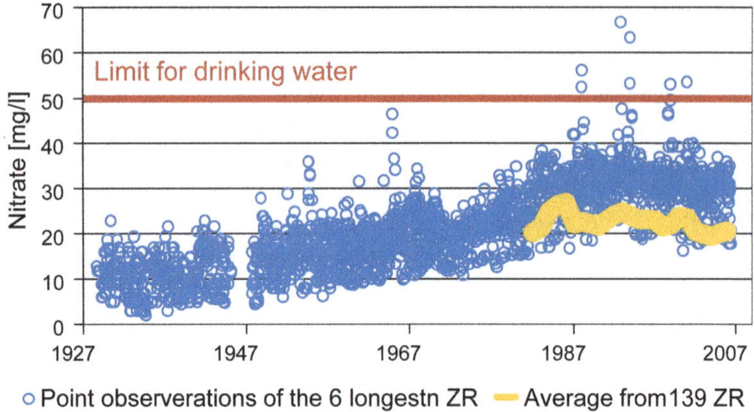

Fig. 22.4 Joint illustration of the longest nitrate observation series

the advent of trace analysis, the number of fully analysed parameters increases to more than 100.

The number of measurements continued to rise and has levelled off in recent years (1995–2004) to a constant level. In 2005, the measurements then decreased since the data are mostly entered into the countries' data storage systems only after a delay. A regionally comprehensive data pool is reliable only after the establishment of the monitoring network in the 1980s. Map 22.1(3) provides an overview of the spatial distribution for the parameter nitrate across long and well-documented time periods. In this presentation, the various data pools can be clearly defined within the overall dataset. BY has only a limited number of time periods (ZR = Zeitreihe) that mostly begin in the 1980s. BW has a cross section of data with some long-term and well-documented time periods (see Fig. 22.4). The first measurements of nitrate are available for the Baden-Württemberg part of the drainage basin already by the end of the 1920s.

22.3 Results

A dataset that comprehensively covers the region of the Upper Danube has existed since the 1970s (quantity) and the 1980s (quality). There are differences in the various sources of data with respect to measurement network and authority and these differences persist even today. The time periods for groundwater level and spring discharges at times exhibit very different patterns with respect to trends, variabilities and seasonal effects. The reason for these differences results from the fact that the measurement sites are filtered from differing groundwater aquifers, which, depending on location, are subject to different factors such as hydrogeological conditions, soil cover, climate, influence of surface waters, water abstraction, etc. It is critical to consider the heterogeneous distribution of the measurement sites in a statistical analysis of the data in order to avoid false conclusions. The choice of measurement sites that can be considered representative of specific areas, for example, forms the basis for future analyses and hence is crucial for the results.

Various relevant properties of the time periods should be emphasised from the quality analyses in particular; the most important of these are (a) frequently changing sampling intervals (the typical intervals are 2–3, 6 or 12 months), (b) substantial gaps in the data (>2 years) and (c) parameters with primarily extremely skewed statistical distributions.

A differentiated approach in the analysis of the data is necessary and helps to more clearly illustrate the dataset based on the groundwater quality data. Figure 22.4 collectively presents all the single measurements of nitrate (blue circles) for the six longest time periods in the quality dataset. The six measurement sites spatially form a group in the regions of Langenau (BW) and Günzburg (BY) and east of the boundary of the Swabian Jura (see also Map 22.1(3)). The development that is typical of regions used for agriculture in Europe is easily apparent (Stockmarr et al. 2005; Visser et al. 2009) with steady increase in nitrate values up until the late 1980s. Area-wide analyses for the Upper Danube basin are possible only to a limited extent, since the compilation of the data pool over the course of the decades has undergone significant change with respect to final depths and type of measurement sites. However, the average trend in the hydrogeological base classes 1 (83 time periods), 4 (30 time periods) and 6 (26 time periods) can be shown for the period from 1983 to 2006, based on 139 pooled time periods with ongoing measurements (see the yellow line in Fig. 22.4).

The GWOW, extraction wells in the water supply system and springs in the measurement network exhibit distinct characteristics that should always be taken into account in the analyses. Map 22.1(4) depicts the distribution of the measurement sites with more than 40 nitrate measurements per site. Especially in BY, many parameters are measured at extraction wells in the water supply system. Pump wells have longer vertical screens and deeper end depths. Compared to GWOW and springs, samples from these measurement sites integrate larger regions of the aquifer. The hydrogeological situation and the configuration of the measurement sites (filter intervals, end depths) and the location within water conservation regions with altered land use both lead to significantly lower nitrate values. Therefore, the single measurements in Fig. 22.4 (end depths around 10 m) cannot be compared directly with the measurements in the 139 time periods (end depths on average 50–70 m).

References

Bayerisches Geologisches Landesamt (2003) Hydrogeologische Raumgliederung von Bayern. GLA Fachberichte No.20. Bayerisches Geologisches Landesamt, Munich

Österreich (1993) Bundesgesetz über den Zugang zu Informationen über die Umwelt. BGBl. 495/93. Österreich, Vienna

Stockmarr J, Grant R, Jørgensen U (2005) Monitoring effectiveness of the EU Nitrates Directive action programmes: approach by Denmark. In: Fraters B, Kovar K, Willems W, Stockmarr J, Grant R (eds) Monitoring effectiveness of the EU Nitrates Directive action programmes, RIVM report 500003007/2005. National Institute for Public Health and the Environment, Bilthoven, pp 107–138

Visser A, Broers HP, Heerdink R, Bierkens MF (2009) Trends in pollutant concentrations in relation to time of recharge and reactive transport at the groundwater body scale. J Hydrol 369:427–439

Chapter 23
Hydropower Plants

Franziska Koch, Andrea Reiter, and Heike Bach

Abstract Hydropower generation has, especially in mountainous areas and in regions with high discharge like the Upper Danube basin, a large proportion of the total renewable energy production. Hydropower as energy supplier will be even more important in the future as predictions assume that the future energy demand will increase. Within the hydropower module of DANUBIA, it is distinct between run-of-the-river power plants mainly situated in rivers and river channels and storage power plants with reservoirs situated especially in regions with high drop heights. The reservoir of a storage power plant is managed by an operation plan mainly depending on hydrological variables. The analysed data includes all larger hydropower plants with a bottleneck capacity of at least 5 MW within the Upper Danube basin. This comprises 118 run-of-the-river power plants with mean annual outputs of 20 and 550 GWh amounting to 13.0 million MWh and 22 storage power plants with mean annual outputs between 25 and 1,000 GWh amounting to 5.7 MWh.

Keywords GLOWA-Danube • Upper Danube • Hydropower • Energy production • Run-of-the-river power plant • Storage power plant

F. Koch (✉)
Department of Geography, Ludwig-Maximilians-Universität München (LMU Munich), Munich, Germany
e-mail: f.koch@iggf.geo.uni-muenchen.de

A. Reiter
Bavarian Research Alliance GmbH, Munich, Germany
e-mail: reiter@bayfor.org

H. Bach
Vista Geowissenschaftliche Fernerkundung GmbH, Weßling, Germany
e-mail: bach@vista-geo.de

W. Mauser, M. Prasch (eds.), *Regional Assessment of Global Change Impacts*, DOI 10.1007/978-3-319-16751-0_23

23.1 Introduction

Predictions assume that the future energy and power demand in Europe are going to rise. This change will be driven on despite increasing efficiencies and energy-saving measures from, for example, growth of population within agglomerations and economic growth (Statistics Austria 2005 and Chap. 66). At the same time, the generation of electricity from renewable energy sources is becoming ever more important. The high alpine regions in particular are predestined for the regenerative energy production from hydropower because of the enormous height differences and because the area is abounding in water. Although the investment expense for the construction of a hydropower plant is quite high and can reach as much as eight times that for a hard coal-fired power plant, hydroelectric power generation is more economical because of low running costs and long operational life spans of the power plants. This proves that hydropower plants operate under efficient energy conversions. Another advantage for using this renewable power source is that it has a very low CO_2 emission rate during construction and while in operation. However, hydropower plants always involve construction of dams that present a great encroachment in the environment and aquatic ecosystems. Therefore, new construction or retrofitting of existing facilities should always consider conservation of environmental, landscape and ecosystem principles; for example, these principles are grounded in the European Water Framework Directive from 2000 and in the amended Renewable Energies Act of 2009.

Hydroelectric power plants in countries with abundant water and strong topographic gradients such as Austria provide for between 60 and 70 % of the total electric power generation; however, even in Bavaria a good proportion of electricity production is from this renewable energy source, at up to 20 %. An estimate of the future available hydropower production is essential for planning the power supply. Therefore, all the larger hydroelectric power plants in the German, Austrian and Swiss parts of the Upper Danube basin were explicitly incorporated and modelled within DANUBIA (see Chaps. 35 and 68).

23.2 Data Processing

Hydropower plants are typically distinguished as either storage power plants that are always associated with reservoir dams or as run-of-the-river power plants at river dams. By definition, the structure of a reservoir dam closes off a valley across its entire breadth and thereby provides a large volume of water storage. In contrast, river dams enclose only the area of flowing water. The key difference of the two types is that reservoir dams store the water for longer periods, whereas run-of-the river power plants involve only short-term damming creating steps of drop height in the river. Because for storage power plants differences in drop heights of up to 1,000 m between the power plant and the reservoir can be used for the generation of energy, this type is usually found in alpine regions. Otherwise, river dams are found in virtually all of the larger rivers within the Upper Danube basin.

The primary purpose of the majority of river dams is power generation. In general, a river dam consists of a weir, a hydroelectric power facility, impounding dams and a lock, if the river is navigable (see Fig. 23.1). In contrast, for reservoir dams the filling area is divided vertically into different zones. The defined levels are based on the type of use, the water supply and the construction of the power plant (see Fig. 23.2).

The water stored in the operation area of a storage power plant is available for generating electricity; depending on the filling level, this area extends from the lower to the upper storage level. The water below the operation area is usually never drained off, except from time to time when the sediment must be dredged. The topmost zone serves as a flood prevention area, and, in the event of flooding, this area is drained via a spillway (Strobl and Zunic 2006). The volume of water stored is managed using controlled discharge that is defined in an operation plan that governs the maximum discharge each month depending on the fill volume of the reservoir. The establishment of an operation plan must be preceded by a comprehensive analysis of the hydrological variables including run-off, precipitation, evaporation, infiltration, groundwater conditions, snow and ice conditions as well as sedimentation patterns in the drainage area of the dam. In addition, the type of use of the dam is an important factor for the operation plan, including not only power production but also other uses such as flood prevention, raising low water levels, recreation, drinking water conservation and irrigation (Maniak 1997). In the case of hydroelectric

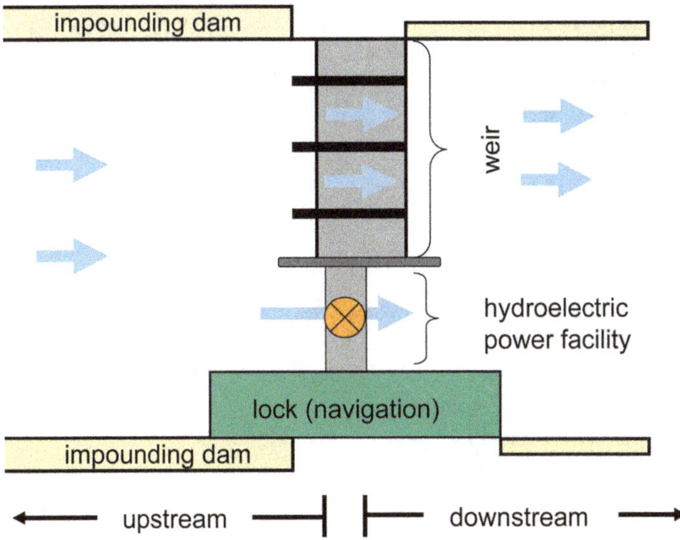

Fig. 23.1 Components of a run-of-the-river power plant (Modified after Strobl and Zunic 2006)

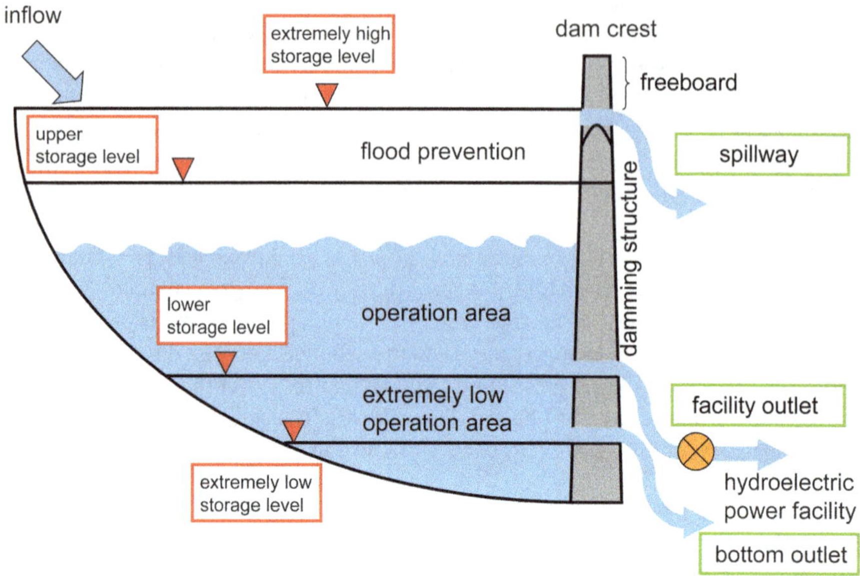

Fig. 23.2 Components and storage levels of a storage power plant (Modified after Maniak 1997)

power generation, the run-off is regulated under the operation plan and the filling level of the reservoir so that in times of elevated power demand, more water can be discharged. In high alpine regions, for example, the inflow from snow and ice melt is stored so that even in the winter months when natural inflow is limited, discharge can be released and electricity produced. This complies in nowadays with the high power demand at this time of the year. Diversions are commonly established that conduct additional water from neighbouring valleys into the reservoir, which increases the efficiency of the power plant.

Run-of-the-river power plants contribute to the generation of baseload electricity based on a constant flow rate. It directly uses the river run-off and is characterised by a usually high run-off rate and a rather low drop height. For energy generation, the best design for this type of power plant is thus the Kaplan Turbine that is shaped like a ship's propeller. One exception among the run-of-the-river power plants are threshold power plants, which can dam up run-off for several hours in order to produce more power during peak demand. Because of their capability to store water over a longer time period, storage power plants can cut into the power network if required and hence can normally supply balancing and peak electricity. Pelton turbines are particularly well suited to make use of the energy from bigger drop heights. Pumped storage plants are a special case of storage power plants. These can drain water from a higher basin into a lower basin during periods of high power demand. In periods of lower demand, typically overnight, the water is pumped back up to the upper basin using other sources of energy in order to be available again for peak periods of demand. Since the management of pumped storage power plants is highly complex and there is no adequate database available for them, they are not simulated in DANUBIA at this time.

23.3 Results

Hydroelectric power plants are mostly characterised according to the so-called annual output and the bottleneck capacity. The mean annual output is represented as the mean energy production for the power plant during a normal year. The bottleneck capacity is the highest feasible long-term output by the power plant, but it only can be reached if the discharge exhibits an ideal value (Fichtner GmbH and Co KG 2003). For run-of-the river power plants, the bottleneck capacity is usually equal to the installed maximum capacity. The design flow (i.e. maximum possible turbine discharge) is selected for each power plant such that the water supply is optimally exploited for energy generation while considering also ecological factors. The design flow at run-of-the river power plants at most central European rivers is achieved or exceeded on approximately 30–60 days per year, at alpine rivers even at 60–115 days per year (Strobl and Zunic 2006; Maniak 1997).

Map 23.1 virtually shows all of the run-of-the-river and storage power plants with a bottleneck capacity of at least 5 MW within the Upper Danube basin. Smaller hydroelectric power plants are thus excluded. At present, the data include 118 run-of-the-river plants with mean annual outputs between 20 and 550 GWh and 22 storage power plants with mean annual outputs between 25 and 1,000 GWh. The overall mean annual output of the run-of-the-river power plants for all facilities included is approximately 13.0 million MWh; for storage power plants this amounts to 5.7 million MWh. This means that in the Upper Danube basin, approximately 70 % of hydroelectric power is generated by the run-of-the-river power plants.

Run-of-the-river power plants within the Upper Danube basin are found on the larger rivers and their tributaries. These include the Danube, the Inn, the Salzach, the Isar, the Lech and the Iller. In the north of the Danube, there are only small hydroelectric power plants that are not simulated in DANUBIA. Sometimes run-of-the-river power plants are located at river canals. The largest run-of-the-river power plants with mean annual outputs greater than 500 GWh such as in Töging, Simbach/Braunau, Schärding and Passau/Ingling and the threshold power plant at Imst are all located on the Inn River, which is characterised by high discharge values. In the central Isar canal, on the Salzach River and on the Danube, the mean annual output at a single power plant is up to 250 GWh. The majority of the power plants on the Iller and the Lech rivers are established in river steps and can achieve up to 100 GWh. The majority of the storage power plants are situated in the high alpine part of the drainage basin in Austria. Among the largest storage power plants and power plant groups with mean annual outputs of at least 500 GWh are the Pradella at Engadin, Prutz at the entrance to the Kaunertal, Sellrain/Silz in Kühtai, Mayrhofen in the Zillertal and Kaprun in the Hohe Tauern range. The two

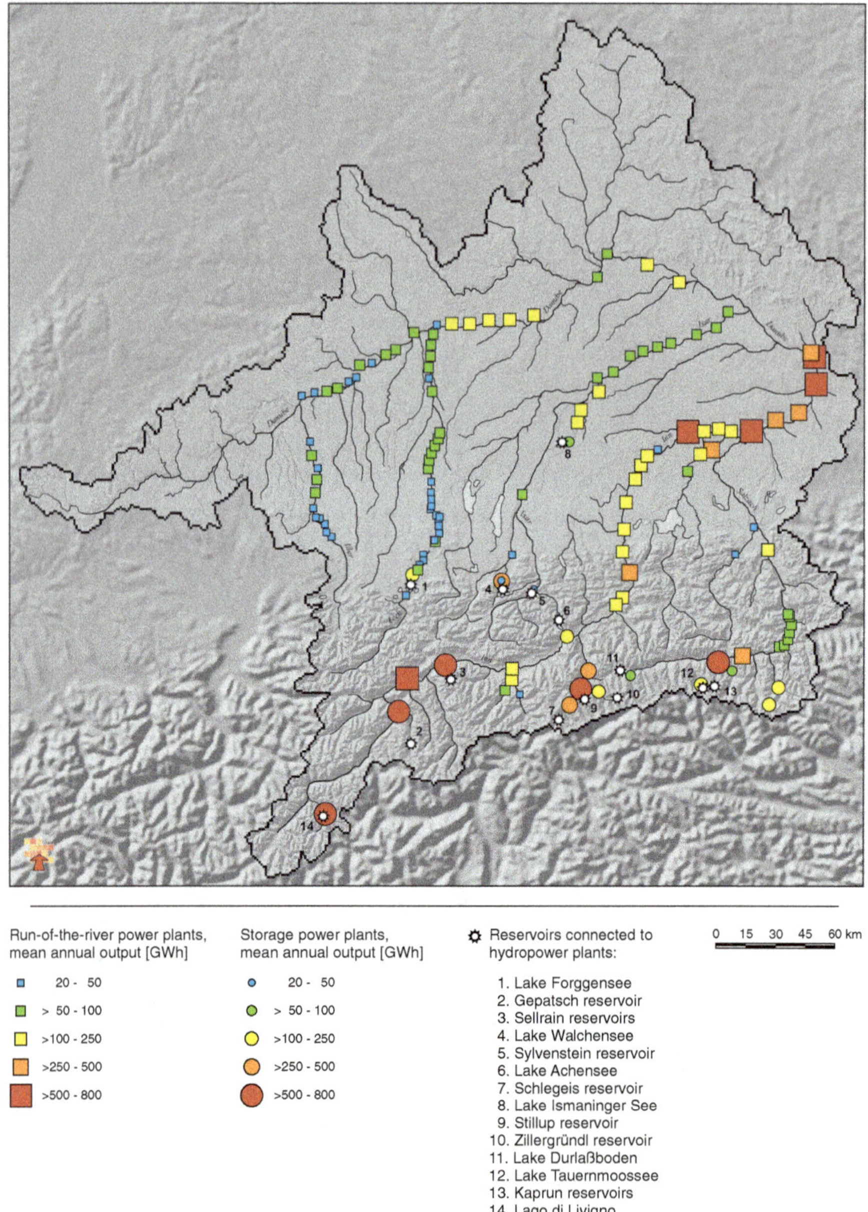

Map 23.1 Hydropower plants (Data sources: DANUBIA – river network, BMLFU – Austrian Federal Ministry for Agriculture, Forestry, Environment and Water Management (2005), Investigations and surveys carried out by the authors, Jarvis et al. (2008); Fichtner and Co (2003))

largest Bavarian facilities are Rosshaupten on the Lech and the Walchensee Power Plant. The most important reservoirs required for the storage power plants are also shown on Map 23.1. The reservoirs are mostly located in close proximity to the sites of the power plants, although in some cases the water is conducted over several kilometres.

References

BMLFU – Austrian Federal Ministry for Agriculture, Forestry, Environment and Water Management (ed) (2005) Hydrological atlas of Austria. Österreichischer Kunst- und Kulturverlag, Vienna

Fichtner GmbH and Co KG (ed) (2003) Die Wettbewerbsfähigkeit von großen Laufwasserkraftwerken im liberalisierten deutschen Strommarkt. Endbericht. Available via DIALOG. http://www.emissionshandel-fichtner.de/pdf/BMWA_Langfassung.pdf. Accessed 24 Nov 2009

Jarvis A, Reuter H I, Nelson A, Guevara E (2008) Hole-filled seamless SRTM data V4. International Centre for Tropical Agriculture (CIAT). Available via DIALOG. http://srtm.csi.cgiar.org. Accessed 19 Sept 2014

Maniak U (1997) Hydrologie und Wasserwirtschaft. Eine Einführung für Ingenieure. Springer, Berlin

Statistik Austria (2005) Volkszählung 2001 – Haushalte und Familien. Statistik Austria, Wien

Strobl T, Zunic F (2006) Wasserbau. Aktuelle Grundlagen – Neue Entwicklungen. Springer, Berlin

Part III
Models

Chapter 24
Groundwater Recharge

Wolfram Mauser and Ralf Ludwig

Abstract The impact of climate change on groundwater recharge is analysed in GLOWA-Danube. The chapter deals with the long-term average spatially distributed groundwater recharge during the climatological period from 1971 to 2000. The chapter first discusses the basic underlying assumptions of the model representation in DANUBIA of water movement in the unsaturated soil, formulates the analytic solution of Philip's equation used and shows in the form of a map the annual groundwater recharge as results of the model run using station data as meteorological driver. The results are discussed.

Keywords GLOWA-Danube • Groundwater recharge

24.1 Introduction

Groundwater recharge, defined as the quantity of water that reaches the groundwater body through percolation from the unsaturated soil zone, is a key variable in the terrestrial water cycle. Its quantitative assessment is critical for estimating the regular, seasonal replenishment of groundwater reserves. In the DANUBIA model approach, groundwater recharge is controlled by hydrological and hydraulic processes within the soil. The dynamics of soil water storage, known as the soil water balance, result from a complex hierarchy of various processes that contribute to either filling or emptying the storage. Storage is replenished through the portion of precipitation or snow melt that infiltrates the soil depending on the local physical conditions of the soil and thus refills the moisture deficits of the unsaturated soil zone. With saturation deficit and a groundwater table near the surface, additional replenishment of the storage can take place through

W. Mauser (✉) • R. Ludwig
Department of Geography, Ludwig-Maximilians-Universität München (LMU Munich), Munich, Germany
e-mail: w.mauser@lmu.de; r.ludwig@lmu.de

W. Mauser, M. Prasch (eds.), *Regional Assessment of Global Change Impacts*, DOI 10.1007/978-3-319-16751-0_24

capillary rise. Factors that contribute to emptying the storage are collectively known as exfiltration. They include evaporation (both evaporation and soil evaporation and plant transpiration and evaporation) and the gravitational emptying that occurs when the moisture level of the ground exceeds a soil-specific threshold known as field capacity.

The water percolating through the soil either vertically reaches the current groundwater table (groundwater recharge) or is conducted as interflow with usually negligible temporal delay to the next receiving waters along existing elevation gradients.

24.1.1 Data Processing

The basis for correctly modelling soil water balance is the assignment of physical parameters of the soils using a soil texture map (see Map 8.1 and Chap. 8). The processes of soil water balance described above must be, on the basis of a suitable soil parameterization, correctly represented in a model of the water movement in unsaturated soils (see following Sect. 24.1.2) in order to simulate groundwater recharge at a plausible level of accuracy.

24.1.2 Model Documentation

Modelling soil water balance was carried out using a method expanded for soils of many layers developed by Eagleson (1978). This method calculates the volumetric water content θ and the matrix potential $\psi(\theta)$ of the root penetration zone of a soil column assumed to be homogenous and consisting of three layers. Depending on the size of the matrix potential, the modelled water content of the soil is available for transpiration by plants or evaporation from the exposed soil surface. Additional physical parameters describing the soil are required in order to calculate θ and $\psi(\theta)$. The change in soil moisture θ and hence the change in the quantity of plant-available soil water are described for homogenous soils by one-dimensional Philip's equation (Philip 1960):

$$\frac{\partial \Theta}{\partial t} = \frac{\partial}{\partial z}\left[D(\Theta)\frac{\partial \Theta}{\partial z}\right] - \frac{\partial K(\Theta)}{\partial z}$$

For this, t is time, z is the depth in the soil, $K(\Theta)$ is the hydraulic conductivity and $D(\Theta)$ is diffusivity in m²/s:

$$D(\Theta) = K(\Theta)\frac{\partial\psi(\Theta)}{\partial\Theta}$$

The partial differential equation developed by Philip describes an analytical approximation solution of the Richards equation (Richards 1931). It applies to one-dimensional, vertical infiltration into a semi-infinite, homogenous soil column. Richards equation is derived from the combination of the continuity equation and Darcy's Law. If Darcy's Law is also applied to unsaturated conditions, the conductivity is no longer just a constant dependent on substrate; instead it also becomes a function of saturation. The hydraulic gradient is thus composed of the gradient in the matrix and the gravitational potential. Philip's equation can either be solved in closed form using specific simplified boundary conditions or through numeric approximations. To solve the equation, discrete solutions for Philip's equation for the partial processes of the soil water movement (infiltration, exfiltration, percolation and capillary rise) must be found. In order to be able to simulate independent processes, the discrete solutions of Philip's equation are linearly superimposed. If the individual layers of the soil column are considered homogenous, the result is the approximation function for the matrix potential $\psi(\theta)$ and the hydraulic conductivity $K(\theta)$, which are based on hydraulic soil parameters that are easy to measure and, more importantly, can be assumed to not vary in time. Approximation functions according to Brooks and Corey (1964) are used, taking into account a functional relationship between matrix potential and soil saturation. Depending on the time and the initial saturation, the analytical solution for Philip's equation leads to equations for infiltration and exfiltration capacities. On the one hand, it is assumed in the first approximation that the soil moisture at greater depths, close to the groundwater table, is seasonally constant, whereby percolation is equal to hydraulic conductivity at this moisture; on the other hand, it is assumed that the groundwater level lies much deeper than the capillary fringe of the soil. Just like exfiltration and infiltration, capillary rise is thus derived from the same soil parameters.

Figure 24.1 schematically illustrates the processes summarised by the unsaturated water movement model of the soil. It is based on a maximum soil thickness of 2 m that is available to the dynamically growing plants as a maximum space for root penetration. This is divided into single layers in the form of a cascade, such that the percolation from one layer into the next can be interpreted as an effective precipitation. Percolation out of the lowest soil layer is transferred to the *Groundwater* model component as groundwater recharge. Depending on slope, a portion of the quantity of water percolating from individual layers is channelled away as interflow and is, for each proxel of the watershed, transferred to the *Rivernetwork* model component together with possible infiltration and saturation surplus formed on the ground surface.

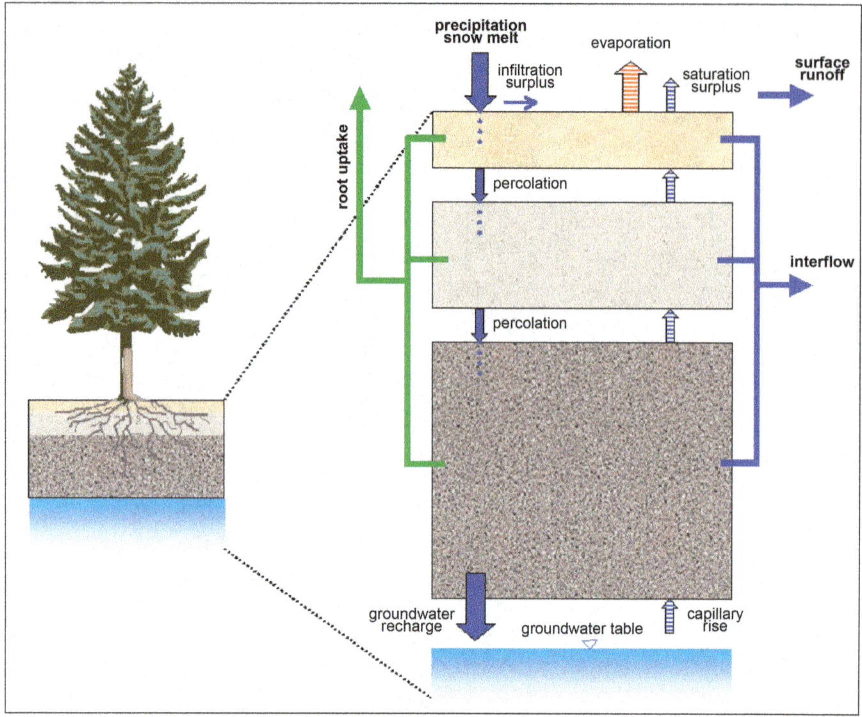

Fig. 24.1 Approach for the calculation of the soil water content in the model component *Soil*

24.2 Results

The mean annual groundwater recharge rate [mm] for the Upper Danube basin for the reference period 1971–2000 is shown on Map 24.1. It is apparent that groundwater recharge is controlled by a variety of factors that lead to several distinct local maxima and minima. Particularly high groundwater recharge is seen in the Allgäu (the result of high precipitation) and in the high-altitude regions of the low mountain ranges (the Black Forest and Bavarian Forest) and the high alpine regions (both caused by relatively high precipitation maxima in association with reduced evaporation and a low storage capacity of the soil). Relative minima occurred in the regions with low precipitation in northern Bavaria but also in the Central Alps as a result of increased evapotranspiration in the forested areas and in the high elevation sites in the Bavarian and Central Alps as a result of the steep relief. Under today's climatic conditions, groundwater recharge, at 554 mm as a long-term average, is a quantitatively significant variable within the Upper Danube basin. This parameter serves the project consortium as a part of the *Groundwater* model component and therein provides the region-wide replenishment of the groundwater reservoir.

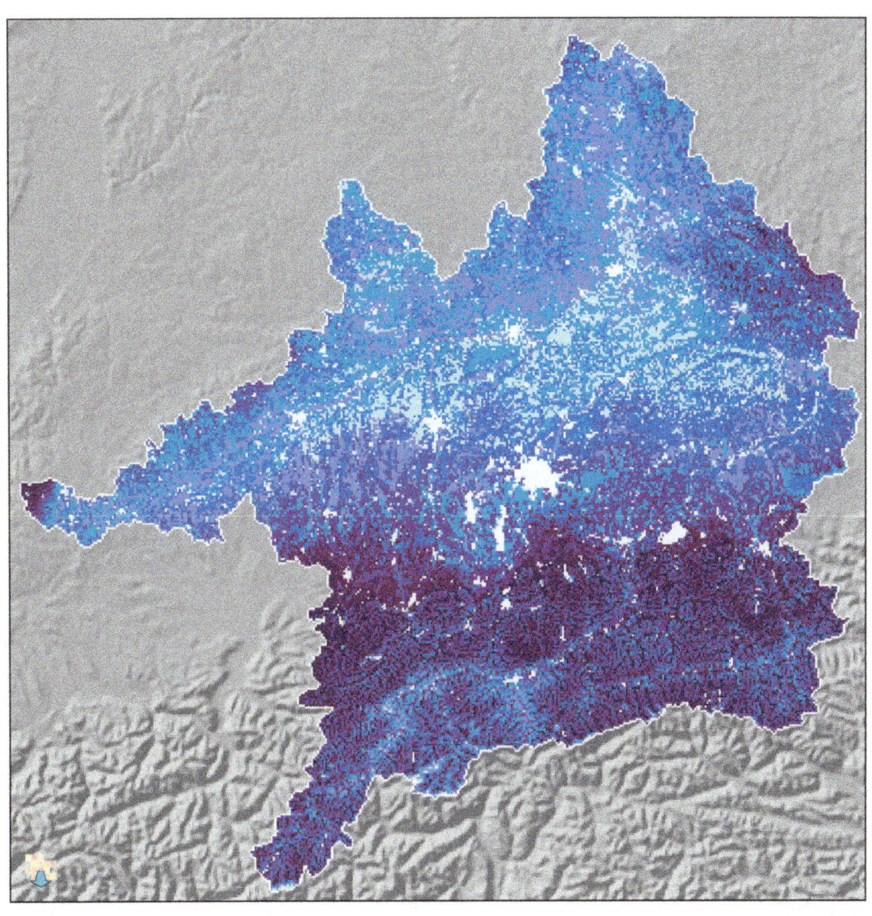

Groundwater recharge per km² in the reference period 1971-2000

0 15 30 45 60 km

[mm/year]	[m³/day*km²]	[l/sec*km²]
0	0 - 2.74	0 - 0.032
> 0 - 100	> 2.74 - 274	>0.032 - 3.17
> 100 - 250	> 274 - 684	> 3.17 - 7.92
> 250 - 400	> 684 - 1,095	> 7.92 - 12.68
> 400 - 550	>1,095 - 1,506	>12.68 - 17.43
> 550 - 700	>1,506 - 1,916	>17.43 - 22.18
> 700 - 900	>1,916 - 2,464	>22.18 - 28.52
> 900 - 1,150	>2,464 - 3,149	>28.52 - 36.44
>1,150 - 1,400	>3,149 - 3,833	>36.44 - 44.36
> 1,400	> 3,833	> 44.36

Map 24.1 Groundwater recharge

References

Brooks RH, Corey AT (1964) Properties of porous media affecting fluid flow. J Irrig Drain Div Am Soc Civil Eng IR2:61–88

Eagleson PS (1978) Climate, soil, and vegetation, 3. A simplified model of soil-movement in the liquid phase. Water Resour Res 14:722–730

Philip JR (1960) General method of exact solution of the concentration dependent diffusion equation. Aust J Phys 13:1–12

Richards LA (1931) Capillary conduction of liquids through porous mediums. Physics 1:318–333

Chapter 25
Run-Off Formation

Wolfram Mauser and Ralf Ludwig

Abstract Run-off formation, the spatial processes that lead to water entering the channel network, is described in its representation in DANUBIA. The representation includes direct run-off and interflow, which are a product of infiltration capacity and water movement in the unsaturated soil layers. As a result the exemplary case of the year 1999 as well as a comparison of the simulated run-off formation and measured discharges at selected gauges for the period 1971–2000 is shown.

Keywords GLOWA-Danube • Direct run-off • Interflow

25.1 Introduction

In addition to groundwater recharge, the formation of run-off, which describes all interrelated processes that lead to water entering the channel network of a watershed, is a significant factor in the discharge behaviour of watersheds. It is quantified here as the sum of direct run-off and interflow on or near the ground surface level. The spatio-temporal variation in the effective soil moisture is a key factor for the formation of run-off, since it controls the ability of the soil to retain and store moisture and is thus the determinative basis for the temporal behaviour of run-off, including the development of hydrological extremes.

Model descriptions of run-off formation should follow an approach that is spatially distributed to represent the spatially variable nature of run-off formation, based on physical principles. The parameters characterising the process at each location in the watershed should further not be calibrated to historical streamflow measurements in order to be able to realistically simulate and map the potential changes in run-off formation that result from future climate change.

Direct run-off is controlled by the hydraulic properties of the soil, the land cover (e.g. sealing) and the patterns of precipitation, while interflow is also controlled by

W. Mauser (✉) • R. Ludwig
Department of Geography, Ludwig-Maximilians-Universität München (LMU Munich), Munich, Germany
e-mail: w.mauser@lmu.de; r.ludwig@lmu.de

W. Mauser, M. Prasch (eds.), *Regional Assessment of Global Change Impacts*, DOI 10.1007/978-3-319-16751-0_25

the soil physics and by topography. If the intensity of precipitation exceeds the soil's capacity for infiltration (see Chap. 8), the result is an above-ground infiltration excess that drains off as surface run-off. If the topmost soil layers become saturated, then saturation excess similarly is fed into groundwater recharge or interflow, which is then transferred to the hydraulic neighbour proxel in the watershed.

25.2 Data Processing

See Chap. 24, Groundwater recharge.

25.3 Model Documentation

See Chap. 24, Groundwater recharge.

At the spatial scale selected for the model, it is not assumed that lateral transport of the generated discharge in the form of sheet flow occurs. It is assumed, since at the selected spatial scale (1 km) each proxel directly connects to the drainage network, that all surface flows as well as the interflow generated by the *Landsurface* component *Soil* during one model time interval (1 h) directly enter the receiving waters of the next downstream neighbour proxel in the drainage network. The water flows in the drainage network are simulated in the model network by the *Rivernetwork* model component.

25.4 Results

Map 25.1 indicates the spatially differentiated formation of run-off within the Upper Danube drainage basin. The significance of land use for the process of run-off formation is obvious from the spatial distribution of the discharge elements. In DANUBIA, each proxel is described only by its dominant land use. In the case of populated areas, this leads to high direct flow in impervious urban regions. To a lesser extent, on heavy soils (e.g. clay soils, see Chap. 8), saturation and infiltration excesses occur and interflow is produced on slopes. Larger amounts of interflow are generated only in regions with marked relief (Alps, Bavarian Forest). They can distinctly shape the patterns of discharge in these areas. Figure 25.1 exemplary depicts the course for the discharge components averaged across the entire region for the 1999 hydrological year. The high level of direct run-off is indicative of the rapid precipitation impulse response in populated areas. The trend in run-off indicates the inertia in the delayed interflow components of flow determined by the soil water dynamics. This is specifically valid for groundwater recharge.

Figure 25.2 is an illustration of the soil water balance. In this case, essential flow components are also shown for the 1999 hydrological year, using deciduous forest

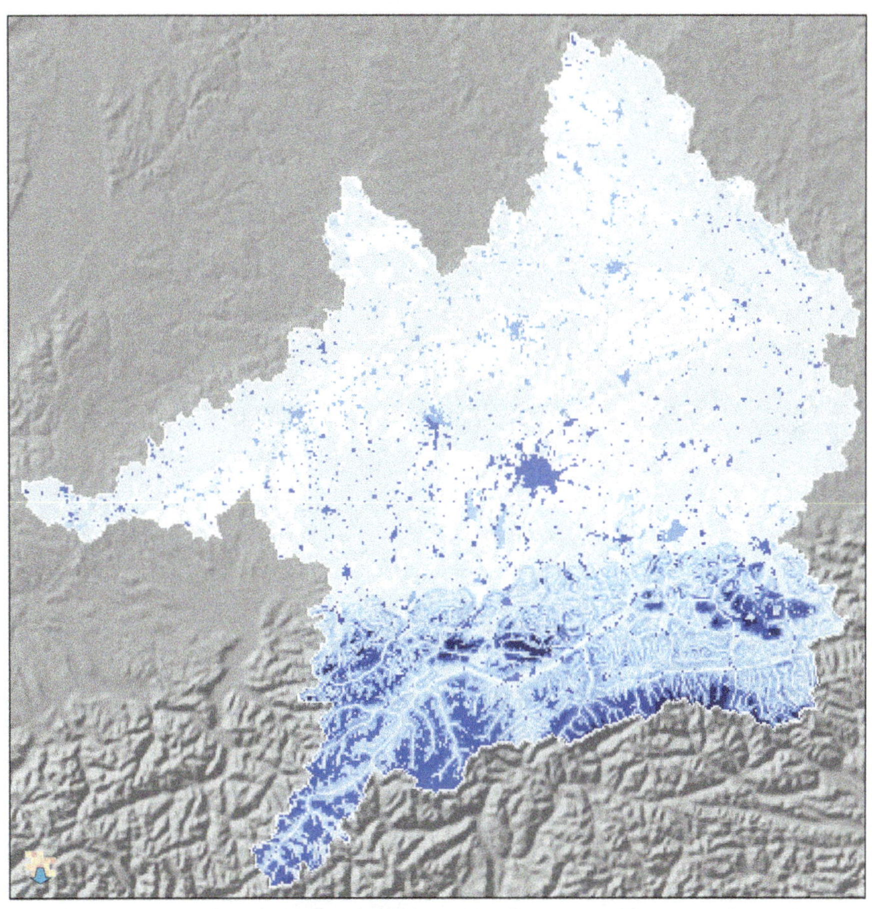

Runoff formation per km² in the reference period 1971-2000

0 15 30 45 60 km

[mm/year]	[m³/day*km²]	[l/sec*km²]
0 - 1	0 - 2.74	0 - 0.032
> 1 - 100	> 2.74 - 274	>0.032 - 3.17
> 100 - 300	> 274 - 821	> 3.17 - 9.51
> 300 - 500	> 821 - 1,369	> 9.51 - 15.84
> 500 - 750	>1,369 - 2,053	>15.84 - 23.77
> 750 - 1,000	>2,053 - 2,738	>23.77 - 31.69
>1,000 - 1,250	>2,738 - 3,422	>31.69 - 39.61
>1,250 - 1,500	>3,422 - 4,107	>39.61 - 47.53
>1,500 - 1,800	>4,107 - 4,928	>47.53 - 57.04
>1,800	>4,928	>57.04

Map 25.1 Run-off formation

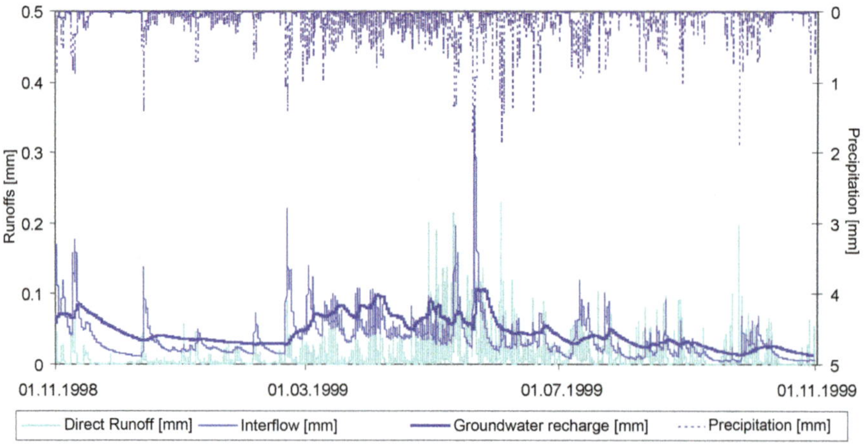

Fig. 25.1 Modelled flow components as mean values for the entire Upper Danube basin for the year 1999

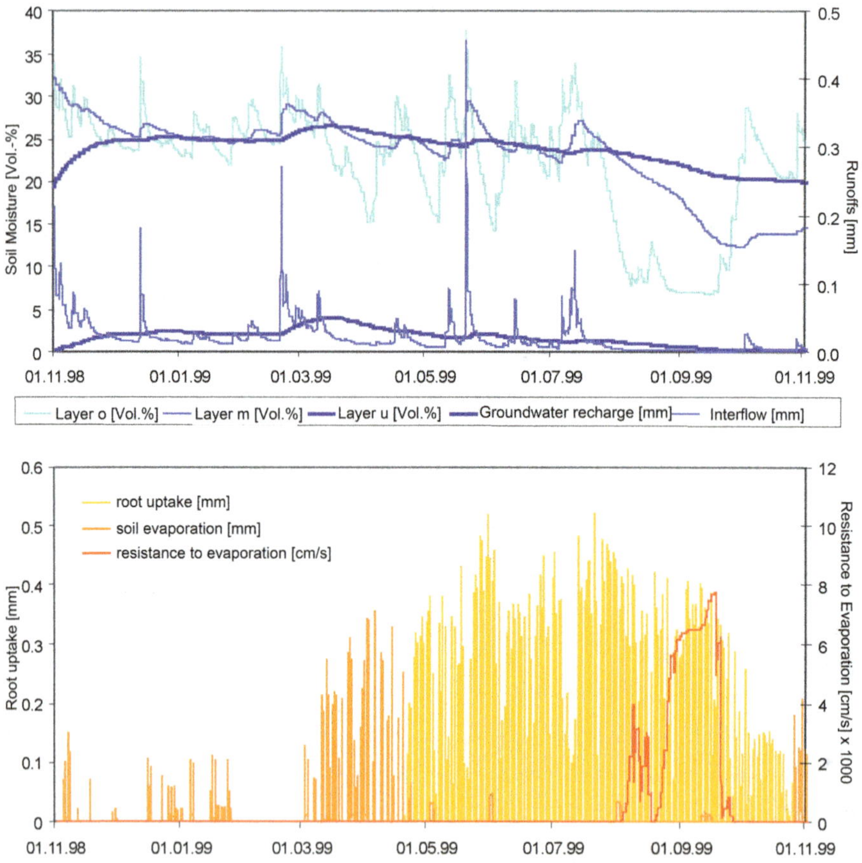

Fig. 25.2 Modelled soil water balance components (position: deciduous forest on loam)

on loam (475 m a.s.l., north of the town of Dillingen, on a slope) as an example. The consideration of soil evaporation and root uptake by plants in the soil water balance is especially apparent in the late summer of 1999. At that time, a longer phase of low precipitation and strong radiation inputs with high evaporation led to a severe drying of the topsoil layers. The topsoil layer is protected from further withdrawal resulting from evaporation by the formation of a resistance to evaporation that is moisture dependent.

25.4.1 Analysis of Long-Term Behaviour

The performance of the water balance component was tested in a 30-year model period (1971–2000) for several validation gauges spread throughout the whole region (see Fig. 25.3). Figure 25.4 presents the statistical relationship between the modelled annual quantities of run-off formation and the measured discharges at each reference gauge. In an ideal case, both should be equal if evaporation from

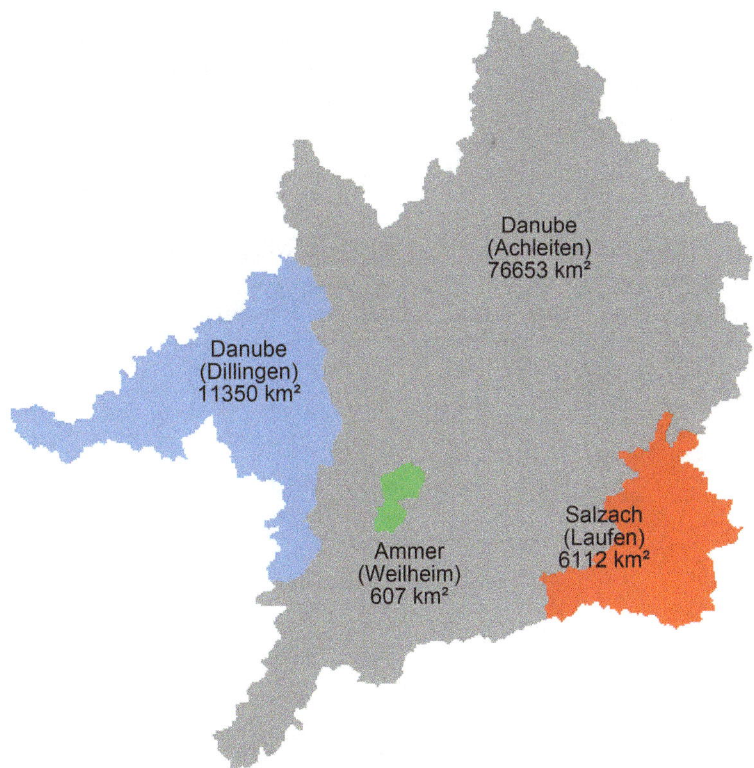

Fig. 25.3 Reference gauges of long-time model time series

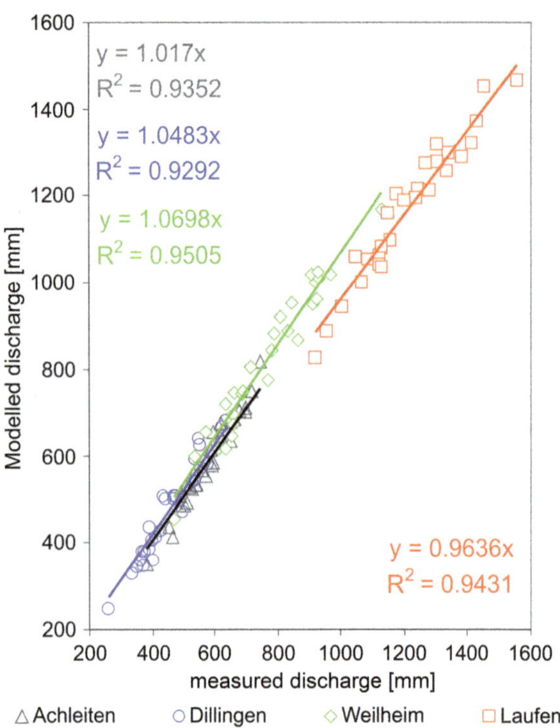

Fig. 25.4 Linear regression and *R^2* of yearly measured discharge and modelled discharge between 1971 and 2000

rivers is neglected. The results confirm good area-wide accuracy in the modelling of annual variation of run-off formation and groundwater recharge based on annually varying rainfall and evapotranspiration patterns. From the figures we conclude that the hydrological processes affecting soil water storage are well represented by the *Soil* model components.

Chapter 26
Groundwater Contour Maps for the Alluvial Aquifers of the Upper Danube Basin

Roland Barthel, Vlad Rojanschi, and Jens Wolf

Abstract Groundwater is the most important source of drinking water in the Upper Danube drainage basin. More than 90 % of drinking water is extracted from wells in shallow quaternary aquifers. The study of the effects of global change on groundwater storage in these aquifers is thus an important step in the attempt to simulate the impact of global change on water resources and water supply. This chapter describes the development of a continuous groundwater table contour map of the uppermost quaternary aquifer in the Upper Danube catchment. These maps show the elevation of the groundwater level above sea level and provide important information needed to assess groundwater movement and boundary conditions. Here we describe the process of collecting base data in various forms from a large variety of sources and the approaches chosen to homogenize and interpolate the data to create the maps presented. The manner in which the resulting datasets are used to conceptualize the groundwater flow model used in DANUBIA is presented.

Keywords GLOWA-Danube • DANUBIA • Groundwater • Upper Danube

R. Barthel (✉)
Department of Earth Sciences, University of Gothenburg,
Göteborg, Sweden
e-mail: roland.barthel@gu.se

V. Rojanschi
Golder Associates Ltd., Calgary, AB, Canada
e-mail: Vlad_Rojanschi@golder.com

J. Wolf
Repository Safety Research Division, Gesellschaft für Anlagen- und Reaktorsicherheit (GRS) gGmbH, Braunschweig, Germany
e-mail: jens.wolf@grs.de

W. Mauser, M. Prasch (eds.), *Regional Assessment of Global Change Impacts*, DOI 10.1007/978-3-319-16751-0_26

26.1 Introduction

Groundwater is the most important source of drinking water in the Upper Danube drainage basin. More than 90 % of drinking water is produced from filtering wells in the water-carrying layers (aquifers) that are close to the surface. These aquifers play a key role in the hydrological cycle as beneficiaries of groundwater recharge and as points of intersection with the rivers. The study of the effects of global change on the groundwater storage in these aquifers is thus an important goal of the GLOWA-Danube project.

As analyses of hydrogeology in the drainage basin indicate (see Chap. 14), the alluvial aquifers (young quaternary sediments in the river valleys; base class 1 in Chap. 14) are the most productive sources of groundwater in the basin. Many significant abstractions come from these aquifers.

Predominantly, hydrogeology makes use of what are known as groundwater contour maps in order to evaluate an aquifer. These maps indicate lines with equal groundwater levels based on a reference level (m a.s.l.). A groundwater contour map provides important information about groundwater movement and flow directions.

A groundwater contour map for the alluvial aquifers is thus a key tool for evaluating the status of and basis for planning and administrating (drinking) water resources. Moreover, groundwater levels and their changes in time and space are useful pieces of information for assessing the ecological roles of groundwater in ecosystems that depend on groundwater (wetlands and riparian zones). However, at the regional scale, this kind of map can only be consulted to evaluate local phenomena (drawdown of individual wells, changes that are specific to location). Local models need to be developed for detailed investigations. However, the groundwater model (*GroundwaterFlow*) prepared within the GLOWA-Danube project and the evaluations based on this model provide the key boundary conditions (lateral tributaries, connection to other aquifers, etc.) for issues of local significance.

26.2 Data Processing

Two types of information are needed to prepare the groundwater model:

1. Static parameters: these describe the subsoil properties that are unchangeable on human time scales – they include data describing the geometry and extent of the groundwater aquifers and aquitards as well as hydraulic parameters that determine the movement and storage of groundwater in the subsoil (see Fig. 26.1).
2. Dynamic parameters: these include all temporally varying variables such as abstractions, groundwater recharge from precipitation and exchange between ground- and surface waters.

Since there is no suitable database for the entire Upper Danube drainage basin at the required extent or at the necessary spatial resolution, the available basic information in many ways must be pooled, interpolated, approximated and above all

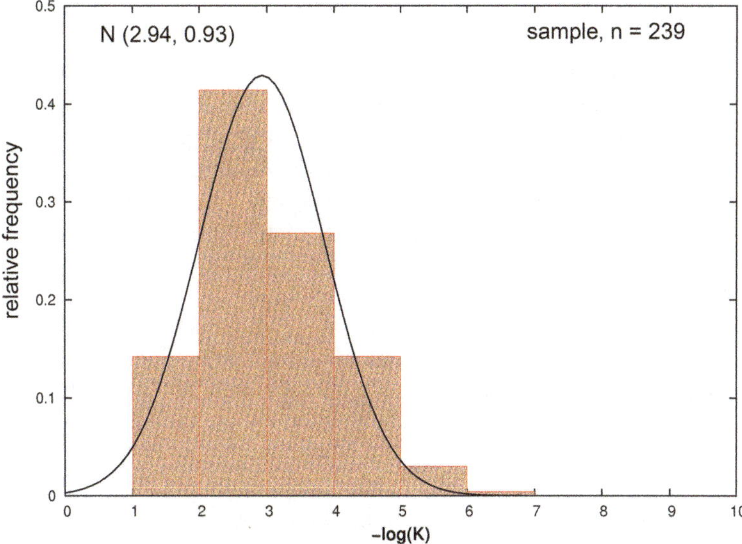

Fig. 26.1 Histogram and adjusted log-normal distribution of measured hydraulic permeabilities

highly abstracted. Therefore, the model shows only limited variability that often does not account for local peculiarities. For example, this is expressed in the summary of the geological properties in the Upper Danube basin into ten hydrological classes (see Chap. 14).

The most important input information for generating a groundwater contour map based on the groundwater model is the groundwater recharge data. In a groundwater model for a complete drainage basin without lateral tributaries, these data provide the entire water input for the groundwater model. In GLOWA-Danube, an averaged groundwater recharge (see Chap. 24) for the period from 1990 to 2000 was calculated from the results of the *Soil* model in the main component *Landsurface*. A transfer function is used to prepare this recharge value for the groundwater model.

In addition to groundwater recharge, the water level in the rivers is another key boundary condition of the model. Like geometry of the aquifer near the surface, this variable is adapted for the DTM 1000 in order to obtain a dataset with hydrological consistency. The presentation of the aquifer geometry on a lower-resolution grid is a special challenge for alluvial aquifers in particular, since their spatial extent is very low compared to the square kilometre grid size chosen for GLOWA-Danube. It is important to compile the following three datasets in an integrative way for a square kilometre grid:

- Aquifer geometry close to the surface
- Geometry of the water body
- Water level in the water body

A detailed presentation of this problem is described in Wolf et al. (2008).

26.3 Model Documentation

The goal of *GroundwaterFlow* within DANUBIA is to calculate the temporal changes in groundwater levels and hence the change in availability of groundwater for the generation of drinking water and for the ecological requirements that depend on natural and temporal changes. The *GroundwaterFlow* model in DANUBIA thus calculates new daily values for all model cells (proxels). Since it is not feasible to present these temporal changes within a single map, it was decided instead to map a stationary model result that is based on inputs averages over the period 1990–2000.

The groundwater model that underlies the *Groundwater* component and its *GroundwaterFlow* model is a finite difference model (FD model). A currently valid international standard was chosen for this problem using *MODFLOW* (Harbaugh et al. 2000). *MODFLOW* is integrated in virtually unchanged form within the Java environment of DANUBIA.

MODFLOW is based on the general groundwater flow equation for an isotropic, inhomogeneous medium in stationary three dimensions:

$$\frac{\partial}{\partial x}\left(T\frac{\partial h}{\partial x}\right)+\frac{\partial}{\partial y}\left(T\frac{\partial h}{\partial y}\right)+\frac{\partial}{\partial z}\left(T\frac{\partial h}{\partial z}\right)=Q$$

x, y, z map coordinates
h groundwater level
Q source or sink term
T transmissivity

In an FD model, this continuous system is transformed using either Taylor series approximation or an integral approach in a system of algebraic equations. The transmissivity T is equal to the product of the thickness filled with water and the hydraulic permeability K of the water-bearing layer. It is constant in confined aquifers and in unconfined aquifers is the product of hydraulic potential (of the groundwater level) h and K.

Thus, K is the key parameter in the stationary model. In general, K is determined using inverse calibration, since there is seldom sufficient data for independent calculation. In the *Groundwater* component, simple zoning was chosen for the alluvial aquifers, and these permeabilities between 10^{-3} and 5×10^{-3} m/s are permitted based on the measurements known from the literature (239 measurements for the entire drainage basin with a mean of 1.2×10^{-3} m/s) (see Fig. 26.1).

26.4 Results

The result of the stationary modelling is the assignment of groundwater levels to the alluvial aquifers. The groundwater contour map is prepared from these results. In Map 26.1, the 10 m isolines and the maximum levels of the groundwater table are presented as means in m a.s.l. for several head watersheds for the decade from 1990 to 2000. Also marked on the map are the regions in which large-scale deep groundwater from the Malm karst raises to the aquifers near the surface. These regions play a key role in evaluating the overall groundwater flow system. A comparison between the measured and calculated groundwater levels is of limited value since the calculated groundwater levels relate to a 1 × 1 km² cell. There can be significant differences in measured groundwater levels at different sites within a cell of this size.

In the context of this limitation, Fig. 26.2 illustrates a comparison between the observed and calculated groundwater levels in the alluvial aquifers. The absolute mean error is 13.2 m or 0.7 % of the total gradient in alluvial aquifers (see Map 7.1 with the maximum value of 2,175 m a.s.l. in the Alps and the minimum of 309 m a.s.l. at the Gauge at Passau-Achleiten).

The resulting contour map reveals results that are comparable to those published by Andres and Wirth (1985).

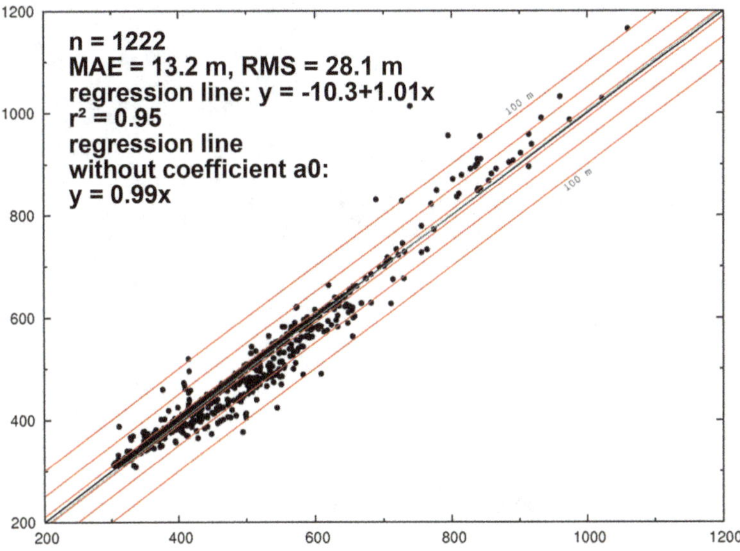

Fig. 26.2 Comparison between 1,222 measured (*x*-axis) and calculated groundwater levels (*y*-axis) in the alluvial aquifers (m a.s.l.)

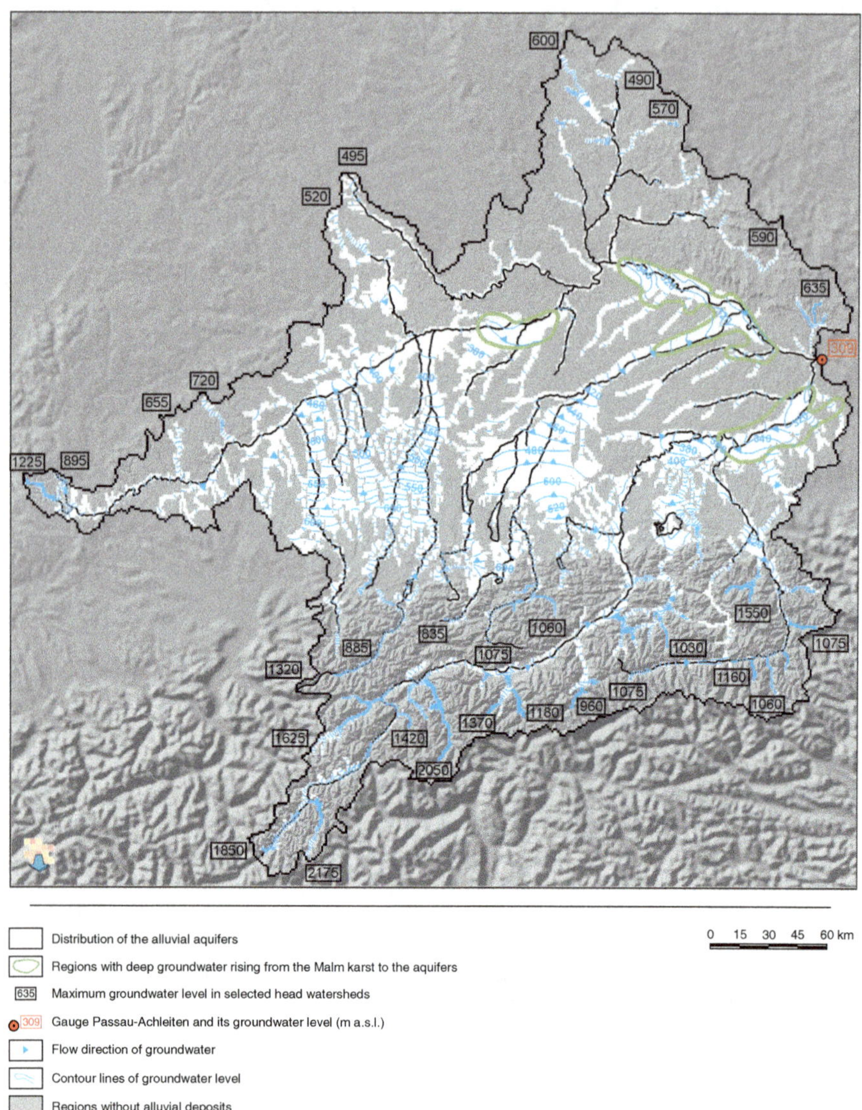

Map 26.1 Modelled groundwater contour maps for the alluvial aquifers of the Upper Danube basin with DANUBIA

Detailed descriptions of the groundwater flow component implemented in DANUBIA and a more specific description of the approaches used to tackle the special challenges of representing the shallow unconfined quaternary aquifer in the numerical model are presented by Barthel et al. (2005, 2008) and Wolf et al. (2008). Barthel (2006) discusses in particular the challenges of coupling the groundwater flow model to the surface water and soil models in DANUBIA.

References

Barthel R (2006) Common problematic aspects of coupling hydrological models with groundwater flow models on the river catchment scale. Adv Geosci 9:63–71

Barthel R, Rojanschi V, Wolf J, Braun J (2005) Large-scale water resources management within the framework of GLOWA-Danube. Part A: The groundwater model. Phys Chem Earth 30(6–7):372–382

Barthel R, Jagelke J, Götzinger J, Gaiser T, Printz A (2008) Aspects of choosing appropriate concepts for modelling groundwater resources in regional integrated water resources management – examples from the Neckar (Germany) and Ouémé catchment (Benin). Phys Chem Earth 33(1–2):92–114

Harbaugh AW, Banta ER, Hill MC, McDonald MG (2000) MODFLOW-2000, The U.S. geological survey modular ground-water model - user guide to modularization concepts and the groundwater flow process. U.S. geological survey, Open-File Report 00-92

Wolf J, Barthel R, Braun J (2008) Modeling ground water flow in alluvial mountainous catchments on a watershed scale. Ground Water 46(5):695–705

Chapter 27
Total Extraction and Total Water Supply per Community

Roland Barthel and Darla Nickel

Abstract The chapter describes a preliminary base version of the *WaterSupply* model which does not have the full functionality of the model version that was finally implemented in DANUBIA. In particular, it does not make full use of the actor concept to simulate human decisions. However, as this preliminary version still forms an essential base module of the later extended model version, it is recommended to read this chapter to get a better understanding of water supply modelling in DANUBIA. This chapter presents the fundamental steps that were taken to conceptualise and implement this base version of *WaterSupply* and explains its main purpose and the interaction with the other models within the DANUBIA system. Four maps exemplarily show selected results of the model which are related to the extraction of raw water and the distribution and consumption of drinking water. The main map shows extraction and consumption aggregated on a community (municipality) level.

Keywords GLOWA-Danube • DANUBIA • Water extraction • Water supply • Upper Danube

27.1 Introduction

The chapter describes a preliminary base version of the *WaterSupply* model which was finally implemented in DANUBIA (see Chap. 28). However, as this preliminary version still forms an essential base module of the later extended model version, it is recommended to read this chapter to get a better understanding of water supply

R. Barthel (✉)
Department of Earth Sciences, University of Gothenburg, Göteborg, Sweden
e-mail: roland.barthel@gu.se

D. Nickel
Deutsches Institut für Urbanistik (Difu), Berlin, Germany
e-mail: nickel@difu.de

W. Mauser, M. Prasch (eds.), *Regional Assessment of Global Change Impacts*, DOI 10.1007/978-3-319-16751-0_27

modelling in DANUBIA. Comprehensive description of this base module and the preparation of data can be found in Nickel et al. (2005) and Barthel et al. (2005a, b).

There are ample supplies of groundwater and surface water in the Danube drainage basin, and only a small fraction of these are used for drinking water or industrial purposes. However, these supplies are distributed unevenly in terms of both quantity and quality. While some regions suffer from water scarcities or must process water at great expense, other regions enjoy a surplus of good quality water (see Chap. 20). Global change – be it, for example, climate change, population change or changes to the patterns of settlement and demand – can lead to a shift or intensification of the regional differences and to conflicts of use. The following related questions serve both the interest of those responsible for the supply of drinking water and water consumers:

- Where is water extracted – today and in the future – for drinking water and for industrial purposes, and where is it used and by what user groups?
- Which communities or regions can supply their own needs from local resources, and which must resort to importing water?
- Are the current structures for the supply of drinking water adequate, or where and under what circumstances might there be supply bottlenecks?

In the GLOWA-Danube project, the overall view on the status of the supply and use situation is the mandate of the *WaterSupply* model within the sub-project Groundwater balance and management and water supply. In addition, the model and excerpts from the results of the model are presented.

27.2 Data Processing

The *WaterSupply* model provides per proxel values for groundwater and surface water extractions. The demand values for individual user groups (household, agriculture, etc.) are input data for the *WaterSupply* model, not results. At the same time, however, they can only be presented from *WaterSupply* with reference to each other or as pooled information and are therefore summarised here as "results". In the following, the term "supply" is used in place of "demand", as this term better describes the activities of the water suppliers. Internal water supply by the industrial sector, even if it is not entirely applicable, is likewise referred to as "supply". In the maps, all proxel values from the months in 1998 are averaged and aggregated over each community area. Pooling to communal level takes into account the fact that the supply of drinking water in every country within the Upper Danube basin is a communal responsibility. It also facilitates a comparison with the data on water extraction and supply from the state statistical offices for the reference years 1995–1999, which are mostly available at the level of political units such as community, district, etc.

27.3 Model Documentation

The *WaterSupply* model belongs to the group of so-called actor models that each represents a user group. The goals include:

- Ascertaining the availability of drinking water with respect to quality and quantity
- The most accurate as possible determination of extraction from groundwater and surface water (status quo) for public drinking water supply and for industrial purposes
- Compliance with limitations on extraction for surface water that are calculated by the *Rivernetwork* model
- Communication of the quantitative and qualitative status of the water bodies to the other actor models

The *WaterSupply* model is based on communication of the modelled water supply companies (hereafter WSC) with the communities they supply. A WSC is a unique object identified by an identification number or code; this object is initialised with specific information. Key pieces of information are the well or river proxel (known as the source proxel) that is available to the WSC for extraction of water and the maximum technical/legal extractable volume of water (=capacity) within that proxel. Similarly, a community has a unique object identifier (community ID). The important attributes of a community include a list of the proxels that lie within the region of the community (consumer proxel) and a list of the WSC from which the community can draw water in order to supply its region. The WSCs are divided into three categories (see Fig. 27.1).

There is a supply hierarchy consisting of three types of suppliers within the entire Upper Danube basin, as shown in Fig. 27.1; the exception is the Swiss portion (Grisons), where drinking water is supplied exclusively through community suppliers. This task of this supply network is above all to ensure supply security in the event of fluctuating demand, as well as to supply regions with water shortages. The mapping of this supply structure is crucial for assigning the extractions to the correct proxels in the *Groundwater* and *Rivernetwork* models.

The *WaterSupply* model operates according to the following scheme, presented here in simplified form: the first step involves aggregating all drinking water demands for each community over the entire community area. These totals are sent to the first assigned WSC that can balance the supply (max. technical/legal extraction

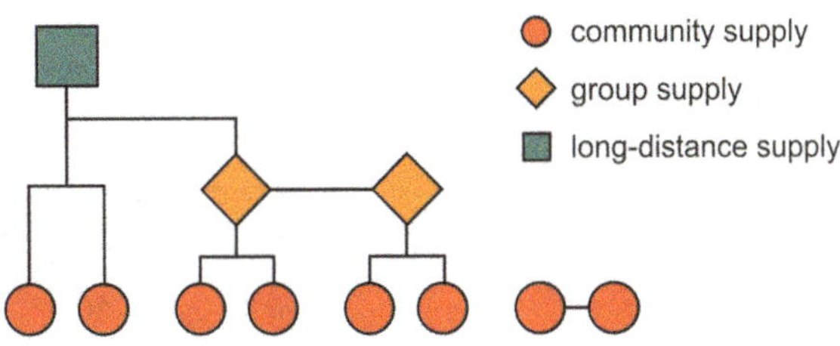

Fig. 27.1 Hierarchies and types of water supply companies in the water supply model

quantity in all source proxels) and demand. A potential deficit is returned to those communities, and they then have the option to relay the unmet demand to a second WCS; this WSC can then similarly balance the supply and demand. The defined allocation of communities to WSC and the source proxels with capacities to WSC ensure that demand is met at least in all reference years (1995–1999) and that extractions take place at the actual extraction sites despite the complex and hierarchical structure of the supply system.

The groundwater and surface water demand of the industrial sector is supplied directly from the proxel in which the demand exists. Surface water is given priority: if the proxel is a river proxel, the demand is supplied from the river, and if it is not a river proxel, the source is groundwater. This falls entirely within the responsibility of the *Economy* model. However, industrial extraction is relayed to the *WaterSupply* model, as the maximum possible extractions per proxel are independent of the type of use. This means that the volume of all surface and groundwater extractions per proxel must be combined within one model and forwarded to the scientific partner model. At the same time, the calculation of the status and elasticity of the water resources necessitates the knowledge of all extractions.

The *WaterSupply* model provides for the fact that the WSC can adapt to conditions that change as a result of global change in order to achieve the goals stated earlier. For example, they can expand or limit their use of specific water resources or increase the water use from another company. This requires the evaluation and communication of the qualitative and quantitative status of the water body. In DANUBIA, defined sections of the water body are divided into a five-level qualitative scale (very good to very poor) where quantitative values underlie the allocation. Assessment of the status of the water body and the adjustment by the WSC to the altered conditions are topics discussed in Chap. 64 of this atlas.

27.4 Results

Four maps are presented which depict the mean water extraction from groundwater and surface water sources for the year 1998 as well as the mean water supplied. The values calculated for the proxels were summed for each community in the drainage basin. Total groundwater extraction (Map 27.1(1)) is composed of

Map 27.1 (continued) e.V. (BGW) Bonn: 112. Statistik Wasser, 2001; EcoGIS, Web-based Map Browser for Environmental Data of Switzerland, BUWAL, 2001–2003, Swiss Agency for the Environment, Forests and Landscape; Operation results of Austrian Waterworks 1999. Statistik DG 1. ÖVGW Austrian Association for Gas and Water, Vienna; Data on withdrawal of water of the Austrian Federal Ministry for Agriculture, Forestry, Environment and Water Management, Strategy Paper "Grundwasserentnahmen", 2004, and "Lage und Abgrenzung von Grundwasserkörpern", 2003; Geographical Information System of Salzburg SAGIS, Office of the Provincial Government of Salzburg, 2004; Data on water extraction and supply, Office of the Upper Austrian Provincial Government, 2004, borderlines: ESRI Deutschland GmbH, Kranzberg, Germany http://ESRI.de, ©BEV – Bundesamt für Eich- udn Vermessungswesen 2014, GG85 ©Swisstopo)

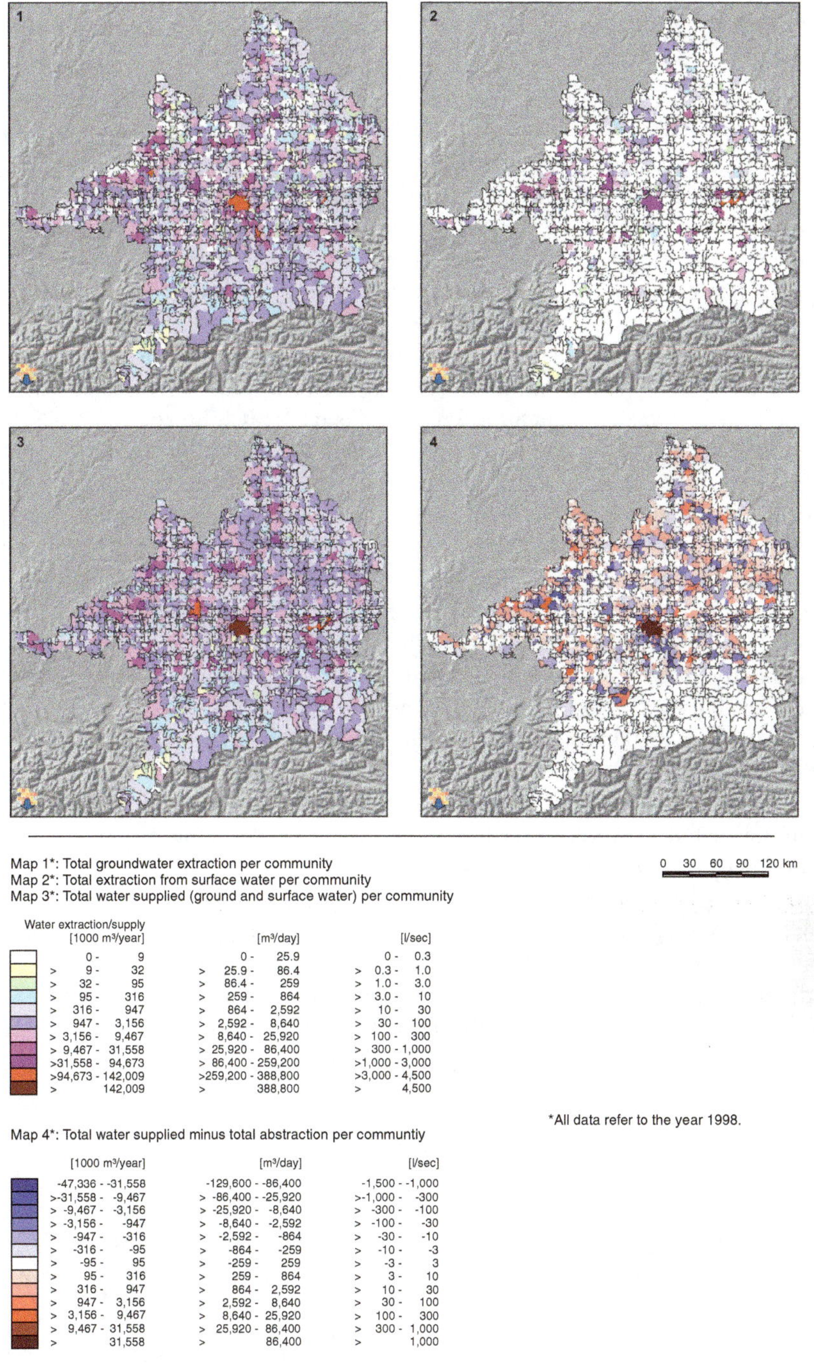

Map 27.1 Total extraction and total water supply per community (Data source: Bavarian State Office for Statistics and Data Processing, Munich 1998; Bavarian State Office for the Environment (LfU), former Bavarian State Office for Water Management 1998; State Statistical Office Baden-Württemberg, Stuttgart 1998; District Offices: Alb-Donau, Biberach, Breisgau-Hochschwarzwald, Heidenheim, Ostalb, Ravensburg, Reutlingen, Schwarzwald-Baar, Sigmaringen, Tuttlingen, Zollemalb, city of Ulm 1998–2003; Federal Association for German Gas and Water Industry

extraction for drinking water and industrial purposes combined. In contrast, the extraction from surface water (Map 27.1(2)) is almost exclusively for industrial purposes. The greatest extraction on both maps is in and around the large cities such as Munich or Augsburg, the northern Alpine fringe as well as along the Swabian Jura in the northwest. Communities with regions where there is no extraction for drinking water and the entire demand is covered by importing water clearly increase from south to north in association with the decreasing availability of water.

The total water supplied (Map 27.1(3)) is also composed of the total output for public drinking water plus the internal supply of the industry. The total water supplied (drinking water plus industrial use) minus the total abstraction (Map 27.1(4)) highlights communities with a net export (communities with blue shading), with a net import (communities with red shading) or with neither import nor export. The white areas in Map 27.1(4) extraction = water supplied) are conspicuous in the southern areas of the basin. These community areas are generally in Switzerland and Austria. They are primarily the result of insufficient data and associated lack of processing for these countries. For example, for the largest Austrian state considered (Tyrol), the extractions are known per aquifer or groundwater unit and not per community. This information is disaggregated to the community using the calculated demand data. Thus, it is already implicitly assumed in the data processing that the abstraction and the demand are equal. A detailed description of the preparation of the database can be found in Chap. 20.

Overall, this map of water extraction and supply at communal level is the first and only detailed overview on the water supply in the drainage basin. Previous overviews are significantly more aggregated and cannot be based on a similarly comprehensive database. However, the mapping and simulation of the status quo is not the primary goal of the project. It is far more important to localise and quantify the changes relative to this reference status in simulated scenarios in order to finally determine the regions in which crises may arise in the future. Likewise, the simulated calculations should support the most efficient, goal-oriented and timely use of preventative measures possible.

References

Barthel R, Nickel D, Meleg A, Trifkovic A, Braun J (2005a) Linking the physical and the socio-economic compartments of an integrated water and land use management model on a river basin scale using an object-oriented water supply model. Phys Chem Earth 30(6–7):389–397. doi:10.1016/j.pce.2005.06.006

Barthel R, Nickel D, Wolf J, Rojanschi V, Trifkovic A, Braun J (2005b) The water supply situation of the alpine region and its representation in the Integrated River Basin Management Model DANUBIA. Landschaftsökologie und Umweltforschung 48:273–283

Nickel D, Barthel R, Braun J (2005) Large-scale water resources management within the framework of GLOWA-Danube – the water supply model. Phys Chem Earth 30(6–7 Spec Issue):383–388. doi:10.1016/j.pce.2005.06.004

Chapter 28
Modelling the Effects of Global Change on Drinking Water Supply: The DeepWaterSupply Decision Model

Roland Barthel and Darla Nickel

Abstract The DeepWaterSupply model described in this chapter presents an extension of the preliminary base version of the WaterSupply model presented in Chap. 27. It is based on the DeepActor approach presented in Chap. 3. Whereas the preliminary model strategy (see Chap. 27) is based on a fixed assignment of extraction sites (supply) and communities (demand) to each water supply company (WSC) and on a fixed maximum withdrawal, the DeepWaterSupply extension allows a dynamic, context-specific adaptation of the infrastructure or a flexible use of resources. The Actors in this deep model are the water supply companies (WSC), who have the main goal to guarantee supply to all connected users. The WSC observe demand and supply. In case the demand exceeds the supply, the WSC takes actions, i.e. it chooses from predefined plans. Whether or not a plan is feasible and how it is performed depends on a variety of dynamic and static variables, including, for example, the status of the groundwater resources. This chapter describes the main components of the DeepWaterSupply model and the typical workflows of decision making. The most decisive factors influencing a WSCs decision to perform certain plans and actions are explained. Results are shown for different climatic and societal scenarios.

Keywords GLOWA-Danube • DANUBIA • Water supply • Upper Danube • Actor-based modelling • Decision modelling

---

R. Barthel (✉)
Department of Earth Sciences, University of Gothenburg,
Göteborg, Sweden
e-mail: roland.barthel@gu.se

D. Nickel
Deutsches Institut für Urbanistik (Difu),
Berlin, Germany
e-mail: nickel@difu.de

W. Mauser, M. Prasch (eds.), *Regional Assessment of Global Change Impacts*, DOI 10.1007/978-3-319-16751-0_28

28.1 Introduction

The *DeepWaterSupply* DeepActor model described in this chapter presents an extension and ultimately a replacement of the *WaterSupply* "shallow" model presented in Chap. 27. It is based on the DeepActor approach presented in Chap. 3. In order to improve understanding of the concept of *DeepWaterSupply*, rereading the text from Chaps. 20 (basis of the data), 27 and 3 is recommended.

In principle, the goals of the modelling, the causes and the effects of the shallow model and the deep model expansion are the same as in the previous version (Chap. 27): the water demands calculated by the actor models are forwarded to the sites of extraction (*Groundwater* and *RiverNetwork* models), mediated by the water supply companies (WSC). The "shallow" model strategy (see Chap. 27) is based on a fixed assignment of extraction sites and communities to the WSC and on fixed maximum withdrawal quantities. If the demand exceeds the capacity assigned to a WSC in the shallow model, a shortage results that cannot be meaningfully interpreted under such a rigid concept. It should however be assumed that in reality changes to conditions result in a dynamic adaptation of the infrastructure or a flexible use of resources. In order to realistically and flexibly depict the relationships that arise even under the conditions of the scenarios, the shallow model approach needs to be expanded with a "deep" and dynamic component. This expansion is activated if a WSC can no longer meet the demands of the users it supplies with the options currently available to it. The *DeepWaterSupply* model accomplishes this using what is known as the DeepActor framework. This framework and the concept underlying the DeepActor modelling in DANUBIA are described in Chaps. 2 and 3.

28.2 Data Processing

An expansion of the data basics described in Chaps. 20 and 27 for the shallow model requires some additional information for the deep model:

Zones: The "zone" concept is tied to the "flag" concept that is described in Chap. 3 (see also Barthel (2011) and Barthel et al. (2010)). A zone represents a "homogenous" groundwater body that responds in a uniform manner. It is believed that any change in boundary conditions (withdrawal, groundwater recharge or climate) equally influences the groundwater supply in a zone. This highly simplifying conceptual assumption is made since determining supply for each proxel is impossible using the current data. Using hydrogeological and hydrological criteria, 405 zones with a mean area of 190 km^2 were defined within the basin. Each zone is assigned characteristic parameters based on its hydrogeological properties (groundwater level, permeability, storage capacity). The parameters include:

Characteristic reaction time: For how long must an indicator be observed in order to determine a significant change?

Weight: What is the significance of each indicator for a specific groundwater body?

The indicators considered for the available groundwater quantities are groundwater level, groundwater recharge and base flow (see below).

28.3 Model Documentation

The expanded *DeepWaterSupply* model is strictly based on the DeepActor framework associated with all DeepActor models and on the multi-actor strategy (MAS) jointly developed by all the socio-economic groups. The technical implementation of the DeepActor framework is described in Chap. 2 (DANUBIA) and the underlying MAS strategy is described in Chap. 3 and in more detail also in Barthel et al. (2008). Actors in this deep model are the WSC. These are divided into one of three types (community, group and long-distance water supply) (see also Chap. 27). The overriding goal of a water supply actor is to guarantee supply to all connected users. For this reason, the actor endeavours to observe demand and supply in the context of a sustainable plan and to counteract foreseeable shortages. The observation of demand and supply takes place by means of an evaluation of the transfer parameters to other actor models (demand) and to the *Soil, RiverNetwork* and *Groundwater* models (supply). Supply is analysed and summarised in a five step analysis ("flags", see above)

A new aspect compared to the shallow model (see Chap. 27) is the calculation of "flags" that represent a categorised disclosure of the quantitative and qualitative status of groundwater (and surface water) resources. The flag information includes the quantity available for extraction, the current amount supplied (the term supply includes all technical, physical, ecological and economic aspects) and the risk of future overexploitation. The fundamental concept of flags for exchange of information between the socio-economic and the scientific models in DANUBIA is explained in Chap. 3. A detailed physical description of the flag calculation is presented in Chap. 64 and Barthel (2011).

The *DeepWaterSupply* model calculates two types of flags (see Table 3.2 in Chap. 3):

GroundwaterQuantityFlag (GQF): the quantitative status of the groundwater resources with respect to month and zone

DrinkingwaterQuantityFlag (DQF): the quantitative status of drinking water supply with respect to month and community supplied

A GQF provides information on the groundwater status in a zone (see Sect. 28.2), whereas a DQF describes the status of the drinking water supply within the supply region of a WSC. Thus, the DQF is relevant for all actors that receive water from the public supply (e.g. households) and the GQF is relevant to all direct users (water suppliers, industry, etc.). There may be significant differences between the status of the groundwater and the status of the drinking water communicated by the WSC, depending on the assumed behaviour of the WSC.

The GQF is calculated from the *groundwaterLevel, inExfiltration* (groundwater in- and outflow from surface waters) and *groundwaterRecharge*, which in turn is calculated from the *Groundwater, Rivernetwork* and *Landsurface* components (see Chap. 2). The GQF is calculated using a weighted average of the GQF from all extraction sites of a WSC. The percentage of each quantity in the total supply is the weight factor.

The DQF also comprises information about the behaviour and the preferences of the individual water supply actors. The actors can react "sensitively" to varying degrees. Thus, the decision as to when a groundwater body can no longer be used to expand capacity or for new infrastructure can be made at different GQF values (3, 4 or 5). With a so-called "insensitive" response, for example, the GQF are not taken into account at all. At the same time, the WSC actor decides whether and how he relays the status value of the GQF in the form of the DQF to the end user. The behaviour of the WSC actor is determined by the scenario definitions. Flags are used both internally in the model and are made available as export parameters (see Sect. 28.4).

An additional important extension involves decision making (=plan selection) in the event that the demand of water-consuming actor cannot be met. The water supply actor has four possible courses of action (=plans):

1. *Default behaviour*: no changes to the extraction and supply structures
2. *Increase capacity of existing extraction sites* – if possible. If not, then:
3. *Exploit a new extraction site*. If not, then:
4. *Crisis management*: unspecified measures such as water delivery in tanker trucks.

The question of which plans can be realised depends on a variety of factors:

- The type of WSC actor and the resources and capacities attributed to it.
- The status of the resource, i.e. the flag values in zones from which the WSC extracts or may extract.
- The sensitivity of the WSC to critical groundwater conditions in a form that can be selected by scenarios. For example, this is controlled by provisions as to which flag values are used as thresholds for discontinuing withdrawal in a zone.

28.4 Results

Beyond the various results of the shallow model (see Chap. 27), the DeepActor model provides a whole series of relevant results related to planning. It should be emphasised in particular that the model results should not be interpreted and used on the local level (e.g. for assessing a well or a single WSC), since the database does not permit this. Interpretation of short-term fluctuations is also not feasible. The results indicate regional tendencies that occur in the mid- to long term. In principle, the following output parameters can be evaluated:

- The spatial and temporal distribution of GQF and DQF flags
- The plan selection by the company

The process of plan selection by one, several or all companies over the course of a model run indicates whether companies are forced to adopt measures that require the construction of infrastructure (plans 2 and 3) or emergency measures (plan 4).

Map 28.1 shows the results of the flag calculations as these have the highest information content from an integrative point of view. A series of eight maps of the spatial distribution of GQF and the DQF is depicted. Considering the presentability, it is somewhat awkward that the flags show only limited variability under the conditions of the past and under the assumption of plausible climate scenarios. The status of the groundwater resources (GQF) in the past was very good except for very few and very brief exceptions (e.g. 1976, 2003). The supply of drinking water in regions with unfavourable hydrogeological conditions has been assured for a long time via the transfer of water. In order to clarify the principles and the goals of the calculations, a "business as usual" climate scenario and a "normal" behaviour by the WSC form the basis of the maps; this approach assumes a continuation of the climatic trend from the past three decades and that a WSC reacts to smaller fluctuations in the groundwater level neither in a manner that is extremely "sensitive" nor extremely "ignorant".

It is clear that a slight deterioration in the status of the groundwater resources (increasing GQF values) can be compensated for by the highly developed supply network; this means that, in general, the status of the supply of drinking water in the drainage basin remains good to very good.

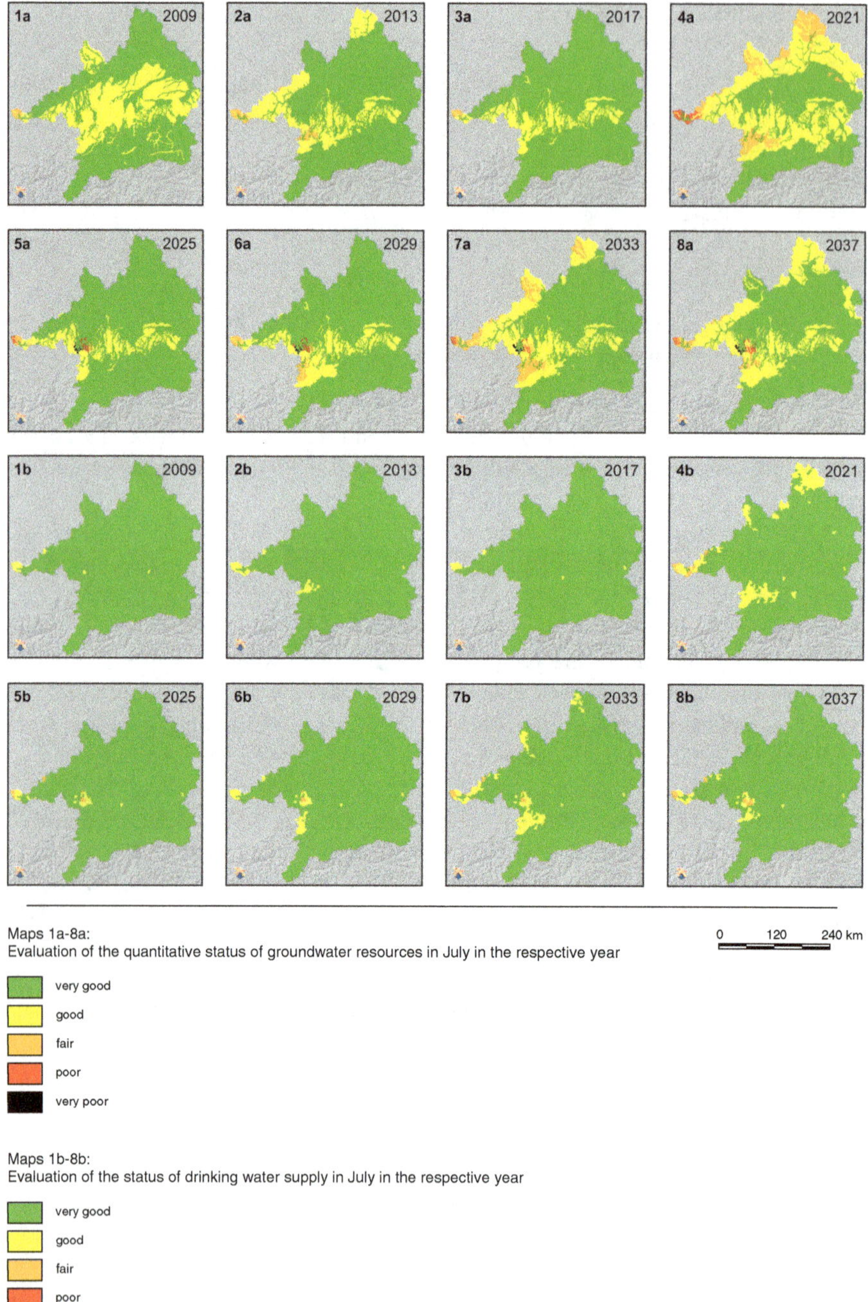

Map 28.1 Test study for the calculation of the quantitative status of groundwater resources and drinking water supply (Data source: Bavarian State Office for Statistics and Data Processing, Munich 1998; Bavarian State Office for the Environment (LfU), former Bavarian State Office for Water Management 1998; State Statistical Office Baden-Württemberg, Stuttgart 1998; District

References

Barthel R (2011) An indicator approach to assessing and predicting the quantitative state of groundwater bodies on the regional scale with a special focus on the impacts of climate change. Hydrogeol J 19(3):525–546

Barthel R, Janisch S, Schwarz N, Trifkovic A, Nickel D, Schulz C, Mauser W (2008) An integrated modelling framework for simulating regional-scale actor responses to global change in the water domain. Environ Model Softw 23(9):1095–1121

Barthel R, Janisch S, Nickel D, Trifkovic A, Hörhan T (2010) Using the multiactor-approach in Glowa-Danube to simulate decisions for the water supply sector under conditions of global climate change. Water Resour Manag 24(2):239–275

Map 28.1 (continued) Offices: Alb-Donau, Biberach, Breisgau-Hochschwarzwald, Heidenheim, Ostalb, Ravensburg, Reutlingen, Schwarzwald-Baar, Sigmaringen, Tuttlingen, Zollemalb, city of Ulm 1998–2003; Federal Association for German Gas and Water Industry e.V. (BGW) Bonn: 112. Statistik Wasser, 2001; EcoGIS, Web-based Map Browser for Environmental Data of Switzerland, BUWAL, 2001–2003, Swiss Agency for the Environment, Forests and Landscape; Operation results of Austrian Waterworks 1999. Statistik DG 1. ÖVGW Austrian Association for Gas and Water, Vienna.; Data on withdrawal of water of the Austrian Federal Ministry for Agriculture, Forestry, Environment and Water Management, Strategy Paper "Grundwasserentnahmen", 2004, and "Lage und Abgrenzung von Grundwasserkörpern", 2003; Geographical Information System of Salzburg SAGIS, Office of the Provincial Government of Salzburg, 2004; Data on water extraction and supply, Office of the Upper Austrian Provincial Government, 2004)

Chapter 29
Surface Water: Discharge Rate and Water Quality

Winfried Willems, Georg Kasper, Peter Klotz, Konstantin Stricker, and Astrid Zimmermann

Abstract Global changes in climate and land use influence both the quality and quantity of surface water. In the surface water sub-project, discharge and water quality parameters for the Danube drainage basin water network up to gauge Achleiten were calculated using the Rivernetwork component and then made available for other models such as Economy, Tourist and Biological. The main component of the Rivernetwork flow model is a Diffusion Analogy Flow approach, developed by the US Geological Survey (DAFLOW) and adapted to the grid-level processing required here. In addition concepts to model headwater regions and small water bodies, sewer systems, lakes, reservoirs and transfer systems are implemented. The main component for modelling water quality is based on the USGS Branched Lagrangian Transport Model (BLTM), and the components considered are water temperature, dissolved oxygen, a sum parameter for oxygen demand, organic nitrogen, nitrate, nitrite and ammonium. The main map depicts the mean routed discharges within the Upper Danube basin for 1995–1999. The accompanying map indicates the largely satisfactory validation of the DAFLOW model based on Nash-Sutcliffe coefficients

Keywords Water flow • Water quality • DAFLOW • BLTM • GLOWA-Danube • DANUBIA

29.1 Introduction

Global changes in climate and land use influence both the quality and quantity of surface water. Changes to the quality of water or the number, duration and intensity of high and low flow phases affect the use and availability of water. This impact has ramifications for industrial, public and agricultural demands, navigation as well as

W. Willems (✉) • G. Kasper • P. Klotz • K. Stricker • A. Zimmermann
Hydrology, Applied Water Resources Management and Geoinformatics IAWG, Ottobrunn, Germany
e-mail: willems@iawg.de

W. Mauser, M. Prasch (eds.), *Regional Assessment of Global Change Impacts*, DOI 10.1007/978-3-319-16751-0_29

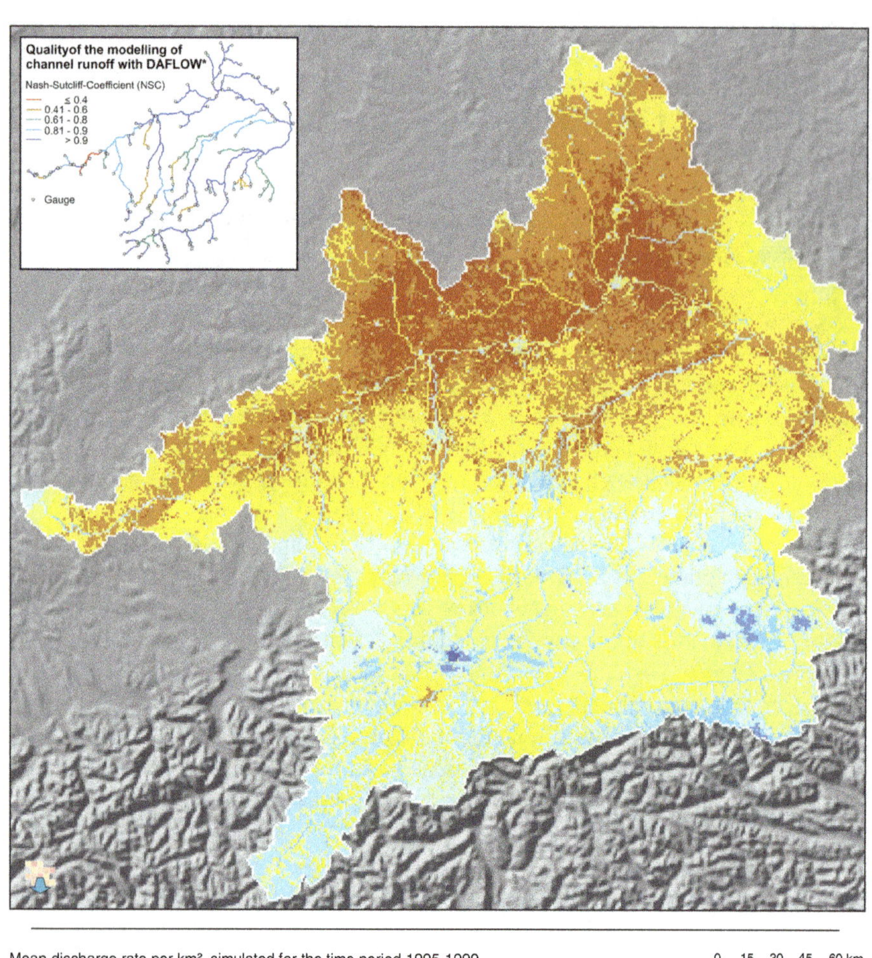

Mean discharge rate per km², simulated for the time period 1995-1999

0 15 30 45 60 km

[m³/sec*km²]	[m³/day*km²]	[l/sec*km²]
0 - 0.0025	0 - 216	0 - 2.5
>0.0025 - 0.005	> 216 - 432	>2.5 - 5.0
> 0.005 - 0.0075	> 432 - 648	>5.0 - 7.5
>0.0075 - 0.01	> 648 - 864	>7.5 - 10
> 0.01 - 0.015	> 864 - 1,296	> 10 - 15
> 0.015 - 0.02	>1,296 - 1,728	> 15 - 20
> 0.02 - 0.025	>1,728 - 2,160	> 20 - 25
> 0.025 - 0.03	>2,160 - 2,592	> 25 - 30
> 0.03 - 0.04	>2,592 - 3,456	> 30 - 40
> 0.04 - 0.05	>3,456 - 4,320	> 40 - 50
> 0.05 - 0.06	>4,320 - 5,184	> 50 - 60
> 0.06 - 0.07	>5,184 - 6,048	> 60 - 70

* Inset map:
Quality of the modelling of channel runoff with DAFLOW for the year 1995.

The discharge data measured at the upstream gauges are used as input data. They are routed to the respective downstream gauges by DAFLOW (see Figure 2.3.1.1).
The Nash-Sutcliff-Coefficient is used as quality criterion, which ranges from negative infinity (no correlation) to 1 (perfect correlation). The results of DAFLOW are predominantly good to very good.

Map 29.1 Discharge rate and quality of the modelling of channel runoff with DAFLOW

on the ecosystems that depend on the water. Flood disasters in recent years indicate that adjusting to discharge conditions is essential when it comes to planning and construction, in order to minimise costly emergency and restorative measures. However, even in the case of less extreme situations, such as the long-term changes in discharge quantities or seasonal shifts in runoff patterns, there are consequences for human and natural systems.

In the surface water sub-project, discharge and water quality parameters for the Danube drainage basin up to gauge Achleiten were calculated using the *Rivernetwork* component and then made available for other models such as *Economy*, *Tourist* and *Biological*. The input data for *Rivernetwork* are principally based on data from the *Landsurface* (surface and interflow), *Groundwater* (in- and exfiltration in water systems) and *Household* (waste water from private households) components.

According to the GLOWA-Danube concept, the lateral movement of water in all proxels is modelled by the *Rivernetwork* component. Headwater regions and water bodies with small drainage basins are mapped using simpler models compared to larger water bodies.

29.2 Data Processing

The central spatial reference for all the modelling within the surface water sub-project is based on the drainage network derived from the elevation model (see Chap. 7). Despite the low resolution of the elevation model for extractions of a river network (1 km edge length), there is a good correspondence between the real and the extracted stream network. Realistic parameterisation is crucial for the applicability of the model explained in further detail below. Thus, the parameterisation strategy must only be drawn from data that are available across the entire drainage basin under consideration. In addition to the elevation model and the dataset drawn from it, key input data include, for example, the flow directions and land use information (see Chap. 9) as well as regionalised values for aggregated discharge quantities related to the gauges (see Chap. 15).

29.3 Model Documentation

29.3.1 Modelling Water Flows

1. Mid-sized and large water bodies: modelling water flows in mid-size to large water bodies (total length approx. 2,000 km) used the DAFLOW (Diffusion Analogy Flow) model developed by the US Geological Survey and adapted to the grid-level processing required here.

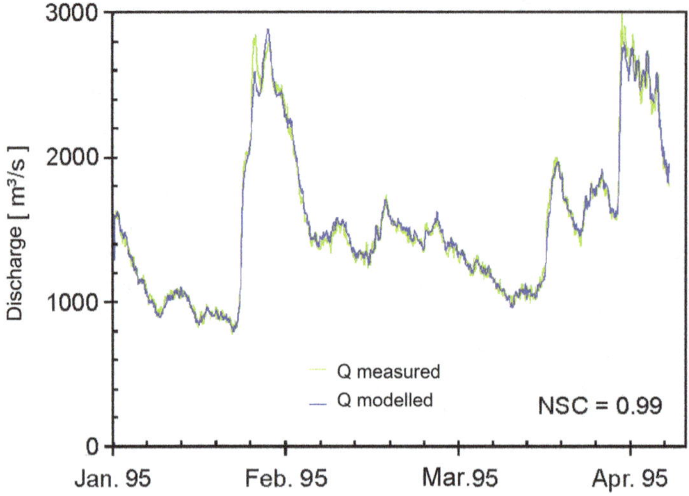

Fig. 29.1 Observed and modelled runoff at the gauge Achleiten, routing of eight upstream gauges

The so-called diffusion analogy forms the basis of the model concept for this method (Jobson 1989; Dyck 1980). The output variables of this model include the runoff values, as well as hydraulic parameters such as water depths and flow speeds; these variables are required for the subsequent modelling of water quality properties. The parameterisation is based on a new strategy (Willems 2004) drawn from regime theory (Leopold 1994). The parameterisation predominantly leads to good or very good mapping of runoff values, as indicated by a comparison of measured and modelled discharges at gauge Achleiten that yielded a Nash-Sutcliffe coefficient of 0.99 (see Fig. 29.1 and the distribution of Nash-Sutcliffe coefficients on the adjacent map). The Nash-Sutcliffe coefficient (NSC) ranges from negative infinity (no correlation) to 1 (perfect correlation).

2. Headwater regions and small water bodies: water flows in headwater regions and small water bodies are modelled within the Rivernetwork object using the Muskingum approach with parameterisation of Cunge (Anderson and Burt 1985). The calculation of the predecessor/successor relationships for all proxels is based on implementing “topological sorting”.
3. Urban sewer systems: the description of water flows in sewer systems requires a modified model approach. The channel network was therefore synthetically generated using a newly developed method for generation (Rödder and Geiger 1996), the data for which included information on the locations of approximately 460 sewage treatment plants (e.g. from GIS_Was, Bavarian State Office for the Environment), approximately 1400 community boundaries and population densities and for which design rules (ATV 1999) were taken into account in a very simplified form.

4. Lakes: to simulate the retention effects of lakes (Chiemsee and Ammersee), the continuity equation was solved numerically and the flow Q was related to the lake water level H using the rating curve $Q = a\,H^b$. The calculation of the coefficients a and b is performed by optimisation. Thus, for example, at lake Chiemsee, a Nash-Sutcliffe coefficient of 0.95 was achieved between the measured and modelled discharge in 1995.
5. Reservoirs: the reservoir model simulates the retention effects of reservoirs such as Sylvenstein, based on simplified operating rules (cp. Ostrowski and Lohr 2002).
6. Transfer systems: The Danube-Main water transfer system transfers water across the natural boundaries of the drainage basin into the catchment area of the Main River. The underlying operating rules that are based, among other factors, on the water levels in the Altmühl, Roth and Brombach lakes as well as the discharges in the Regnitz, Altmühl and Danube rivers, are simulated here in a simplified manner. Since the gauge Hüttendorf/Regnitz that is relevant for management is outside of the Danube basin, the discharge data from it are generated by regression of discharge values for proxels that lie close to the basin boundary and are thus highly correlated with Hüttendorf ($r^2 = 0.87$).

29.3.2 Modelling Water Quality

1. Mid-sized and large water bodies: in mid-sized and large water bodies of the Upper Danube basin, modelling water quality is done using the Branched Lagrangian Transport Model (BLTM) (Jobson and Schoellhamer 1987). The following seven parameters are modelled: water temperature, dissolved oxygen, a sum parameter for oxygen demand, organic nitrogen, nitrate (see Fig. 29.3), nitrite and ammonium.

   The original BLTM method was modified to include a corrective developed for winter temperature; the effect of this modification is clear in Fig. 29.2. The "Shuffled Complex Evolution" method based on the evolution strategy was used to calibrate the model (Duan and Qingqyun 1992).
2. Headwater regions and small water bodies: the boundary conditions for the deterministic modelling of water quality, i.e. the hydrographs for the water quality parameters at the "initial nodes" of the drainage network modelled with BLTM, are prepared using an internally developed statistical method. This method is based on the time-variant expansion of the multivariate statistical regionalisation model drawn from the Group Method Data Handling procedure (GMDH, Ivachnenko and Ivachnenko 1995) suggested by Kishi (2000). In this approach, the water quality parameters (target variables) are modelled as linear combinations of regional characteristics (explanatory variables). The GMDH algorithm serves in this method for the most appropriate selection of the explanatory variables (here: size of the drainage basin, population density, percentage coniferous forest, annual precipitation).

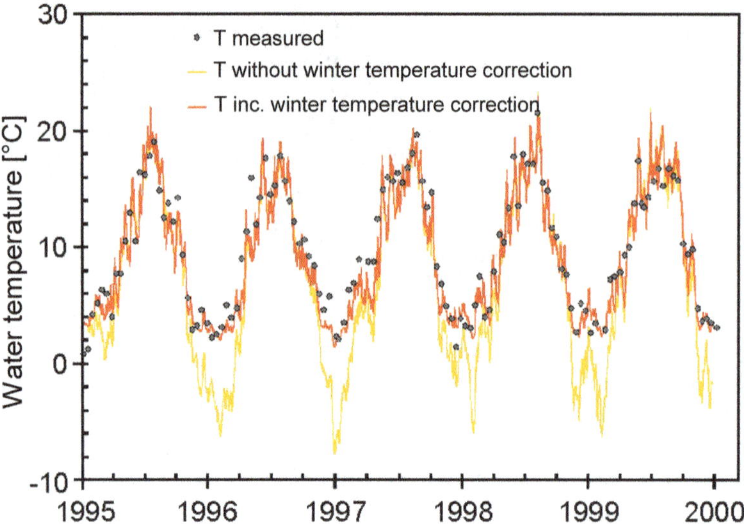

Fig. 29.2 Observed and modelled water temperature at the gauge Fischen/Ammer

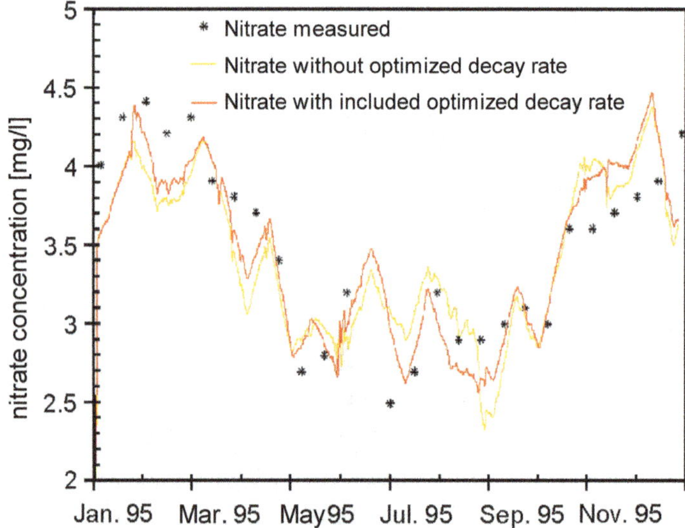

Fig. 29.3 Nitrate concentration at the gauge Passau-Kachlet

29.4 Results

The main map depicts the mean routed discharge rate within the Upper Danube basin for 1995–1999. In the spatial distribution of the discharge rate, the generation of discharge is shaped by the process modified by translation and retention effects of the channels (see Chap. 25). For the drainage outlet at Achleiten, a mean discharge of 20.9 l/s*km^2 (1,590 m^3/s) was calculated for the modelled period. This value adequately matches the measured mean discharge of 18.5 l/s*km^2 (1,420 m^3/s) for the period from 1901 to 2000. The accompanying map indicates the largely satisfactory validation of the DAFLOW model for the year 1995 based on Nash-Sutcliffe coefficients. Both maps are based on a model run in which only the partial models 3.1a and 3.1b were implemented. Thus, the sections for the lakes and reservoirs that were not modelled show greater deviations on the accompanying map.

References

Anderson MG, Burt TP (eds) (1985) Hydrological forecasting. Wiley, New York, p 605

ATV (1999) Hydraulische Bemessung und Nachweis von Entwässerungssystemen. Arbeitsblatt A 118. ATV, Hennef

Duan Q, Qingqyun V (1992) The shuffled complex evolution (SCE-UA) method. Department of Hydrology and Water Resources, University of Arizona Tucson

Dyck S (1980) Angewandte Hydrologie. Teil 2. Verlag für Bauwesen, Berlin

Ivachnenko AG, Ivachnenko GA (1995) The review of problems solvable by algorithms of the group method data handling (GMDH). S Mach Perc 5:527–535

Jobson HE (1989) Users manual for an open-channel stream flow model based on the diffusion analogy, U.S. Geological Survey WRI, 89–4133. Department of the Interior, U.S. Geological Survey, Reston, p 73

Jobson HE, Schoellhamer DH (1987) Users manual for a branched Lagrangian transport model, U.S. Geological Survey WRI, 87–4163. Department of the Interior, U.S. Geological Survey, Reston, p 80

Kishi RT (2000) Modellierung stofflicher Parameter mit Hilfe raumbezogener Daten, vol 9. Schriftenreihe des Institutes für Siedlungswasserwirtschaft, Universität Karlsruhe

Leopold L (1994) A view of the river. Harvard University Press, Cambridge, MA/London

Ostrowski M, Lohr H (2002) Modellgestützte Bewirtschaftung von Talsperrensystemen. Wasser und Abfall 13:40–45

Rödder A, Geiger WF (1996) Berechnungsgrundlagen für Schmutzfrachtberechnungen zur regionalen Darstellung des Stoffaustrags aus Kanalisationen. In: Beichert J, Hahn HH, Fuchs S (eds) Stoffaustrag aus Kanalisationsnetzen. Hydrologie bebauter Gebiete. DFG Forschungsbericht, Weinheim

Willems W (2004) DANUBIA software-documentation. GLOWA-Danube Papers, Techn. Rel. No. 12, Rel.: 0.92_13, Munich, Unpublished

Chapter 30
Mean Snow Cover Duration

Markus Weber and Michael Kuhn

Abstract The temporary storage of precipitation in short term in the form of snow cover has an important function in the water balance of the Upper Danube drainage basin. Among other factors, snow cover depends on the climate variables temperature and precipitation; thus, it is a good indicator for climate change and its impacts for water balance and economic activities. The formation and development of a snow cover is the short-term result of processes of formation and depletion that are controlled by local weather conditions. These processes have a very direct and highly sensitive dependence on the climate parameters radiation, temperature, humidity, wind and precipitation. As a significant location factor, local snow cover duration SD is defined as the number of days with snow cover within a specified period. The results obtained here with the model *Snow* for the spatial distribution of snow cover duration on the 1×1 km^2 grid are quite detailed and potentially even more accurate than statistical extrapolations from observational data on similar maps. The physical-based model *Snow*, which is integrated within DANUBIA, is shortly documented.

Since in general the months from July to October are snow-free at elevations below 3,000 m a.s.l., only the period from November up to and including June is used for the map of snow cover duration. The resulting distribution is plausible. It illustrates not only a distinct correlation between the persistence of snow cover and elevation but also regional dependencies.

Keywords GLOWA-Danube • Snow • Snow cover • Snow cover duration

---

M. Weber (✉)
Commission for Geodesy and Glaciology of the Bavarian Academy of Sciences and Humanities, Munich, Germany
e-mail: Wasti.Weber@kfg.badw.de

M. Kuhn
Institute of Meteorology and Geophysics, University of Innsbruck, Innsbruck, Austria
e-mail: Michael.Kuhn@uibk.ac.at

W. Mauser, M. Prasch (eds.), *Regional Assessment of Global Change Impacts*, DOI 10.1007/978-3-319-16751-0_30

30.1 Introduction

The temporary storage of precipitation both short term in the form of snow cover and long term as glacial ice has an important function in the water balance of the Upper Danube drainage basin. This importance increases as a result of the general decrease in atmospheric temperature with increasing elevation.

The temporary changes in surface cover that arise from snow cover are the subject of this project but are of interest to other disciplines as well, including meteorology, hydrology, agronomy and forestry as well as tourism, because these areas constantly depend on data on the available and predicted snow amounts. Moreover, this variable plays a key role in the water and energy sectors. Among other factors, snow cover depends on the climate variables temperature and precipitation; thus, it is a good indicator for climate change and its impacts for water balance and economic activities.

The formation and development of a snow cover is the short-term result of processes of formation and depletion that are controlled by local weather conditions. These processes have a very direct and highly sensitive dependence on the climate parameters radiation, temperature, humidity, wind and precipitation. Therefore, in the context of scenario calculations with altered weather conditions, snow conditions can no longer be deduced from observations but instead need to be calculated hourly for each proxel using a model that accurately simulates the processes of snow cover formation and depletion using variables for the meteorological conditions. Both the water equivalent for the currently available snow cover and the quantity of meltwater formed (per proxel) are forwarded via the *Surface* and *LandSurfaceControllers* interfaces as input data for other DANUBIA models such as *Tourist* (snow depths in ski areas) and *Rivernetwork* (meltwater discharge).

The model results can only be validated by local observations in the context of the reference runs, which recalculate past weather patterns. In contrast, general changes from climate changes are studied using derived climate variables that remain constant under stationary conditions. The oft-studied duration of snow cover variable is just such a parameter for snow cover. Map 31.1 prepared from the results of the reference run for 1995–1999 provides a basis for analysing the scenario calculations on climate changes.

30.2 Data Processing

As a significant location factor, local snow cover duration SD is defined according to DIN 4049–3 as the number of days with snow cover within a specified period. A snow cover day is a day on which there is snow cover. Normally, SD is derived from measurements or observations of snow depths (KLIWA 2005).

The measurement of snow depth is extremely error-prone. A measurement site is only rarely representative of a larger area. Local rearrangement by wind or avalanches creates large variations in the depths of the layer without the snow mass within a larger area appreciably changing. For this reason, comprehensive and high-resolution mapping of snow cover duration within a larger region simply on the basis of local spot measurements is highly inaccurate. This is particularly true for Alpine regions. Extrapolation to the other points using statistical models (Wielke et al. 2004) only considers some of the processes that contribute to snow cover formation and depletion.

The results obtained here with the model *Snow* for the spatial distribution of snow cover duration on the 1×1 km^2 grid are quite detailed and potentially even more accurate than statistical extrapolations from observational data on similar maps. However, direct comparison is difficult since the model data set is limited to date to four successive winter seasons; this sample size is rather small for a climatology series. In this respect, Map 30.1 cannot be compared with "real" climate maps that are based on 30 years of observations. However, the advantage is the high spatial resolution that cannot be achieved with conventional analyses of station data.

30.3 Model Documentation

The *Snow* model developed for DANUBIA by the Glaciology Working Group is an advancement of the model *PEV* designed according to Escher-Vetter (2000) for the calculation of meltwater production in glaciated drainage basins. The hourly increase in the water equivalent for snow cover is determined using the aggregate state and quantity of precipitation calculated from the atmospheric temperature and humidity close to the ground.

As the key process for the depletion of snow cover, snow melt is calculated using the energy available for this process at the surface, S. This value is typically calculated as the result of the balance of radiation fluxes R, turbulent fluxes H (sensible heat) and HE (latent heat, evaporation and condensation) between the atmosphere and the surface, as well as the usually small thermal conductive flux in and out of the snow cover G, according to the equation:

$$S = R - H - HE - G$$

The radiation fluxes out of the atmosphere are derived externally in this case from measurements and prepared as input data. The energy radiated from the surface and the turbulent sensible and latent heat flux are calculated internally by the model and depend on atmospheric temperature, moisture, pressure and wind. Empirical parameterisation according to Weber (2009) was used; this method is particularly suitable to the special conditions for snow and ice. An important key variable for calculating the energy balance is the surface temperature of the snow cover, which either must be assumed to be 0 °C under melting conditions or a lower value, so that energy fluxes at the surface offset each other.

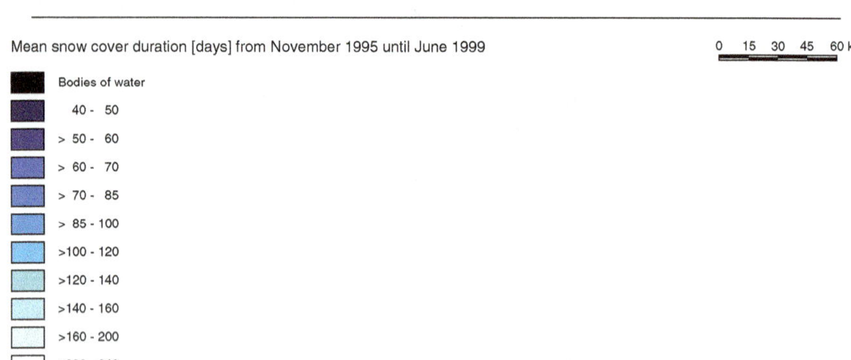

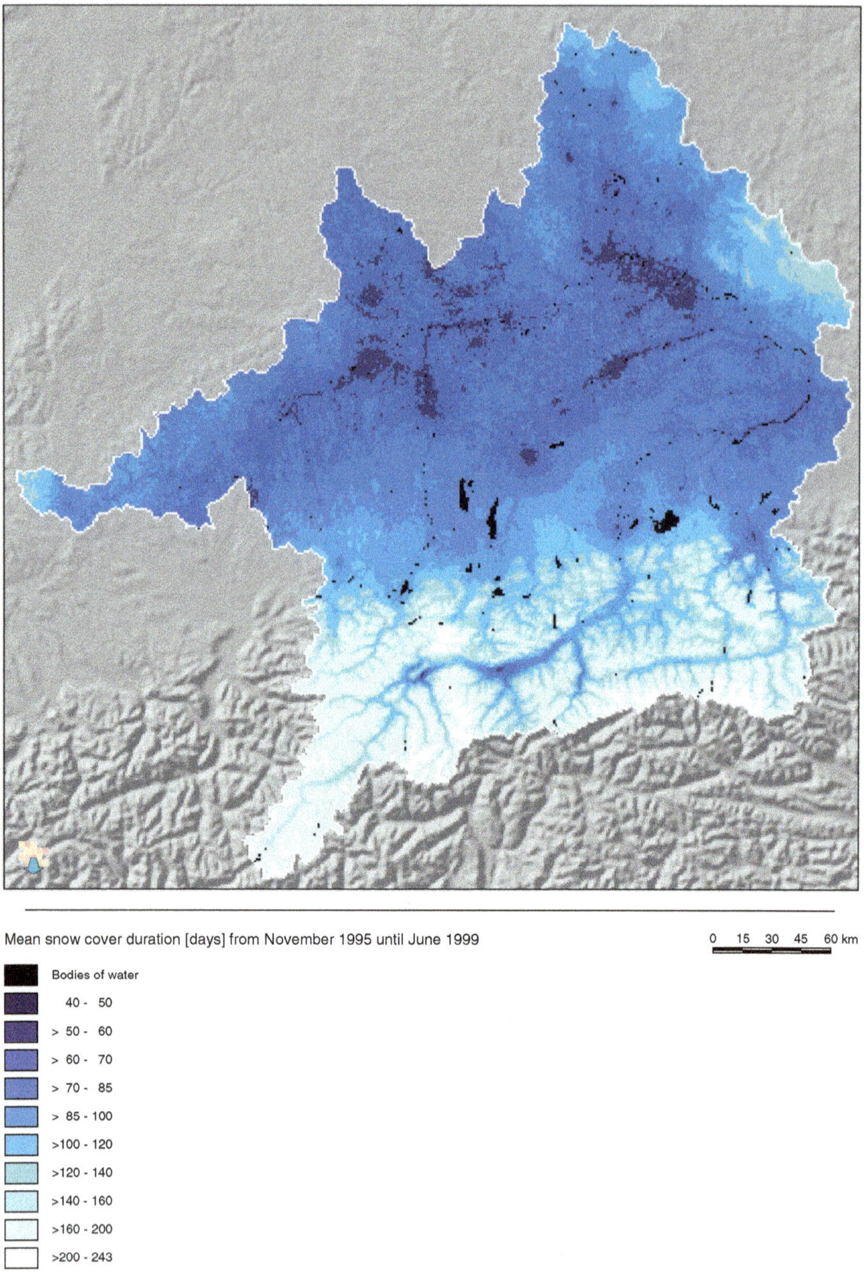

Map 30.1 Mean snow cover duration (water equivalent of snow accumulation calculated by DANUBIA 1995–1999)

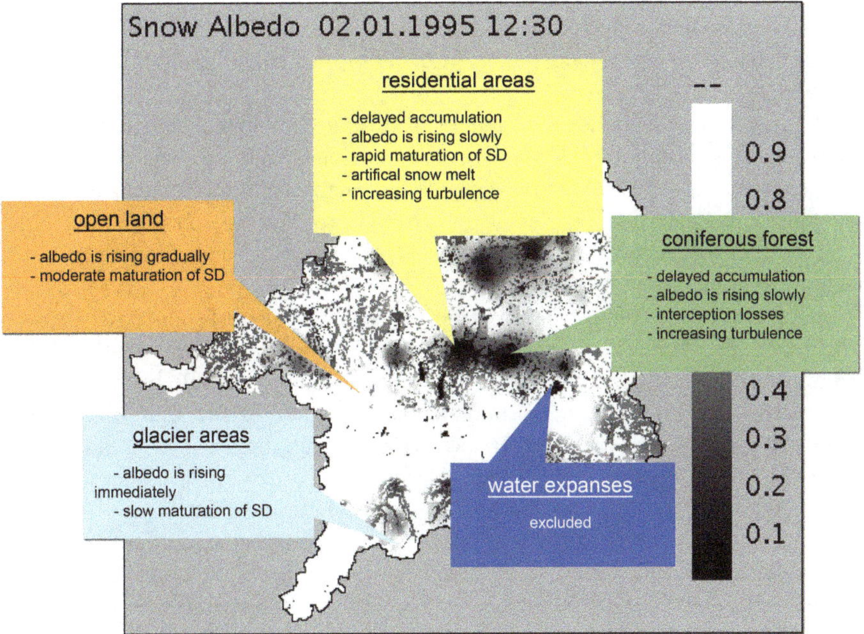

Fig. 30.1 Land use categories and processes included in the snow model

The heterogeneous land use within the drainage basin necessitates a further differentiation for the calculation of accumulation and ablation. While the physics of the energy exchange is similar across the various surfaces, there are differences in the characteristic parameters. An overview of the different land use types in the model and the general consequences for the treatment of the processes is illustrated in Fig. 30.1.

The *Snow* model integrated within DANUBIA does not provide snow depths; it only generates the quantities of water stored in the snow cover. Thus, observational data on the water equivalent in the snow cover are needed to validate the model results (see Chap. 10).

30.4 Results

Since in general the months from July to October are snow-free at elevations below 3,000 m a.s.l., only the period from November up to and including June is used for the map of snow cover duration. A snow cover day is counted if the daily mean water equivalent of the snow cover is more than 1 mm. As explained above, the means depicted for 1995–1999 are only representative for this period and are thus not directly comparable with long-term climate periods. More important is the demonstration of the possibility to generate high-resolution maps of snow cover duration based on model calculations instead of statistical interpolation of observational data.

The resulting distribution is plausible. It illustrates not only a distinct correlation between the persistence of snow cover and elevation but also regional dependencies on precipitation quantities or on the type and depth of the topsoil. In the Upper Danube drainage basin, complete snow cover is expected only on 40–50 days each year at elevations below 600 m a.s.l. in both lowland regions and even in areas in the centre of Munich or in Innsbruck.

In contrast, in the Alpine foothills there are 70–120 days of snow cover depending on the elevation. There is snow cover over the entire period only in the highest summits of the drainage basin and even only on the northern side of the mountain chain. The Wetterstein mountain range is one exception, where significantly more solid precipitation falls compared to the surrounding regions, as a result of the western exposure, in the direction of the preferred track of the low-pressure centre. In spring, the water equivalent of the snow cover on the Zugspitzplatt is up to 1,400 mm, which is equal to a snow cover thickness of 5–6 m. These special circumstances of accumulation are ultimately the cause for the local presence of some glacier remnants (Schneeferner and Höllentalferner glaciers) on the Zugspitze at elevations of only 2,600–2,800 m a.s.l., even though the most important glacial areas are found near the main Alpine ridge, significantly higher than 3,000 m a.s.l.

Despite the limitation of the sample to only four winter periods, the spatial distribution of snow cover duration in Map 30.1 corresponds well to the published results from KLIWA (2005), for example. It can be expected that the mean persistence of snow cover will decrease noticeably at all elevations with further global warming.

References

Escher-Vetter H (2000) Modelling meltwater production with a distributed energy balance method and runoff using a linear reservoir approach results from Vernagtferner, Oetztal Alps, for the ablation seasons 1992 to 1995. Zeitschrift für Gletscherkunde und Glazialgeologie 36:119–150

KLIWA (2005) Langzeitverhalten der Schneedecke in Baden-Württemberg und Bayern. In: KLIWA-Berichte, H 6. Arbeitskreis KLIWA (Landesanstalt für Umweltschutz Baden-Württemberg, Bayerisches Landesamt für Wasserwirtschaft, Deutscher Wetterdienst). ISBN 3-937911-18-9. www.kliwa.de/download/KLIWAHeft6.pdf

Weber M (2009) Mikrometeorologische Prozesse bei der Ablation eines Alpengletschers. Bayerische Akademie der Wissenschaften. In: Abhandlungen der Mathematisch-Naturwissenschaftlichen Klasse, H. 177, S. C.H.Beck, München, p 258. ISBN 978-3-7696-2564-6

Wielke L-M, Haimberger L, Hantel M (2004) Snow cover duration in Switzerland compared to Austria. Meteorol Z 13:13–17

Chapter 31
Future Changes in the Ice Reservoir

Markus Weber, Monika Prasch, Michael Kuhn, and Astrid Lambrecht

Abstract In summer, meltwater from glaciers supplies a reliable contribution to runoff in the alpine headwater regions. In a warming climate with reduction of the glacier surface and depletion of the ice reservoir, this contribution is increasing in an initial stage, but decreasing on the long run.

The observations of glacier retreat over the past century and an increase in global temperatures are indications of the impressive dynamics of the processes that have continued to accelerate as a result of interactions and feedback, especially in the last two decades. These processes are calculated in precise detail in DANUBIA using the SURGES (Subscale Regional Glacier Extension Simulator) glacier sub-model, so that even the future extent of the glaciers and discharge can be studied under the various climate scenarios.

Similar to Map 12.1, Map 31.1 presents the mean available ice mass in mm of water equivalent per proxel for the IPCC regional climate trend and the baseline climate variant in the model years 2030 and 2060. The results reproduced here confirm estimates that suggest the ice reservoir in the Eastern Alps may completely disappear as early as the second half of the century.

Keywords Glacier • SURGES • Climate trend • IPCC • Ice dynamics • GLOWA-Danube • DANUBIA

M. Weber (✉) • A. Lambrecht
Commission for Geodesy and Glaciology of the Bavarian Academy of Sciences and Humanities, Munich, Germany
e-mail: Wasti.Weber@kfg.badw.de; Astrid.Lambrecht@keg.badw.de

M. Prasch
Department of Geography, Ludwig-Maximilians-Universität München (LMU Munich), Munich, Germany
e-mail: m.prasch@lmu.de

M. Kuhn
Institute of Meteorology and Geophysics, University of Innsbruck, Innsbruck, Austria
e-mail: Michael.Kuhn@uibk.ac.at

W. Mauser, M. Prasch (eds.), *Regional Assessment of Global Change Impacts*, DOI 10.1007/978-3-319-16751-0_31

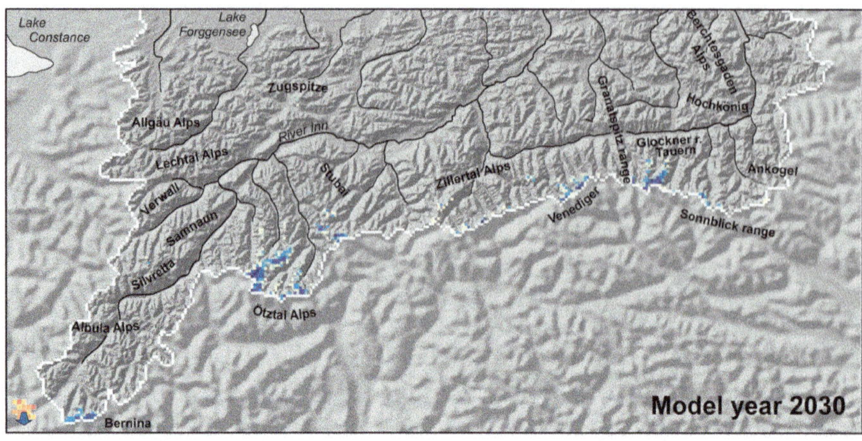

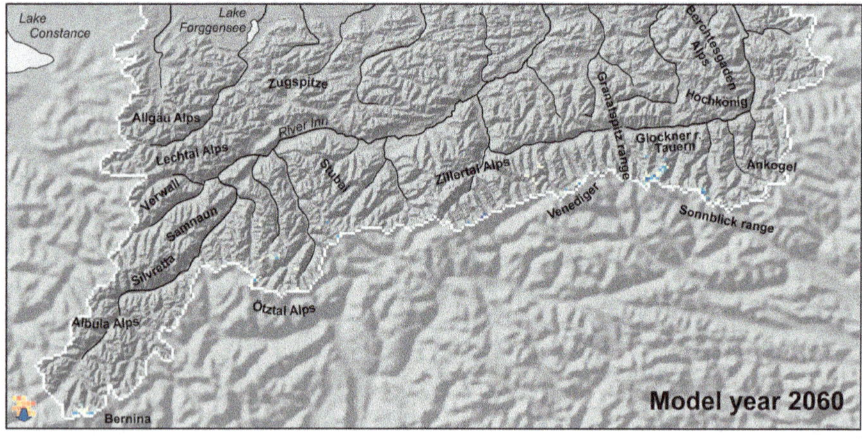

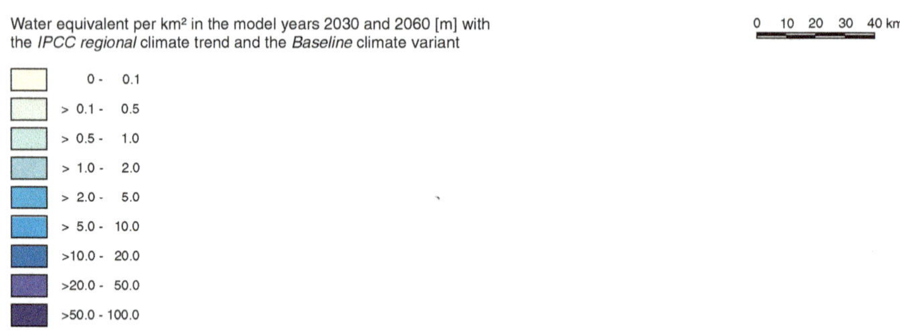

Map 31.1 Future changes of the ice reservoir (cadastre of glaciers (Institute of Meteorology and Geophysics at the University of Innsbruck IMGI (University of Innsbruck 2008); Commission for Geodesy and Glaciology of the Bavarian Academy of Sciences and Humanities, Munich; Department of Geography, University of Zurich); DANUBIA – Digital terrain model and derived river network; Results of the statistical climate generator (see Chap. 49): IPCC regional climate trend and baseline climate variant)

31.1 Introduction

In summer, meltwater from glaciers supplies a reliable contribution to runoff in the alpine headwater regions. In a warming climate with reduction of the glacier surface and depletion of the ice reservoir, this contribution is increasing in an initial stage, but decreasing on the long run.

The observations of glacier retreat over the past century and an increase in global temperatures by 0.8 K are indications of the impressive dynamics of the processes that have continued to accelerate as a result of interactions and feedback, especially in the last two decades. These processes are calculated in precise detail in DANUBIA within the glaciology sub-project in the context of the treatment of land surface processes, so that even the future extent of the glaciers and discharge can be studied under the various climate scenarios. The results reproduced here by way of example from the modelling based on the moderate *IPCC regional* climate trend and the *baseline* climate variant (see Chap. 47) confirm estimates that suggest the ice reservoir in the Eastern Alps may completely disappear as early as the second half of the century.

31.2 Model Documentation

Alpine glaciers are objects that can change their shape via the following three main processes:

- Accumulation of snow
- **Ablation** of snow and ice
- **Flow of ice** into the valley

The first two processes take place almost exclusively on the glacier surface; however, ice movement affects the whole ice body with higher velocities at the surface. The main source of accumulation is snowfall; the percentage of snowfall depends on the air temperature **T** near the surface. As a result of a general decline in **T** with increasing elevation, snowfall increases at higher elevations. Accumulation also takes place locally from avalanches or drifting snow.

The most important process for ablation is melt. The meltwater feeds the river network with discharge. Thawing occurs if the snow or ice surface is warmed to 0 °C and if there is energy available for the heat of fusion. This latter condition is equivalent to a positive energy balance at the surface. Its components include short- and long-wave radiation (e.g. global radiation **G**) and the turbulent heat flow between the atmosphere and the surface.

Calculation of both snow accumulation and the snow and ice melt at the surface of the glacier is achieved with the SURGES (Subscale Regional Glacier Extension Simulator) glacier sub-model (Weber et al. 2010; Prasch et al. 2013). As a result of the high dependence of the driving variables on relief and the rather crude localisation of the glaciated areas on the DANUBIA grid, the model algorithm is applied on the subscale. Thus, the ice bodies are approximated on the grid units using a stepwise model that is derived from area, elevation and ice thickness distributions. The preparation of this dataset is explained in more detail in the text description accompanying Map 12.1.

31.2.1 Subscale Extrapolation

Figure 31.1 illustrates the extrapolation of the basic meteorological variables prepared using the supporting programme to the level of the proxel and in relation to their elevation dependency. Accumulation and ablation are then calculated independently for each individual level with the appropriate parameterisation adjusted to the problem. In this way, snowfall in the upper regions and rain or thaw in lower-lying regions can be calculated simultaneously within a 1×1 km^2 unit. The potential difference between snow and ice surfaces is similarly important, since ice melt is up to four times more efficient than snow melt because of a lower albedo.

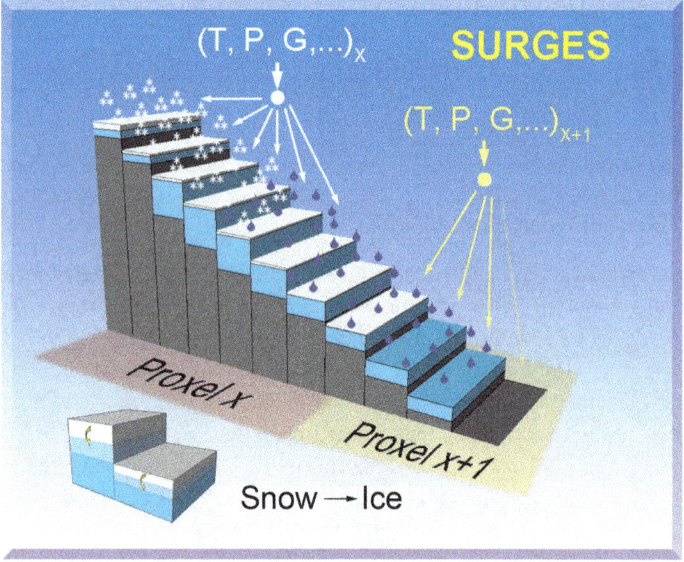

Fig. 31.1 Schematic function of the subscale glacier model component SURGES and its integration in DANUBIA

31.2.2 Change in Mass and Surface Area

If the snow cover at one level outlasts a budget year, then it gradually changes into ice, which is added to the ice body. In contrast, snow-free levels can thaw completely and uncover the ground below. This reduces the surface area of the glacier. If more ice melts than forms within a hydrological year, the overall glacier mass balance is negative.

Depending on the climatic boundary conditions, the mass balance is usually positive for the uppermost levels and negative for the lowest levels on an annual basis. In part, the movement of ice ensures the exchange of ice from higher to lower levels. If the overall balance of the glacier is negative, this process slows down the decrease in surface area, but the loss of mass is increased as a whole. However, if the overall glacier mass balance is positive, this mechanism leads to an advance of the glacier. SURGES simulates this mechanism using an empirical parameter for the redistribution of ice. Since the movement of ice is driven by gravity, this factor loses importance in periods with ongoing loss of mass. Thus, the currently observed diminution in the area increases disproportionately with the loss of mass.

31.3 Results

Similar to Map 12.1, Map 31.1 presents the mean available ice mass in mm of water equivalent per proxel for the *IPCC regional* climate trend and the *baseline* climate variant in the model years 2030 and 2060. Under these conditions, the ice reservoirs disappear quite drastically. The Alps within the drainage basin are virtually ice-free within 50 model years.

However, it should be noted that this is not a “forecast” for the status of the glaciers by a deadline; instead, this is a case study on the behaviour of a glacier based on the current state of the glacier in 2000 under a climate scenario that has not yet been reached. Figure 31.2 is a detailed depiction of the change in the surface area of the glacier for a section of the Ötztal glacier based on contour maps. On the one hand, the figure highlights that the relatively low subscale resolution has limited suitability for realistic modelling of the temporal changes of glacier areas. On the other hand, the figure also provides an overview of the dynamics of the future retreat of the glaciers and the extent of the ice region under conditions with regularly prolonged periods of mass loss, in a way that has never before been observed, with the exception of 2003.

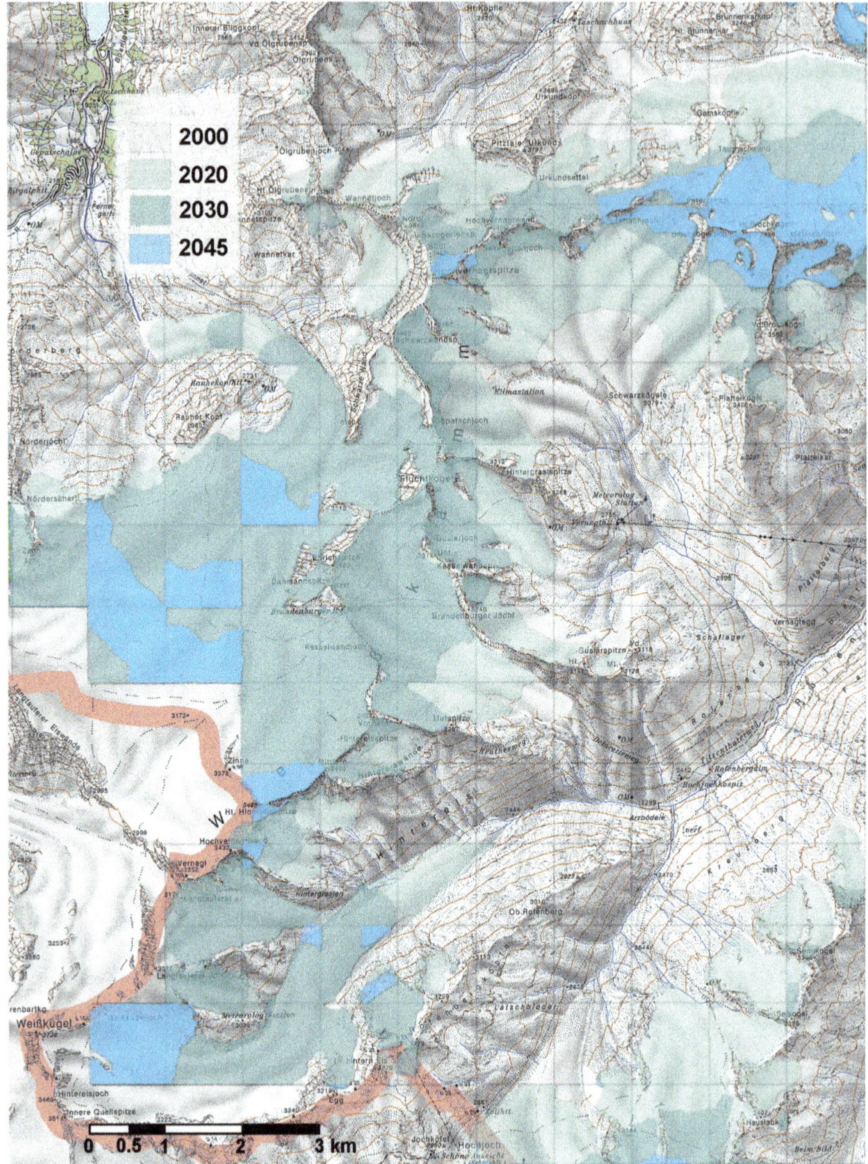

Fig. 31.2 In detail map section of the modification of glacier areas in the Ötztal valley (*IPCC regional* climate trend and *baseline* climate variant, modelled with SURGES) (data sources: 1:50,000 map of the Hydrological Atlas of Austria (BMLFU – Austrian Federal Ministry for Agriculture, Forestry, Environment and Water Management 2005) B4.3; contour line scheme 50 m from data of the Austrian Glacier Inventory, Institute of Meteorology and Geophysics, Universtiy of Innsbruck (Kuhn et al. 2013))

References

BMLFU – Austrian Federal Ministry for Agriculture, Forestry, Environment and Water Management (ed) (2005) Hydrological atlas of Austria. Vienna

Kuhn M, Lambrecht A, Abermann J (2013) Austrian glacier inventory 1998. PANGAEA. doi:10.1594/PANGAEA.809196

Prasch M, Mauser W, Weber M (2013) Quantifying present and future glacier melt-water contribution to runoff in a Central Himalayan river basin. Cryosphere 7:889–904. doi:10.5194/tc-7-889-2013

University of Innsbruck (2008) Austrian glacier inventory. Institute University of Innsbruck, Innsbruck

Weber M, Braun L, Mauser W, Prasch M (2010) Contribution of rain, snow- and icemelt in the Upper Danube discharge today and in the future. Geogr Fis Dinam Quat 33(2):221–230

Chapter 32
Precipitation and Temperature

Barbara Früh, Volkmar Wirth, Josef Egger, Andreas Pfeiffer, and Janus W. Schipper

Abstract Climatological studies indicate that climate change lead to an increase in the mean global temperature of around 0.5 °C until the end of the twentieth century. This warming impacts the atmospheric humidity, wind, radiation, and precipitation. However, the magnitude of changes is not equally distributed over the globe but differs markedly with regions, making a regionalization of the global information essential. The GLOWA-Danube project follows such a downscaling approach with the focus on the drainage basin of the Upper Danube River.

Since the horizontal resolution of global climate simulations (approx. 200 km) is still too coarse for studies on regional or even local scales, the results of the global projections are post-processed using the regional climate model MM5 to downscale the information to 45 km. The empirical-statistical method AtmoMM5 subsequently downscaled the MM5 results to the desired resolution of 1 km.

In order to test the quality of the meteorological data used in DANUBIA, the time series of different proxels within the period from 1991 to 2000 were compared with corresponding periods from the station measurements. The station measurements are interpolated with respect to distance and direction on a regular grid. These gridded fields in general follow the pattern of the observed precipitation and 2 m air

B. Früh (✉)
Deutscher Wetterdienst DWD, Offenbach, Germany
e-mail: Barbara.frueh@dwd.de

V. Wirth
Institute for Atmospheric Physics, Johannes Gutenberg University Mainz, Mainz, Germany
e-mail: vwirth@uni-mainz.de

J. Egger
Meteorological Institute Munich, Ludwig-Maximilians Universtiät München (LMU), Munich, Germany
e-mail: j.egger@lrz.uni-muenchen.de

A. Pfeiffer
Institute of Atmospheric Physics, DLR, Oberpfaffenhofen, Germany
e-mail: a.pfeiffer@dlr.de

J.W. Schipper
Süddeutsches Klimabüro, Karlsruher Institut für Technologie, Karlsruhe, Germany
e-mail: schipper@kit.edu

W. Mauser, M. Prasch (eds.), *Regional Assessment of Global Change Impacts*, DOI 10.1007/978-3-319-16751-0_32

temperature. The scaling of the regional model MM5 with AtmoMM5 causes a distinct reduction of the root mean square error between the observations and MM5 or AtmoMM5, respectively.

Keywords GLOWA-Danube • MM5 • AtmoMM5 • Upper Danube • Precipitation • Temperature • Downscaling

32.1 Introduction

Our climate is changing as a result of the effects of global change; these climate changes are visible in the long-term averages of the weather conditions. Studies indicate that there has already been an increase of around 0.5 °C in the mean global temperature over the past 50 years (IPCC 2001). The responses to this increase include changes to the atmospheric water vapour content, wind conditions, radiation balance and precipitation. However, these changes are not manifested equally over the globe but regionally differ markedly instead. Thus, global warming may even lead to a cooling effect in certain regions. The regionalization of global climate change is therefore an important area of research. The GLOWA-Danube project is an example of a study of these effects, specific to the drainage basin of the Upper Danube River.

For most people, climate change is most readily noticed by a change in temperature or precipitation, because both temperature and precipitation have direct impacts on the lives of humans. These factors interact not only with aspects of the natural environment, such as the soil, groundwater, natural vegetation and agriculture, but also with socioeconomic conditions such as water demands by agriculture and private homes, industry and the tourism sector (see Chaps. 39, 41 and 46).

Even putative negligible changes in the climatological (i.e. long-term) averages of temperature and precipitation can entail significant consequences in the complex system. Serious ramifications on various aspects of human living can also be expected in case of major deviations from the climatological averages. Long-lasting heat waves or droughts increase mortality rates, bathing lakes experience elevated concentration of bacteria, power stations may not be supplied with enough cooling water, plants desiccate, and, in extreme cases, the supply of drinking water for the population may be at risk both in terms of quantity and quality.

32.2 Data Processing

To date, simulations of global climate have not had sufficient horizontal resolution to allow for realistic, detailed studies of the effects of climate change on regional or even local scales. In particular, prominent features of an orographically strongly structured landscape such as the Alps are not captured within a global climate model. In order to achieve this regionalization, the results of the global simulations were additionally processed by using them as forcing for a regional climate model at physically consistent, higher resolutions.

To cover the current regional climate and to validate the overall model approach, data from past global meteorological analyses were fed into a regional model at first. The three-dimensional meteorological data used is provided by the "European Centre for Medium-Range Weather Forecasts" (ECMWF) for long time series, in high quality, and in a highly consistent temporal format known as reanalyses. The results of the simulations from the regional model for the period 1991–2000 were used for the maps presented here.

32.3 Model Documentation

The mesoscale atmosphere model *MM5* (Grell et al. 1994) was used to calculate the meteorological parameters for DANUBIA. Since the resolution of *MM5* (at present 45 km, possibly eventually up to 15 km) is lower than the 1×1 km^2 proxel grid of DANUBIA, the meteorological parameters from *MM5* need to be scaled.

In principle, operating *MM5* on a horizontal resolution of 1 km would be desirable. However, since the climate scenarios in DANUBIA should cover long time intervals, the required demand for computer resources exceeds the nowadays realistically feasible amount. Therefore, as a compromise, a combination of dynamical-physical downscaling using *MM5* (e.g. from the typical resolution of a global climate model of about 200 km to 45 km) and a subsequent empirical-statistical downscaling of the *MM5* results to the desired resolution (from 45 km to 1 km) was chosen. This method ties in the advantages of both a dynamical-physical procedure and a statistical-empirical approach. This combination of scaling algorithm of *MM5* with the statistical-empirical approach is called *AtmoMM5*.

The scaling method used in *AtmoMM5* first generates monthly climatologies based on station measurements (see Chap. 10); these are representative for the average conditions for each month during 1991–2000. The next step interpolates these station measurements to the 1×1 km^2 proxel grid. To do so, these point measurements are related to the area using an interpolation method that is weighted by distance and direction (Shepard 1968). The dependence of precipitation on orographic elevation is taken into account using PRISM (**P**arameter-elevation **R**egression on **I**ndependent **S**lope **M**odel, Daly et al. 1994) which has already been applied by Schwarb et al. (2001) to the Alpine region. A local elevation regression was determined to take into account the orography for the air temperature. This process leads to climatological fields for precipitation and temperature at a 1×1 km^2 resolution.

The scaling algorithm consists of two distinct steps: first, the local subscale variability is derived from the high-resolution climatological observations. Then a bias correction is performed, which adjusts the low-resolution model climatology to the observed one (Früh et al. 2006). For temperature, the two steps are carried out additive; for precipitation multiplicative.

In order to test the quality of the meteorological data used in DANUBIA, the time series of different proxels within the period from 1991 to 2000 were compared with corresponding periods from the station measurements. When interpreting this comparison, it should be noted that the DANUBIA results are based on a 1×1 km^2 proxel area and hence present mean values for an area, whereas the station measurements are specific to a point location. The meteorological parameter studied in this first step includes precipitation P and atmospheric temperature T. In the following, the observed time series *Obs* are compared with the results of the mesoscale atmosphere model *MM5*, the scaled model *AtmoMM5* and the *AtmoStations* model. *AtmoStations* interpolates the station measurements directly using an interpolation method that is weighted by distance and direction. Figure 32.1 illustrates the precipitation P in 1995 at the Regensburg station. It is clear that the precipitation from *AtmoStations* in general follows the pattern of the observed precipitation. In some periods, the precipitation is overestimated by the simulation compared to the station measurements (e.g. end of April, middle of July and December), and in some periods, it is underestimated (e.g. end of May to the beginning of June and end of July to the beginning of August).

To assess the changes in the simulation results for precipitation that results from scaling, the root mean square error is calculated. It is defined as

$$\mathrm{RMSE} = \sqrt{\frac{1}{N}\sum_{i=1}^{N}\left[X(\mathrm{Obs}) - X(\mathrm{Model})\right]^2}$$

where X stands either for the daily mean temperature or the total daily precipitation. The term "model" in the above equation refers either to *MM5*, *AtmoMM5* or *AtmoStations*. N is the number of days with measurements. Squaring the deviations gives more weight to large differences. A perfect match between the observed and scaled values is given by the value zero for RMSE.

Downscaling (*MM5* → *AtmoMM5*) causes a reduction of the RMSE between daily total precipitation from the observations Obs and *AtmoMM5* to 2.78 mm compared to 3.25 mm for Obs and *MM5*. As expected, the RMSE is lower (0.92) for *AtmoStations*, since the station measurements Obs are direct input of the interpolation model. The slight temporal shift in the precipitation results (Fig. 32.1) can be attributed to the fact that the precipitation measurements take place at the appointed climate time and not at midnight (see Chap. 10).

Figure 32.2 shows the corresponding results for the 2 m air temperature. The observations Obs are reproduced quite well by *AtmoMM5*. The RMSE was reduced from 2.01 to 1.36 by applying the scaling algorithm.

Figure 33.1 depicts the seasonal dependence of the RMSE for precipitation. For all models used, the RMSE is highest in the summer months (May to August). This is related to the common summertime convective precipitation events that are difficult to simulate in all cases: in the *MM5* model, convective precipitation is generated via a subscale parameterization. In *AtmoStations* and the precipitation climatology that forms the basis for the scaling, precipitation measurements are interpolated to the proxel; this process smoothes steep

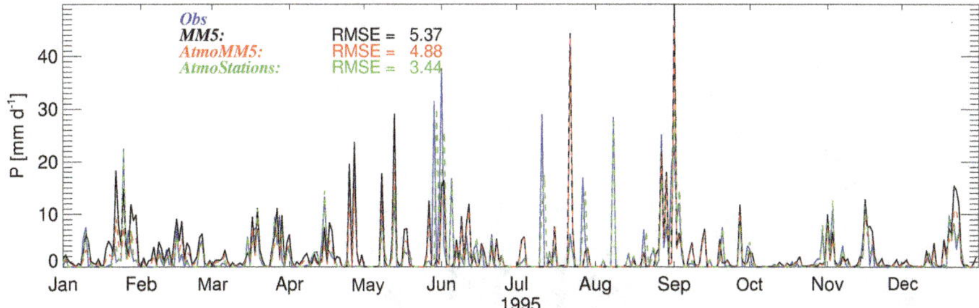

Fig. 32.1 Precipitation (*P*) from the observations Obs (*blue*), the simulations of MM5 (*black*) and AtmoMM5 (*red*) together with AtmoStations (*green*) and their root mean square errors (*RMSE*) in relation to the observations in Regensburg in 1995

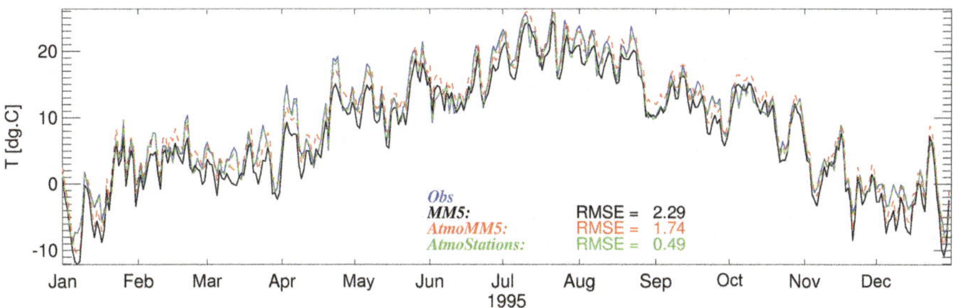

Fig. 32.2 2 m air temperature (*T*) from the observations Obs (*blue*), the simulations of MM5 (*black*) and AtmoMM5 (*red*) together with AtmoStations (*green*) and their root mean square errors (*RMSE*) in relation to the observations in Regensburg in 1995

spatial gradients that arise from the narrowly restricted spatial extent of the convective events.

The RMSE for the 2 m air temperature (see Fig. 32.4) has a significantly weaker seasonal dependence than the RMSE for precipitation (see Fig. 32.3). The RMSE for the 2 m temperature is also very much reduced as a consequence of the scaling (Table 32.1).

32.4 Results

Figure 32.5 shows a specific precipitation episode on January 18, 1995, between 13 and 24 UTC as an example of a simulation from *AtmoMM5*. The hourly precipitation fields (from upper left to lower right) serve as input data for DANUBIA.

The dynamics of the precipitation system that moved across the drainage basin is easy to discern. The influence of the orography can be seen by the higher intensity of precipitation at the northwestern fringe of the Alps.

Both maps (Maps 32.1 and 32.2) illustrate the climatologies for precipitation and temperature for the decade from 1991 to 2000. The first Map 32.1 presents the

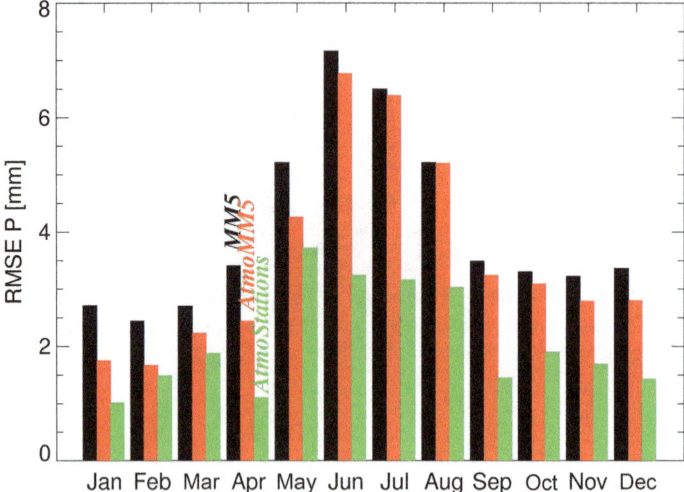

Fig. 32.3 Monthly root mean square error (*RMSE*) between the daily sum of precipitation (*P*) from the observations (*Obs*) and the DANUBIA models at the Regensburg station for the time period 1991–2000. *Black columns* show the RMSE between Obs and MM5, *red* ones that between Obs and AtmoMM5 and *green* ones that between Obs and AtmoStations

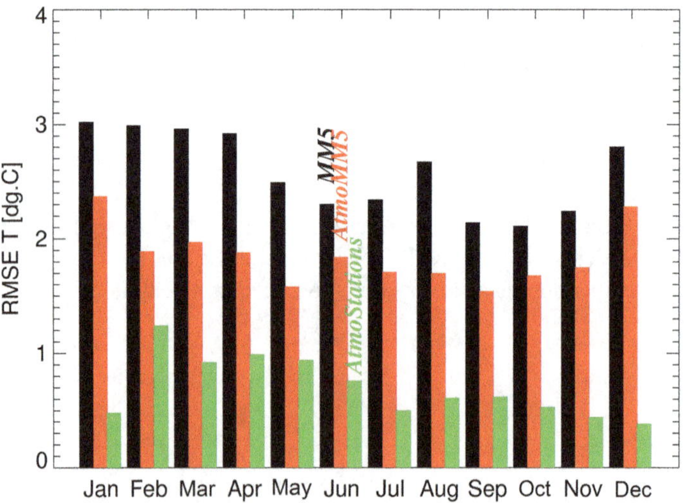

Fig. 32.4 Monthly root mean square error (*RMSE*) between the daily mean air temperature (*T*) from the observations (*Obs*) and the DANUBIA models at the Regensburg station for the time period 1991–2000. *Black columns* show the RMSE between Obs and MM5, *red* ones that between Obs and AtmoMM5 and *green* ones that between Obs and AtmoStations

Table 32.1 The mean values and the RMSE for the daily temperature and precipitation averaged across the five stations at Regensburg, Augsburg, Munich, Garmisch-Partenkirchen and Hohenpeißenberg for the time period between 1991 and 2000

	T (°C) Mean	T (°C) RMSE	P (mm) Mean	P (mm) RMSE
Obs	8.7		2.7	
AtmoStations	8.3	0.9	2.7	1.7
MM5	7.0	2.8	3.1	3.5
AtmoMM5	8.7	2.1	2.9	3.4

The mean values from *AtmoStations* and *AtmoMM5* correspond very well with the observed values Obs. The scaling reduces the RMSE for both temperature and precipitation

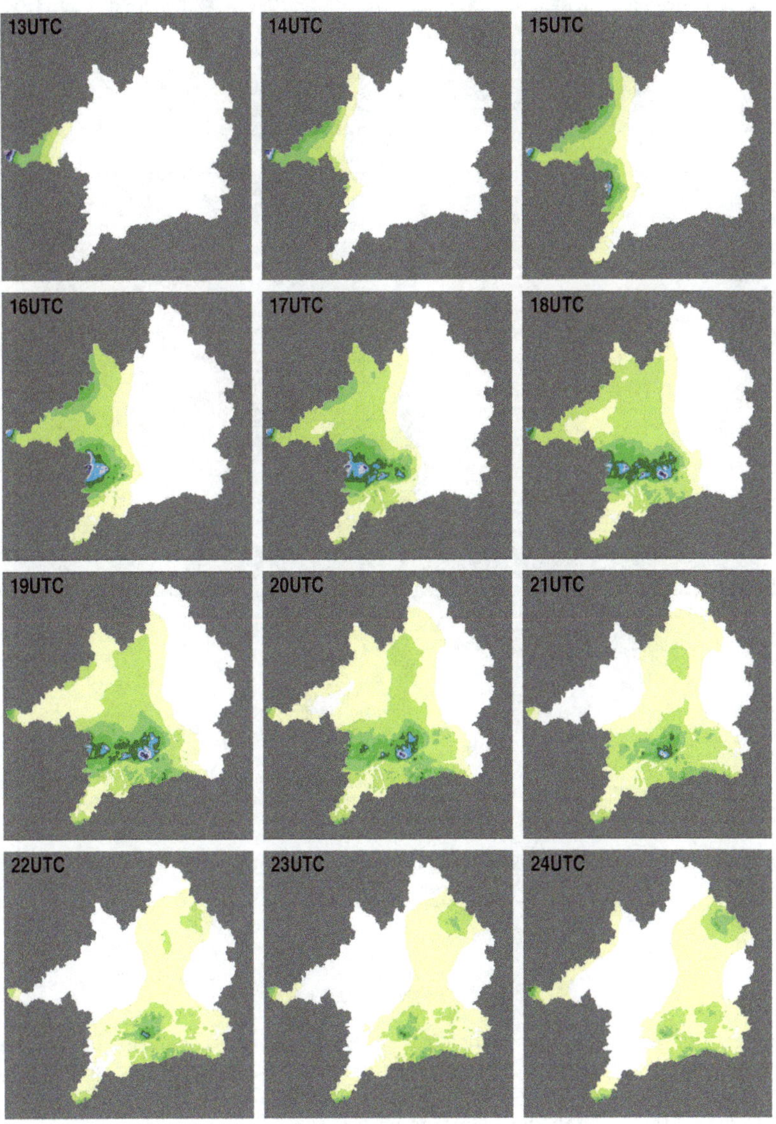

Fig. 32.5 Precipitation on January 18, 1995, 13–24 UTC (hourly fields simulated by AtmoMM5)

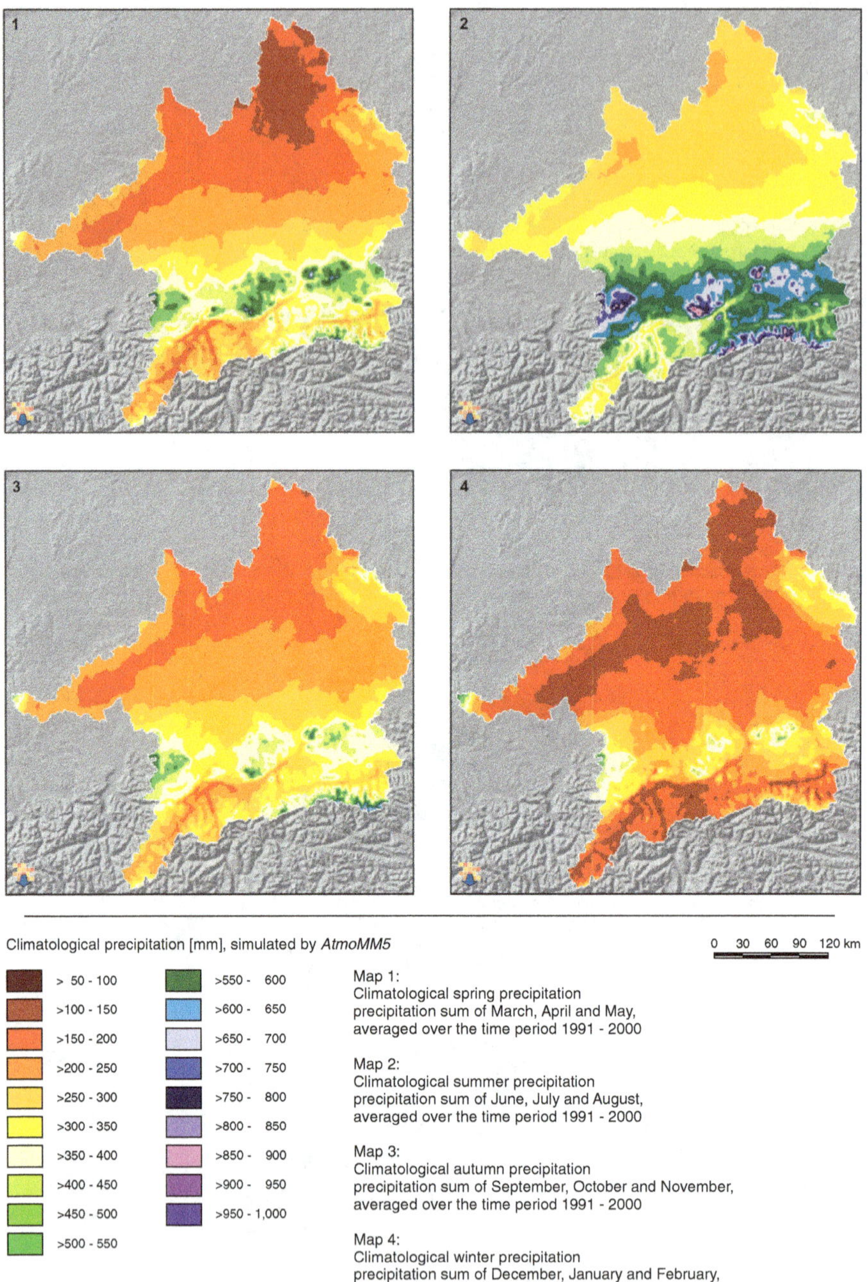

Map 32.1 Climatological precipitation (Data source: Deutscher Wetterdienst (DWD); Austrian Federal Ministry for Agriculture, Forestry, Environment and Water Management (BMLFU); Zentralanstalt für Meteorologie (ZAMG), boundary and initial conditions of MM5 simulations: ERA-40 data of ECMWF (Simmons and Gibson 2000))

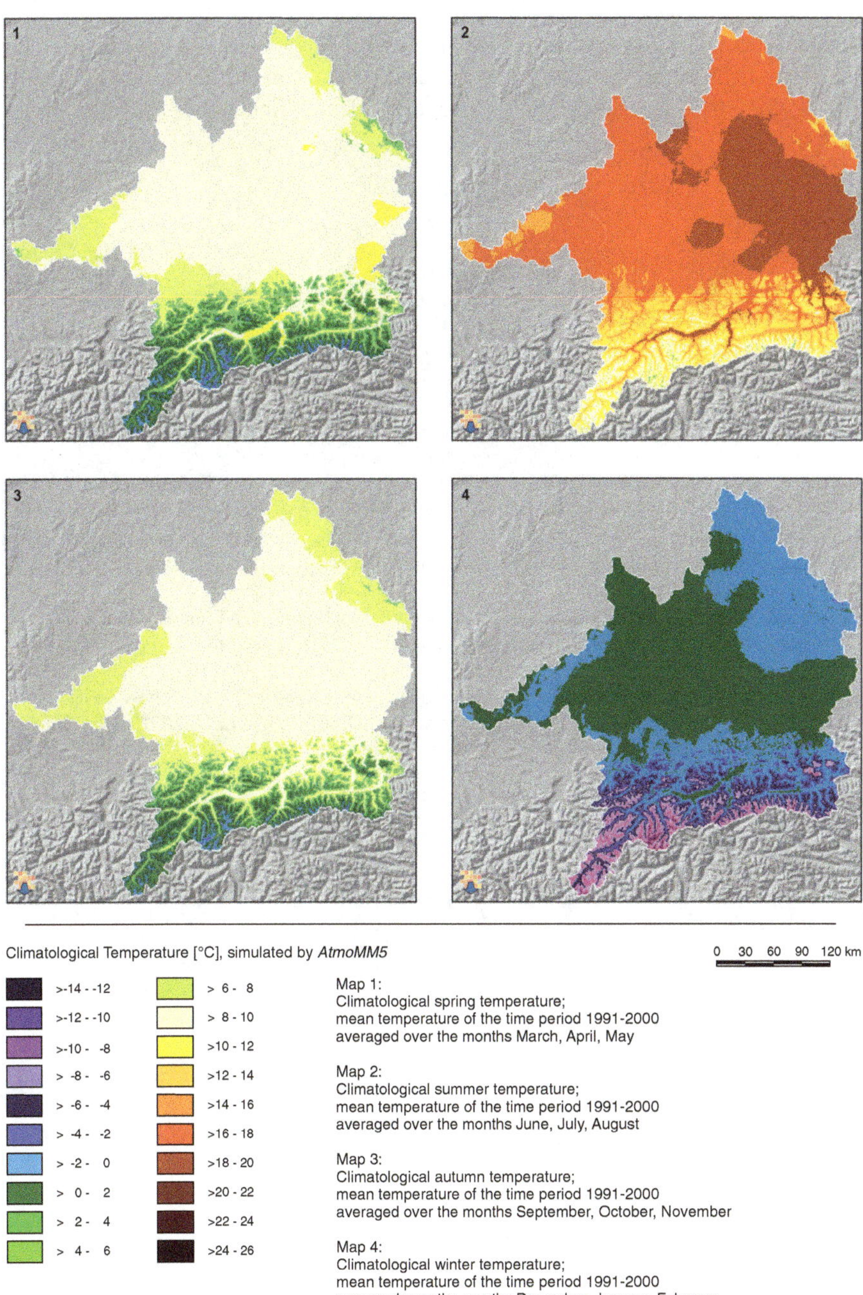

Map 32.2 Climatological temperature (Data source: Deutscher Wetterdienst (DWD); Austrian Federal Ministry for Agriculture, Forestry, Environment and Water Management (BMLFU); Zentralanstalt für Meteorologie (ZAMG), boundary and initial conditions of MM5 simulations: ERA-40 data of ECMWF (Simmons and Gibson 2000))

seasonal sum of precipitation, while the second Map 32.2 presents the seasonal mean air temperature. The seasons are defined as winter (December–February), spring (March–May), summer (June–August) and fall (September to November).

References

Daly C, Neilson RP, Phillips DL (1994) A statistical topographic model for mapping climatological precipitation over mountainous terrain. J Appl Meteorol 33:140–158

Früh B, Schipper JW, Pfeiffer A, Wirth V (2006) A pragmatic approach for downscaling precipitation in Alpine scale complex terrain. Meteorol Z 15:631–646

Grell GA, Dudhia J, Stauffer DR (1994) A description of the fifth-generation Penn State/NCAR mesoscale model (MM5). NCAR Technical Note NCAR/TN-398+STR, Boulder, USA, p 138

IPCC (2001) Climate change 2001: synthesis report. A contribution of Working Groups I, II, and III to the third assessment report of the Intergovernmental Panel on Climate Change (Watson RT, The Core Writing Team, eds). Cambridge University Press, Cambridge/New York, p 398

Schwarb M, Daly C, Frei C, Schär C (2001) Mean annual and seasonal precipitation in the European Alps 19711990. Hydrological Atlas of Switzerland. Plates 2.6 and 2.7, Federal Office for Water and Geology, Bern

Shepard D (1968) A two-dimensional interpolation function for irregularly-spaced data. In: Proceedings of 23rd ACM national conference. Brandon/Systems Press, Princeton, pp 517–524

Simmons AJ, Gibson JK (2000) The ERA-40 project plan. ERA-40 Project Report Series, No. 1, Reading, UK. Available at http://old.ecmwf.int/publications/library/do/references/list/192. Accessed 20 May 2015

Chapter 33
Two-Way Coupling the PROMET and MM5 Models

Florian Zabel, Wolfram Mauser, and Thomas Marke

Abstract Regional climate models still lack in an adequate description of the land surface. Here, hydrological land surface models historically developed from small-scale catchment models with higher spatial resolution than regional climate models. Hence, they have a more detailed view on land surface processes. Therefore, a basic idea was to substitute a land surface module within the regional climate model with the improved hydrological land surface model. For this purpose, we adopted the Mesoscale Model 5 (MM5) and the hydrological land surface model PROMET for coupling the two models in both directions. As a result, we conclude that the bilateral coupling improved simulation results in both the atmosphere model and the hydrological land surface model.

Keywords Model coupling • MM5 • PROMET • Atmosphere • Land surface • GLOWA-Danube

33.1 Introduction

The land surface is a key component of the climate system (IPCC 2007). Its properties are spatially heterogeneous (e.g. land use, terrain, soil) and influence energy and mass fluxes at the land surface through processes such as plant transpiration and soil evaporation, heating, reflection of short-wave radiation and emission of long-wave radiation.

The various vegetation covers on the land surface, as well as the impervious surface areas such as occur in cities, are primary factors that can vary significantly in space. Although the heterogeneity of the land surface can mostly be well

F. Zabel (✉) • W. Mauser
Department of Geography, Ludwig-Maximilians-Universität München (LMU Munich), Munich, Germany
e-mail: f.zabel@lmu.de; w.mauser@lmu.de

T. Marke
Institute of Geography, University of Innsbruck, Innsbruck, Austria
e-mail: thomas.marke@uibk.ac.at

W. Mauser, M. Prasch (eds.), *Regional Assessment of Global Change Impacts*, DOI 10.1007/978-3-319-16751-0_33

documented with remote-sensing methods, accounting for this heterogeneity in climate models presents a scientific challenge. Most climate models have features known as land surface modules (LSMs) that are responsible for modelling the land surface processes. The energy and mass fluxes at the land surface are typically calculated at the resolution of the respective climate model and represent the lower boundary conditions for the atmospheric simulations in a specific region. Newer LSMs often use a subscale approach in which mass and energy fluxes at the land surface are calculated individually for the most frequently occurring land use classes according to the proportional areas for the resolution of a pixel, and these values are then aggregated. However, this approach does not account for subscale orography. The characteristics of the land surfaces, which have a spatial dimension that is smaller than the pixel size of the climate models, are not taken into account as a result of the limited spatial resolution of the very computationally intensive climate models. This also affects a number of meteorological processes such as the complex feedback mechanisms between the land surface and the atmosphere, which often take place at much smaller scales than can be simulated with the current climate models. These subscale processes and feedback mechanisms that need to be considered are a focus of current research efforts and can contribute to a better understanding of the complex land surface-atmosphere system and hence also to improving future climate scenarios. For this purpose, a method was developed in which the hydrological model PROMET (see Chap. 4; Mauser and Bach 2009) is coupled with the regional climate model *MM5* (Pfeiffer and Zängl 2009).

33.2 Coupling Area

The coupling area in this study represents a section of the *MM5* simulation area, in which the energy and mass fluxes at the land surface are calculated at high resolution by PROMET instead of by *MM5* itself. The simulation area of the regional model *MM5* is driven by reanalysis data or global climate model data at its margins (see Chap. 51). The coupling area extends across more than $1{,}170 \times 1{,}170$ km^2.

It encompasses Central Europe with a total of 18 countries from the North Sea to the Mediterranean. Figure 33.1 illustrates the topography of the coupling area and the Upper Danube drainage basin. The landscape is characterised by large, fertile plains, low mountain ranges and the Alps, which represent a climatic boundary for the temperate zone to the Mediterranean climate.

33.3 Land Use Classification

Land use patterns in Europe are largely shaped by anthropogenic influences. Agriculture makes up approximately 45 % of the land use. Cities and settlements seal large areas, especially in densely populated areas. A detailed mapping of land

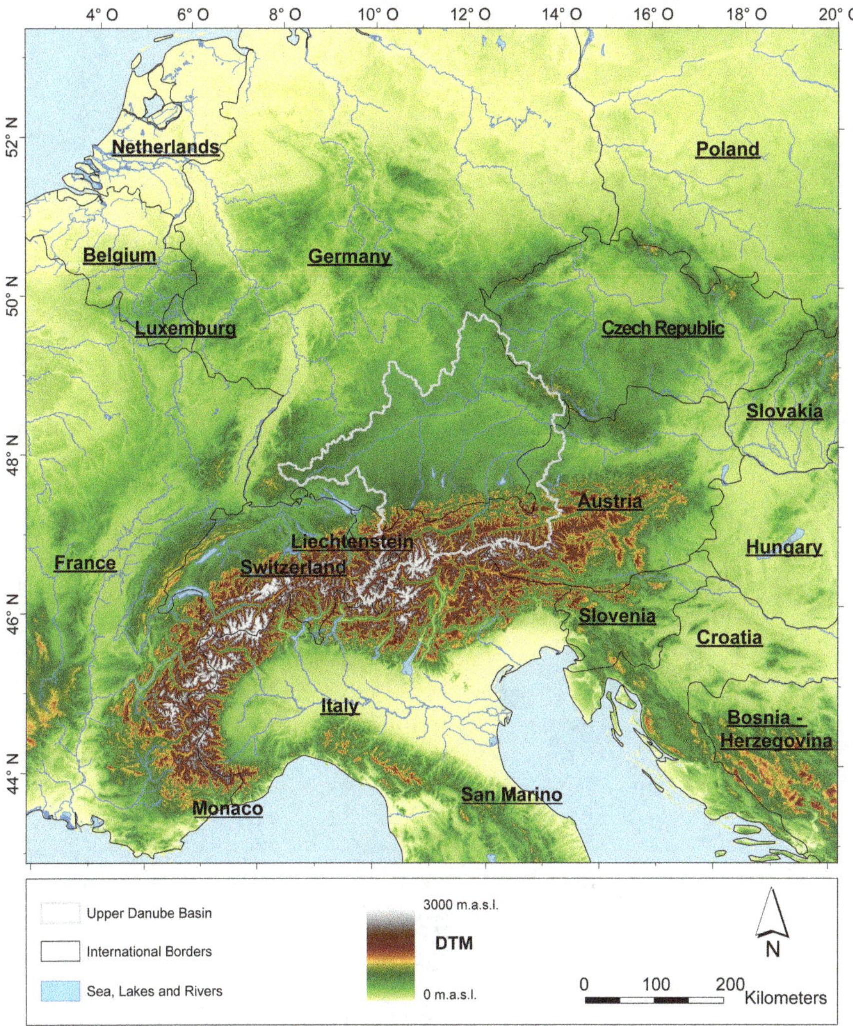

Fig. 33.1 Topography of the coupling area used for the two-way coupling (Data source: Elevation data from SRTM (Jarvis et al. 2008))

use at a spatial resolution of 1×1 km^2 was required for modelling land surface processes with PROMET (see Chap. 4) within the coupling area.

Figure 33.2 depicts the land use and cover for the coupling area at a resolution of 45×45 km^2, as it is originally specified in *MM5*. Agricultural areas dominate this land use/cover classification; these areas can be summarised into the class “dryland, cropland and pasture”. While coniferous forests are predominately found in the Alps, deciduous forests are situated north, west and especially southeast of the Alps.

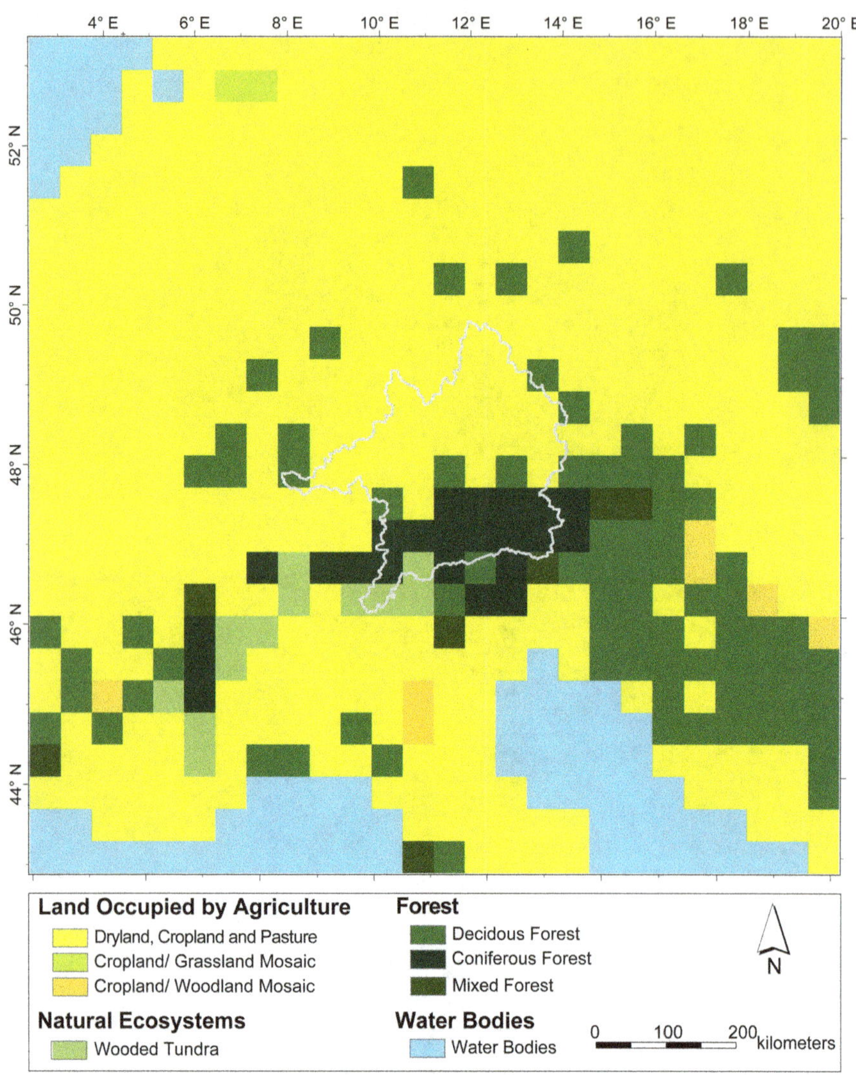

Fig. 33.2 Land use classification of the MM5 model (45×45 km^2) for the coupling area

In order to create a detailed map of the land surface for the simulation area at the resolution of PROMET (1×1 km^2), a detailed land use/cover map was generated based on the CORINE 2000 land use/cover classification system; this map also incorporates the heterogeneity of the agricultural areas. This was achieved using high-resolution MERIS Normalized Differenced Vegetation Index (NDVI) data, combined with statistical data from EUROSTAT (Zabel et al. 2010).

Figure 33.3 depicts the land use classification as it is used in PROMET at a 1×1 km^2 resolution. The significantly higher resolution compared to the land use from the meteorological model *MM5* leads to a much more precise mapping of reality, in which the heterogeneity of the land surface is apparent.

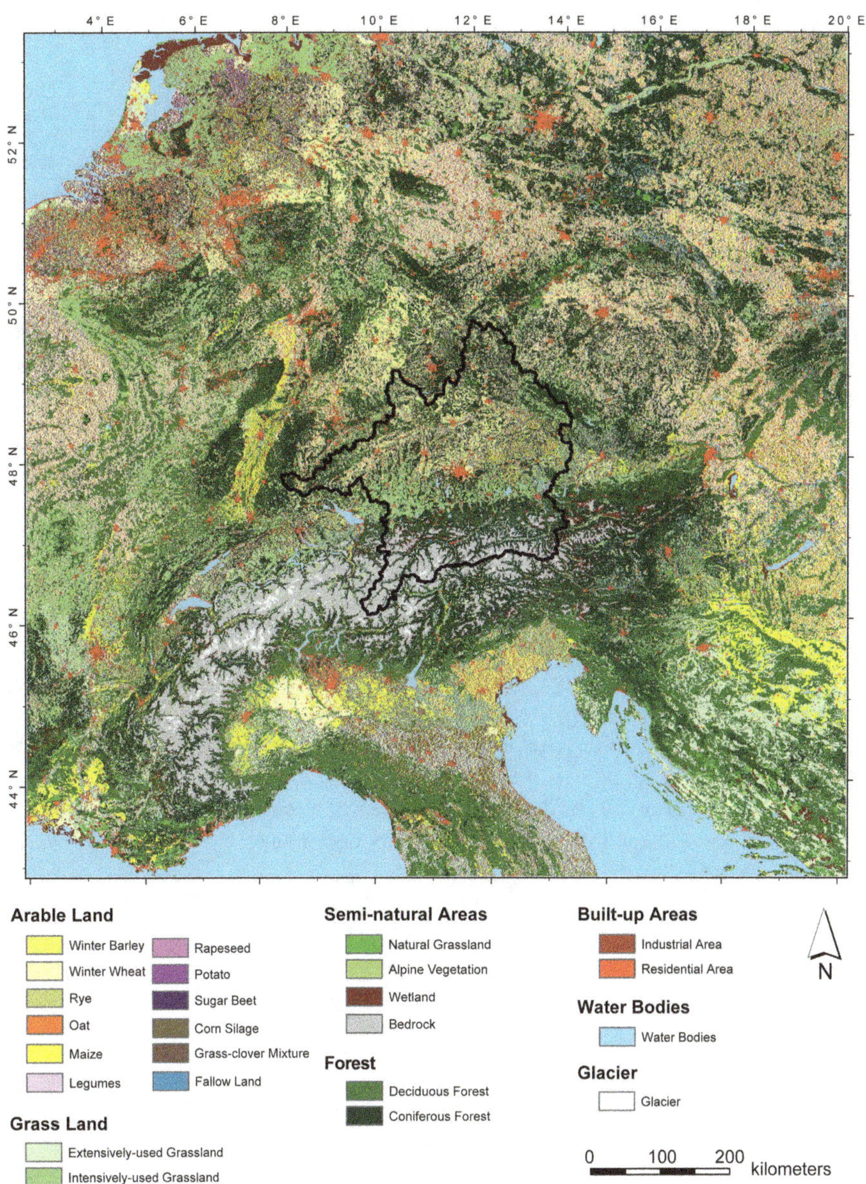

Fig. 33.3 Land use classification for the PROMET model at a spatial resolution of 1×1 km^2 for the coupling area

It is not only cities, rock or glacier that cannot be depicted at a 45×45 km^2 resolution; agricultural categories are also able to be subdivided at higher spatial resolution into the key types of agricultural use. Given the high percentage of agricultural area in Central Europe, this is especially important since different phenological developments within various types of cultivations can have significant effects on temporal dynamics of the land surface, such as transpiration or other energy fluxes (Zabel et al. 2010).

33.4 Two-Way Coupling

PROMET has already been successfully run with data from regional climate models (*REMO, MM5*) (Marke 2008; Marke et al. 2011). This has generated the regional climate variants from the regional climate models *REMO* and *MM5* (*downscaled and bias corrected*), which are described in detail in Chap. 51. This one-way coupling was developed within the GLOWA-Danube project to enable the PROMET model to be driven by data from regional climate models. To do so, the meteorological data from regional climate models must be downscaled to the resolution of the hydrological models and then subjected to a bias correction (see Chap. 51).

While within the one-way coupling approach both the regional climate model (RCM) and the hydrological model (see Fig. 33.4) each use their own land surface data (land use, DTM, soil) to calculate the energy fluxes at the land surface, which induces inconsistencies in the coupled model system, two-way coupling systematically excludes such inconsistencies. This is achieved by sharing the same land surface model in both models (see Fig. 33.4). In this case, the land surface serves as the link between the atmosphere and hydrology, through which mass and energy are exchanged.

In the two-way coupling method presented here, the energy fluxes at the land surface are now calculated with PROMET at a spatial resolution of 1×1 km^2; the meteorological variables are then prepared every 9 simulation minutes by *MM5* at 45×45 km^2 and then downscaled by the model interface SCALMET (Marke 2008) to 1×1 km^2.

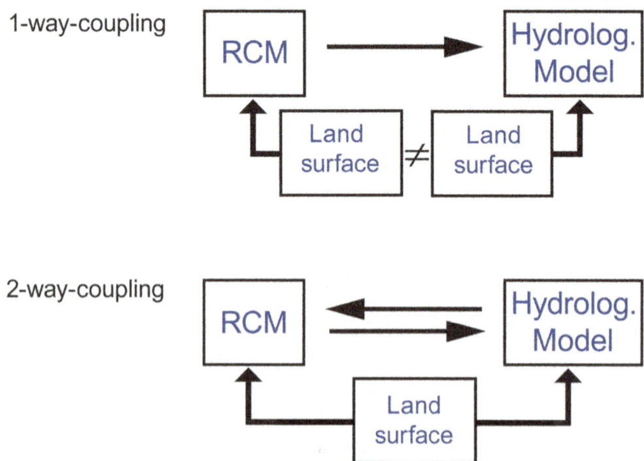

Fig. 33.4 Scheme of the one-way and two-way coupling

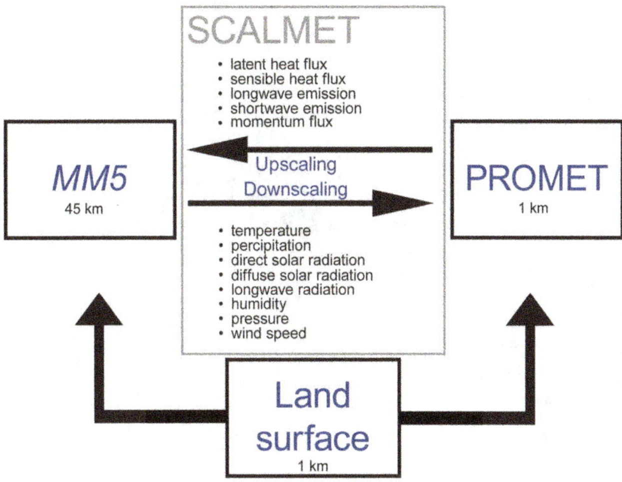

Fig. 33.5 Methodology of the two-way coupling

In contrast to the one-way coupling approach, the scaling within the two-way coupling approach is applied without empirical scaling functions or bias correction (see Chap. 51). Instead, it uses physical models or quasi-physical methods such as elevation gradients that are derived from the meteorological data from *MM5* for each coupling time step. Energy and mass thus always remain preserved. The energy fluxes at the land surface calculated by PROMET are then passed to *MM5*, in order to complete the energy balance within the coupled land-atmosphere system. This requires upscaling the energy fluxes from 1 to 45 km (see Fig. 33.5). There are no scaling issues in this case, since the energy fluxes exhibit linear behaviour and can simply be averaged.

33.5 Results

Coupled and uncoupled model simulations were conducted from 1996 to 1999 using ERA-40 reanalysis data from the ECMWF.

The mean annual atmospheric temperature from the station data within the Upper Danube basin is 7.0 °C for the 4-year simulation period, while the results of the *MM5* simulation with the land use classification generated internally by *MM5* show a mean atmospheric temperature for the basin of 6.1 °C. In the fully coupled simulation, mean atmospheric temperature is 6.9 °C. Over the entire simulation period, the atmospheric temperature from the fully coupled simulation is approximately 0.8 °C warmer than the temperature from the uncoupled simulation and is thus closer to the interpolated measurements from station data. The 3-h temperature patterns are relatively similar over the course of the year. Figure 33.6 shows an example of this for the year 1999.

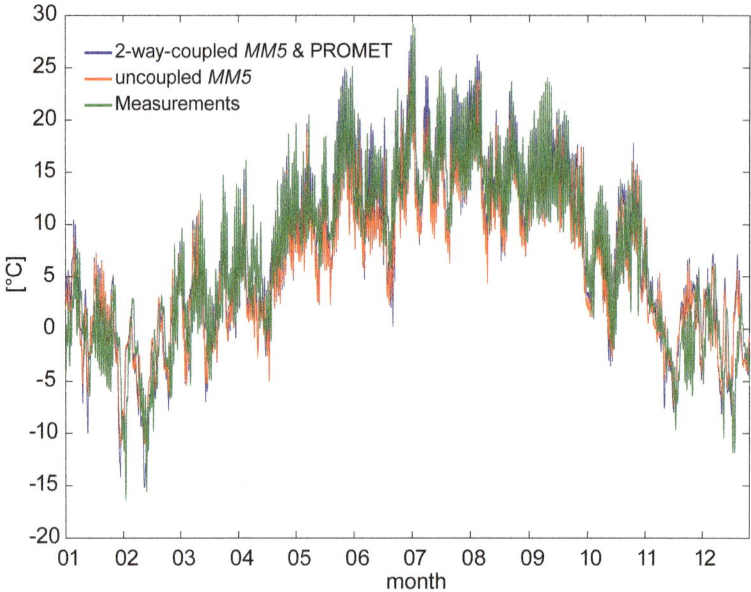

Fig. 33.6 Modeled three-hourly near surface air temperature of uncoupled MM5 (*red*) and fully coupled (*blue*) simulations in comparison with measurements (*green*). All data is averaged over the Upper Danube basin

Both for the fully coupled and the uncoupled simulations, the model results are predominately driven by the ERA-40 reanalysis data that determines the principle course of the temperature.

However, it should be noted that atmospheric temperature during the summer months in the uncoupled simulation is too cool compared to the temporally and spatially interpolated data from the climate stations (Pfeiffer and Zängl 2009). This effect is particularly evident at the daily maxima. In order to explain this temperature pattern, Fig. 33.7 shows the mean monthly diurnal variation in atmospheric temperature in June and in December for the model simulations as a mean across the simulation period from 1996 to 1999.

In this figure, it is clear that the effects of the high-resolution land surface on the simulated atmospheric temperature in the Upper Danube basin in winter are marginal, but that in summer there are significant differences. In winter, energy input at the land surface is low. As a result, differences between the coupled and uncoupled model systems are low, while in summer, feedback effects change mechanisms between the land surface and the atmosphere which lead to greater variations in atmospheric temperature. Thereby, the relation of latent and sensible heat changes, resulting in an increased diurnal temperature in July in case of the tow-way coupling (see Fig. 33.7) (Zabel and Mauser 2013). For this reason, the simulated temperatures from two-way coupling are closer to the measured climate station data.

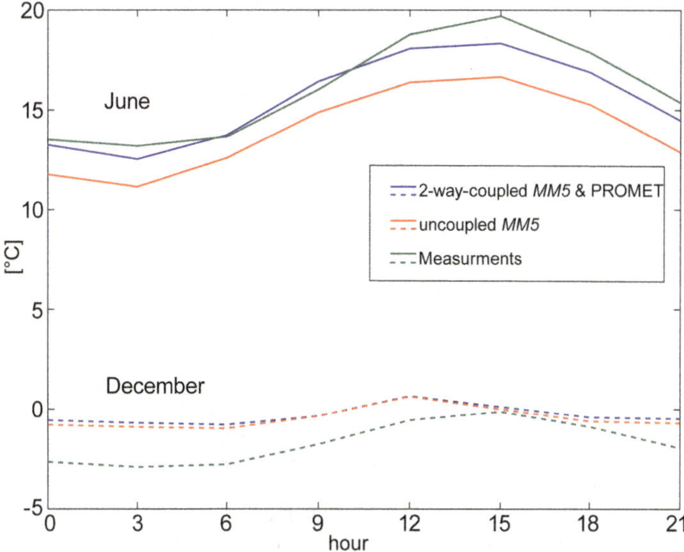

Fig. 33.7 Daily course of the near surface air temperature, averaged over the Upper Danube basin and the years 1996–1999

One cause for this may be the consideration of impervious surfaces such as cities or for rock in the two-way coupling approach, while the *MM5* resolution does not resolve these areas. Since impervious surfaces warm up more intensely the surrounding air masses in turn also heat up more intensely. Besides the additional consideration of small-scale land use and cover due to the increased spatial resolution in the hydrological land surface model PROMET, other important feedback effects between the land surface model and the atmospheric model were identified. Particularly, the more detailed consideration of soil information and soil processes as well as the more sophisticated handling of snow pack in the alpine areas had great impacts on the improved simulation of land-atmosphere interactions in the two-way coupling approach (Zabel et al. 2012).

33.6 Conclusion

Since higher-resolution climate models to date are computationally too intensive, the two-way coupling involving internal downscaling and upscaling offers an opportunity to account for the heterogeneity of the land surface in the climate simulations; this approach involves calculating the land surface processes at a higher resolution than the atmospheric processes operate at. Driving the *MM5* regional model at the boundaries of the model domain using ERA-40 reanalysis data decisively determines the model behaviour. Two-way coupling changes the results of the regional climate model, due to an improved land surface, higher spatial resolution and improved parameterisation of land surface properties in PROMET. It was demonstrated that the diurnal variation in atmospheric temperature yielded better results due to land-atmosphere interactions caused by the changed radiation and energy balance at the land surface. As a result of the improved simulation of feedback effects, precipitation patterns changed in the atmospheric model which in turn improved the simulation results of the water balance and runoff in the Upper Danube catchment (Zabel and Mauser 2013).

References

IPCC (eds) (2007) Climate change 2007 the physical science basis. In: Solomon S, Qin D, Manning M, Chen Z, Marquis M, Averyt KB, Tignor M, Miller HL (eds) Contribution of the working group I to the fourth assessment report of the Intergovernmental Panel of Climate Change. Cambridge University Press, Cambridge/New York, p 996

Jarvis A, Reuter HI, Nelson A, Guevara E (2008) Hole-filled seamless SRTM data V4. International Centre for Tropical Agriculture (CIAT). Available via DIALOG. http://srtm.csi.cgiar.org. Accessed 19 Sept 2014

Marke T (2008) Development and application of a model interface to couple regional climate models with land surface models for climate change risk assessment in the Upper Danube watershed. Dissertation, LMU Munich

Marke T, Mauser W, Pfeiffer A, Zängl G (2011) A pragmatic approach for the downscaling and bias correction of regional climate simulations: evaluation in hydrological modeling. Geosci Model Dev 4:759–770

Mauser W, Bach H (2009) PROMET Large scale distributed hydrological modelling to study the impact of climate change on the water flows of mountain watersheds. J Hydrol 376:362–377

Pfeiffer A, Zängl G (2009) Validation of climate-mode MM5-simulations for the European Alpine region. Theor Appl Climatol 101:93–108

Zabel F, Mauser W (2013) 2-way coupling the hydrological land surface model PROMET with the regional climate model MM5. Hydrol Earth Syst Sci 17:1–10

Zabel F, Hank TB, Mauser W (2010) Improving arable land heterogeneity information in available land cover products for land surface modelling using MERIS NDVI data. Hydrol Earth Syst Sci 14:2071–2084

Zabel F, Mauser W, Marke T, Pfeiffer A, Zängl G, Wastl C (2012) Inter-comparison of two land-surface models applied at different scales and their feedbacks while coupled with a regional climate model. Hydrol Earth Syst Sci 16:1017–1031

Chapter 34
Mean Number of Storm Days

Boris Thies, Thomas Nauss, Christoph Reudenbach, Jan Cermak, and Jörg Bendix

Abstract Precipitation events are the main driving force for hydrological processes; for this reason, correctly compiling the distribution of precipitation in the study area is given high priority. Therefore, three models for assessing precipitation were implemented in DANUBIA: a mesoscale atmosphere model, an interpolation model based on station data and a satellite-supported rainfall retrieval. The satellite-based derivation of precipitation takes place using data from the European Meteosat system. In a first step, the boundaries of the raining cloud areas are delineated. Second, the precipitation rate is assigned considering the precipitation processes identified before.

Comparing monthly mean precipitation in the study area for 1999 based on the atmospheric model, the interpolation method and the satellite-based technique reveal shortcomings in identifying stratiform precipitation for the satellite method. On the other hand, weather models have slight weaknesses in calculating convective precipitation. Further results show the average number of storm days from May to September between 1995 and 1999 derived using the satellite retrieval technique. The frequency distribution indicates the expected midsummer maximum in July and reveals an increase in thunderstorm frequency caused by orography within the drainage basin.

Keywords GLOWA-Danube • DANUBIA • Storm days • Upper Danube

34.1 Introduction

Precipitation events are the main driving force for hydrological processes; for this reason, correctly compiling the distribution of precipitation in the study area is given high priority. Therefore, three models for assessing precipitation were

B. Thies (✉) • T. Nauss • C. Reudenbach • J. Bendix
Department of Geography, University of Marburg, Marburg, Germany
e-mail: boris.thies@geo.uni-marburg.de; nauss@geo.uni-marburg.de; reudenbach@geo.uni-marburg.de; bendix@geo.uni-marburg.de

J. Cermak
Department of Geography, University of Bochum (RUB), Bochum, Germany
e-mail: jan.cermak@rub.de

W. Mauser, M. Prasch (eds.), *Regional Assessment of Global Change Impacts*, DOI 10.1007/978-3-319-16751-0_34

implemented in DANUBIA: a mesoscale atmospheric model (*AtmoMM5*), an interpolation model based on station data (*AtmoStations*) and a satellite-supported precipitation retrieval (*AtmoSat*).

A method based on radar data was not implemented, since systematic errors (clutters) caused by the topography of the drainage basin in the area of the Alps would limit a comprehensive application to the Alpine foreland. The satellite-based precipitation retrieval *AtmoSat* is thus the only method that is based on comprehensive, area-wide data datasets (satellite images) that permit direct mapping of the spatial patterns of precipitations; for this reason, it was primarily used to validate the other two models (*AtmoMM5*, *AtmoStations*). Moreover, *AtmoSat* can be applied to the complete Meteosat series of satellites; this allows for the generation of long precipitation time series and hence facilitates an analysis of the regional effects of climate change in the Upper Danube basin.

In addition to compiling area-wide precipitation information, *AtmoSat* can also be used for capture and real-time observation of specific weather events such as thunderstorms (Nauss et al. 2004), since these are associated with deep, convective cloud cores that can be identified with a high degree of certainty in the thermal sensor channels. The high spatio-temporal variability of storms requires a quasi-continuous measurement system that is provided by the 30- and 15-min repetitions of the Meteosat systems.

34.2 Data Processing

The data from the European Meteosat system forms the basis for establishing the precipitation retrieval technique; at an elevation of approximately 36,000 km, this system is in a geostationary orbit over the point of intersection between 0° longitude and the equator that is synchronous to the rotation of the earth. This position permits recording of the entire hemisphere facing the sensor (Europe, Africa) at a spatial resolution of only a few kilometres and a temporal resolution on 30 or 15 min since the introduction of Meteosat-8. With the conversion to Meteosat-8 since the beginning of 2004, the spectral resolution of the Spinning Enhanced Visible and Infrared Imager (SEVIRI) sensors on board has also significantly increased. The METEOSAT Visible and Infra-Red Imager (MVIRI) sensor on satellites up to Meteosat-7 only had three available channels; SEVIRI has 12 channels within the visible spectrum, the near-infrared range, the water vapour absorption bands and the thermal infrared range.

Various preprocessing steps are required to derive precipitation information; these steps are implemented within the operational process chains of the precipitation/remote sensing subproject. The data received from Meteosat (5×5 km^2) are first calibrated, then transferred to the DANUBIA map projection (1×1 km^2) using a bilinear interpolation and finally forwarded to the retrieval integrated within DANUBIA.

34.3 Model Documentation

The derivation of precipitation in *AtmoSat* takes place using the two-part Advective-Convective Technique (ACT) (Nauss et al. 2004; Reudenbach et al. 2001, 2007) that is integrated within DANUBIA. The first part serves to delineate the boundaries of the raining cloud areas in the satellite image, and the second is for assigning a precipitation rate, which in ACT is based on three-dimensional cloud model calculations and accounts for the precipitation processes identified before. The remote sensing part of the ACT consists of two modules that are used to identify precipitation events in convective clouds (e.g. thunderstorms) and stratiform clouds (e.g. from nimbostratus clouds that arise in conjunction with convective clouds). Precipitation from convective clouds is calculated on the basis of the measured cloud surface temperature in water vapour (WV, ~6.5 μm) and infrared channel (IR, ~11 μm). Tjemkes et al. (1997) demonstrated that high-altitude (storm) clouds with cloud tops near reaching above the tropopause are warmer in the WV channel than in the IR channel, so that positive WV-IR temperature differences suggest the existence of high-altitude, convective clouds. In this way, a physically based identification of storm cells is possible. In contrast to the derivation of convective precipitation, the identification of stratiform precipitation areas for the period 1995–2003 is not directly physically based, since for this period, only data from Meteosat MISR with the three channels are available. The identification is based on a combination of cluster analysis of the WV and IR signals and an analysis of the compactness of the cloud fields, by means of which a decision is ultimately made whether or not a precipitation situation is present.

With the advent of Meteosat-8 SEVIRI in the beginning of 2004, this second module of the ACT could be replaced by the physically based Rain Area Delineation Scheme (RADS) (Nauss 2006), which was likewise implemented within *AtmoSat*. The key innovation of the concept model of RADS is that the areas of precipitation are derived from information about the vertical depth of the clouds and the size of the cloud droplets.

A sufficient vertical depth is not only required to form sufficiently large droplets according to the dynamics of cloud systems but is also a controlling factor for evaporation of drops falling beneath the clouds, which in turn has a direct impact on the necessary droplet size. The semi-analytical cloud retrieval algorithm (SACURA) (Kokhanovsky et al. 2003 and Nauss et al. 2005) was used to derive the cloud parameters that require the enhanced spectral resolution of SEVIRI; this algorithm is based on measurements of reflection in the visible spectrum and the near-infrared range.

34.4 Results

Figure 34.1 shows a comparison of the monthly mean precipitation in the study area for 1999 based on *AtmoMM5*, *AtmoStations* and the ACT in *AtmoSat*. The monthly deviations are largely attributable to the fact the satellite method contains

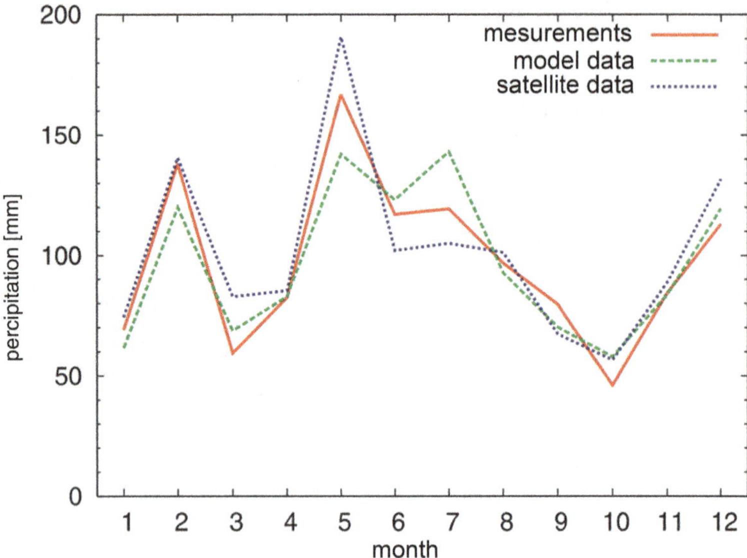

Fig. 34.1 Comparison of mean monthly areal precipitation for the year 1999 based on spatio-temporal interpolated measurements (AtmoStations), atmospheric model calculations (AtmoMM5) and satellite data (AtmoSat)

shortcomings in identifying stratiform precipitation because of the requirement to use the old Meteosat MISR sensor; on the other hand, weather models have slight weaknesses in calculating convective precipitation, and, in addition, the calculated distribution of precipitation is highly dependent on the resolution selected. The interpolation models based on station data ultimately tend to reduce peak precipitation rates while at the same time expanding the area of precipitation. However, the deviations should not belie the fact that all three models represent the respective status of research and on average allow for quite reliable statements about the distribution of precipitation.

Map 34.1 shows the average number of storm days from May to September between 1995 and 1999 derived using the ACT. A day was identified as a storm day if, within a period of 24 h, at least one significant convective cloud core that protruded into the region of the tropopause was observed over the respective position, since this type of system will almost certainly lead to a storm.

Unlike the climatological definition of a storm day that is spoken of if an observation of thunder is heard for a location, the satellite-based measurement of the location of the storm core must match the observation location. The number of storm days calculated with the satellite method thus appears to be somewhat lower compared to subjective observations.

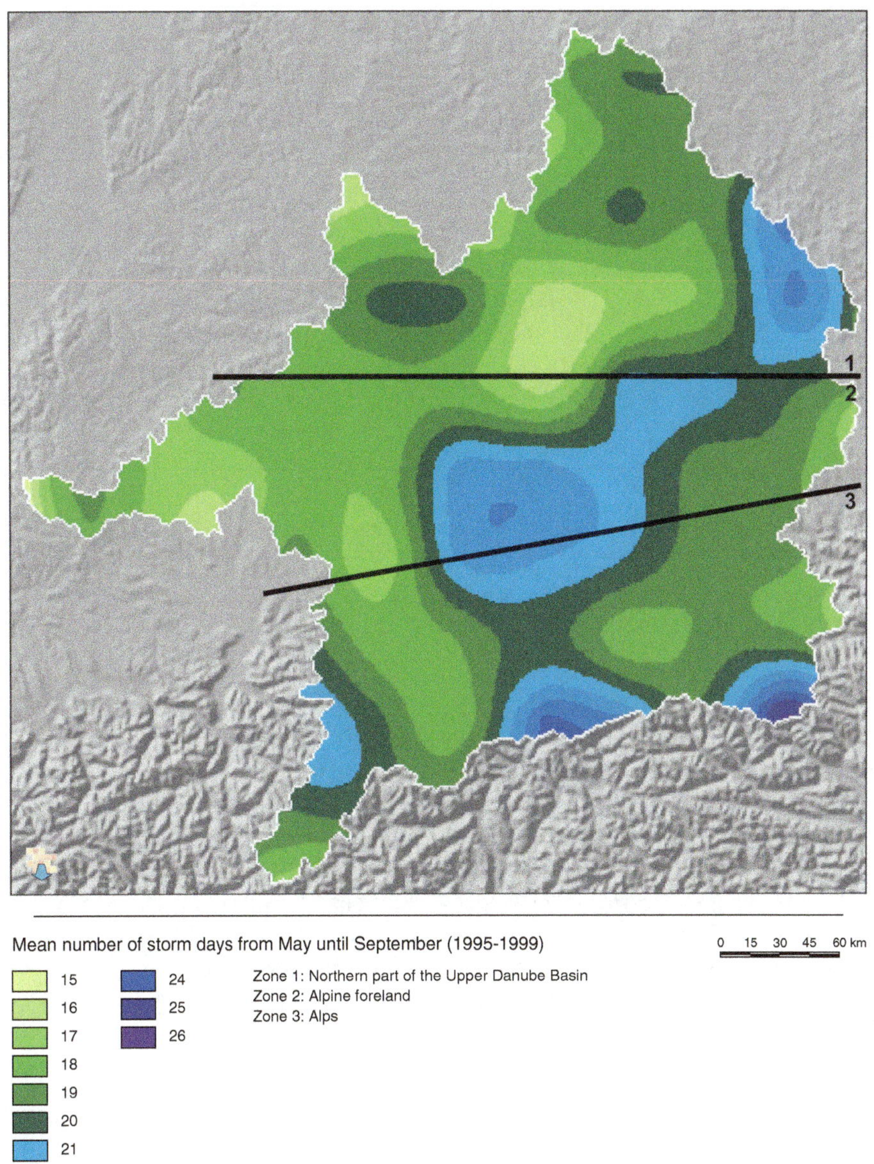

Map 34.1 Mean number of storm days (Data source: Meteosat-7 satellite data 1995–1999)

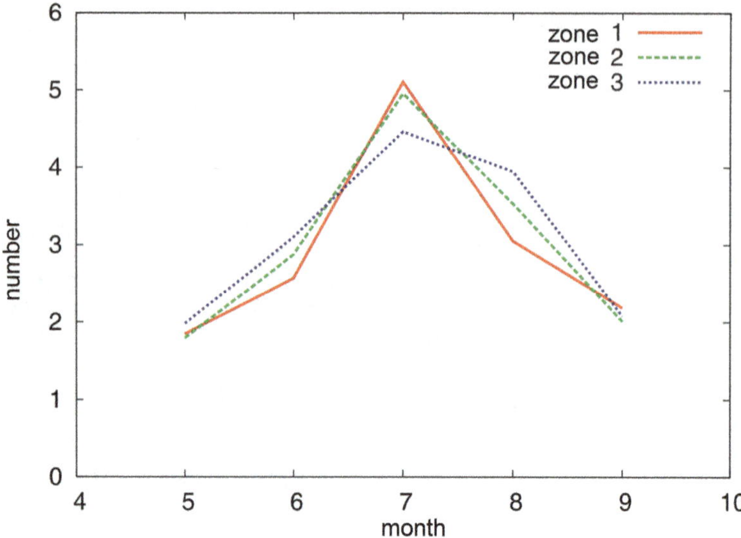

Fig. 34.2 Mean number of storm days for the months May to September in the Alps (*zone 3*), the Alpine foreland (*zone 2*) and the northern part of the Upper Danube basin (*zone 1*) based on satellite data of the years 1995–1999

The map basically indicates the increase in the frequency of thunderstorms caused by orography within the drainage basin. In the east, these orographic factors are the Bavarian Forest and the Bohemian Forest, in the south there are the Alps and in the northwest there are the Swabian Jura and in part also the Franconian Jura. In the lee of the Black Forest and the Swabian and Franconian Jura, there is a region with distinctly lower likelihood of storms. On average for the years 1995–1999, 14–26 storm days occurred in the Upper Danube basin between May and September. When compared to the climatological definition of a storm day, by which up to 30 storm days were counted in the Bavarian lowlands between January and December, the data from the satellite system can be considered realistic (Fig. 34.2).

References

Kokhanovsky AA, Rozanov VV, Zege EP, Bovensmann H, Burrows JP (2003) A semianalytical cloud retrieval algorithm using back-scattered radiation in 0.4–2.4 μm spectral region. J Geophys Res 108:AAC 4-1–AAC 4-19

Nauss T (2006) Das Rain Area Delineation Scheme RADS. Ein neues Verfahren zur satellitengestützten Erfassung der Niederschlagsfläche über Mitteleuropa. Marburger Geographische Schriften 143

Nauss T, Reudenbach C, Cermak J, Bendix J (2004) Operational identification and visualisation of cloud processes for general aviation using multispectral data. In: Proceedings of the Eumetsat Satellite conference, Prague, p 196–172

Nauss T, Kokhanovsky AA, Nakajima TY, Reudenbach C, Bendix J (2005) The intercomparison of selected cloud retrieval algorithms. Atmos Res 78:46–78

Reudenbach C, Heinemann G, Heuel E, Bendix J, Winiger M (2001) Investigation of summertime convective rainfall in Western Europe based on a synergy of remote sensing data and numerical models. Meteorol Atmosph Phys 76:23–41

Reudenbach C, Nauss T, Bendix J (2007) Retrieving precipitation with GOES, Meteosat and Terra/MSG at the tropics and midlatitudes. In: Levizzani V, Bauer P, Turk FJ (eds) Measuring precipitation form space – EURAINSAT and the future. Advances in global change research, vol. 28, XXVI. Springer Science & Business Media, Dordrecht, p 722

Tjemkes SA, van de Berg L, Schmetz J (1997) Warm water vapour pixels over high clouds as observed by Meteosat. Contrib Atmos Phys 70:15–21

Chapter 35
Energy: Simulation of Hydropower Generation and Reservoir Management

Franziska Koch, Andrea Reiter, and Heike Bach

Abstract Within the hydropower module of DANUBIA, all large run-of-the-river and storage power plants in the Upper Danube basin are included. Potential and kinetic energy are used for the generation of hydroelectric power which depends mainly on discharge and drop height. For each hydropower plant, the discharge and the power output were simulated depending on a discharge-power output function for each modelled time step of 1 h. For annual comparability, the annual output was calculated whereof the hydraulicity was derived for the years 1960–2006. Moreover, a monthly based reservoir operation plan was included for each reservoir power plant considering basic reservoir operation rules. A validation was carried out for all hydropower plants, which are, with a coefficient of determination of 0.99, very well reproduced in DANUBIA. The mean simulated annual output of the run-of-the-river and reservoir power plants from 1971 to 2000 is shown, divided in the six subbasins Inn, Salzach, Isar, Lech, Iller and the remaining area of the Danube basin.

Keywords GLOWA-Danube • Hydropower model • Mean annual output • Reservoir operation plan

35.1 Introduction

Hydropower plays an enormous role within the Upper Danube basin. As described in Chap. 23, the modelling of DANUBIA includes virtually all of the large run-of-the-river and storage power plants. In this chapter, their mode of operation within

F. Koch (✉)
Department of Geography, Ludwig-Maximilians-Universität München (LMU Munich), Munich, Germany
e-mail: f.koch@iggf.geo.uni-muenchen.de

A. Reiter
Bavarian Research Alliance GmbH, Munich, Germany
e-mail: reiter@bayfor.org

H. Bach
Vista Geowissenschaftliche Fernerkundung GmbH, Weßling, Germany
e-mail: bach@vista-geo.de

W. Mauser, M. Prasch (eds.), *Regional Assessment of Global Change Impacts*, DOI 10.1007/978-3-319-16751-0_35

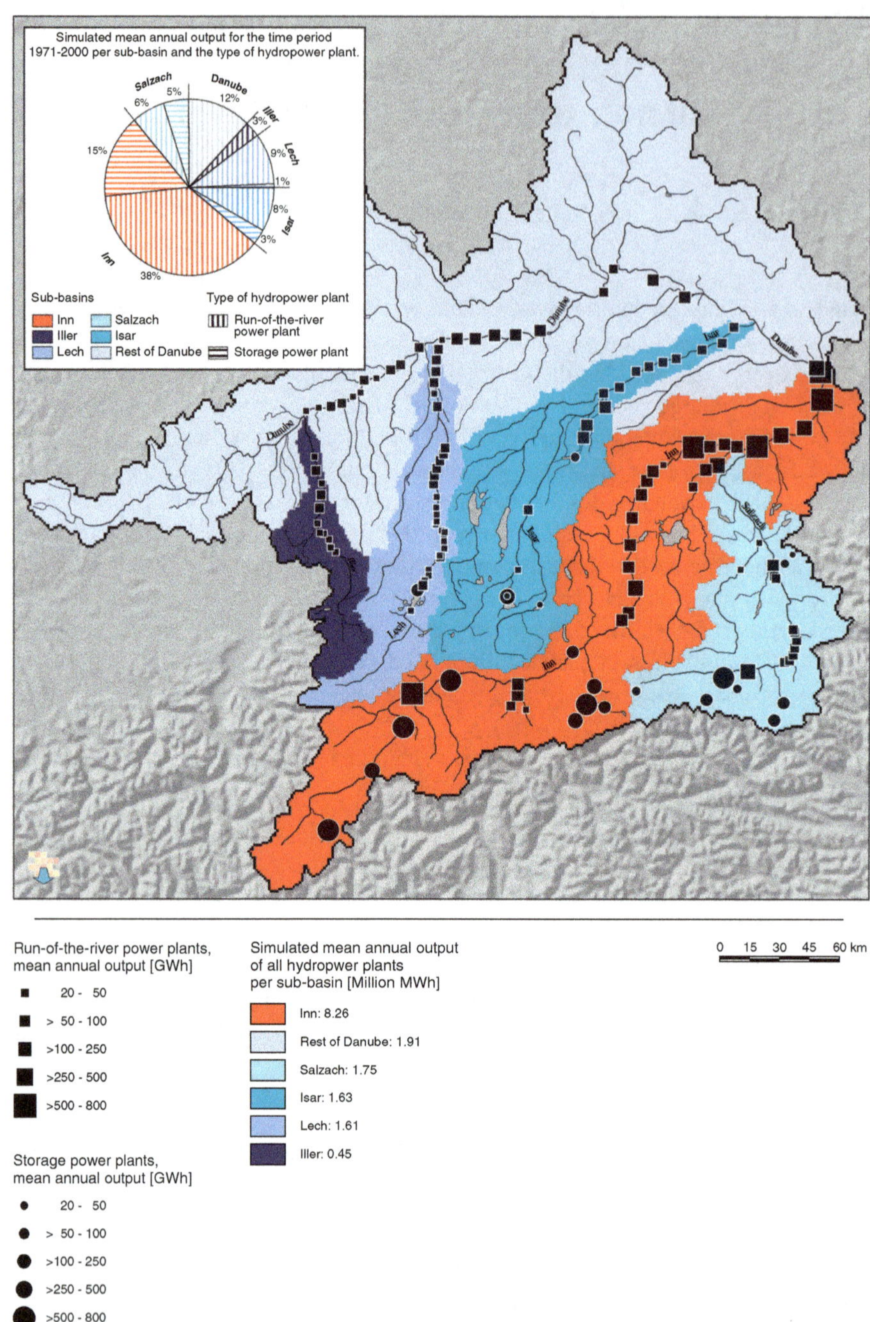

Map 35.1 Simulation of hydropower generation and reservoir management (Data source: DANUBIA – river network, BMLFU – Austrian Federal Ministry for Agriculture, Forestry, Environment and Water Management (2005), Investigations and surveys carried out by the authors, Jarvis et al. (2008), Fichtner and Co (2003))

DANUBIA is demonstrated. In addition, the management of reservoirs is dealt with. The largest reservoirs that are primarily used for energy production are located in the high alpine regions. For those, the operation plans for reservoir management are oriented towards hydroelectricity (see Chap. 23). The discharge of natural inflow over the course of a year is sometimes shifted significantly as a result of storage, so that the correct modelling of discharge and power output for reservoirs becomes a major priority.

35.2 Data Processing

See Chap. 23, Hydropower Plants.

35.3 Model Documentation

35.3.1 Simulated Annual Output

To calculate the hydropower generation in DANUBIA and to validate the implemented hydropower plants, the following parameters are needed:

- Discharge Q [m^3/s]
- Overall efficiency η
- Drop height H [m]
- Maximum/bottleneck capacity P_{max} [MW]
- Optimal discharge Q_{opt} [m^3/s]
- Minimum discharge Q_{min} [m^3/s]
- Maximum discharge Q_{max} [m^3/s]
- Annual output W_R [MWh]
- Year of commissioning [a]

Potential and kinetic energy are used for the generation of hydroelectric power, since the power output P is largely dependent on both discharge Q and drop height H. The drop height refers to the difference in elevation between the upper water level and the lower water level at the axis of the turbines in the power plant itself. In modern hydroelectric facilities, the overall efficiency η of a hydropower plant is usually between 0.8 and 0.9. This also accounts for loss from energy conversion within the entire facility such as from the inlet and outlet of the turbines. The following formula is used to calculate the power output of hydroelectric power plants (Strobl and Zunic 2006):

$$
\begin{aligned}
&\mathrm{P} = \eta\ p\ Q\ g\ H\ [\mathrm{W}] \\
&\eta \quad \text{overall efficiency} \\
&p \quad \text{density of the water} \left[\mathrm{kg/m^3}\right] \\
&Q \quad \text{discharge} \left[\mathrm{m^3/s}\right] \\
&g \quad \text{gravitational acceleration} \left[\mathrm{m/s^2}\right] \\
&H \quad \text{drop height} [\mathrm{m}]
\end{aligned}
$$

In DANUBIA, the discharge and the power output that results therefrom are calculated for each simulated time interval (i.e. hourly) and each implemented power plant. If the power outputs generated over the course of a year are summed, the result is the simulated annual output W_J of a plant, which represents the mean energy production for the year.

Figures 35.1 and 35.2 schematically illustrate the formation of the model of energy production as a function of discharge for run-of-the-river and storage power plants in DANUBIA. The turbines start running at a specified minimum discharge Q_{min} based on the legally regulated minimum flow rates, and so too the run-of-the-river power plants implemented in the model begin with energy generation. The power output increases with increasing discharge up to the optimal discharge Q_{opt}, at which the maximum output P_{max} and the maximum turbine discharge are ultimately reached (see Fig. 35.1). The run-of-the-river power plants also produce energy up to a specified maximum discharge Q_{max}, although this is reduced since the water level increases downstream and hence the drop height is reduced. In extreme flood situations, the energy production is stopped beyond

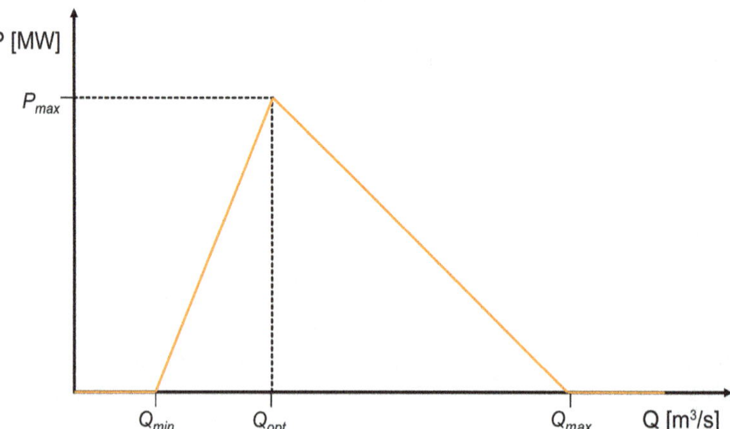

Fig. 35.1 Schematic representation of the relationship between discharge and power output of a run-of-the-river power plant (Koch et al. 2011)

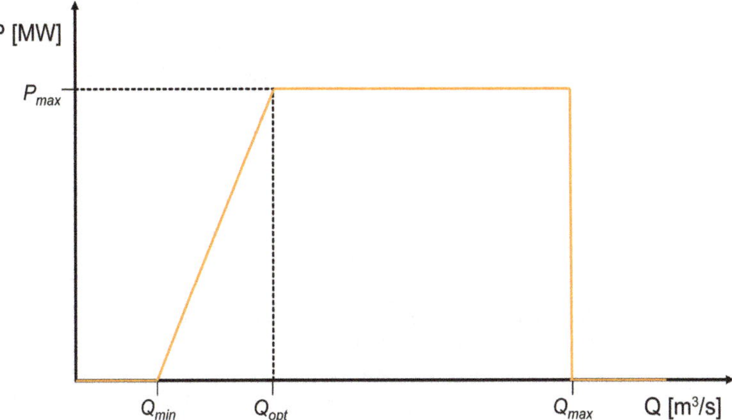

Fig. 35.2 Schematic representation of the relationship between discharge and power output of a storage power plant (Koch et al. 2011)

Q_{max} in order to simulate the protective mechanism against flood damage within the facility (Koch et al. 2011).

The functioning within storage power plants is identical, with an increase in power output up to the maximum discharge (see Fig. 35.2). However, with an increasing discharge, the maximum capacity is maintained at a constant level that is above the installed flow rate, since the elevation difference is usually such that the downstream gradient increase is not a significant factor. Beyond a specified discharge, production of energy is largely discontinued due to flood protection.

Validation was carried out for the period from 2000 to 2006 for all hydropower plants that began operation prior to year 2000. Thus, all but four power plants were considered in the validation. The mean annual output values provided by the power plant operators were compared with the simulated mean outputs (see Fig. 35.3). Overall, the hydropower plants are reproduced very well in DANUBIA with a coefficient of determination of more than 0.99 and, hence, can be used without hesitation for future simulations.

35.3.2 Reservoir Management

In general, run-of-the-river power plants immediately use the river run-off. In contrast, storage power plants retain the inflow over a longer period in reservoirs and release it in a delayed manner. Therefore, for each storage power plant with a long-term annual storage, a monthly operation plan to control the release of discharge and the stored volume has been implemented within DANUBIA. The following

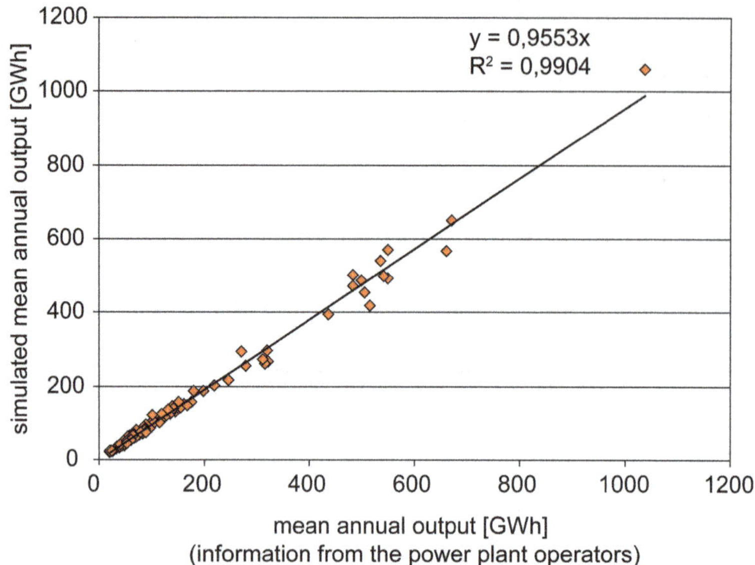

Fig. 35.3 Validation of the simulated annual output of the hydropower plants for the time period 2000–2006 (Modified after Koch et al. 2011)

basic reservoir operating rules have been defined for this (Ostrowski and Lohr 2002; Willems 2007):

- Accounting for maximum and minimum output capacities
- Accounting for maximum and minimum storage provisions
- Normal discharge output
- Flood discharge
- Automatic output increase or reduction
- Maximum change in discharge per time interval

Figure 35.4 illustrates the example of the operation plan for the Gepatsch reservoir in the form of a look-up table. The storage space within a reservoir typically can be divided using lamellae defined in the operation plan and which specify a certain fill volume (see Chap. 23). Each month, the discharge output is defined with regard to the fill volume and is interpolated between the lamella (Willems 2007). In the case of the Gepatsch reservoir, the discharge output given the same fill volume is higher in the winter months from November to March than in the summer months from April to October. This result is based on the fact that in summer, snow and ice melt inflows are increasingly stored for energy production during winter.

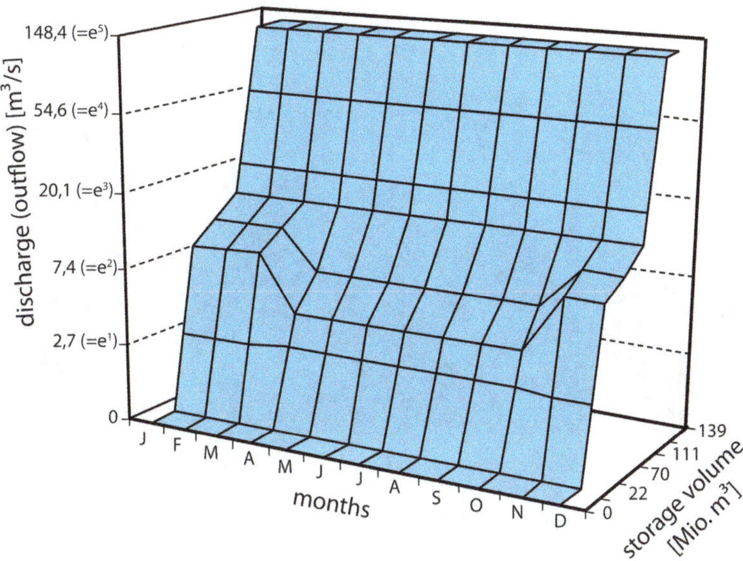

Fig. 35.4 Look-up table for the operation plan of the Gepatsch reservoir under normal conditions (Modified after Koch et al. 2011)

35.4 Results

Hydraulicity serves as a relative measure for hydroelectric power generation; this variable represents the relationship of the current output to its long-term mean for each year. Figure 35.5 depicts both simulated hydraulicity and simulated annual output for 1960–2006. Simulated annual output also increased because more hydroelectric power plants continue to be built and connected to the power network. An excess is generated if hydraulicity is above 100 %, while if it is below, less than average is generated. Thus, in years of low hydraulicity such as in the extremely hot and dry year 2003, there are significant losses of output.

The simulated mean annual output from 1971 to 2000 for all storage and run-of-the-river power plants of the subbasins of Iller, Lech, Inn and Salzach rivers and the remaining area of the Danube basin is shown in Map 35.1. The map clearly shows that over half of the mean simulated annual output is generated in the Inn catchment (approximately 9.3 million MWh). The partial drainage basins of the Lech, Isar, Salzach and the remaining Danube area generate much less, with 1.6–2.4 million MWh each. In the relatively small drainage catchment of the Iller, approximately 0.45 million MWh is produced. The simulated mean annual output in storage power plants takes on a greater significance in the high alpine Inn and Salzach basins. In contrast, in the Isar and Lech basins, only a small fraction of the total simulated mean annual output is generated by storage power plants (see the pie chart in Map 35.1). Only run-of-the-river power plants are found on the Iller and the rest of the Danube.

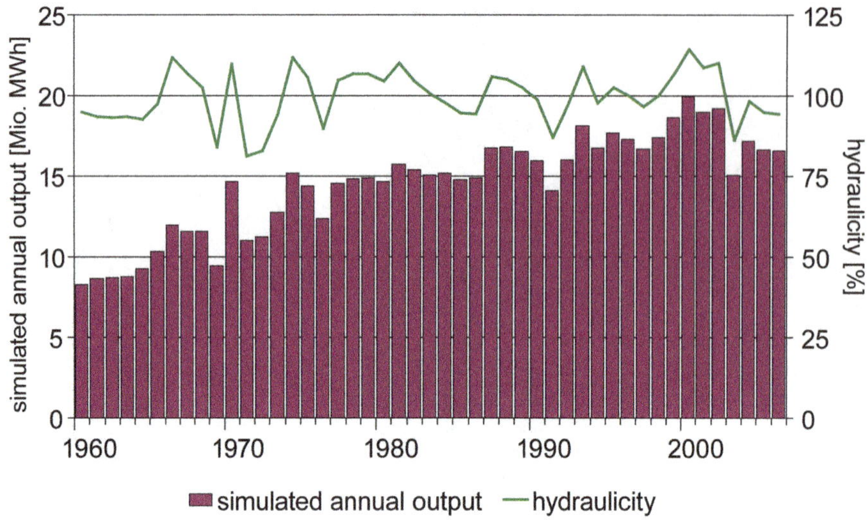

Fig. 35.5 Hydraulicity and simulated annual output of the hydroelectric power generation within the Upper Danube basin for the time period 1960–2006

References

BMLFU – Austrian Federal Ministry for Agriculture, Forestry, Environment and Water Management (ed) (2005) Hydrological atlas of Austria. Österreichischer Kunst- und Kulturverlag, Vienna

Fichtner GmbH, Co KG (ed) (2003) Die Wettbewerbsfähigkeit von großen Laufwasserkraftwerken im liberalisierten deutschen Strommarkt. Endbericht. Available via DIALOG. http://www.emissionshandel-fichtner.de/pdf/BMWA_Langfassung.pdf. Accessed 24 Nov 2009

Jarvis A, Reuter HI, Nelson A, Guevara E (2008) Hole-filled seamless SRTM data V4. International Centre for Tropical Agriculture (CIAT). Available via DIALOG. http://srtm.csi.cgiar.org. Accessed 19 Sept 2014

Koch F, Prasch M, Bach H, Mauser W, Appel F, Weber M (2011) How will hydroelectric power generation develop under climate change scenarios? A case study in the Upper Danube Basin. Energies 4(10):1508–1541. doi:10.3390/en4101508

Ostrowski M, Lohr H (2002) Modellgestützte Bewirtschaftung von Talsperrensystemen. Wasser und Abfall 4(1–3):40–45

Strobl T, Zunic F (2006) Wasserbau. Aktuelle Grundlagen – Neue Entwicklungen. Springer, Berlin

Willems W (2007) Teilprojekt Oberflächengewässer. Integration wassermengen- und wasserqualitätsrelevanter technischer Strukturen und Detektion von Überlastungen in oberirdischen Gewässern. In: Mauser W, Strasser U (eds) GLOWA-Danube. Abschlussbericht Phase 2. Department für Geographie und geographische Fernerkundung. Ludwig-Maximilians-Universität München, Munich, pp 161–204

Chapter 36
CO_2 Fluxes and Transpiration

Victoria I.S. Lenz-Wiedemann, Tim G. Reichenau, Christian W. Klar, and Karl Schneider

Abstract Plants play a key controlling role within the hydrological cycle. For analysing global change impacts on water resources in the Upper Danube basin, coupled and process-based modelling of vegetation water and carbon fluxes is needed. The model component *Biological* is part of the simulation system DANUBIA and calculates the processes of carbon assimilation and transpiration for various vegetation categories (e.g. grassland, winter wheat, sugar beet and maize). To best depict the complex interplay of water, carbon and nitrogen fluxes in agroecosystems, decisions on crop management are included in the modelling. Additionally, meteorological and pedological model input data are provided by other dynamically coupled DANUBIA model components. Modelling of photosynthesis and transpiration takes into account not only the predicted increases in air temperature and atmospheric CO_2 concentration but also the availability of water and nitrogen. Maps of transpiration totals for one hydrological year are presented for several agricultural land uses in the Upper Danube basin. Local conditions, characteristics of the different vegetation categories and differences in management are shown. In this way, spatial and temporal changes in plant water demand and supply under global change conditions and altered cultivation practices are assessed at the regional scale.

Keywords GLOWA-Danube • DANUBIA • Global change • Upper Danube • Hydrology • Ecosystem model • Transpiration • Photosynthesis

36.1 Introduction

Vegetation plays a key controlling role within the hydrological cycle, because plants extract water from the soil and release it into the atmosphere through the process of transpiration. Evapotranspiration is the term applied to the joint action of plant

V.I.S. Lenz-Wiedemann (✉) • T.G. Reichenau • K. Schneider
Institute of Geography, University of Cologne, Cologne, Germany
e-mail: victoria.lenz@uni-koeln.de; tim.reichenau@uni-koeln.de; karl.schneider@uni-koeln.de

C.W. Klar
Forschungszentrum Jülich, Jülich, Germany
e-mail: c.klar@fz-juelich.de

W. Mauser, M. Prasch (eds.), *Regional Assessment of Global Change Impacts*, DOI 10.1007/978-3-319-16751-0_36

transpiration, interception evaporation within moist plant populations and soil evaporation.

Transpiration is of particular significance since it accounts for the greatest portion by far of evapotranspiration within plant populations, is actively controlled by the plants and is closely associated with CO_2 uptake (assimilation) by vegetation. Both the release of water and CO_2 exchange take place via the plant stomata.

During periods of adequate water supply, the stomata are controlled by photosynthesis. Transpiration in this case is determined by assimilation, which (based on a defined leaf surface area) depends in particular on radiation, air temperature and nutrient availability. In contrast, in situations when water is scarce, opening of the stomata is controlled by the quantity of water available. Hence, actual transpiration is lower than potential transpiration, and at the same time, assimilation is reduced. In addition to this direct association between assimilation and transpiration, there are a number of other interactions between plant growth and water demand. Thus, changes in plant growth determine the leaf surface area and therewith evaporation as well as transpiration and interception.

Calculations of past and future water demand and water supply therefore require coupled and process-based modelling of water and carbon fluxes for all vegetation categories in the Upper Danube drainage basin (see Chap. 9). In agricultural ecosystems (e.g. maize, winter wheat, grassland), it is important to consider the type of agricultural land use and its management, since these influence the water balance.

The *Biological* model component was developed to enable adequate mapping of the interactions between water demand and plant growth under varying climatic conditions. *Biological* models the plant physiological control of CO_2, water and nitrogen fluxes as well as plant growth.

The modelling of plant growth and biomass production based on the modelling of assimilation and transpiration as explained in this chapter is to be found in Chap. 37.

36.2 Data Processing

Biological uses the results from the DANUBIA *Atmosphere*, *RadiationBalance*, *Soil*, *SNT* (*Soil Nitrogen Transformation*) and *Farming* model components as input variables. The first three model components produce hourly data for the environmental, meteorological (air temperature, relative humidity, radiation, atmospheric CO_2 concentration, wind speed, atmospheric pressure) and pedologic (water, nitrate and ammonium content of the specific soil layers, soil temperature) parameters that are relevant for plant growth.

The *Farming* model supplies decisions on agricultural management (land use, dates and quantities for fertiliser application and sowing, cutting and harvesting dates). The cutting and harvesting dates are dynamically calculated by *Farming* depending, among other factors, on the modelled biomass and on the development stage. In addition, *Biological* uses physical parameters of the soil (wilting point, field capacity, saturated water content, bulk density) derived from the soil textures presented in Chap. 8.

36.3 Model Documentation

A model approach based on plant physiology is necessary to ensure the validity of the modelled fluxes of CO_2 and water even under altered environmental conditions. The method implemented in the model Genotype-by-Environment interaction on CROp growth Simulator (GECROS) (Yin and van Laar 2005) is taken for coupled modelling of photosynthesis and transpiration in *Biological*. The key aspects of this method are described below. *Biological* uses the approach of Farquhar et al. (1980) for modelling the biochemical processes of photosynthesis. Assimilation is calculated based on radiation, air temperature and atmospheric CO_2 concentration. Yin and van Laar (2005) have expanded this model of photosynthesis for C_4 plants and added the dependency of photosynthesis on nitrogen in photosynthetically active leaves.

Modelling potential transpiration is based on the Penman-Monteith equation and considers the water vapour saturation deficit in the air and the calculated rate of photosynthesis. If insufficient soil water is available to plants, the potential transpiration is correspondingly reduced (actual transpiration). The reduction of transpiration and hence of latent heat flux influences leaf temperature and therewith affects photosynthesis via the temperature dependency of its biochemical reactions.

Therefore, modelling of photosynthesis and transpiration in *Biological* takes into account not only the predicted increases in air temperature and atmospheric CO_2 concentration but also the availability of water and nitrogen. *Biological* uses hourly time increments in order to account for the non-linear responses of photosynthesis and transpiration to short-term changes in environmental conditions. Both processes are first modelled at leaf scale, differentiating between sunny and shaded leaves. Then the results are scaled up to the plant canopy, taking into account the relative areas of sunny and shaded leaves.

The fundamental process descriptions for CO_2 and water exchanges in *Biological* are equally applicable for all categories of vegetation, although a distinction is made between C_3 and C_4 plants depending on the type of carbon fixation.

C_4 plants use water more efficiently than C_3 plants, since they require less water for biomass production. As the most significant example of a C_4 crop plant in Europe, maize takes up vast acreages in the Upper Danube basin.

The various significant plant physiological variables for modelling of photosynthesis and transpiration are specifically parameterised for each vegetation category. Data from the literature (Yin and van Laar 2005) and field measurements (Lenz 2007; Lenz-Wiedemann et al. 2010) form the basis for this parameterisation in *Biological*.

Measurements of CO_2 and water exchange at the canopy scale (eddy covariance and Bowen ratio methods) and summary biomass measurements are used to validate the model.

36.4 Results

In order to illustrate the strong temporal dynamics and the patterns of the modelled fluxes specific to different vegetation categories, an example of the daily and annual course in CO_2 uptake and transpiration is shown in Fig. 36.1. The dependency of

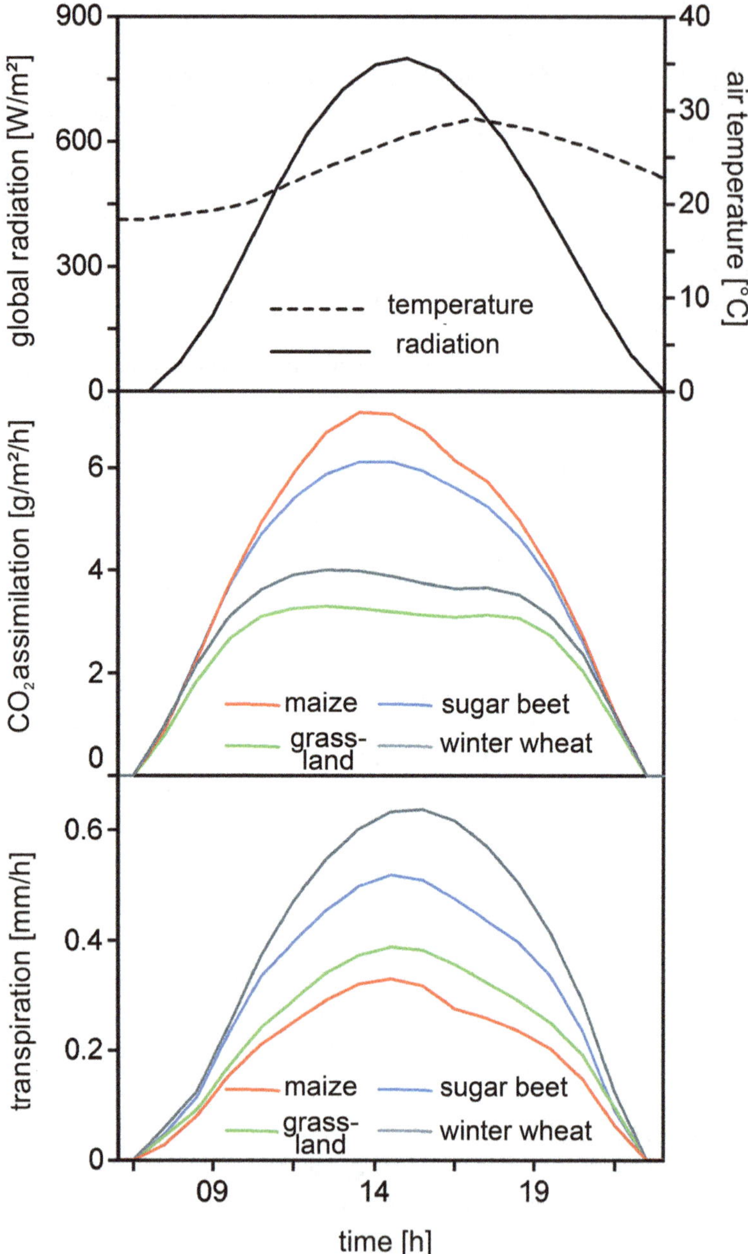

Fig. 36.1 Modelled CO_2 assimilation and transpiration of different vegetation categories: example for the dependency of the diurnal course on global radiation and air temperature

CO_2 uptake and transpiration on the meteorological input variables global radiation and air temperature is shown for grassland, winter wheat, maize and sugar beet. The figure clearly depicts the different responses of the vegetation categories to environmental conditions. Figure 36.2 shows the annual course of CO_2 uptake and transpiration for the same vegetation categories. The annual changes in the modelled fluxes are essentially controlled by the different phenological patterns for each vegetation category and its management (sowing, harvest and cutting dates). The transpiration totals shown in Map 36.1 for selected agricultural vegetation categories for the hydrological year 1994/1995 reflect local conditions, characteristics of the different

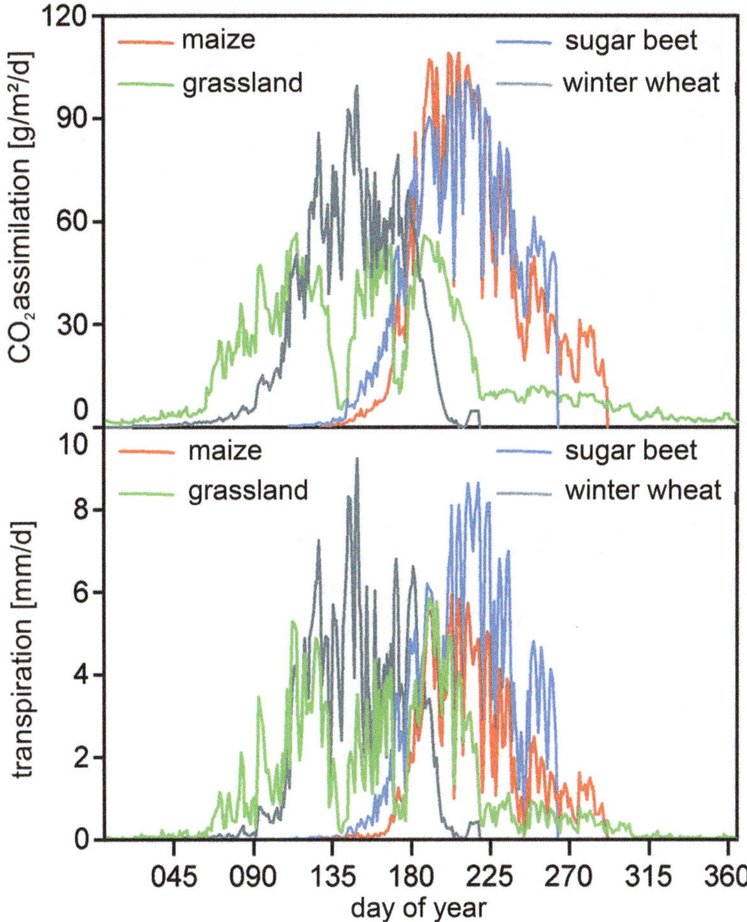

Fig. 36.2 Modelled CO_2 assimilation and transpiration of different vegetation categories: example of the annual course

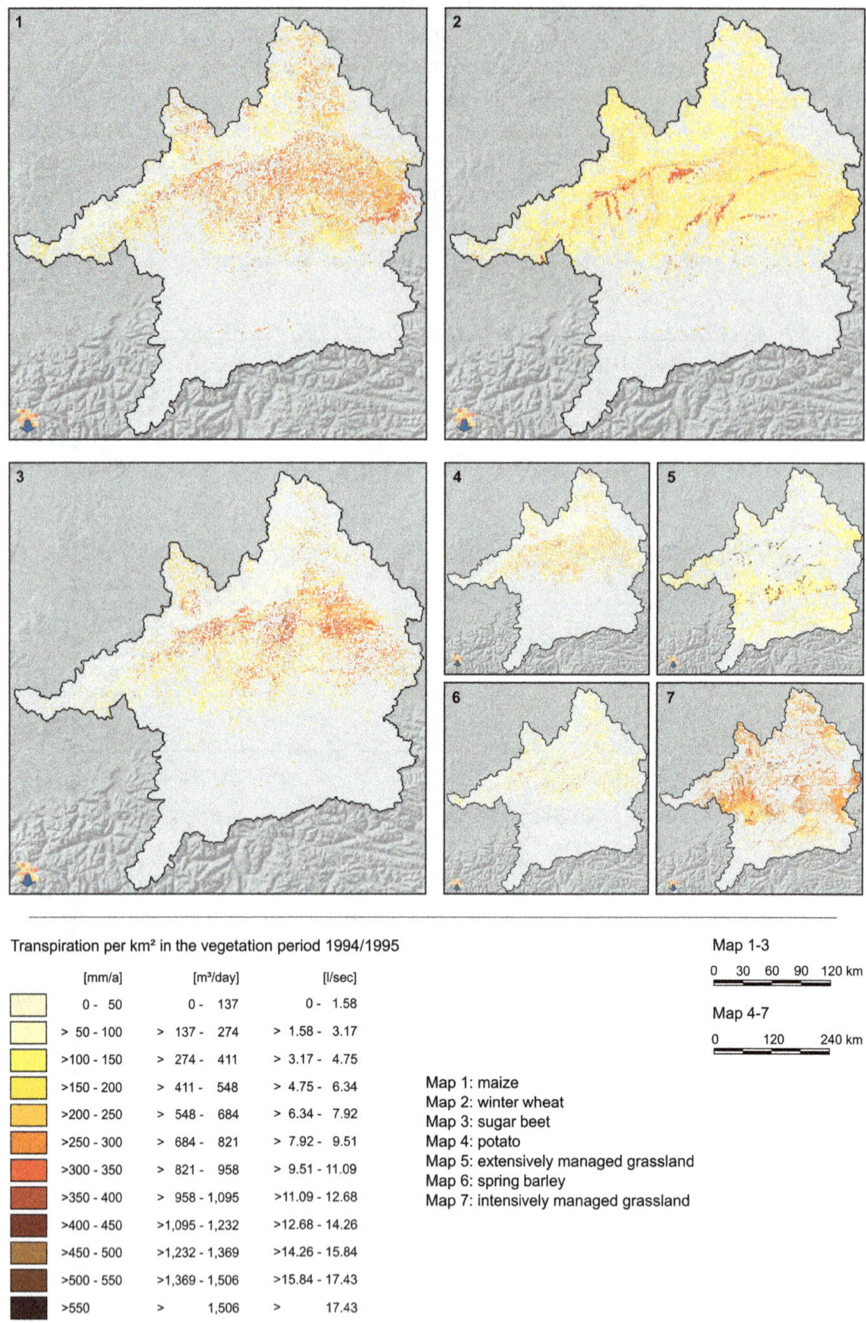

Map 36.1 Modelled transpiration sums (dynamically coupled model run, data source of management data: KTBL Association for Technology and Structures in Agriculture (2000/01), Bavarian State Ministry of Food, Agriculture and Forestry (1996))

vegetation categories and differences in management. These model results were calculated with a subscale land use map that accounts for several vegetation categories per proxel. The influence of small-scale heterogeneity can be better captured at this subscale resolution, and hence, more accurate statements about transpiration can be made.

Low totals for transpiration are predominantly found on the Karst soils in the north-western part of the study area, especially for maize, potatoes and sugar beet. In addition, low totals are found on the grassland in the Alps as a result of lower temperature and shorter growing seasons. Maximum transpiration values are found in areas with intensive agriculture. Forested land also shows high transpiration totals as a result of the long growing period (not shown as a map).

References

Bavarian State Ministry of Food, Agriculture and Forestry (ed) (1996) Bayerischer Agrarbericht, Munich

Farquhar GD, von Caemmerer S, Berry JA (1980) A biochemical model of photosynthetic CO_2 assimilation in leaves of C3 species. Planta 149:78–90

KTBL Association for Technology and Structures in Agriculture (2000/01) Taschenbuch Landwirtschaft 2000/01, Landwirtschaftsverlag GmbH, Münster

Lenz VIS (2007) A process-based crop growth model for assessing Global Change effects on biomass production and water demand – a component of the integrative Global Change decision support system DANUBIA. PhD thesis, University of Cologne

Lenz-Wiedemann VIS, Klar CW, Schneider K (2010) Development and test of a crop growth model for application within a Global Change decision support system. Ecol Model 477:314–329

Yin X, van Laar H (2005) Crop systems dynamics. An ecophysiological simulation model for genotype-by-environment interactions. Academic Publishers, Wageningen

Chapter 37
Plant Growth and Biomass Production

Victoria I.S. Lenz-Wiedemann, Tim G. Reichenau, Christian W. Klar, and Karl Schneider

Abstract In the Upper Danube basin, plant growth and biomass production strongly influence water, carbon and nitrogen fluxes. The model component *Biological* is part of the simulation system DANUBIA and calculates plant growth and biomass production for various vegetation categories (e.g. grassland, winter wheat, sugar beet and maize). An ecohydrological model approach is needed to account for the interactions of water, carbon and nitrogen fluxes in the soil-plant-atmosphere system. Meteorological and pedological model input data are provided by dynamically coupled DANUBIA model components. When analysing global change effects on agroecosystems, it is crucial to consider agricultural decisions such as type of use and management. *Biological* uses this information as input data from the coupled *Farming* actor model. In turn, the modelled biomass production and yield data are needed by *Farming* for the selection of crops to be cultivated. For the model validation analysis, measured and modelled values were compared for several test fields covering a wide range of meteorological and pedological conditions. On the district scale, agricultural statistics were used for validation. Maps of biomass production for 1 year are shown for selected field crops and managed grassland in the Upper Danube basin. Within DANUBIA, spatial and temporal changes in plant growth and biomass production for the past, present and future are assessed at the regional scale.

Keywords GLOWA-Danube • DANUBIA • Global change • Upper Danube • Hydrology • Ecosystem model • Plant growth • Biomass

V.I.S. Lenz-Wiedemann (✉) • T.G. Reichenau • K. Schneider
Institute of Geography, University of Cologne, Cologne, Germany
e-mail: victoria.lenz@uni-koeln.de; tim.reichenau@uni-koeln.de; karl.schneider@uni-koeln.de

C.W. Klar
Forschungszentrum Jülich, Jülich, Germany
e-mail: c.klar@fz-juelich.de

W. Mauser, M. Prasch (eds.), *Regional Assessment of Global Change Impacts*, DOI 10.1007/978-3-319-16751-0_37

37.1 Introduction

Plant growth and biomass production govern to a large extent the water and matter regime within the Upper Danube drainage basin. Especially in agricultural ecosystems, biomass production and the expected yield are crucial for decisions on land use and management. The *Biological* model component simulates the effects of changed environmental conditions and land use on plant growth and biomass production for seminatural and agricultural ecosystems as well as their effects on the fluxes of water, carbon (C) and nitrogen (N).

Calculations of CO_2 uptake and transpiration (see Chap. 36) form the basis for modelling plant growth and biomass production. When analysing the effects of changed environmental conditions on plant growth and biomass production, it is necessary to consider numerous other processes. Thus, for example, biomass production and also yield are determined not only by the phenological pattern but also by the distribution of assimilated C and incorporated N among the different plant organs (allocation).

Climate change therefore leads to adjustments in agricultural decisions such as type of use and management (e.g. sowing date, fertiliser plan). The close interdependency of natural and socioeconomic processes is implemented by coupling the *Biological* model component and the *Farming* actor model (see Chap. 40). The management decisions modelled by *Farming* influence biomass production. The resulting yield in turn affects the selection of crops to be cultivated.

37.2 Data Processing

See Chap. 36.

37.3 Model Documentation

Biological models the water and nutrient fluxes of the vegetation in seminatural and agricultural ecosystems within DANUBIA. A variety of interactions determine the fluxes of water, C and N in the soil-plant-atmosphere system. Thus, for example, photosynthesis is influenced by the concentration of nitrogen in the leaves. Higher concentrations of nitrogen promote photosynthesis. Increased uptake of carbon boosts biomass production, which raises nitrogen demand. Nitrogen uptake is coupled to water uptake and depends on root length density (root length per unit soil volume). An ecohydrological model approach is needed to account for these interdependencies. In this way, the response of plants to meteorological conditions and to the availability of water and nitrogen is depicted.

Biological combines the two process-based plant growth models Crop Environment Resource Synthesis (CERES) (Jones and Kiniry 1986) and Genotype-

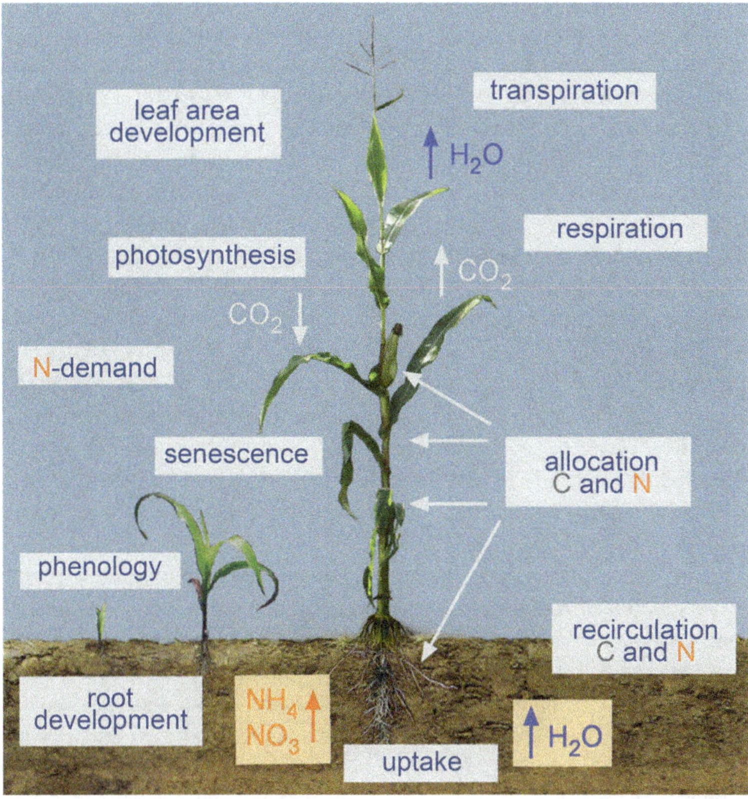

Fig. 37.1 Modelled processes in *Biological*

by-Environment interaction on CROp growth Simulator (GECROS) (Yin and van Laar 2005). The key processes modelled in *Biological* are presented in Fig. 37.1. Uptake of water and N is modelled according to CERES approaches. In contrast, the following processes are modelled using GECROS concepts: photosynthesis and transpiration (see Chap. 36), allocation of C and N, respiration (CO_2 release), N demand, N fixation in legumes, leaf area development and senescence (leaf and root decay).

The formation of root biomass and its decay are calculated according to GECROS. The development of root length density as a determining variable for water and N uptake is modelled according to CERES. Water and N uptake and root length density are modelled soil layer specific. Soil water, nitrate and ammonium content, which are dynamically modelled by *Soil* and Soil Nitrogen Transformation (SNT) (see Chaps. 24 and 38), are used as input along with field capacity, wilting point, saturated water content and bulk density, all of which are soil texture specific. N uptake is differentiated as either nitrate or ammonium uptake.

The GECROS phenology model was expanded to model phenological development according to Streck et al. (2003) in order to include vernalisation in addition to temperature and day length. Vernalisation is a key factor for winter wheat and winter rapeseed especially in the context of climate change. Moreover, modelling of germination and emergence (the appearance of sprouts or leaves at the ground surface) was complemented according to CERES. This approach ensures that an adjustment of the sowing date (on the part of the *Farming* actor model) as a result of a predicted climate change is adequately depicted by the model.

Dynamic modelling of the C and N allocation is needed in order to predict future plant growth and biomass production. Therefore, allocation is calculated based on the functional equilibrium approach according to GECROS. Thus, during times of insufficient water or N supply, root biomass production is increased, and correspondingly, with low available radiation, production of above-ground biomass is promoted. Modelling of leaf area development and senescence is physiologically based, according to GECROS. Leaf area is calculated based on the dynamic allocation of C and N in the leaves. Senescence is triggered when the concentration of N in the leaves or roots drops to a critical value.

All process descriptions in *Biological* are equally applicable for all vegetation categories. A specific differentiation is achieved by assigning the considered plant to functional groups (e.g. winter or summer crop) as well as by plant-specific parameterisation of the model input variables. *Biological* implements plant growth for a total of 22 different vegetation categories. Process-based growth models have been implemented for winter wheat (a representative for winter cereals), spring barley (a representative for summer cereals), winter rapeseed (for oil seeds), peas (for legumes), maize, potato, sugar beet and managed grassland. The yields for the other vegetation categories are calculated from empirically derived regression equations based on the modelled yields for plants with comparable phenology and management. For example, spring wheat yield is derived from the value modelled for spring barley. The regression equations were calculated from long-term statistics for agricultural districts (BLSD 2008 and SLBW 2008). For forests, only transpiration and net assimilation are modelled based on a predefined leaf area index (LAI) development.

Validation of the model results was carried out independent of scale. Measured and modelled values were compared for several test fields that cover a wide range of meteorological and pedological conditions. The measured data consist of LAI (leaf area per unit ground area) as well as biomass and C and N content of the individual plant organs, all for different plant types. Validation data are taken in part from field measurements by the working group (e.g. Lenz 2007; Lenz-Wiedemann et al. 2010) and in part from the literature (Kenter 2003 and McVoy et al. 1995). On the district scale, agricultural statistics were used for validation. In addition to the central task of modelling transpiration, *Biological* also serves within DANUBIA to provide harvest yields to the *Farming* actor model. The model results relating to the interaction between soil and plants (root growth, uptake of water and N, release of organic C and N back to the soil) are used by *Soil* and *SNT*. Results relating to the interaction between the atmosphere and vegetation (LAI, leaf temperature, plant height) are imported by *RadiationBalance* and *Surface*.

37.4 Results

The comparison of measured and modelled biomass and leaf area development is shown in Fig. 37.2 using the example of a sugar beet field. At the beginning of growth, the model marginally overestimates the photosynthetically active leaf area, while later deviations are greater. The mean square error between the modelled and measured LAI is 0.54 m^2/m^2. Measured values and model results for green leaf biomass agree well at some points of time but show significant deviations at others. This results in a mean square error of 147 g/m^2. The modelled biomass of dead brown leaves (44 g/m^2) is almost exactly equivalent to the last measured value (46 g/m^2). The mean square error for the beet body is 86 g/m^2. A corresponding validation with comparable results was carried out for six additional sugar beet fields, three spring barley fields, two maize fields, three winter wheat fields and two potato fields (Lenz 2007).

Map 37.1 shows the spatial distribution of biomass production in 1995 for the Upper Danube basin. The model results are presented for selected field crops (total biomass without roots) and managed grassland (biomass at cutting date). The influence of soil characteristics on biomass production can easily be traced on the maps.

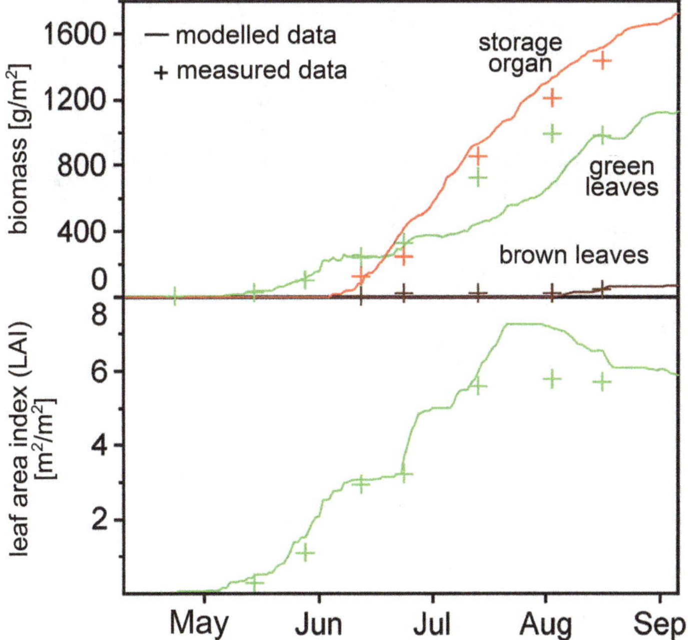

Fig. 37.2 Comparison of measured and modelled biomass (in g dry matter per m^2) and leaf area development (m^2 green leaf area per m^2 ground surface) for sugar beet (2005 in Feienberg, measurements taken by the authors) (With permission from Lenz-Wiedemann et al. 2010)

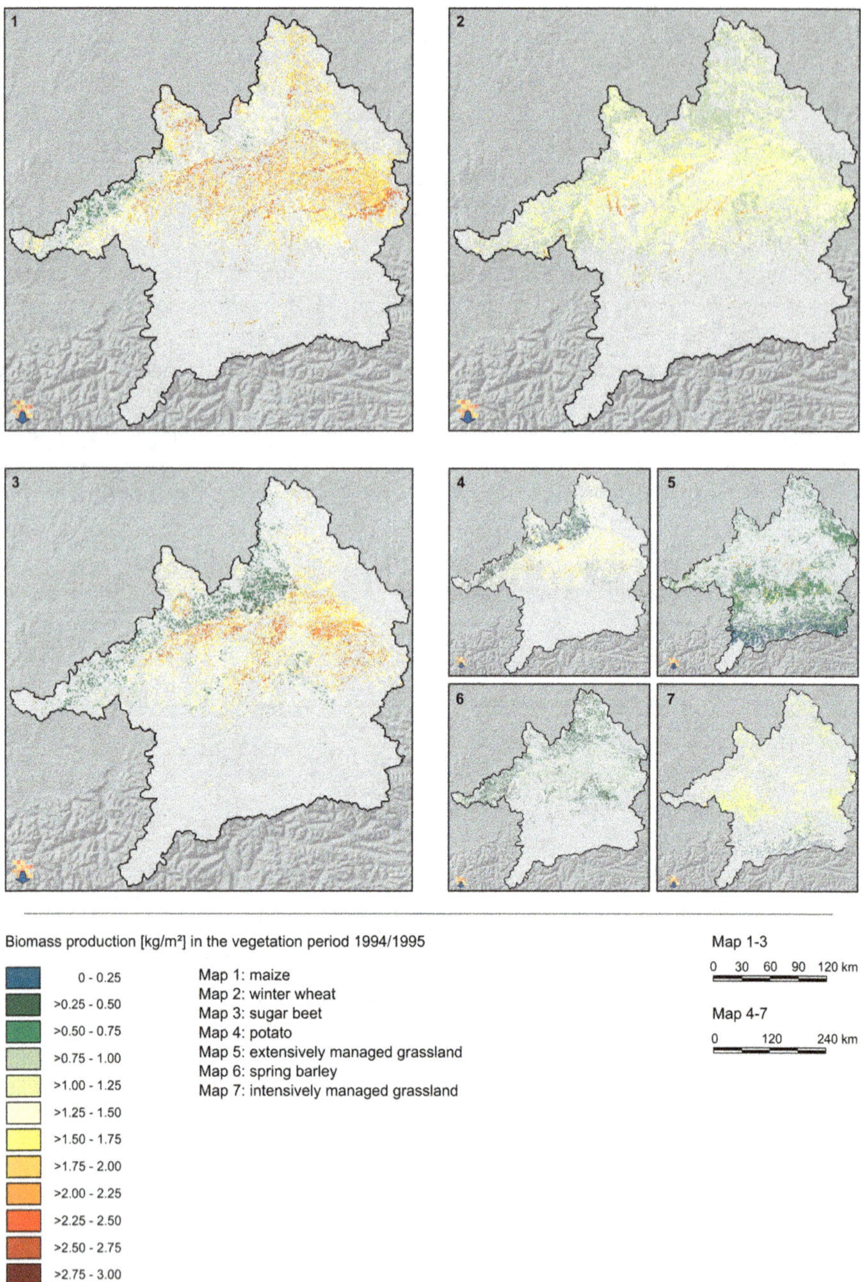

Map 37.1 Modelled biomass production (dynamically coupled model run, data source of management data: KTBL Association for Technology and Structures in Agriculture (2000/01), Bavarian State Ministry of Food, Agriculture and Forestry (1996))

Thus, the intensively cultivated regions on loess soils in the Dungau area show high biomass production, while the regions less favourable to agriculture in the Allgäu exhibit lower biomass production.

References

Bavarian State Ministry of Food, Agriculture and Forestry (ed) (1996) Bayerischer Agrarbericht, Munich

BLSD (2008) Datenbank GENESIS online. Bayerisches Landesamt für Statistik und Datenverarbeitung, München

Jones CA, Kiniry JR (eds) (1986) CERES-Maize. A simulation model of maize growth and development. Texas A&M University Press, College Station

Kenter C (2003) Ertragsbildung von Zuckerrüben in Abhängigkeit von der Witterung. PhD thesis, Georg-August-University, Göttingen

KTBL Association for Technology and Structures in Agriculture (2000/01) Taschenbuch Landwirtschaft. Landwirtschaftsverlag GmbH, Münster

Lenz VIS (2007) A process-based crop growth model for assessing Global Change effects on biomass production and water demand. A component of the integrative Global Change decision support system DANUBIA. PhD thesis, University of Cologne

Lenz-Wiedemann VIS, Klar CW, Schneider K (2010) Development and test of a crop growth model for application within a Global Change decision support system. Ecol Model 221:314–329, Elsevier

McVoy CW, Kersebaum KC, Arning M, Kleeberg P, Othmer H, Schröder U (1995) A data set from north Germany for the validation of agroecosystem models: documentation and evaluation. Ecol Model 81:265–300

SLBW (2008) Struktur und Regionaldatenbank. Statistisches Landesamt Baden-Württemberg, Stuttgart

Streck NA, Weiss A, Baenziger PS (2003) A generalized vernalization function for winter wheat. Agron J 95:155–159

Yin X, van Laar H (2005) Crop systems dynamics. An ecophysiological simulation model for genotype-by-environment interactions. Wageningen Academic Publishers, Wageningen

Chapter 38
Nitrate Leaching

Tim G. Reichenau, Christian W. Klar, Victoria I.S. Lenz-Wiedemann, Peter Fiener, and Karl Schneider

Abstract Nitrate leaching has a significant influence on plant nitrogen supply and groundwater quality. Spatially detailed information on nitrate leaching is required to assess land-use management options and to develop effective groundwater resource protection measures. Comprehensive information for decision support can be derived based on spatiotemporally dynamic modelling.

Coupled simulations with the DANUBIA simulation system (plant growth, balances of carbon, nitrogen, water, energy) were performed for several test sites and for the Upper Danube catchment. Model validation results for soil mineral nitrogen show good correspondence between the model results and field measurements without a site-specific calibration.

For the spatially explicit analysis on the catchment scale, land-use and cultivation practice (timing, fertilisation) were set according to best-practice recommendations. Modelled nitrate concentrations in the leachate from the vadose zone were analysed for the period 1995–2000. A comparison of simulated and measured data proved the consistency of the model results.

In general, nitrate concentrations above 50 mg l^{-1} were calculated mainly for regions characterised by intensive agriculture or peatland soils. The lowest nitrate concentrations occur in forested areas and regions with little arable land. Overall, the results reveal the potential for dynamic and comprehensive modelling of nitrogen leaching with DANUBIA.

Keywords Nitrate leaching • Plant growth • Nitrogen balance • DANUBIA • GLOWA-Danube • Ecosystem model

T.G. Reichenau (✉) • V.I.S. Lenz-Wiedemann • K. Schneider
Institute of Geography, University of Cologne, Cologne, Germany
e-mail: tim.reichenau@uni-koeln.de; victoria.lenz@uni-koeln.de; karl.schneider@uni-koeln.de

C.W. Klar
Forschungszentrum Jülich, Jülich, Germany
e-mail: c.klar@fz-juelich.de

P. Fiener
Institut für Geographie, Universität Augsburg, Augsburg, Germany
e-mail: peter.fiener@geo.uni-augsburg.de

W. Mauser, M. Prasch (eds.), *Regional Assessment of Global Change Impacts*, DOI 10.1007/978-3-319-16751-0_38

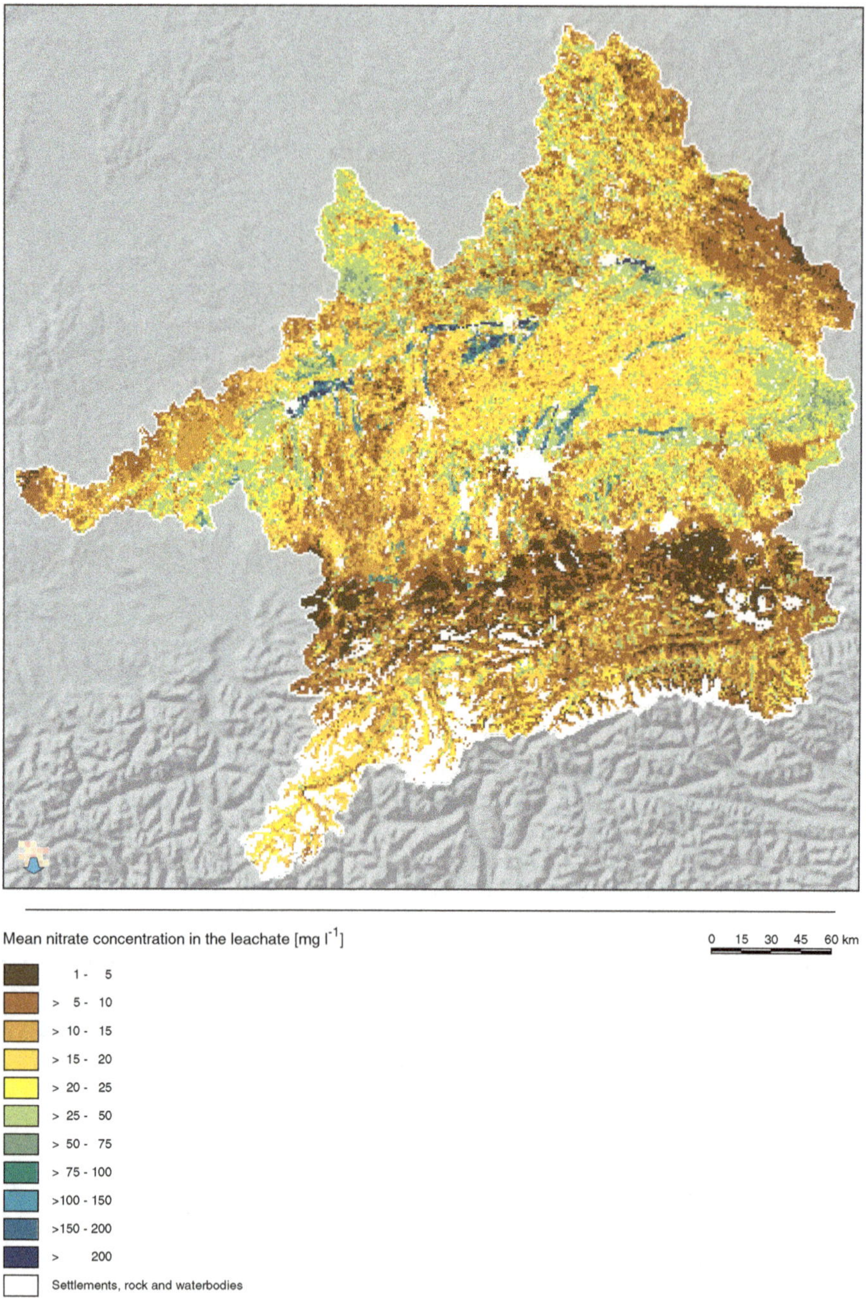

Map 38.1 Nitrate leaching

38.1 Introduction

Nitrate leaching, defined as the process of nitrate percolation from the root zone, is one of the most important factors influencing groundwater quality. Moreover, nitrate leaching reduces N availability in soils and hence potentially affects plant growth. The majority of nitrate in groundwater originates from diffuse leaching from agricultural soils. However, in particular cases also point emission from farms and industry might be important.

Spatially detailed information on nitrate leaching is required to assess land-use management options and to develop effective groundwater resource protection measures. Comprehensive information for decision support can be derived based on spatiotemporally dynamic modelling. DANUBIA provides such a tool to analyse optimised, region-specific management options under climate change conditions.

38.2 Preparation of the Data

See Chaps. 8 and 21.

38.3 Description of the Model

In DANUBIA, the turnover and storage of soil nitrogen in arable land and grassland is modelled with a dynamic and process-oriented approach, while for forests, an empirical approach based on Feldwisch et al. (1999) and Block et al. (2000) is applied. Dynamic modelling is performed with the model-component SNT (Soil Nitrogen Transformation) which is closely linked to the process-oriented plant growth model-component Biological (see Chaps. 36 and 37). SNT is conceptually based on CERES Maize 2.0 (Jones and Kiniry 1986). This modelling concept on the one hand is suitable for regional applications because of its limited data demand; on the other hand, it considers the major processes potentially affected by land-use and climate change. The combination of regional applicability and process-oriented is major prerequisite for SNT to be used for Global Change Research in the Upper Danube catchment.

For arable land and grassland, SNT considers the following processes (see Fig. 38.1):

- Nitrogen mineralisation and immobilisation
- Nitrification
- Denitrification
- Hydrolysis of urea
- Nitrate redistribution within the soil profile

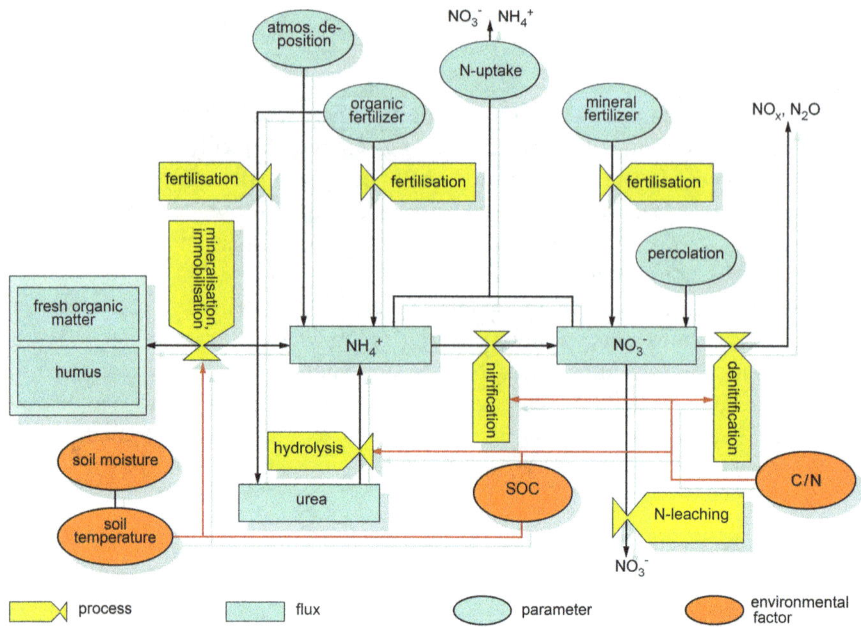

Fig. 38.1 Schematic diagram of SNT including modelled soil nitrogen storages, fluxes and turnover processes (Reprinted from Klar et al. (2008) with permission from Elsevier)

With the exception of the hydrolysis of urea, which is assumed to be limited to the top soil layer, all processes are calculated specifically for each soil layer. Vertical nitrate redistribution between soil layers is coupled to the modelled water fluxes. Considering the spatial resolution (1×1 km^2) of the modelling approach, lateral fluxes are neglected.

All N fluxes are calculated as a function of soil moisture and soil temperature as well as depending upon the physical and chemical characteristics (field capacity, saturated water content, wilting point, bulk density, soil organic carbon content and C:N ratio) of the soil. Nitrate redistribution within the soil profile is proportional to water percolation rates and nitrate concentrations within the individual soil layers. Nitrate leaving the lowest soil layer (max. 2 m) is equivalent to the amount of nitrate leaching from the vadose zone.

The nitrogen status and transformation within the soil is driven by nitrogen inputs (fertilisation, atmospheric deposition) and nitrogen losses (nitrate leaching, uptake by plants, denitrification). SNT accounts for the dynamics within the mineral (nitrate and ammonium) and organic (nitrogen in fresh organic matter and in the humus layer) nitrogen reserves (see Fig. 38.1).

The most important input parameters for SNT are (1) the nitrogen uptake by plants and (2) the N inputs from fertiliser application. Uptake of nitrogen is calculated in Biological as a function of root length density, soil water content, availability of mineral nitrogen (N_{min}) and current demand by plants. The modelling of N-demand is based on the concept of functional equilibrium between carbon and nitrogen in plants (Brouwer 1962). N-demand is modelled dynamically based on carbon assimilation (Lenz-Wiedemann et al. 2010). Due to the process-based simulation of plant growth, the model is able to appropriately address N-uptake under conditions of climate change. For modelling future land-use and land management scenarios, data on the quantity and composition of fertilisers are provided by the Farming component (see Chap. 39).

All processes are modelled on a daily basis, with the exception of nitrate redistribution, which is calculated hourly. Nitrate leaching is given as mass per unit area (kg ha^{-1}) or as nitrate concentration in the percolating water (N_c) from the lowest soil layer (mg l^{-1}). Nitrate leaching for a proxel is calculated as the area weighted mean of the total leaching from the land-use classes on the proxel. Modelled nitrogen availability within and nitrate leaching from soils are transferred to the DANUBIA components Biological and Groundwater.

38.4 Presentation of the Results

Modelled and measured annual dynamics of N_{min} content in three soil layers are exemplarily shown for two test fields (winter wheat and spring barley) in Fig. 38.2. In the top soil layer, the response of N_{min} content to fertiliser application is obvious for spring barley. Also, the dynamics without external N input (here: winter wheat) are correctly depicted. A comprehensive, field-related validation (Klar et al. 2008) showed good correspondence between the model results and the measurements without performing a site-specific calibration. This is an important prerequisite for the spatial transfer of the model to the scale of a drainage basin.

Map 38.1 shows the modelled mean nitrate concentration (mg l^{-1}) in water percolating from the lowest soil layer for 1995–2000. Data regarding soil management, especially fertiliser application, were estimated based on land-use data and information from literature (KTBL 2000/01) in combination with district-specific agricultural statistics (BStMLF 1996). Areas of settlements, rock and water bodies were not modelled and are masked in the map (white areas).

Overall, nitrate concentrations below the threshold of 50 mg l^{-1} (German Drinking Water Directive) can be found in almost the entire catchment (>95 % of area; see Fig. 38.3). Higher concentrations often correspond to regions that are characterised by intensive agriculture (e.g. Passau and Dungau districts); in addition, particularly high nitrogen concentrations are modelled for peatland soils (e.g. the Donaumoos and the Erdinger Moos). Here peak values of up to 440 mg l^{-1} were calculated. These very high concentrations are backed up by measurements in

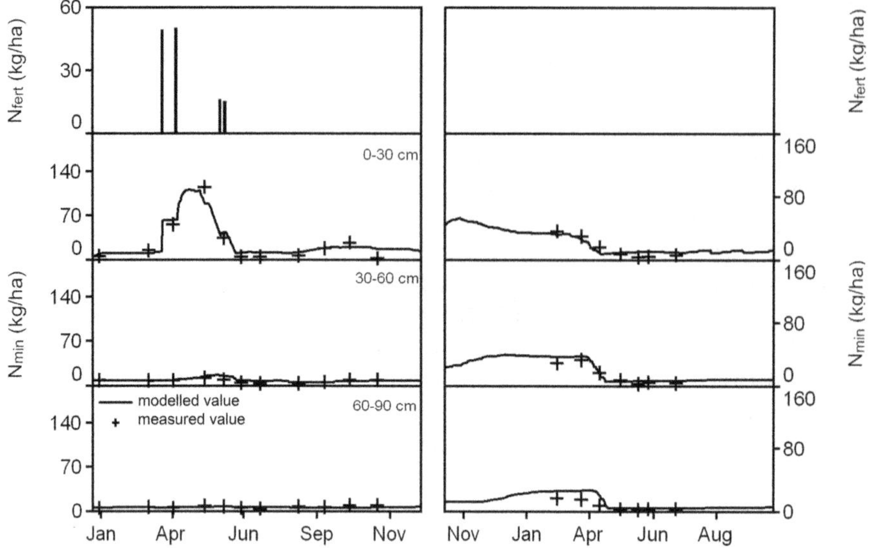

Fig. 38.2 Modelled and measured N_{min} dynamics in three soil layers under summer barley (*left*) and unfertilised winter wheat (*right*); applied fertiliser amounts given in topmost row (Reprinted from Klar et al. (2008) with permission from Elsevier)

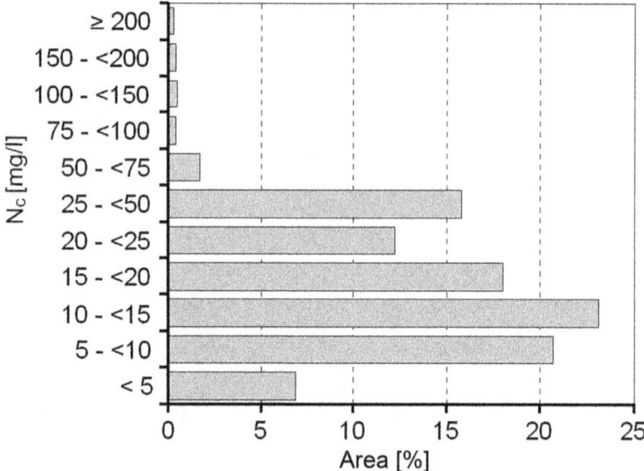

Fig. 38.3 Areal proportion of different nitrogen concentration classes in the Upper Danube catchment

the Donauried region where similar concentrations were observed (Briemle and Lehle 1991). In contrast, the lowest nitrate concentrations are modelled in forested areas and regions with little arable land, such as the Bavarian Forest and the Northern Alps.

Despite remaining conceptual uncertainties (e.g. processes of denitrification during percolation through the unsaturated zone between soil and groundwater), a

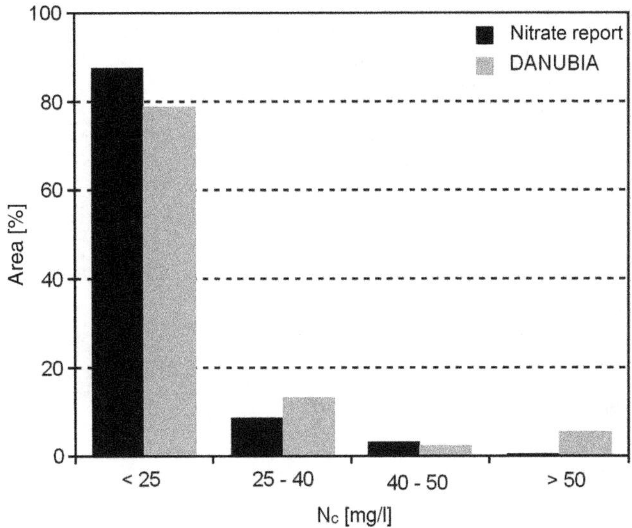

Fig. 38.4 Modelled areal proportion of different nitrogen loading classes compared to data taken from the nitrate report (LfU 2008) in the administrative region of Upper Bavaria for the year 2000

comparison of the modelled nitrate concentrations with measured data proves the consistency of the model results. This is illustrated in Fig. 38.4 which shows a comparison of modelled and measured N load classes for the administrative region of Upper Bavaria (approx. 17 500 km^2).

Overall, the results reveal the potential for dynamic and comprehensive modelling of nitrogen leaching within DANUBIA.

References

Block J, Eichhorn J, Gehrmann J, Kölling C, Matzner E, Meiwes KJ, Wilpert V, Wolff KB (2000) Kennwerte zur Charakterisierung des ökochemischen Bodenzustandes und des Gefährdungspotentials durch Bodenversauerung und Stickstoffsättigung an Level II-Waldökosystem-Dauerbeobachtungsflächen. Arbeitskreis C der Bund-Länder-Arbeitsgruppe "Level II", BML (Bundesministerium für Ernährung, Landwirtschaft und Forsten), 167 p

Briemle G, Lehle M (1991) Einfluss von Bewirtschaftung und Standort auf den Stickstoffhaushalt von Moorböden. Gemeinsamer Versuchsbericht der LVVG Aulendorf und der Universität Hohenheim, Aulendorf

Brouwer R (1962) Nutritive influences on the distribution of dry matter in the plant. Neth J Agric Sci 10:361–376

BStMLF Bayerisches Staatsministerium für Ernährung, Landwirtschaft und Forsten (ed) (1996) Bayerischer Agarbericht. München

Feldwisch N, Frede H-G, Hecker F (1999) Verfahren zum Abschätzen der Erosions- und Auswaschungsgefahr. In: Frede H-G, Dabbert S (eds) Handbuch zum Gewässerschutz in der Landwirtschaft. Ecomed, Landsberg

KTBL (Kuratorium für Technik und Bauwesen in der Landwirtschaft) (2000/01) Taschenbuch Landwirtschaft. Landwirtschaftsverlag GmbH, Münster

Jones PK, Kiniry JR (eds) (1986) CERES-Maize. Simulation model of maize growth and development. Texas A&M University Press, College Station

Klar CW, Fiener P, Neuhaus P, Lenz-Wiedemann VIS, Schneider K (2008) Modelling of soil nitrogen dynamics within the decision support system DANUBIA. Ecol Mod 217:181–196

Lenz-Wiedemann VIS, Klar CW, Schneider K (2010) Development and test of a crop growth model for application within a Global Change decision support system. Ecol Mod 477:314–329

LfU, Bayerisches Landesamt für Umwelt (ed) (2008) Nitratbericht Bayern. Berichtsjahre 2000–2004. LfU, Augsburg

Chapter 39
Agricultural Land Use and Drinking Water Demand

Alexander Wirsig, Tatjana Krimly, and Stephan Dabbert

Abstract The assessment of agricultural water demand and agricultural land use is a common element in quantitative model-based policy analysis. Economic, political and pre-existing climatic and soil conditions shape the structure of agricultural land use which determines drinking water demand and nutrient input from agriculture. These factors vary strongly across regions. A nonlinear process analytical model based on agricultural statistics on district level was used to calculate parameters such as farm incomes, optimum cultivation plan, animal husbandry, quantities of fertiliser and the demand for drinking water at district level. To produce a more detailed picture with a higher spatial resolution, the regional optimization model was combined with an agricultural actor model that simulates farmers' crop management decisions. An allocation tool was developed to allocate the different farm actors on a spatial unit of 1 × 1 km. For the analysis modelled yield changes from interdisciplinary global change scenarios were used.

Keywords Global change • Regional optimization model • Global change scenarios • Agricultural production • GLOWA-Danube

39.1 Introduction

Regional models have been developed as tools to support political decision-making. Agro-economic regional models serve the specific decisions that relate to the agricultural sector and represent the agricultural actors in interdisciplinary projects. Economic and political conditions in agricultural production are determinative for

A. Wirsig (✉)
MBW Marketing- und Absatzförderungsgesellschaft für Agrar- und Forstprodukte aus Baden-Württemberg mbH, Stuttgart, Germany
e-mail: wirsig@mbw-net.de

T. Krimly • S. Dabbert
Production Theory and Resource Economics, Universität Hohenheim, Stuttgart, Germany
e-mail: T.Krimly@uni-hohenheim.de; dabbert@uni-hohenheim.de

W. Mauser, M. Prasch (eds.), *Regional Assessment of Global Change Impacts*, DOI 10.1007/978-3-319-16751-0_39

the factor inputs, production and agricultural incomes (Dabbert et al. 1999). These shape the structure of land use along with the natural conditions, such as precipitation. In the context of different global change scenarios in the Upper Danube drainage basin, the agro-economic regional model facilitates an analysis and evaluation of the agricultural water demand and agricultural land use that affects water quality and quantity, from a production-related and socioeconomic point of view.

39.2 Data Processing

The areas of arable land and grassland aggregated to district level for each year generated within the DANUBIA Land Use Database (see Chap. 9) serve as basis for the calculations within the agricultural sector model for optimal agricultural land use. A downscaling procedure distributes these district-based results to the level of the proxel, which is used as a standard within DANUBIA.

39.3 Model Documentation

The *Farming* model simulates land use, drinking water demand, nutrient input from agriculture and farmers' crop management decisions. As a part of the main component *Actor*, *Farming* is exchanged with the *WaterSupply, Biological, SNT, Soil* and *Atmosphere* simulation models. The agricultural economics model *Farming* in DANUBIA consists of the following basic components (see Fig. 39.1):

1. The agricultural sector model ACRE, which operates based on statistical data available at district level
2. The *Deepfarming* agricultural actor model that simulates farmers' crop management decisions per proxel

1. The agricultural sector model calculates an optimum annual plan for the subsequent fiscal year, which maps economic parameters such as farm incomes,

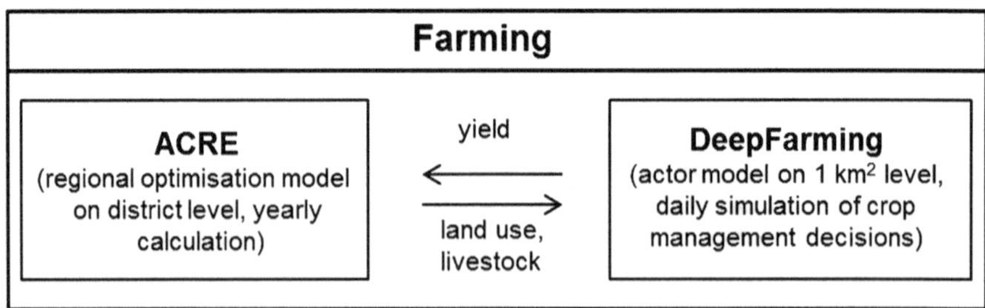

Fig. 39.1 Overview on the agro-economic model component farming

optimum cultivation plan, animal husbandry, quantities of fertiliser and the demand for drinking water at district level. The agricultural sector model ACRE (Agro-eConomic pRoduction model on rEgional level) is a nonlinear process analytical model based on agricultural statistics on district level. Its temporal resolution is 1 year. ACRE is based on the method of "positive mathematical programming" (PQP) (Howitt 1995; Howitt and Mean 1999), using a modification developed by Röhm and Dabbert (2003) to account for different production variants (e.g. different production intensities). The nonlinear optimisation model describes spatial heterogeneity within the districts by approximation. The objective function variable of the model is the total gross margin of the model region. The gross margin is calculated as the difference between the revenue and variable production costs. Optimised land use at district level is an important result within DANUBIA. ACRE was developed by Winter (2005) in the standard language GAMS (General Algebraic Modelling System) and further improved by Henseler et al. (2009). The regional farm approach was used to aggregate the individual farms to the districts. To do so, the factor capacities of the farms occurring in a district are added, and it is assumed that the entire district is represented by a single farm. The accuracy of the reproduction by the model depends on its calibration (i.e. accurate modelling of the base period). ACRE is calibrated using statistical data for the reference year 1995. The statistical data include information on the extent of production of plants and animals and on production output as well as data on socioeconomic conditions. Unrealistic constant marginal yields, an overspecialisation of the model with limited response options and volatile changes of the organisation within scenario calculations are all disadvantages of linear regional farm models that are overcome by using a nonlinear regional farm model (Röhm and Dabbert 2003). PQP also enables the production process and the interrelationship between the individual production methods to be presented in detail. The model formulates a total of 12 production methods for animal husbandry, 15 for crops and 2 for grasslands. Sales activities emerge from the model, taking the interrelations among the animal husbandry practices into account. Data from other model components of DANUBIA are imported to simulate climate and socioeconomic scenarios, for example, data on changes in crop yield that are calculated from *Biological*. ACRE calculates data that describe changes in the production structure (e.g. land use) and the socioeconomic situation (e.g. regional total gross margin). Figure 39.2 gives a simplified overview on the product flows in ACRE . Food crops, forage crops and industrial crops are cultivated. Marketable products from plant and animal production are sold at producer prices, and the revenues (price × quantity) are incorporated into the total gross margin. Forage crops are used as fodder for animal production. The animals provide manure; mineral fertiliser and feed concentrates can be purchased.

2. Daily management decisions are simulated using an agricultural farm actor for each proxel. In *Deepfarming* an actor represents a farm of a certain type, which was deduced from statistical data and the knowledge of agricultural experts. Overall 28 farm types with different production directions were deduced. A farm actor attempts to implement the optimum cultivation plan specified by ACRE for

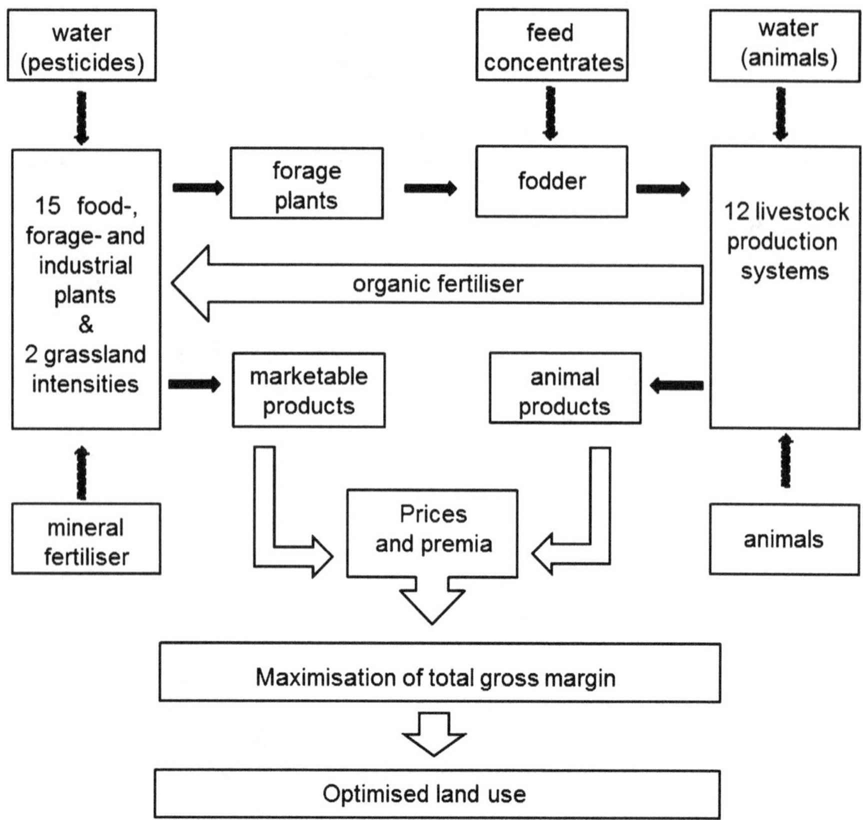

Fig. 39.2 Agricultural sector modelling in ACRE

the proxel taking natural factors such as precipitation, passability of the field and soil temperature into account. To do so, the actor makes decisions on a daily basis about the crop management activities arising within various timeframes, such as sowing, fertilising and harvesting and cutting, respectively, while taking into account the prevailing natural conditions; each result and altered agricultural land use, respectively, is transferred to other models in DANUBIA (*Soil* and *Biological*). The real yields calculated from *Biological* within DANUBIA are balanced with the expected yields from ACRE. If the real yields deviate for several years from the expected yields based on ACRE, these changes are accounted for in the future cultivation plans within ACRE. A further explanation of *Deepfarming* is given in Chap. 40.

An allocation tool was developed to allocate the different farm actors on the proxels. This tool includes an internal rule-based decision matrix. The rules were developed with the aid of spatial information (e.g. "Landwirtschaftliche Vergleichszahl (LVZ)" which is an agricultural comparative figure, community

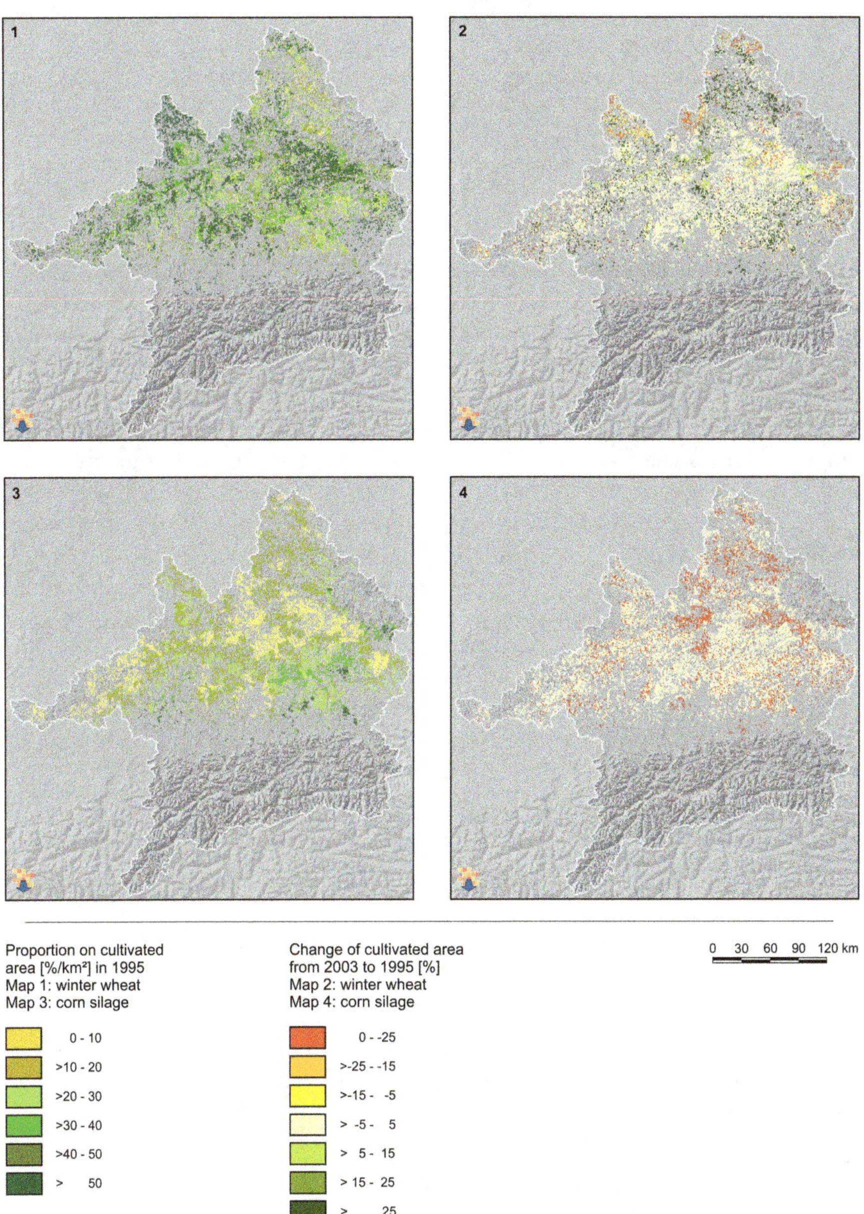

Map 39.1 Agricultural land use (Data sources: © Bavarian State Office for Statistics and Data Processing 1995; Federal Statistical Office 1998; © State Statistical Office, Baden-Württemberg 1995; © STATISTIK AUSTRIA 1995; Federal Statistical Office 2003; EEA, European Environment Agency 2005)

statistics) and expert knowledge from interviews (e.g. local conditions, crop rotations), according to a disaggregation approach developed by Herrmann et al. (2003).

39.4 Results

Changes to income and agricultural land use compared to the reference situation arise as an interdisciplinary result from the yield changes modelled in DANUBIA based on local climate and soil conditions. Maps 39.1–(1–4) illustrate the simulated cultivation of corn silage and winter wheat compared to the starting situation for 1995. Although the extent of winter wheat cultivation at the level of the proxel shows both increases and decreases, the cultivation of corn silage is marked by a single decrease. The validation of the model for the German part of the basin was conducted using an ex post analysis. For 1999, a geometric absolute percentage forecast error (GAPE) of 7.68% was calculated for the cost-related approach (Winter 2005). At this error measure, a quantitative assessment of the accuracy of the forecast at district level took place. According to Hazell and Norten (1986), values for comparable mean absolute percentage forecast error (MAPE) of up to 10% are considered good.

References

Dabbert S, Herrman S, Kaule G, Sommer M (eds) (1999) Landschaftsmodellierung für die Umweltplanung. Springer, Berlin

Hazell P, Norton R (1986) Mathematical programming for economic analysis in agriculture. Macmillan Publishing, New York

Henseler M, Wirsig A, Herrmann S, Krimly T, Dabbert S (2009) Modeling the impact of global change on regional agricultural land use through an activity-based non-linear programming approach. Agric Sys 100(1-3):31–42

Herrmann S, Schuster H, Zárate M (2003) Disaggregation agrarökonomischer Daten als Grundvoraussetzung für eine disziplinübergreifende Modellkopplung – Das Beispiel GLOWA-Danube. Poster presented at the 43rd. Jahrestagung der Gesellschaft für Wirtschaftsund Sozialwissenschaften des Landbaus, Stuttgart-Hohenheim, 29 Sept – 1 Oct 2003

Howitt RE (1995) Positive mathematical programming. Am J Agric Econ 77(2):329–324

Howitt RE, Mean U (1999) A positive approach to microeconomic programming models. Workingpaper 6, Department of Agricultural Economics, University of California, Davis

Röhm O, Dabbert S (2003) Integrating agri-environmental programs into regional production models: an extension of positive mathematical programming. Am J Agric Econ 85(1):256–267

Winter T (2005) Ein Nichtlineares Prozessanalytisches Agrarsektormodell für das Einzugsgebiet der Oberen Donau. Dissertation, Universität Hohenheim

Chapter 40
Actor Model for Farmers' Crop Management Decisions: The *DeepFarming* Model

Tatjana Krimly, Josef Apfelbeck, Marco Huigen, and Stephan Dabbert

Abstract Plant growth and farmers' crop management decisions are strongly influenced by the climatic conditions. To simulate the interplay between crop management activities, crop growth and weather and its changes due to long-term development of climate change, the actor model *DeepFarming* was developed. *DeepFarming* has a high spatial and temporal resolution and is very closely linked with the agricultural sector model *ACRE* und the plant growth model *Biological. ACRE* delivers data on the yearly cultivation plan and *Biological* on the daily development stage of the plants to the actors. *DeepFarming* represents 28 different actor types, which were derived from statistical data on types of farm and their management practices at district level and are allocated to the drainage basin using specific rules. Key task of the actors is to make decisions on the timing of the crop management such as sowing, fertilising and harvesting for each crop. Relevant decision parameters are information on the daily weather conditions, soil saturation level and the development stage of the plants. As a result, the initialisation of the actors in the drainage basin is presented.

Keywords DeepFarming model • Crop management decisions • Actor type • Climate • Allocation • GLOWA-Danube

40.1 Introduction

Climate change is a significant trigger for changes in agriculture, since the process of plant growth and the crop management are to a large extent dependent on climatic conditions. The frequency of precipitation, its amount and duration, the length of frost periods and the daily sunshine duration are all variables that have much greater significance for plant growth than changes to the annual mean temperature

T. Krimly (✉) • J. Apfelbeck • M. Huigen • S. Dabbert
Production Theory and Resource Economics, Universität of Hohenheim,
Stuttgart, Germany
e-mail: T.Krimly@uni-hohenheim.de; josefapfelbeck@gmx.net;
marco_huigen@hotmail.com; dabbert@uni-hohenheim.de

W. Mauser, M. Prasch (eds.), *Regional Assessment of Global Change Impacts*, DOI 10.1007/978-3-319-16751-0_40

and CO_2 concentration (Rosenzweig et al. 2000). Different crops are affected to varying degrees by changes to climatic conditions, because of their specific demands on the climate. Therefore, for the individual farm, the effects of climate change also depend on the role that individual crops play in a farm's crop rotation and production type.

Modelling these complex relationships requires high spatial and temporal resolution (Rosenzweig et al. 2000; Seneviratne et al. 2006). Regional agro-economic models have limits in this regard, since they use (annual) means and, therefore, are not able to adequately reproduce the climatic variability and agricultural management activities. Therefore, the goal here is to estimate the effect of climate changes and the variability of the weather on the vegetation period for different crops, the crop rotation in various types of farms and the yields that result from crop management activities that are delayed or forsaken as a result of weather; this is accomplished using a spatially and temporally high-resolution DeepActor model (see Chap. 3).

40.2 Data Processing

As already mentioned in Chap. 39, the changes to land use calculated by ACRE (Agro-eConomic pRoduction model on rEgional level) at district level are transformed to the level of the proxel via the decision behaviour of the actors in the *DeepFarming* model.

40.3 Model Documentation

The model *Farming* basically consists of two components:

1. The agricultural sector model ACRE
2. The DeepActor model *DeepFarming*

The focus in this chapter is on the actor model *DeepFarming* that intensifies the coupling and data exchange especially with the *Biological* and *WaterSupply* models.

A total of 28 different actor types (farm types) were defined for the *DeepFarming* model (see Table 40.1). These types each have a specific crop rotation, setting of arable land and grassland and animal husbandry. The actor types were derived from statistical data at district level for 1995 with respect to production type, land use and the number of farms in each production category. The agricultural statistics distinguish five production types: cash cropping types, fodder growing types, meat production types, permanent crop types and mixed farm types.

Table 40.1 Actor types of the *DeepFarming* model grouped by production direction and distinguishing features within the groups

Actor type (n=28)	Distinguishing features
Cash cropping (n=4)	Crop rotation
Meat production (n=3)	Fattening pigs or lactating sows, number of animals, crop rotation
Fodder growing (n=13)	Arable land and/or grassland; dairy cattle, fattening bulls, sheep, horses or suckler cows; number of animals; crop rotation; intensity of grassland use
Permanent crop hop (n=1)	–
Mixed (n=7)	Grassland or no grassland, dairy cattle and fattening bulls or fattening pigs and lactating sows, number of animals, crop rotation

Specific rules were applied in the allocation of the 28 actor types within the drainage basin for 1995. Significant allocation factors for each actor type are the specified crop rotation, the setting of arable land and grassland and the suitability of a proxel for the cultivation (1 = very good, 2 = good, 3 = poor) of individual crops. To initialise the actors, in a first step a randomly selected proxel was assessed for its setting of arable land and/or grassland. In a second step, the actor types were identified that have the matching setting of arable land and/or grassland. In a third step that used the suitabilities for cultivation on the respective proxel, the sum of the suitability for cultivation resulting from the crop rotation was deduced for these selected actor types and the actor type with the lowest sum was localised on the proxel. In order to adhere to the number of farms within a production type per district that is reported by the agricultural statistics, the sums of the suitabilities for cultivation are weighted. The weight factor incorporates the already localised actor types and the number still to be localised. There is a total of 58,984 proxels with agricultural use in the drainage basin, and each was occupied with an actor type.

The actors obtain their cultivation plan for the coming harvest year from ACRE (see Chap. 39). The land use calculated at district level by ACRE must now be interpreted by the actors. The bases to adjust the actors to a changed land use are rules that are mainly based on expert knowledge; they are applied individually for each actor type. The actors assess whether ACRE specifies an expanded, reduced or unchanged extent of cultivated area for a specific crop type for the actor's district and then follows this signal. For example, if ACRE indicates a decrease to the cultivated area of a crop, the actor in turn will attempt to reduce cultivation of this crop and to cultivate the freed-up land with another crop for which ACRE has decreed an expanded area. Within these rules for each actor type, specific crop rotation upper and lower limits must be observed. The basis for these limits are the crop rotation rules developed by Könnecke (1967), which have been adjusted to account for today's options with respect to plant protection and plant breeding. Actors with animal husbandry must also make sufficient land available for forage. This is ensured using a minimum requirement on forage area per animal.

If an actor's cultivation plan for the coming harvest year is fixed, it adaptively implements the plan using daily decisions under the given climatic conditions. The key task of the actors is to make management decisions about cultivation for each crop (sowing, fertilisation and harvest). These decisions involve both the knowledge from the previous years about climatic conditions and changes and the pattern of plant growth, as well as the current daily information about the weather and development stages of the plants. Parameters that are relevant to the decisions include precipitation and the mean daily temperature from the *Atmosphere* model, the soil saturation from the *Soil* model and the development stage of the plants from the *Biological* model. The actor makes an important decision about the sowing of a crop, for which the start of growth is triggered by the *Biological* model. Each crop has a certain start date for sowing. From this date on, the crop can be sown or planted if the field is passable, soil temperature is not too low and there is no heavy rain. If these conditions are not fulfilled, the actor tries to sow or plant the next day. The temporal pattern of plant growth and hence also a possible maturity date are determined by the sowing date. For fertilising and harvesting, a certain development stage of the crops must be reached which is transferred on a daily time step by *Biological*. If this development stage is reached and the field is passable, soil temperature is not too low and there is no heavy rain, the management action can be carried out. Overall, each crop type has specific natural and crop-specific demands related to cultivation, which the actor must observe when making management decisions. Figure 40.1 depicts an example of this type of decision using sowing of winter wheat. Similar decision-making algorithms were designed for fertilisation

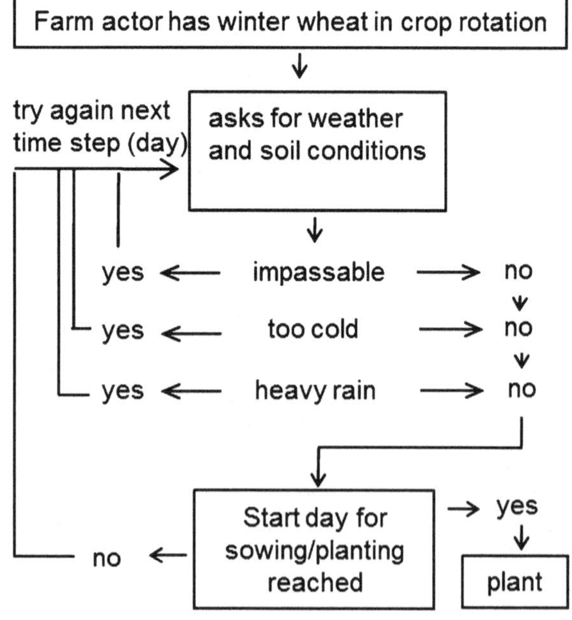

Fig. 40.1 Decision algorithm for sowing of winter wheat

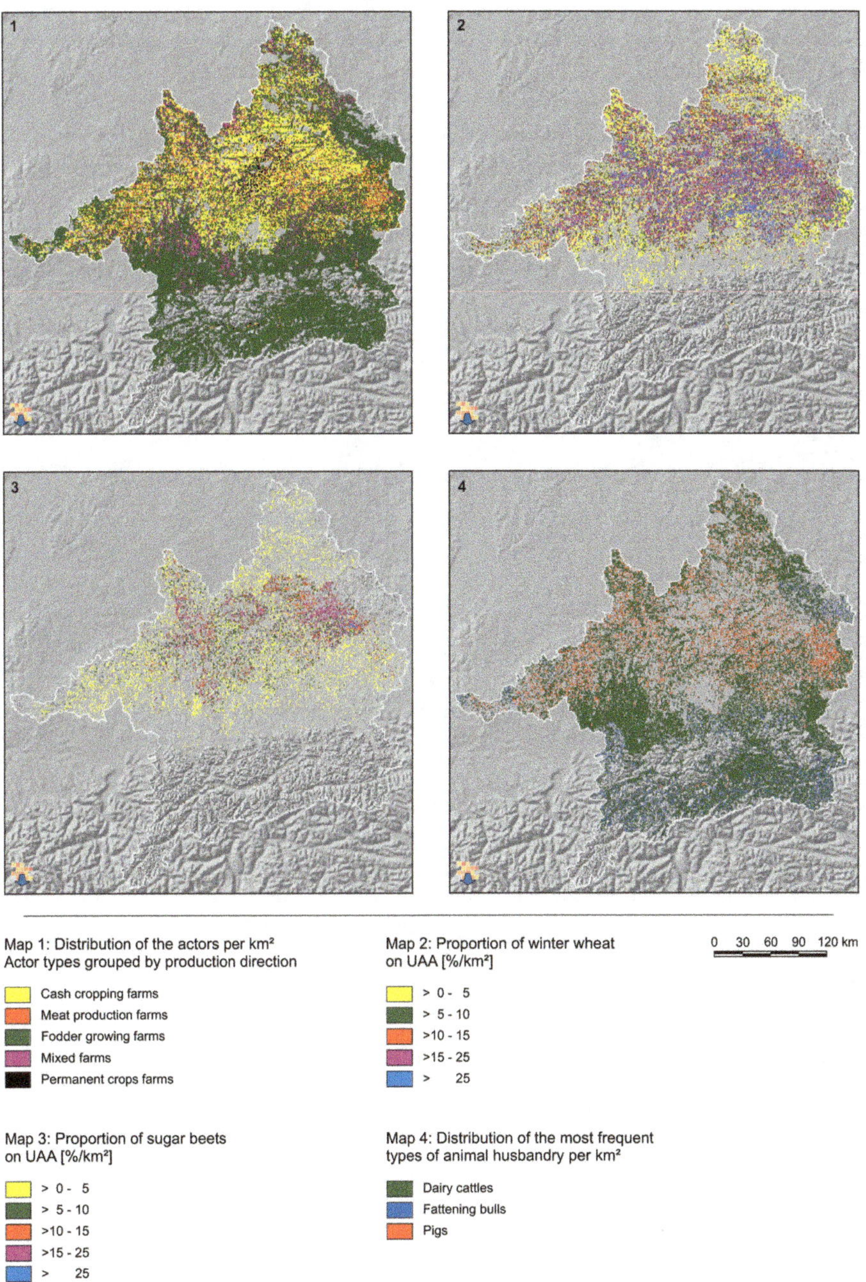

Map 40.1 Distribution of actor types, agricultural land use and animal husbandry (*DeepFarming* model component). (The maps present the modelling results of the initialisation of the actors. These are based on data from the agricultural statistics of 1995 relating to the number of farms of the specific production types within a district and the suitability of a proxel for the cultivation of different crops.) (Data sources: ©Bavarian State Office for Statistics and Data Processing 2002; ©State Statistical Office, Baden-Württemberg 1996, 1997, 2000, 2005)

and harvest. If delayed and forsaken management measures that arise as a result of weather cause changes to yields for individual crops compared to what is expected based on the planning in ACRE, then these changes are summarised at district level and forwarded to ACRE for planning the following harvest year.

The actors also make decisions about the irrigation of different crops. These decisions are influenced by the degree of soil saturation and the availability of groundwater, which are variables supplied by the models *Soil* and *WaterSupply*. The quantities of groundwater required for irrigation are transferred to the *WaterSupply* model.

40.4 Results

The results of the initialisation of the actors are presented in Maps 40.1(1–4). Map 40.1(1) shows the distribution of the actor types within the drainage basin, grouped into the 5 production types. Fodder growing farms are largely found in the Alps, the Alpine Foothills and the Eastern Bavarian Uplands which are dominated by grasslands. In the regions favourable to crop cultivation south of the Danube, there are mostly cash cropping farms that are replaced by meat production farms with decreasing proportions of arable land in the Tertiary Hills. The greatest extent of winter wheat cultivation is found in the regions with cash cropping and meat production (Map 40.1(2)), while cultivation of sugar beets is concentrated in cash cropping farms west of the Eastern Bavarian Uplands (Map 40.1(3)). Dairy cattle and fattening bulls are largely found in the grassland regions in the southern half of the basin as well as in the Eastern Bavarian Uplands, and pig farming is mostly located in the Tertiary Hills and in the area around the Swabian Jura (Map 40.1(4)).

References

Könnecke G (1967) Fruchtfolgen. VEB Deutscher Landwirtschaftsverlag, Berlin

Rosenzweig C, Iglesias A, Yang XB, Epstein PR, Chivian E (2000) Climate change and U.S. agriculture: the impacts of warming and extreme weather events on productivity, plant diseases and pests. Center for Health and the Global Environment, Harvard Medical School, Boston

Seneviratne SI, Lüthi D, Litschi M, Schär C (2006) Land-atmosphere coupling and climate change in Europe. Nature 433:205–209

Chapter 41
Water Demand by Private Households and the Public Sector

Andreas Ernst, Silke Kuhn, Carsten Schulz, Nina Schwarz, and Roman Seidl

Abstract In this section, the shallow (statistical) *Household* model is described and results from runs are presented. In later stages of the development of DANUBIA, this model was replaced by the DeepHousehold actor model. The shallow model calculates the water demand by private households and the public service sector. The calculation of water demand is based on 25 categories of households. The household types are distinguished by the number of persons living in the household as well as their monthly net incomes in a 5×5 matrix. Data stem from representative written and telephone surveys of a total of 1,317 people within the German population of the Upper Danube catchment. The computations include the domestic water demand per household type based on the ten types of use (e.g. showers, toilets, tooth brushing etc.) taking into account annual variations, the water price and price elasticity and the aggregation of water demands by household types for the individual proxels taking into account an urban-rural bias factor and an innovation factor. The latter two were necessary to account for considerable deviations on a spatial level and over time.

Keywords Domestic water use • Water use • Household types • Shallow model • GLOWA-Danube

A. Ernst (✉) • S. Kuhn • C. Schulz
Center for Environmental Systems Research (CESR), University of Kassel, Kassel, Germany
e-mail: ernst@usf.uni-kassel.de; der.schulz@web.de

N. Schwarz
Department of Computational Landscape Ecology, Helmholtz-Centre for Environmental Research – UFZ, Leipzig, Germany
e-mail: nina.schwarz@ufz.de

R. Seidl
Institute for Environmental Decisions, ETH Zürich, Zürich, Switzerland
e-mail: roman.seidl@env.ethz.ch

W. Mauser, M. Prasch (eds.), *Regional Assessment of Global Change Impacts*, DOI 10.1007/978-3-319-16751-0_41

41.1 Introduction

The *Household* model calculates the water demand by private households and the public service sector. This model allows the mapping of private water demands under changing climatic conditions and social circumstances. In addition, regions with particularly high water demands related to extremely hot summer temperatures and possible saving potentials can be identified with the spatial resolution of the model. This kind of information is especially helpful in the development of long-term strategies by and decision makers in policy and administrations.

41.2 Data Processing

The calculation of water demand is based on 25 categories of households. The household types are distinguished by the number of persons living in the household as well as their monthly net incomes (5×5 matrix). For all of the 25 household types, the demand in litres per person per day is available for 10 different use types. These data stem from representative written and telephone surveys of a total of 1,317 people within the German population of the Upper Danube basin. The exact consumption for the 10 use types for each household type was determined using these data together with statistical data from the literature (Abke 2001).

The 10 Types of Water Use
washing machines, dishwashers, showers, baths, toilet flushing, cleaning, dishwashing by hand, tooth brushing, washing hands, miscellaneous

The splitting of water demand into the individual use types allows the mapping of annual variations in water demand, for example. The household types exhibit typical patterns of water consumption; these are graphically represented in Fig. 41.1.

The following statements apply with respect to the typical consumption patterns:

1. The higher the household income, the more water tends to be consumed
2. The more people live in a household, the higher the absolute consumption of water in the household, although the per capita consumption decreases with increasing number of household members

The reason for this pattern is that for some of the types of water use, the consumption value per week is approximately the same, regardless of how many people live in the household. In addition, 1-person households have a machine usage that leads to higher water consumption compared to households with more than one person: thus, for example, a 4-person household uses the washing machine only

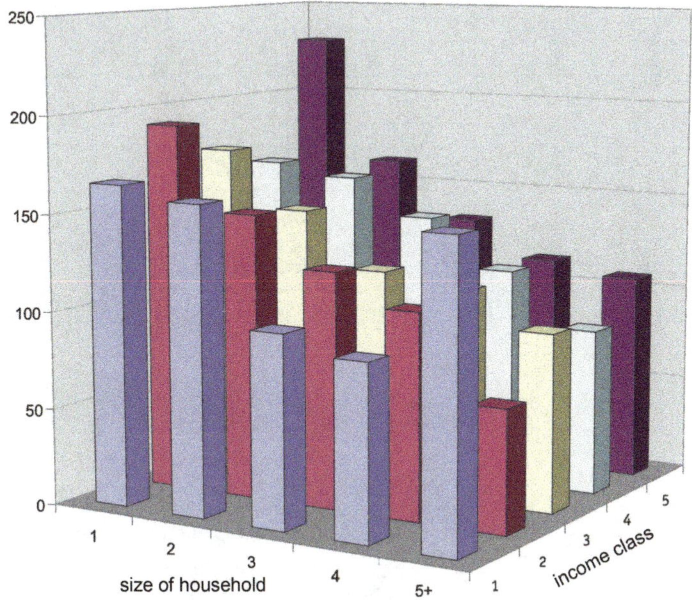

Fig. 41.1 Per person water demand of the 25 household types (matrix 5 × 5); the *left* y-axis shows the water demand in litres per person per day; the *right* z-axis shows the net income classes (1: < 1,100€; 2: 1,100<2;000 €; 3: 2,000<2,900 €; 4: 2,900<5,000 €; 5: > 5,000 €); a 5 or more person household consists of 5.3 persons on average. *Missing bars* indicate that there is no empirical data on water demand from the water metre for these household types

3 times and not 4 times as often as a 1-person household. In a 4-person household, a dishwasher is used only 2.2 times as often as in a 1-person household.

41.3 Model Documentation

The calculation of the domestic water demand by the model takes place in three steps:

1. The calculation of water demand per household type based on the 10 types of use (e.g. showers, toilets, tooth brushing etc.) taking into account annual variations
2. The incorporation of the water price and its price elasticity
3. The aggregation of the water demands by household types for the individual proxels taking into account an urban-rural bias factor and an innovation factor (see Fig. 41.2)

The splitting of water demand into individual use types allows the model to account for seasonal fluctuations: if the monthly mean temperature increases above 10 °C, we assume an increased demand for water for showers and baths (garden watering is considered in a further development of DANUBIA). The total demand

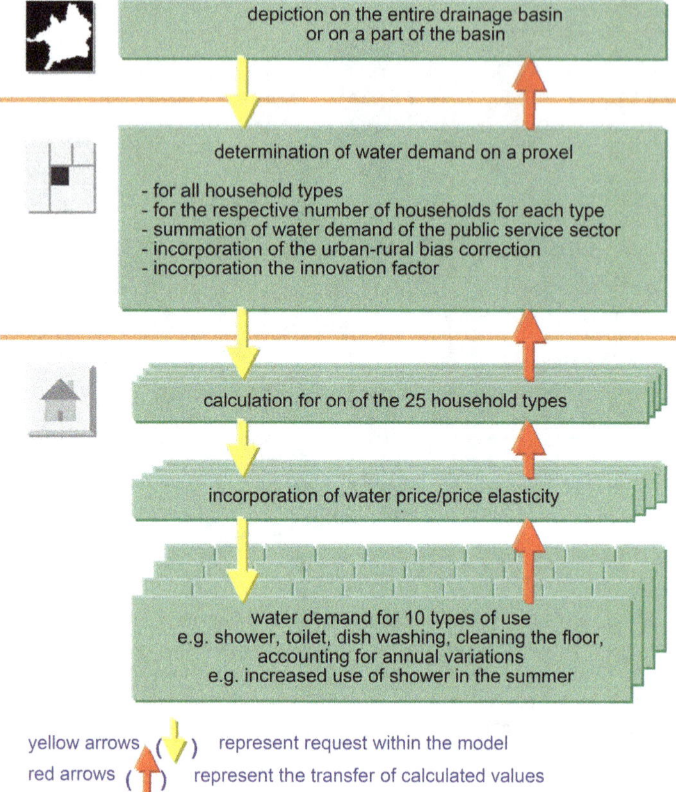

Fig. 41.2 Steps of calculation in the *Household* model (the symbols on the *left* show the calculation level: household, proxel or Upper Danube basin)

for water per household type based on the 10 use types includes – in a second step – the water price (Fig. 41.2).

Finally, in a third step, the water demands for all households of a proxel are summed up. The baseline data on how many households of which types are located within each proxel are supplied by the *Demography* model.

The urban-rural bias correction accounts for the fact that urban proxels have a statistically higher mean per capita consumption. This fact is the result of higher per capita consumption in cities compared to rural areas (due to a higher number of single-person and 2-person households, as well as more small businesses that are included in the statistical water demand). This bias correction is calculated using regression equations.

An innovation factor is also applied. It is based on statistical consumption data for 1991, 1995, 1998 and 2001 and represents the fact that water consumption in recent years have declined continually as a result of technical innovations (e.g. washing machines with water-saving programmes).

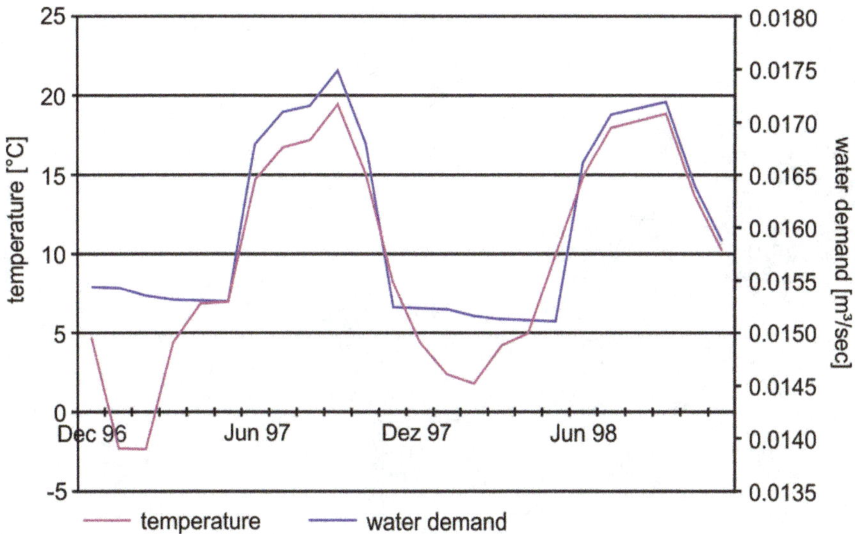

Fig. 41.3 Modelled water demand of a Munich proxel (ID 94185) for the years 1997/1998 as a function of mean daily temperature

In Fig. 41.3, the trend of water consumption for 2 years for one proxel in Munich is depicted. Since the seasonal variations do not influence the consumption in the colder months, it can be easily seen that the water consumption decreases slightly overall.

A model of water consumption of the public sector is integrated in the *Household* model. It is a simple additive model that adds 8 % to the water consumption of private households. This water demand is termed as the category "miscellaneous" by the BGW (Bundesverband der Gas- und Wasserwirtschaft = Federal Association for German Gas and Water Industry). It includes, for example, public swimming pools, schools, opera houses, fountains, central government departments, armed forces, universities and hospitals. The water demand in this category statistically amounts to approximately 8 % of the domestic water demand.

41.4 Results

Map 41.1 presents the modelled water consumption by private households for the month of April in 1998. The data are from the reference run that was carried out with DANUBIA in December 2004. The validation of a modelled water demand is based on statistics for water consumption for the communes in the states of Baden-Württemberg and Bavaria on the mean per capita daily consumption for 1998. Since the consumption by the public service sector is not separately included in the statistics, the validation for the *Household* model can only be carried out for the domestic water demand.

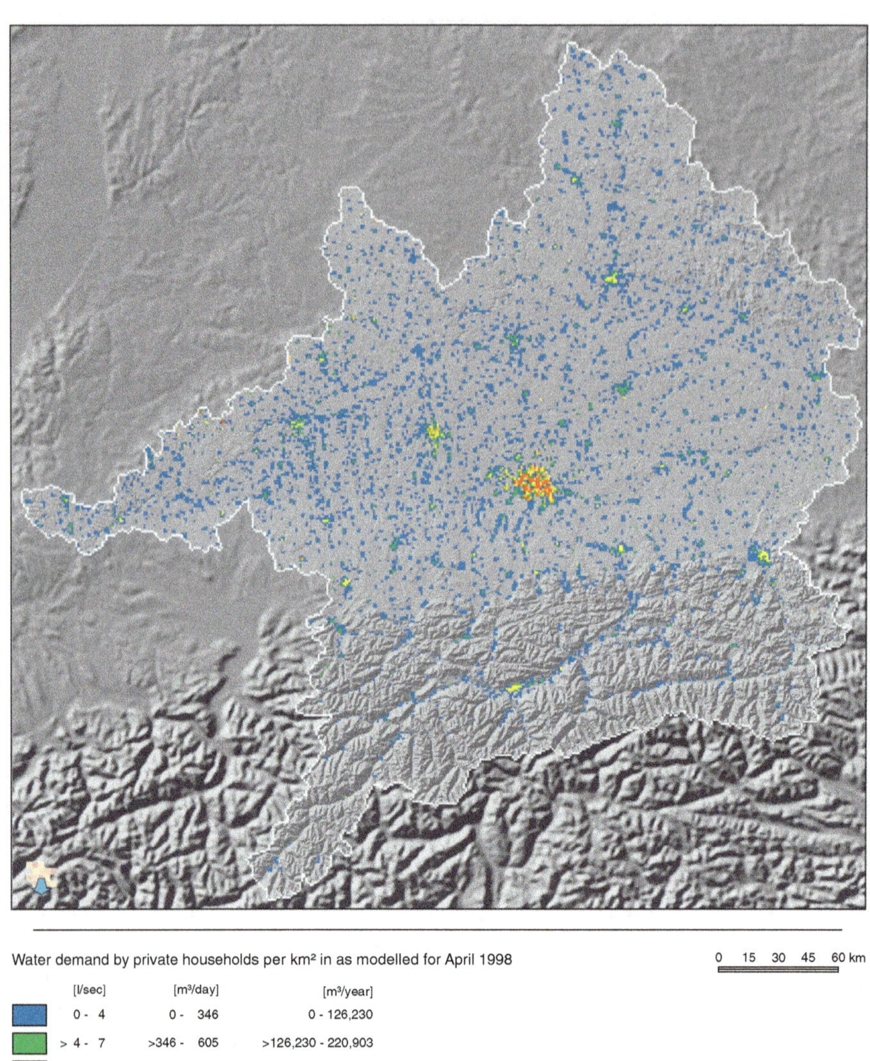

Map 41.1 Water demand by private households and the public sector modelled by DANUBIA Household component (reference run)

The summed-up consumption over all proxels quite well matches the total consumption within the basin, although every single proxel deviates from the statistical data, sometimes even considerably. The statistics have fluctuations from 57.7 to 301.1 (mean = 129.13; standard deviation = 24.1) litres per person per day. One reason for this is that in some communes, many private wells are used as sources of

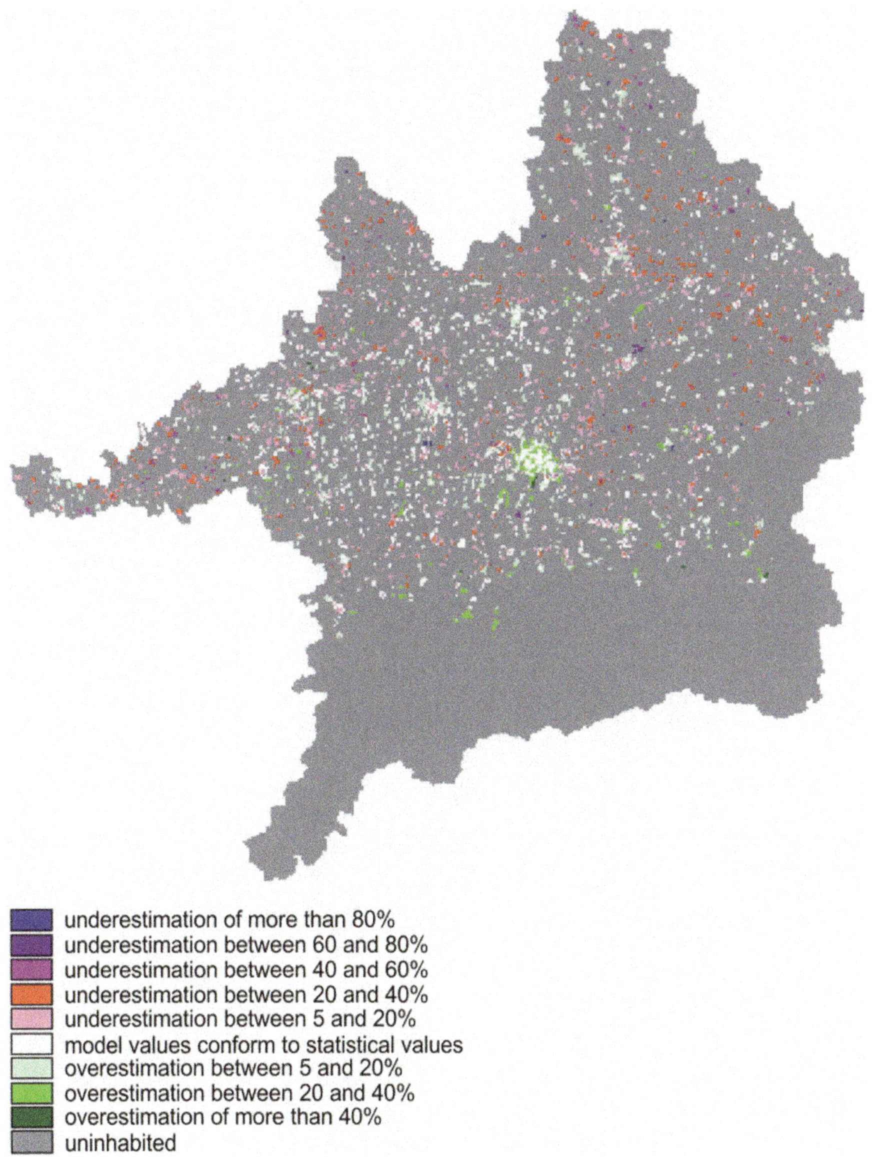

Fig. 41.4 Percent deviation of modelled water demand in relation to the statistical demand of private households for the German part of the basin in the year 1998

drinking water, and in others, the consumption of water by small businesses is disproportionately high. However, the model has a range of only from 121.9 to 234.5 l per person per day with a standard deviation of 10.2. Figure 41.4 shows the deviations of the modelled consumptions compared to the statistics for water consumption by private households in the German communes of the Upper Danube basin for 1998 in percent.

The *Household* model calculates a total demand of 17.66 m^3/s for all households in the basin for 1998. Thus, the model overestimates the domestic water demand in 1998 over all proxels by 1 %.

Reference

Abke W (2001) Wasserversorgung. In: Lechner K, Lühr HP, Zanke UCE (eds) Taschenbuch der Wasserwirtschaft. Parey, Berlin

Chapter 42
Modelled Domestic Water Demand 2: The *DeepHousehold* Decision Model

Andreas Ernst, Silke Kuhn, and Roman Seidl

Abstract This section presents the *DeepHousehold* decision model of domestic water demand, a multi-agent model in which the actors' decision processes are described explicitly. As such it replaces the shallow Household model. The aggregation of actors to five types according to a lifestyle classification is described. Lifestyles pool groups of individuals that have similar values, socio-economic status, behaviours and common aesthetic preferences. The lifestyle classification used in this model also provides geo-referenced relative quantities of population for each lifestyle type. Thus, there are five actors for each inhabited proxel in the catchment ($N = 9{,}115$), giving a total of $9{,}115 * 5 = 45{,}575$ actors. The characteristics of the modelled actors together with their preferences, sensitivity to external events and others are given. Finally, the range of possible applications and types of analyses with the model are depicted, including the definition of test regions, singling out specific water uses, or the behaviour of lifestyle groups.

Keywords Agent-based model • Domestic water use • Decision-making • Lifestyles • GLOWA-Danube

A. Ernst (✉) • S. Kuhn
Center for Environmental Systems Research (CESR),
University of Kassel, Kassel, Germany
e-mail: ernst@usf.uni-kassel.de

R. Seidl
Institute for Environmental Decisions, ETH Zürich, Zürich, Switzerland
e-mail: roman.seidl@env.ethz.ch

W. Mauser, M. Prasch (eds.), *Regional Assessment of Global Change Impacts*, DOI 10.1007/978-3-319-16751-0_42

42.1 Introduction

The *DeepHousehold* (DHH; Ernst et al. 2008) model developed by the environmental psychology subproject is a multi-actor model in which the agent, i.e. the actors, are described explicitly (see Chap. 3). As such it replaces the modelling of water consumption carried out with the *Household* model (see Chap. 41).

42.2 Data Processing

Since not all households (i.e. approx. 4 million) in the basin can be modelled individually, they are aggregated into types. The categorisation is based on an existing lifestyle classification. Lifestyles comprise groups of individuals that have similar values, socioeconomic situations, behaviours and common aesthetic preferences (Fig. 42.1).

The basis of our classification is the so-called Sinus-Milieus® (Sinus Sociovision 2007) from commercial market research. Based on the lifestyle classification according to Sinus, there are five milieu groups. These are assumed as actor types in the model.

DeepHousehold thus includes the following five actor types:

- Post-materialists
- Leading milieus (includes modern performers and established persons)
- Traditional milieus (includes tradition rooted, GDR nostalgics, conservatives)

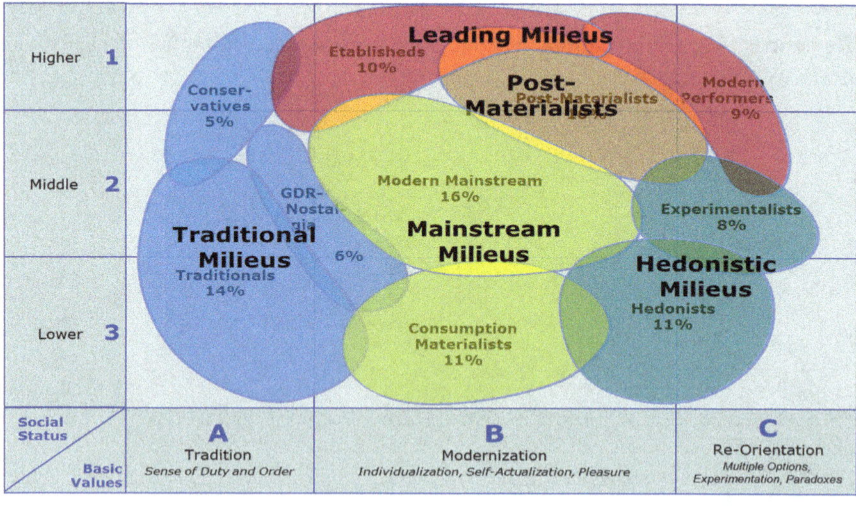

Fig. 42.1 Classification of the ten Sinus-Milieus® by social status and basic orientation (Source: Sinus Sociovision 2007). In addition the Sinus-Milieus® are divided in four milieu groups, called segments (Sinus Institute)

- Mainstream milieus (includes mainstream middle class, consumerist-materialists)
- Hedonistic milieus (includes hedonists, experimentalists)

The data on the geographical localisations of the Sinus-Milieus® are commercially available data stemming from microm® (microm 2007). These data were converted into DANUBIA coordinates in our subproject. Accordingly, for each inhabited proxel, the percentage of each of the 5 milieu groups is known.

42.3 Model Documentation

DHH is designed so that one modelled actor represents all households of a specified type (i.e. one of the 5 milieu groups) on a proxel. Thus, there are 5 actors for each inhabited proxel in the basin (N=9,115), for a total of 9,115 * 5=45,575 actors.

The characteristics of each actor type in *DeepHousehold* are stored within its profile. In addition to an actorID and the localisation on a proxel, the profile contains information on value orientation (modernity: conservative vs. modern) and importance of the environment, price (price sensitivity), behaviour of others and future mindedness (importance of future consequences of current activity). Moreover, the profile includes data on age, income and number of individuals living in each household.

In the shallow *Household* model – the precursor to DHH – the sociodemographic features income and household size made up the 25 household types (see Chap. 41). These are now stored as a part of the characteristics of the actors in DHH.

In order to account for the fact that not all members of a milieu group earn exactly the same amount of money and they are not all precisely of the same age, the variables age, household size and income show a distribution of values. For example, an actor in the traditional milieu group consists of 10 % from age class 1 (20–40 years of age), 20 % from age class 2 (40–60 years) and 70 % age class 3 (over 60 years).

42.3.1 Water Use Types

The 10 water use types modelled in *DeepHousehold* differ only marginally from those used in the shallow *Household* model. The use types in DHH were selected based on the effectiveness, representativeness and innovation sensitivity of their characteristics.

The 10 Types of Water Use in *DeepHousehold*
washing machines, dishwashers, use of rainwater, showering, bathing, tooth brushing, dishwashing by hand, toilet flushing, cleaning floors, cooking

42.3.2 Plans

For each of the 10 water use types, the actors use so-called plans. Plans summarise typical behaviour patterns of each of the use types. The majority of plans are related to the frequency of the water use, but some are based on other features, such as shower length. In the following, the plans within the model are described using the example of showering as the use type.

First, four habit patterns are determined using the data from the empirical studies. They are at the same time both as different but also as representative as possible. These are taking a shower once per week, every other day, daily or twice a day.

Because DHH also has to describe extreme situations in which no more water is supplied, the option of not showering was also included as one plan. Accordingly, there are five shower frequency plans: no showering, showering once per week, showering every other day, showering daily or showering twice a day.

42.3.3 Events

Everyday water use behaviour is largely characterised by habit, that is, water use is an automatic behaviour. The habits are empirically determined and implemented for the five actor types in DHH. Thus, calculated decisions are required for the model only in the case of a deviation from one's typical habit. The departure from a habit (i.e. the deviation from a plan that the actor implements in order to adopt another plan) is triggered by specific conditions known as events. Thus, the current frequency of taking a shower is only evaluated if at least one of three *thresholds* is exceeded for *warning flags*, *temperature* or *price of water*.

This is the case if:

- *WaterSupply* yields a drinking water quantity flag greater than 1 (see Chap. 3)
- The daily mean temperature exceeds 10 °C
- The price of water increases by more than 5 % compared to the previous time interval

42.3.4 Attributes

All plans have *attributes*. Attributes describe the value of a plan with respect to three dimensions: price of water, environmental consciousness and modernity. The values range from −1 to 1. A value of 1 for environmental consciousness means that less water is consumed and the plan is thus quite environmentally friendly. The plan to shower daily is given a value of −0.6, and the plan to shower once per week is assigned a value of 0.4.

The values for the dimension modernity are based on which behaviour is considered to be trendy. For the example of showering, this means that it is modern to shower more often. The specific values are currently based on expert ratings since related empirical data are missing up to now.

42.3.5 The Decision Process

Based on their profiles, the actors have varying sensitivities to the different events. For an actor in the traditional milieu group with a very price-sensitive profile, an increase in the price of water has greater significance than for an actor from the post-materialist milieu who is less price-sensitive, but more environmentally conscious and modern. Accordingly, there is a slight decrease in the water consumption for the traditional milieu for certain water use types following an increase in the price of water by over 5 % after some time interval (water price event). In the post-materialist milieu, this event does not lead to any change in behaviour because of the weak sensitivity to price. One actor represents all households in its milieu group within a proxel. However, not all individuals that belong to the same milieu group make exactly the same decisions. Therefore, each actor type contains percentage values representing the relative quantity of different decisions. Thus, for each use type, following an event, the percentage to which each plan is carried out is computed.

When an actor selects a plan, the associated action is invoked. In the example of the increase in water price, this may lead to a reduction in the frequency of showering in some actors and hence to lower total water consumption.

For each actor, the percentage at which each plan is chosen is stored. All events are also stored.

The following summarises the calculations and result outputs that DHH is capable of. It is possible to model the water demand for the entire basin in specifically defined test regions and also based on specific proxels. In addition to total water consumption, the water consumption for each of the 10 use types can be calculated separately. This can be done either for all households or for each of the actor types. Furthermore, the influence of various constraints (i.e. events and scenarios) can be calculated per use type and per actor. The decisions (i.e. the selection of plans by the actors) can be reproduced over time (in monthly intervals). The presentation of the results can be on a spatial or temporal scale.

The first validation runs indicate that the total water consumption calculated by the model was approximately 8 % lower than the total water demand for the basin based on statistics. At the moment, this seems a satisfactory approximation. Further validations are planned.

42.4 Results

Maps 42.1(1–4) illustrate the modelled water consumption for the taking a shower use for August 2035 in litres per household per day for two milieus in two

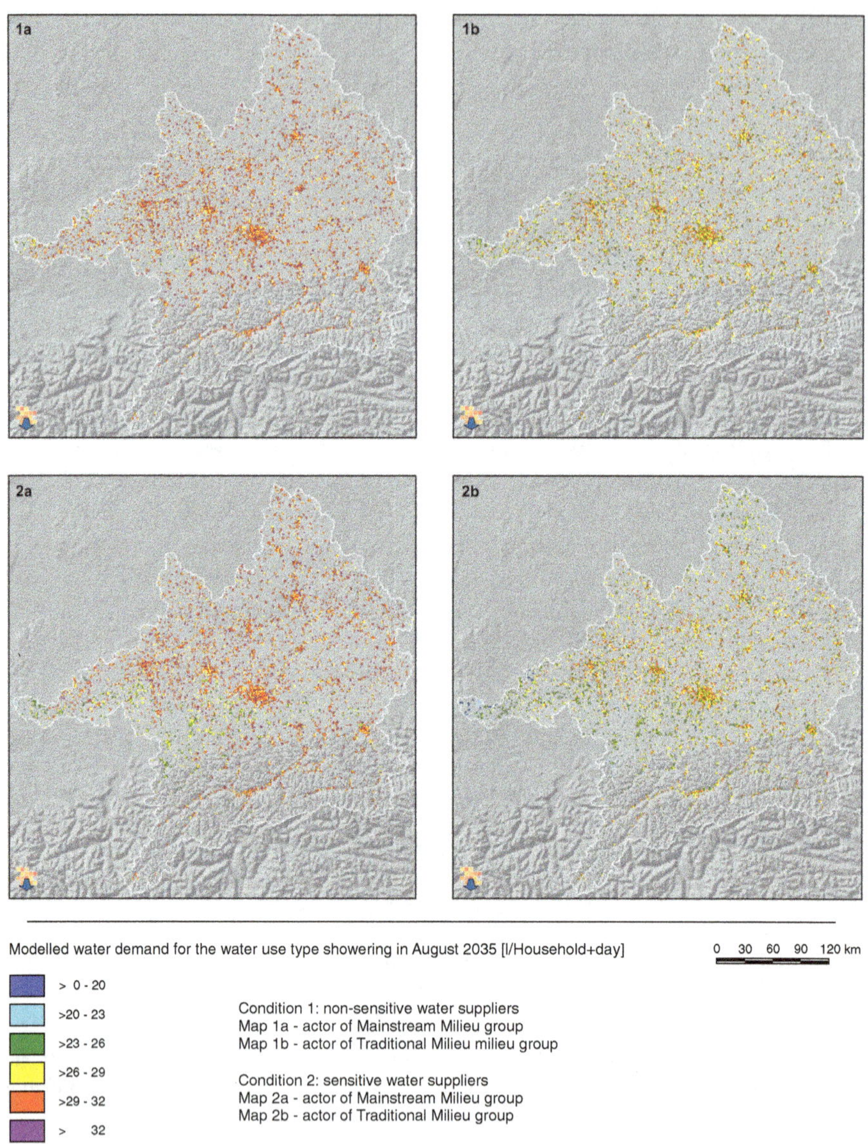

Map 42.1 Test study for the calculation of water demand with the water use type showering (Water demand modelled by *DeepHousehold* model component for two actors of different milieu groups and different *WaterSupply* actor types)

different water supply scenarios (see Chap. 28). One is a "business-as-usual" climate scenario that assumes a continuation of the climate trend over the past three decades. It includes the results from a pilot study on trends in drinking water supply (see Chap. 28).

The actors in the mainstream milieu group consume more water per day under both conditions (1 and 2) for showering use. Since the households for both actors are of the same size (average of 2.2 individuals), the observed differences can be attributed to differing habits of the actors. If the water supplier responds sensitively (condition 2), there is a reduction in the water consumption for both actors in parts of the basin. This response is particularly perceivable west of Munich (in the middle of the basin) and is based on the fact that warning flags are shown on the proxels in this region with a value greater than 1.

References

Ernst A, Schulz C, Schwarz N, Janisch S (2008) Modelling of water use decisions in a large, spatially explicit, coupled simulation system. In: Edmonds B, Hernández C, Troitzsch KG (eds) Social simulation: technologies, advances and new discoveries. Information Science Reference, Hershey, pp 138–149

microm Micromarketing-Systeme und Consult GmbH (2007) MOSAIC Milieus®. http://www.microm-online.de

SinusSociovision (2007) Milieulandschaft 2007. http://www.sinus-sociovision.de/

Chapter 43
Diffusion of Water-Saving Technologies in Private Households: The Innovation Module of *DeepHousehold*

Nina Schwarz, Silke Kuhn, Roman Seidl, and Andreas Ernst

Abstract The Innovation model is a supplementary module in the *DeepHousehold* model within the Environmental Psychology subproject. It focuses on water saving through increased efficiency, manifested by the purchase of newer, more efficient technologies. The factors that influence the extent of adopting water-saving innovations were investigated, and an estimate was calculated as to how the diffusion of these innovations in the near future might look. Three innovations were chosen: shower heads, toilet flushes and rainwater use. The Innovation model simulates the decision process of domestic households based on two empirical studies: a quantitative, written study and supplementary telephone interviews. The five *Household* actors generally decide on the adoption or rejection of new technologies using two different decision algorithms: a deliberative decision rule, based on "Theory of Planned Behavior", and the so-called take-the-best heuristic. If no clear decision is made, the *Household* actors imitate the behaviour of their social network. Three innovation scenarios were developed in contrast to a baseline: an information campaign, subsidies and regulation. Simulation runs show that the regulation has the greatest impact on all three innovations. However, information campaigns also led to a spread of the water use innovations. Financial incentives had no additional impact on the spread for either water-saving shower heads or for 2-volume toilet

N. Schwarz (✉)
Department of Computational Landscape Ecology, Helmholtz-Centre for Environmental Research – UFZ, Leipzig, Germany
e-mail: nina.schwarz@ufz.de

S. Kuhn • A. Ernst
Center for Environmental Systems Research (CESR), University of Kassel, Kassel, Germany
e-mail: ernst@usf.uni-kassel.de

R. Seidl
Institute for Environmental Decisions, ETH Zürich, Zürich, Switzerland
e-mail: roman.seidl@env.ethz.ch

W. Mauser, M. Prasch (eds.), *Regional Assessment of Global Change Impacts*, DOI 10.1007/978-3-319-16751-0_43

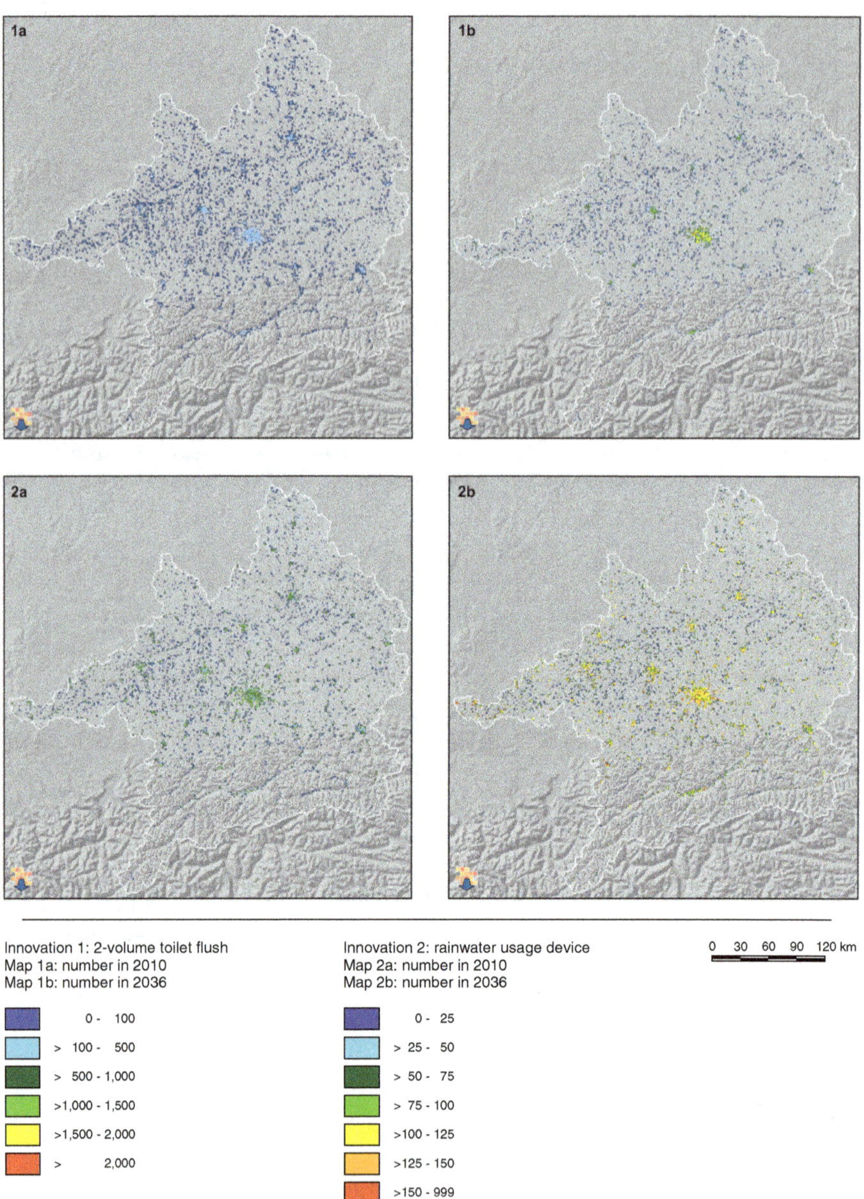

Maps 1a and 1b show the temporal spreading of the innovation of 2-volume toilet flushes at two different dates (2010 and 2036).
Maps 2a and 2b show the temporal spreading of the innovation rainwater usage device at two different dates (2010 and 2036).

As the installation of a rainwater usage device is only relevant in case of building a new house, this innovation is spreading much more slowly.

Map 43.1 Test study for the calculation of diffusion of 2-volume toilet flushes and rainwater usage devices in private households (innovation module in the *DeepHousehold* model component)

flush; just like the *Baseline* scenario, they led to a moderate increase. In the case of rainwater usage devices, both the *Information* scenario and the *Subsidies* scenario led to a notable effect in approximately equal size.

Keywords Innovation diffusion • Water use • Agent-based modelling • GLOWA-Danube

43.1 Introduction

Savings of two kinds can be realised in many areas of the environment: either through changes in behaviour or through increased efficiency, which can be manifested by the purchase of newer, more efficient technologies. Especially in private households, the more efficient use of resources has a great impact on the environment, since the sum of the saved resources is very high as a result of the high number of users. The adoption of water-saving technology is thus a key factor in future water demands.

The hot summer of 2003 made it clear that when it comes to water use, the water supply can be influenced by local and regional climatic factors to such an extent that bottlenecks in the supply of drinking water can arise. The likelihood of such extreme events is increasing with climate change. Therefore, saving drinking water resources is imperative.

For this reason, with the example of water use, the factors that influence the extent of water-saving innovations were investigated and an estimate was calculated as to how the diffusion of these innovations in the near future might look (Schwarz 2007; Schwarz and Ernst 2009). The model is called the Innovation model and is implemented as a supplementary module in the *DeepHousehold* model within the Environmental Psychology subproject (see Chap. 42).

43.1.1 *Innovation Research*

Because the decision to purchase products such as modern toilet flushes, shower heads, etc. is ultimately a decision about the adoption of innovations, understanding the extent of water use innovations draws from the scientific results of diffusion research. For *DeepHousehold*, it was decided that a separate Innovation model should be developed that uses the differentiation of actors according to their lifestyles. This has not been realised to date in conventional diffusion models. The typology of the actors was implemented using Sinus-Milieus® and milieu groups (microm 2007 and Sinus Sociovision 2007). Data for the geo-referenced basin are resolved at the proxel level. A detailed description is provided in Chap. 42.

Table 43.1 Technologies included in the innovation module

Technology	Type of technology
Shower	Standard shower head
	Water-saving shower head
Toilet flushing	Push flushing
	Standard flushing
	Stop button
	2-volume toilet flush
Rainwater use	Rainwater use devices
	No rainwater use devices

The selection of innovations is based on the savings potential of the technology. Three innovation areas were chosen: shower heads, toilet flushes and rainwater use (Table 43.1).

43.2 Data Processing

One particularity of the Innovation model is that it presents not only why individuals decide for or against an innovation but also represents the decision process itself. The components of the decision process are based on two empirical studies: a quantitative study ($N=272$) of the significance of the features of innovations and the lifestyles of the participants and supplementary telephone interviews ($N=12$) in which the decision process was surveyed in detail. It could be determined from the empirical data that membership in a milieu group has an impact on the adoption of water use innovations. Moreover, a key criterion in the decision process is the normative pressure from other significant individuals. An actor in the leading milieus group represents a high share of adopters – these are individuals who adopt an innovation. At the same time, this milieu is oriented only little towards its social surroundings, whereas other milieu groups are guided by their peers in their decisions about buying water-saving innovations. Therefore, social networks are integrated as one element of the Innovation model. With the programme TooDaReD (Tool for DANUBIA Result Data) developed within the subproject, various social networks can be generated between the actors, for example, small-world networks or networks that are specific to a lifestyle.

Furthermore, the results of the study indicate that, depending on the lifestyle, other innovation characteristics are responsible for the adoption of innovations. In our study, seven such innovation characteristics were relevant: environmental protection, luxury, ease of use, compatibility with customs, compatibility with existing infrastructure, financial resources and authorisation to install the technology in the house. From the telephone interviews, it was concluded that in addition to these features, the financial savings arising from the innovation is also a key factor in the adoption. Accordingly, this feature was also implemented in the model.

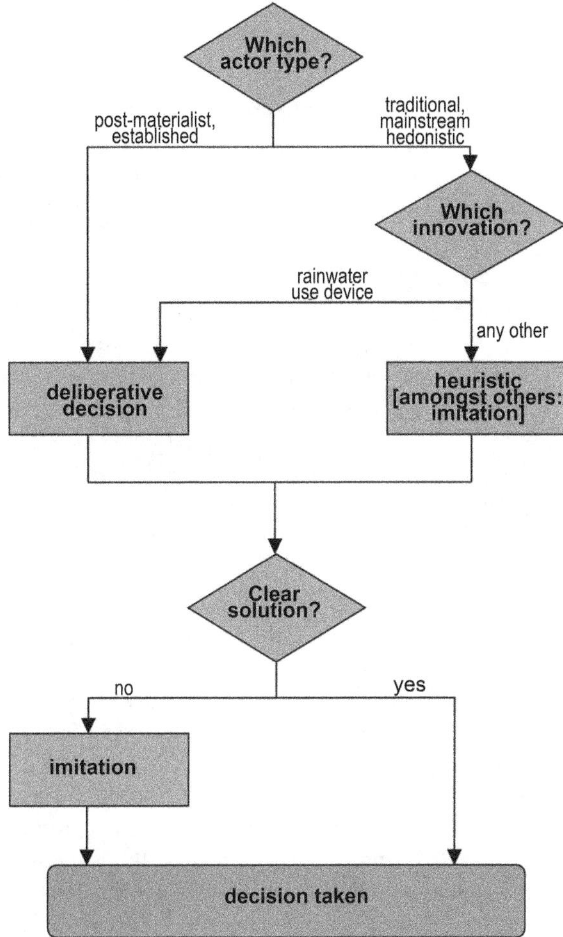

Fig. 43.1 Decision process on an actor concerning adaption or refusal of a water use innovation

43.3 Model Documentation

Once the magnitudes of the influence of the various factors were determined from the empirical studies, decision rules were created for the model based on these. Thus, for example, the significance which the leading milieus ascribe to their surroundings is set to low, while for other milieus it is higher. Furthermore, the actor types use different innovation characteristics in the formation of their attitudes.

The five *Household* actors generally decide on the adoption or rejection of new technologies using two different decision algorithms: one is a deliberative decision rule, based on the theory of planned behaviour (Ajzen 1991), and the other is the so-called take-the-best heuristic (Gigerenzer et al. 1999). If no clear decision is made, the *Household* actors imitate the behaviour of their social network (Fig. 43.1).

43.4 Results

To validate the sub-model, both independent statistical data (GfK 2004, 2006; Mall 2006) and data from lifestyle-specific surveys within the project were used. The validation of the sub-model indicated satisfactory to very good results.

Four innovation scenarios were developed to test the model within DANUBIA.

The 4 Innovation Scenarios

The *Business-as-usual* scenario is the baseline scenario.

The *Information* scenario contains a simulated intervention to increase environmental consciousness.

The effects of subsidies are studied in the *Subsidies* scenario.

The *Regulation of environmental policy* scenario assumes that only environmentally friendly options for toilet flushing and shower heads are available.

It could be shown from the simulation runs that the regulation of environmental policy has the greatest impact on all three innovations. However, information campaigns also led to a spread of the water use innovations. Financial incentives had no additional impact on the spread for either water-saving shower heads or for 2-volume toilet flush; just like the *Baseline* scenario, they led to a moderate increase. In the case of rainwater usage devices, both the *Information* innovation scenario and the *Subsidies* scenario led to a notable effect in approximately equal size.

Figure 43.2 indicates the spread of the water-saving shower head innovation in the four innovation scenarios for 2006–2020. It is clear that the *Regulation of environmental policy* scenario has the largest effect, followed by the *Information* scenario. Both the *Baseline* and the *Subsidies* scenarios show a similar trend.

Figure 43.3 illustrates the spread of the water-saving shower head for the two milieu groups (post-materialist and mainstream) in the *Baseline* scenario from 2006 to 2036 within a *Business-as-usual* climate scenario (see Chap. 28).

In the figure, it is clear that the post-materialist actor already has more water-saving shower heads (6 %) at the onset (2006) than the mainstream actor (0 %). By the end of the simulation under a *Business-as-usual* climate scenario, approximately 11 % of the post-materialist actors and approximately 3 % of the mainstream actors own the innovation. Among other factors, these differences can be attributed to the fact that the profile of the post-materialist actor places higher value on modernity and environmental consciousness than the mainstream actor (see also Chap. 42).

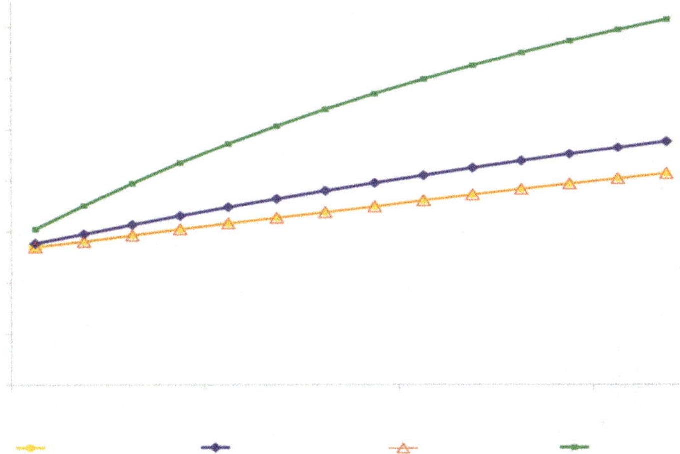

Fig. 43.2 Diffusion of water-saving shower head in the four innovation scenarios

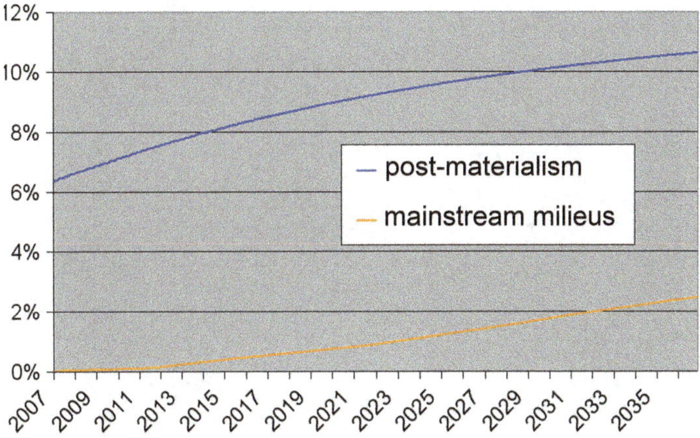

Fig. 43.3 Diffusion of water-saving shower head with actors from two different milieu groups (post-materialist and mainstream milieus) over a time period of 30 years, with assumed *Business-as-usual* climate scenario

References

Ajzen I (1991) The theory of planned behavior. Organ Behav Hum Decis Process 50:179211

GfK (2006) Sanitärstudie 2006: Eine schriftliche Befragung bei 2.000 repräsentativ ausgewählten privaten deutschen Haushalten. Porvided by Vereinigung Deutsche Sanitärwirtschaft e.V

GfK – Gesellschaft für Konsumforschung (2004) Baden und Duschen in Deutschland 2004. Provided by Grohe AG

Gigerenzer G, Todd PA, ABC-Research-Group (1999) Simple heuristics that make us smart. Oxford University Press, New York

Mall GmbH (2006) Pressemitteilung: Wirtschaftsfaktor Regenwasser. Available via DIALOG. http://mall.info/Pressemitteilung-Wirtschaftsfa.7110.0.html. Accessed 7 Jan 2008

microm Micromarketing-Systeme und Consult GmbH (2007) MOSAIC Milieus®. http://www.microm-online.de. Accessed 7 Jan 2008

Schwarz N (2007) Umweltinnovationen und Lebensstile, eine raumbezogene, empirisch fundierte Multi-AgentenSimulation. Metropolis, Marburg

Schwarz N, Ernst A (2009) Agent-based modeling of the diffusion of environmental innovations—an empirical approach. Technol Forecast Soc Chang 76(4):497–511

SinusSociovision (2007) Milieulandschaft 2007. http://www.sinus-sociovision.de/. Accessed 7 Jan 2008

Chapter 44
Modelling Risk Perception and Indicators of Psychosocial Sustainability in Private Households: The Risk Perception Module in DeepHousehold

Roman Seidl, Silke Kuhn, Michael Elbers, Andreas Ernst, and Daniel Klemm

Abstract In this section, we present the risk perception module for the *DeepHousehold* component. As part of the decision support system DANUBIA, it represents the households' anticipation of potential long-term consequences and water-related risks due to global change. Basically, the majority of households in southern Germany are not concerned yet by major changes regarding risks related to water supply or floods. However, recurring events may increase stress and concern. Depending on patterns in the simulated natural and social environment (e.g. high temperature, floods, risk perception patterns of other agents), the agents of *DeepHousehold* can respond in two ways. One is behavioural (e.g. switching between habitual water use and deliberate actions; see Chaps. 41 and 42) and the other psychological (long-term stress) which, however, may impact on behaviour and weaken individuals' health. We address the response of households to climate-related risks using different combined climatic and societal scenarios. This risk perception module contributes to the agent-based model *DeepHousehold* based on empirical data from qualitative and quantitative studies, an indicator concept derived from systems theory and psychological models of the consideration of future consequences and defence mechanisms.

R. Seidl (✉)
Institute for Environmental Decisions, ETH Zürich, Zürich, Switzerland
e-mail: roman.seidl@env.ethz.ch

S. Kuhn • A. Ernst
Center for Environmental Systems Research (CESR), University of Kassel, Kassel, Germany
e-mail: ernst@usf.uni-kassel.de

M. Elbers
Micromata GmbH, Kassel, Germany
e-mail: m.elbers@micromata.de

D. Klemm
wusoma GmbH, Munich, Germany
e-mail: mail@danielklemm.de

W. Mauser, M. Prasch (eds.), *Regional Assessment of Global Change Impacts*, DOI 10.1007/978-3-319-16751-0_44

Keywords Agent-based model • Global change • Households • Risk perception • Social response • Spatially explicit • Water risks • GLOWA-Danube

44.1 Introduction

The overall climate change problem involves a highly abstract threat potential without the individual being specifically affected by it, yet. Nevertheless, potential and actual climate changes and the environmental changes that are consequences thereof constitute a potential threat and are thus perceived as risks (Lorenzoni et al. 2007) (Map 44.1).

Therefore, *DeepHousehold* (DHH) considers the psychological main and side effects of a changed environment from the point of view of psychosocial sustainability using a risk perception module (Seidl 2009). Psychosocial sustainability entails a long-term consideration of psychological needs, which include, for example, emotional stress and dissatisfaction. The *RiskPerception* module was implemented within the already existing multi-actor simulation (MAS) of *DeepHousehold*. It ties together psychological variables and the indicators derived from the systems theory of Bossel (1999).

44.2 Data Processing

A bottom-up approach in the form of semi-standardised, approximately 1-h interviews of personal climate change considerations was chosen for this topic. The knowledge gathered from these interviews was used to develop a quantitative survey study. A total of 240 individuals completed that survey.

The results indicate differences between the milieu groups (microm Micromarketing-Systeme und Consult GmbH 2007; SinusSociovision 2007), especially in the variables consideration of future consequences (CFC; see Strathman et al. 1994) and psychological hygiene (PsyHyg) (Guldberg et al. 1993). The traditional milieus together with the post-materialists have the highest CFC values and thus place relatively high weight on long-term consequences of current actions. At the lower end of the scale is the Hedonist milieu group. The mainstream milieus show the most mental hygiene (processes such as repression or rationalisation). The post-materialist and the Hedonist milieus differ significantly from the Mainstream milieu group.

Various profiles were developed and implemented for the DHH actor types based on these empirical data. The differing sensitivities of the milieu groups (see also Chap. 42) with respect to the indicator dimensions were determined in close collaboration with Sinus-Sociovision®, the company providing the lifestyle classification.

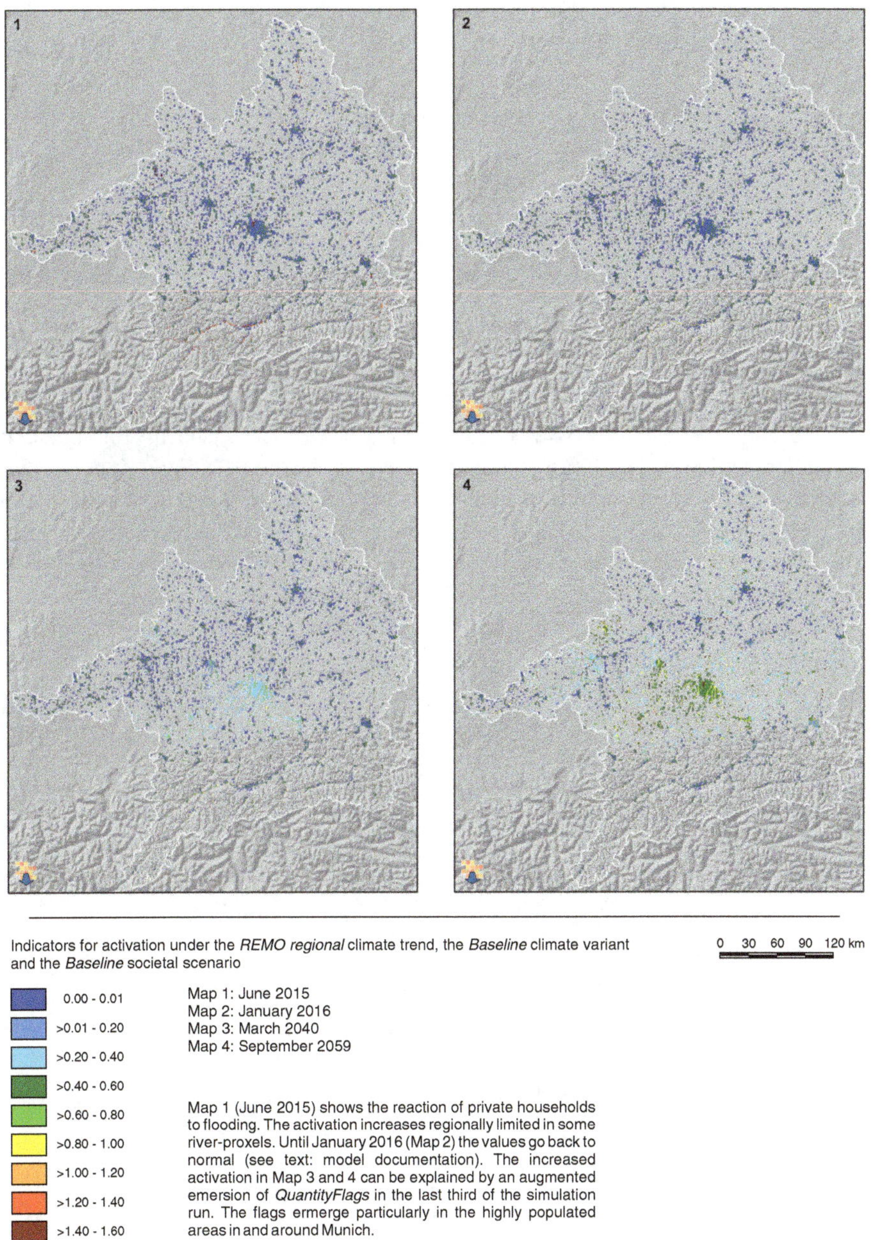

Map 44.1 Modelling risk perception and indicators of psychosocial sustainability in private households

44.3 Model Documentation

Figure 44.1 provides an explanation of the model concept. This figure shows only part of the more complex overall model. The central factors of the model are described in more detail below according to their sequence.

Specific environmental events (1) such as a drinking water shortage or a flood influence the indicators (3) and ultimately the agents' activation (6). This can happen directly through immediate experience (2) or indirectly through the media or closely associated individuals (social network) (5). Personal psychological factors (4) also influence the degree of activation. Aspects of sustainability are considered using the concept of indicator dimensions (3) utilised in the *RiskPerception* model.

The concept follows Bossel's (1999) orientor theory. Based on fundamental environmental attributes, Bossel identified indicators (the so-called orientors) that a system must satisfy (in our case a household) in order to successfully survive in its environment. Table 44.1 shows these orientors vis-à-vis the respective properties of the environment (adapted from Bossel 1999, and Krebs and Bossel 1997). For example, a water shortage affects the *Freedom* indicator since everyday activities such as showering are limited by it. Similarly, the *Security* indicator is affected by flooding.

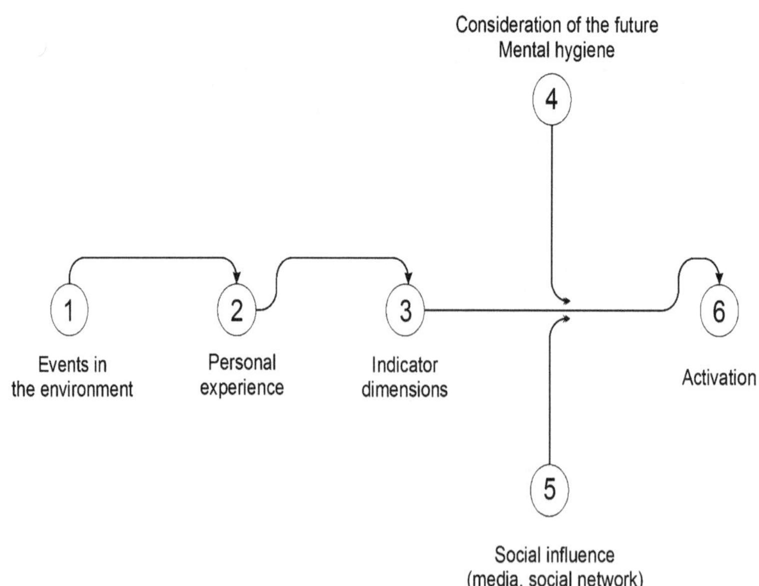

Fig. 44.1 Overview of the set of implemented concepts and their causal structure

Table 44.1 The fundamental properties of an environment and the system's respective orientor dimensions

Properties of the environment	Orientors
Resource scarcity: resources (energy, matter and information) required for a system's survival are not immediately available when needed	Effectiveness: the system should on balance, i.e. over the long term, be effective (not necessarily efficient)
Variety: many qualitatively different processes and patterns of environmental variables occur in the environment	Freedom of action: the system must have the ability to cope in various ways with the challenges posed by environmental variety
Insecurity: variation around normal environmental state	Security: the system must be able to protect itself from the detrimental effects of environmental variability
Change: in the course of time, the normal environmental state may change gradually or abruptly to a permanently different normal environmental state	Adaptability: the system should be able to learn, adapt, and self-organise to generate more appropriate responses to challenges posed by environmental change

Note 1: A drinking water scarcity not only is an example of a scarce resource problem, pushing on the orientor effectiveness, but also changes the environmental variety by introducing a reduction of freedom of action. The variability around the environmental normal state refers to impending events, which can occur under normal circumstances. The security orientor, therefore, maps the relative security or insecurity of the environment. The orientor called adaptability is affected if there is a qualitative change, i.e. the system settles on a new normal state

All indicator dimensions are characterised by the fact that satisfaction occurs only after minimum fulfilment of all indicators. A system is threatened if one indicator is below its minimum fulfilment values. This cannot be compensated by overfulfilment of another indicator. The overall situation of the system can be described by the notion of variable viability (Bossel 1999), which comprises the current status of the indicator dimensions and hence of the whole system. When comparing the effects of different scenarios, this value can provide an estimation which of the different scenarios proves sustainable. If there are no environmental events, then the indicators stay the same. The differences between the milieus result from perceptions and information processing differentiated by milieu (see the section on data preparation).

Activation is defined as the expression of cognitive emotional impairment of the psychological system. As a summary index, the activation (6) is derived from the psychologically moderated overall situation of the indicators. Social network and media influence as well as internal psychological processes (4) (consideration of the future, mental hygiene) enhance or reduce the activation. At every step, an actor compares its activation level with that of its collaborators (=other actors known to the actor). Depending on whether the activation level of the other actors is higher or lower than its own one, the actor increases or decreases its own activation level.

44.4 Results

The calculations that are shown below are based on a GLOWA-Danube scenario that is based on the *REMO regional* climate trend, the *Baseline* climate variant, and the *Baseline* societal scenario (see Chaps. 47, 48, 49, 50, 51 and 52). Figure 44.2 illustrates the temporal pattern of the modelled activation (for each milieu group) and the warning flags for flooding and drinking water shortage (*FloodFlag, QuantityFlag*). Obviously, flood flags are common, but appear in specific regions and at specific times. Therefore, the influence of flood flags in the pooled depiction of activation is difficult to detect. In the last third of the simulation, the level of *QuantityFlags,* which signal a decrease in drinking water resources, is higher. Accordingly, the activation of the actors increases significantly.

Figure 44.3 below illustrates the indicator types for two different actor types in July 2060. The leading and traditional milieus were singled out since they differ in their indicator sensitivities. It is noticeable that for the leading milieus actor type, the indicators *Effectiveness*, *Freeaction,* and *Adaptability* are affected to a greater extent, whereas the *Security* indicator is less affected than for the traditional milieus. Overall, however, the restrictions under these conditions (the *Baseline* societal scenario) are minimal.

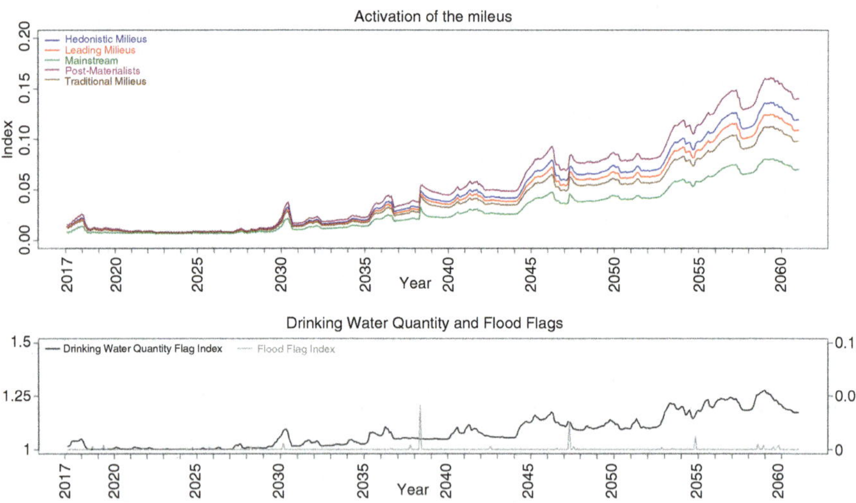

Fig. 44.2 Timeline of the activation values for each of the milieu groups (*upper graph*) and the drivers *FloodFlag* and *QuantityFlag*

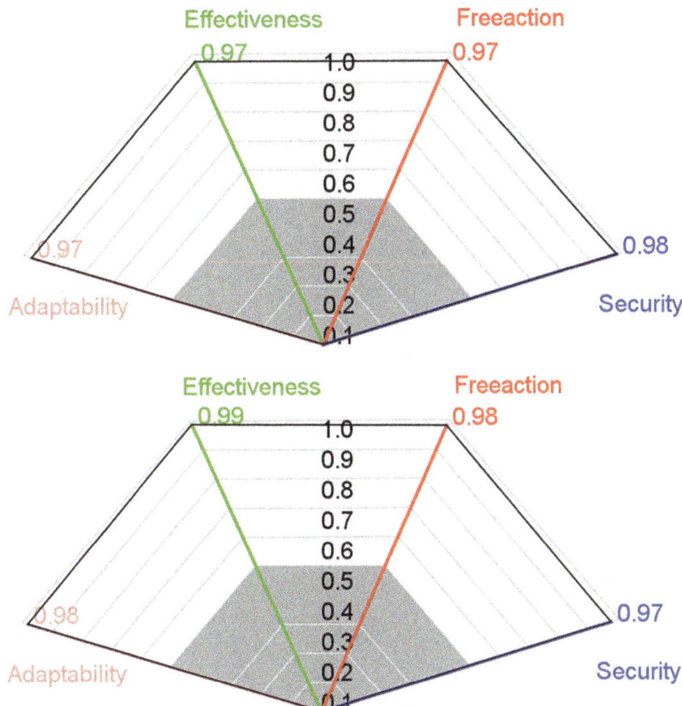

Fig. 44.3 Comparison of two actor types at one time step regarding their indicator values: societal scenario baseline; time step 07/2060. Aggregation over the whole area and all agents. Further explanations in the text

References

Bossel H (1999) Indicators for sustainable development: theory, method, applications. A report to the Balaton Group. IISD, Winnipeg

Guldberg CA, Hoglend P, Perry JC (1993) Scientific methods for assessing psychological defenses. Nord J Psychiatry 47(6):435–446

Krebs F, Bossel H (1997) Emergent value orientation in self-organization of an animal. Ecol Model 96:143164

Lorenzoni I, Nicholson-Cole S, Whitmarsh L (2007) Barriers perceived to engaging with climate change among the UK public and their policy implications. Glob Environ Chang 17:445–459

microm Micromarketing-Systeme und Consult GmbH (2007) MOSAIC Milieus®. http://www.microm-online.de. Accessed 15 Jan 2015

Seidl R (2009) Eine Multi-Agentensimulation der Wahrnehmung wasserbezogener Klimarisiken [A multi-agent simulation of the perception of water related climate change risks]. Metropolis, Marburg

SinusSociovision (2007) Milieulandschaft 2007. http://www.sinus-sociovision.de/

Strathman A, Gleicher F, Boninger DS, Edwards CS (1994) The consideration of future consequences – weighing immediate and distant outcomes of behavior. J Pers Soc Psychol 66:742–752

Chapter 45
Environmental Economy: Industrial Water Abstraction

Matthias Egerer and Markus Zimmer

Abstract This chapter presents the methods employed for modelling the water use in different industrial applications and in the energy industry. The district values are allocated to the proxels with the help of remote sensing. Water abstraction by the industrial sector is modelled as the dependent variable in a system of regression equation analyses in RIWU, which forms the basis for the DANUBIA model *Economy*. There are a total of 1,610 proxels with industrial development. Taken together, approximately 34 m^3 of water per second were supplied in January 1995. This is equivalent to approximately 2.9 million m^3 of water per day. According to the model calculations, a total of approximately 35 % of the water abstraction is sourced from groundwater.

Keywords GLOWA-Danube • DANUBIA • Industrial water abstraction • Economy • Upper Danube

45.1 Introduction

Within the Upper Danube drainage basin, supply of water for industrial purposes is a significant form of anthropogenic water use. Figure 45.1 illustrates the relative proportions of the key use types for Bavaria, Baden-Württemberg and Austria. The use of water by the industrial sector includes abstractions by companies in the manufacturing, mining and quarrying sectors. Water supply to the industrial sector is thus clearly distinguished from water use for energy supply (i.e. thermal power

M. Egerer (✉)
Bavarian Ministry of Economic Affairs and Media, Energy and Technology, Munich, Germany
e-mail: Matthias.Egerer@stmwi.bayern.de

M. Zimmer
Center for Energy, Climate and Exhaustible Resources, Ifo Institute – Leibniz Institute for Economic Research, University of Munich, Munich, Germany
e-mail: zimmer@ifo.de

W. Mauser, M. Prasch (eds.), *Regional Assessment of Global Change Impacts*, DOI 10.1007/978-3-319-16751-0_45

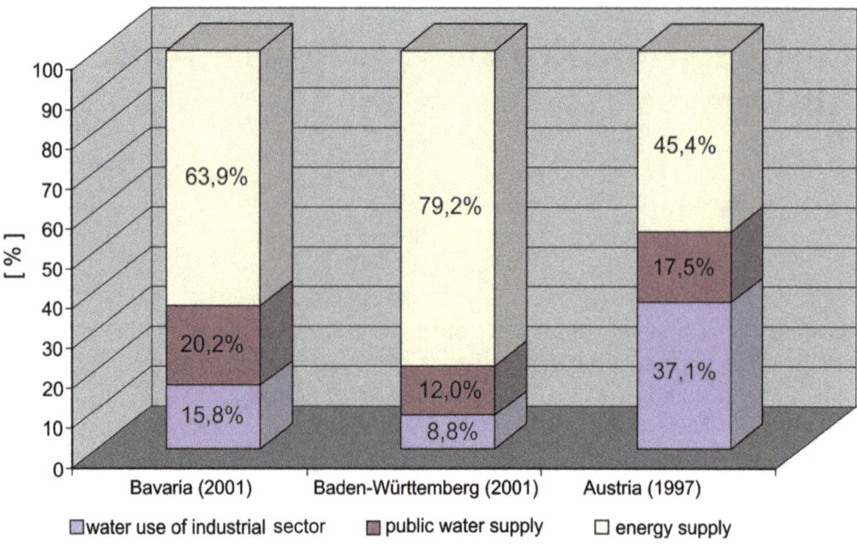

Fig. 45.1 Development of water extraction divided by sector (in %)

stations) for public supply. This distinction is largely based on the different uses of the water by the two sectors. Whereas water in industrial applications is used in different ways in the sector-specific production processes, in the energy industry, water is used exclusively for cooling the power generating equipment. Modelling of the water use that is distinguished in this way in DANUBIA is presented in this chapter. Water supply for the energy sector has not yet been covered within DANUBIA.

Water used in the manufacturing sectors within the regions around Bavaria and Baden-Württemberg is primarily used for cooling (approximately 74 % in Bavaria and 62 % in Baden-Württemberg) as well as for production-specific purposes (approximately 20 and 33 %). Sectors that have particularly high water consumption include the chemical industry and paper manufacturing. The water used for industry is mostly sourced from internal sources (approximately 84 and 92 %). Especially for large production companies, supply from the public network is only a minor source and is chiefly used for personnel purposes. Due to quality aspects, only companies in the food industry primarily use water from the public water supply. The above statistics are based on the year 2001 (Bavarian State Office for Statistics and Data Processing 2003; State Statistical Office Baden-Württemberg 2004).

Water abstraction by the industrial sector is modelled in the DANUBIA model Economy as a dependent variable. The regional economic simulation model RIWU (Regional Industrial Water Use) forms the basis for this model. Both the theoretical basis of the model and the model itself, as well as the statistical data upon which it is based, are presented in detail in the explanatory notes in Chap. 17.

The calculation of proxel values from the district values for industrial water abstraction generated by RIWU is presented below.

45.2 Preparation of the Data

All data that are used by several models in DANUBIA are transferred to the level of the proxel. It was therefore necessary to allocate the district values for industrial water abstraction calculated by RIWU – just as for all variables generated by the model and interchanged with other groups – to the level of the proxel.

The method for disaggregation of values that are given at the scale of administrative boundaries to the extremely small grid size of square kilometres is rather uncommon from the standpoint of the economics discipline and required some pioneering work. However, there is little point in adhering to administrative regions in the context of the present interdisciplinary project. Therefore, a general method was developed to allocate the district values to the proxel with the help of remote sensing. Chapter 16 describes the basic approach to using remote sensing.

Using the geographic information system from the hydrology/remote sensing subproject, each proxel was assigned a value indicating the proportional amount of the land falling in the use category "industry". For the assignment of the industrial water abstraction values, it was assumed that the industrial sector companies abstract water only in proxels that are industrially developed according to the remote sensing data. First, the district values calculated by RIWU are disaggregated into these proxels proportional to the percentage of this land use category per district. Subsequently, the distinction is made whether the abstraction on each proxel is from ground or surface water sources, where each proxel is assigned only one or the other source. This method begins in each model time step (RIWU uses a month as a unit) with the first proxel and works through the following algorithm for each proxel:

If the proxel is one with industrial water demand and there is also a river present (see Chap. 7), the sum of the abstractions for the eight adjacent proxels is added to the abstraction value for this proxel, if these proxels themselves are not "river proxels". This approach takes into account that abstraction from rivers for industrial purposes is frequently concentrated in large, interdependent industrial complexes in which the individual companies do not need to be located immediately adjacent to the river. The "summed" proxels are then blocked out and ignored in further run-throughs to avoid being counted twice. In the above example from Fig. 45.2, there is thus a summed demand of 0.4 m^3/sec on "river proxel" 5. Proxel 9 is skipped.

If the water demand by the industrial sector is on a proxel with no river, it is determined whether one of the adjacent eight proxels is a "river proxel" with industrial water demand. If the answer is yes, a null value is assigned to the proxel with no river and its demand is ascribed to the river proxel. If none of the eight neighbouring proxels is a river proxel, then groundwater demand is assigned (see proxels 6 and 11 in the lower portion of Fig. 45.2 as an example). In this way, either surface- or groundwater abstraction is clearly assigned to each proxel. It is clear that this pragmatic

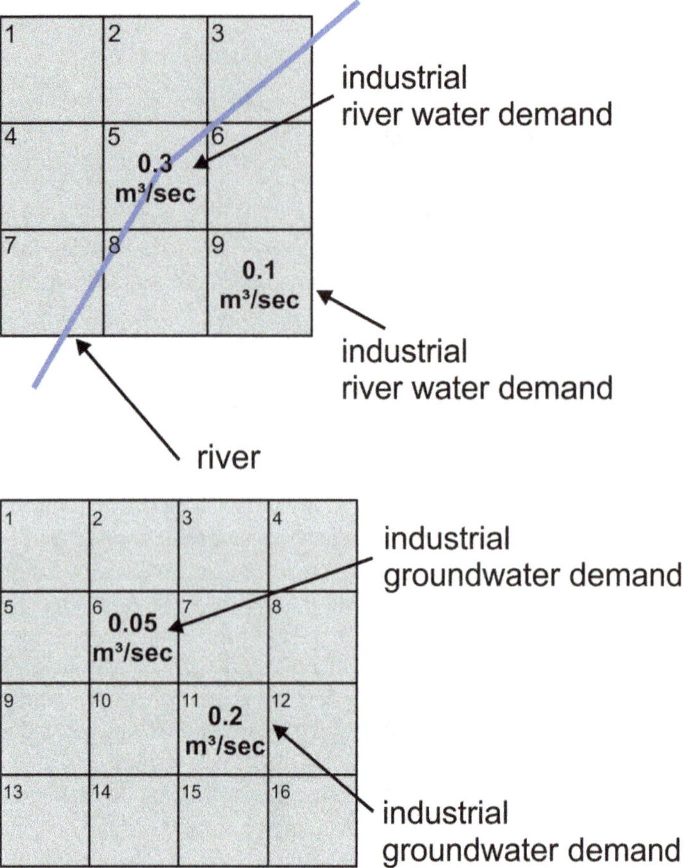

Fig. 45.2 Surface water and groundwater extraction

approach cannot completely match reality. Overall, groundwater abstraction is slightly overestimated.

45.3 Description of the Model

In RIWU, water abstraction by the industrial sector is modelled in an econometric equation system. It should be noted here that the statistical data and hence also the model only apply to the water abstracted for each company's own purpose. The negligible external sourcing of water from the public supply network is not taken into account. In addition, differences in the extent of water use in the sense of single, multiple or recycled use is not considered. Only the magnitude of the abstraction from nature is estimated. Seasonal differences in water abstraction were not considered.

45.4 Presentation of the Results

Map 45.1 depicts the proxel values calculated using the approach described above for January 1995, the month for which the model is calibrated. A total of 1,610 proxels are industrially developed. Taken together, approximately 34 m^3 of water per second were extracted in this month. This is equivalent to approximately

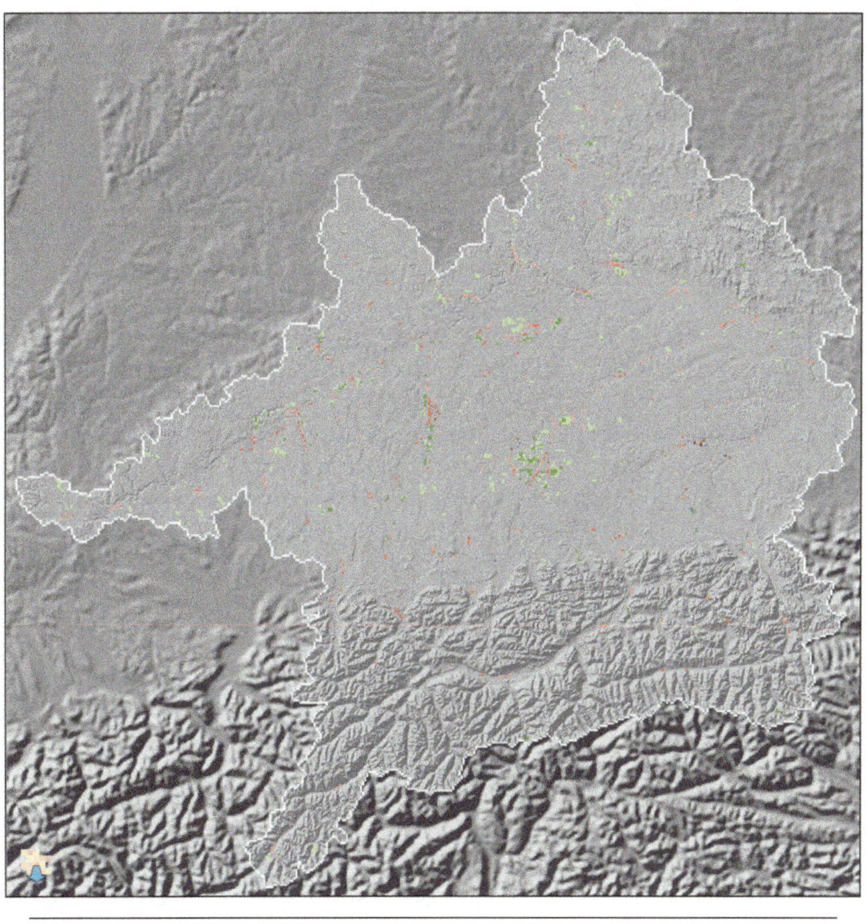

Industrial water extraction per km² as modelled for January 1995

0 15 30 45 60 km

from rivers	from groundwater	[m³/year]	[m³/day]	[l/sec]
		0 - 365,250	0 - 1,000	0 - 11.57
		> 365,250 - 1,826,250	> 1,000 - 5,000	> 11.57 - 57.87
		>1,826,250 - 3,652,500	> 5,000 - 10,000	> 57.87 - 115.74
		>3,652,500 - 10,957,500	>10,000 - 30,000	>115.74 - 347.22
		> 10,957,500	> 30,000	> 347.22

Map 45.1 Industrial water extraction (DTM: Jarvis et al. 2008)

2.9 million m^3 per day. The mean absolute error of the model, taken as the deviation of the values calculated by the model from the values from statistics, was 24 %. This relatively high mean error is primarily attributable to the district of Altötting, in which the value from statistics deviated quite clearly, by more than 200 %. This error can be attributed to the importance of this district in the overall industrial water abstraction. More than 40 % of the water supply for industrial use in Bavaria is directed to the chemical companies in this district.

According to the model calculations, a total of approximately 35 % of the water abstraction is sourced from groundwater. Thus, compared to values from statistics, the groundwater percentage is somewhat overestimated. In Bavaria, the percentage of abstraction from groundwater is only approximately 30 %; in Baden-Württemberg the value is only 23 % (Bavarian State Office for Statistics and Data Processing 2003; State Statistical Office Baden-Württemberg 2004). In Austria, companies use up to approximately 30 % groundwater (Statistics Austria, 1994, Wasserverbrauch der Großgewerbebetriebe 1994, Personnel communication).

In a reference test in DANUBIA that modelled the years 1995–1999, abstraction of water for industrial purposes increased altogether by around 3–35 m^3/s in December 1999. This can be attributed to the increasing economic output by the industrial sector, which slightly overcompensates for the effects of the increasing efficiency in water use that are also assumed in the model. A look at the statistics from Bavaria and Baden-Württemberg indicates that industrial water abstraction in this period increased in Bavaria by 4.5 % but decreased in Baden-Württemberg by approximately 7 %. Overall in both states, industrial water abstraction remained almost the same (Bavarian State Office for Statistics and Data Processing 2003; State Statistical Office Baden-Württemberg 2004). Increasing economic output and efficiency balanced each other out.

References

Bavarian State Office for Statistics and Data Processing, Munich (2003) Wasserversorgung und Abwasserentsorgung des Verarbeitenden Gewerbes in Bayern 2001. München

Jarvis A, Reuter HI, Nelson A, Guevara E (2008) Hole-filled seamless SRTM data V4. International Centre for Tropical Agriculture (CIAT). Available at DIALOG. http://srtm.csi.cgiar.org. Accessed 19 Sept 2014

State Statistical Office Baden-Württemberg (2004) Wasserwirtschaft in Baden-Württemberg. Statistisches Landesamt Baden-Württemberg, Stuttgart

Chapter 46
Tourism Research: Water Demand by the Tourism Sector

Mario Sax, Jürgen Schmude, and Alexander Dingeldey

Abstract Within the context of a study of the scenarios of global change, a quantification of the water demands by the tourism sector that were as precise as possible and at the best possible spatial resolution was undertaken. Whether the water demand by the tourism sector is reconcilable in the sustainable long term can be estimated from this quantification. The facilities were located on points and associated with unique proxels in order to be able to carry out the spatial distribution of the water demand by the tourism sector with DANUBIA (In this way, the spatial distribution of the tourism facilities could be transferred to the model. The tourism research subproject basically follows a supply-oriented approach for creating a model. This means that elements of the tourism infra- and suprastructure facilities are treated as actors in the model that can be influenced by the environment and are responsible for the tourism water demand. At present, the content of the Tourism model consists of three subcomponents: the "DeepTourism" component, the "Attractiveness model" and the "Water consumption model". The demand for drinking water by the tourism sector is quite heterogeneous is space and time. Particularly high demands occur in the Alps and as a whole in the region of Munich. Overall, it can be concluded that there is a slight overestimate in the calculation of the demand for drinking water by the tourism sector.

Keywords Climate change • Tourism facilities • Water use • Tourism sector • Tourism industry • Actor model • GLOWA-Danube

---

M. Sax (✉)
Wilhelm-Hauff-Str. 51, 84036 Landshut, Germany
e-mail: sax.mario@googlemail.com

J. Schmude
Department of Geography, Ludwig-Maximilians-Universität München (LMU Munich), Munich, Germany
e-mail: j.schmude@lmu.de

A. Dingeldey
BWL Reiseverkehrsmanagement, Duale Hochschule Baden-Württemberg, Ravensburg, Germany
e-mail: Dingeldey@dhbw-ravensburg.de

W. Mauser, M. Prasch (eds.), *Regional Assessment of Global Change Impacts*, DOI 10.1007/978-3-319-16751-0_46

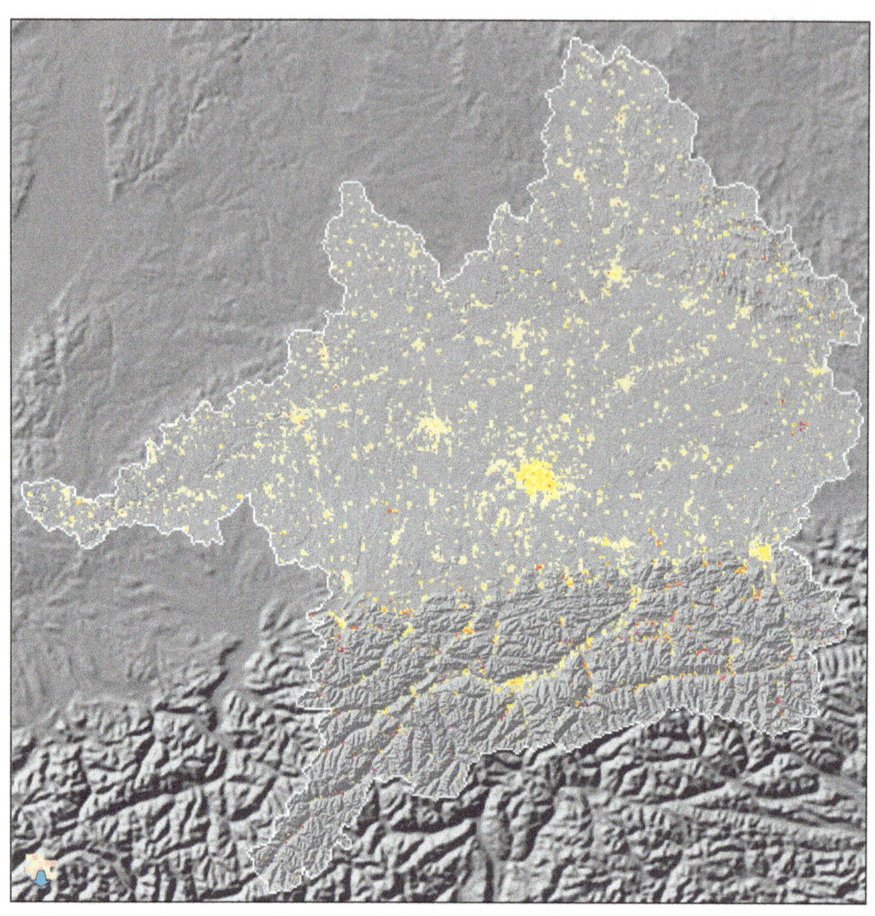

Water demand by the tourism sector per km² in 1998

0 15 30 45 60 km

[m³/year]	[m³/day]	[l/sec]
0 - 2,000	0 - 5	0 - 0.06
> 2,000 - 5,000	> 5 - 14	>0.06 - 0.16
> 5,000 - 15,000	> 14 - 41	>0.16 - 0.48
>15,000 - 25,000	> 41 - 68	>0.48 - 0.79
>25,000 - 40,000	> 68 - 110	>0.79 - 1.27
>40,000 - 75,000	>110 - 205	>1.27 - 2.38
>75,000 - 150,000	>205 - 411	>2.38 - 4.76
> 150,000	> 411	> 4.76

Map 46.1 Water demand by the tourism sector in 1998 (Data sources: Bavarian State Office for Statistics and Data Processing 2004, Federal Statistical Office 2004, State Statistical Office Baden-Württemberg, 2004, Statistics Austria 2003, Federal Statistical Office 2003, EEA, European Environment Agency 2005, Chap. 9 of this book)

46.1 Introduction

Within the context of a study of the scenarios of global change, a quantification of the water demands by the tourism sector that were as precise as possible and at the best possible spatial resolution was undertaken. Whether the water demand by the tourism sector is reconcilable in the sustainable long term can be estimated from this quantification. At the same time, other project groups receive this demand calculation to determine the price of drinking water, for example. The presentation of the results in the form of a map highlights the key points of water demand by the tourism sector. Thus, the map of the values for the water demand by the tourism sector is applicable to outside interest groups.

46.2 Data Processing

The calculations are based on the tourism infrastructure facilities presented in Chap. 18, which are each assigned a specific water demand calculated in the investigations of the first and second phases of the project (2001–2007) and which are calculated in a simulation. A description of the data collection is not provided here, since it was already given in Chap. 18. The necessary water quantities for various infra- and suprastructure facilities were compiled to quantify the water demand by the tourism sector. The facilities were located on points and associated with unique proxels in order to be able to carry out the spatial distribution of the water demand by the tourism sector with DANUBIA. In this way, the spatial distribution of the tourism facilities could be transferred to the model.

46.3 Model Documentation

The tourism research subproject basically follows a supply-oriented approach for creating a model. This means that elements of the tourism infra- and suprastructure facilities are treated as actors in the model that can be influenced by the environment and are key for the tourism water demand. Thus, the *Tourism* model is based on the DeepActor concept that applies to all the social science models and which facilitates the development of models that can respond to the simulated environmental conditions of DANUBIA. At present, the content of the *Tourism* model consists of three subcomponents (see Fig. 46.1): the "*DeepTourism*" component, the "Attractiveness model" and the "Water consumption model".

The "*DeepTourism*" component portrays the spatial extent and the basic mode of operation of the tourism infra- and suprastructure within the study area (e.g. opening, closing, water or snow use) in DANUBIA. For the current *Tourism* model, water-intensive infrastructure facilities (golf courses, swimming pools, ski areas

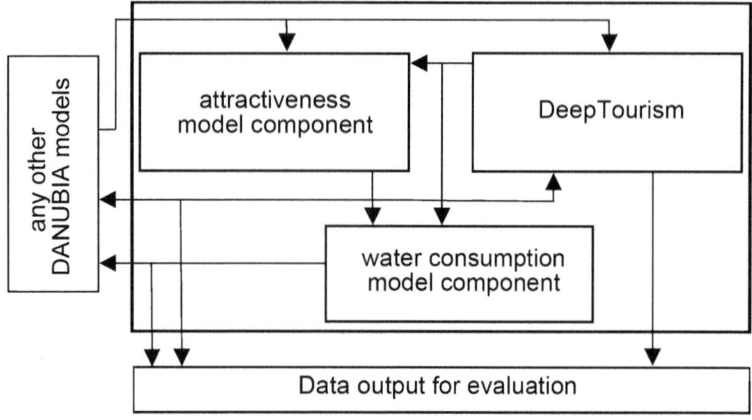

Fig. 46.1 Tourist model (Dingeldey 2008, modified)

with and without artificial snow production as well as hotels and gastronomy facilities) are modelled, taking into account the seasonal opening dates for each. Each tourism actor (e.g. a ski area) has a set of specific plans (e.g. "open ski area" or "close ski area") from which the actor selects the appropriate one using a decision system. A day was chosen as the temporal resolution because the model can then better respond to extreme values. Below, the decision rules are shown for the example of an actor for a ski area with no snow production.

IF			
Current date	$\geq$	Season beginning (December 15)	AND
Current date	$\leq$	Season end (April 30)	AND
Natural snow cover	$\geq$	30 cm	AND
Mean opening date	$\geq$	break even operation days	
THEN "open ski area"			

With the "Attractiveness model" component, both the number of day visitors and overnight visitors are calculated. The key feature of this component is that the calculations take place at the communal level. The distribution of the calculated values of the tourism demands is at the level of the individual proxel using a population distribution based on this grid and which was prepared in the first project phase by the remote sensing group. The number of same-day visitors is calculated using a gravitation method (Klaasen et al. 1979). Studies which involve the daily flow of traffic (Harrer 1995; Maschke 2005) are incorporated into the calculation. These provide information about the activities of the same-day visitors. An absolute attractiveness is determined for each proxel that depends on land use (e.g. residential building) in order to calculate the number of same-day visitors. The capacity for scenarios of this model component is ensured in that the population values from the *Demography* model that underlie these calculations as well as the operating statuses (e.g. "open" or "closed") of the gastronomy and swimming pool actors are read in dynamically via interfaces during a simulation.

In addition to same-day visitors, the number of bed nights is also calculated in the "Attractiveness model" component. The modelling of the demand by overnight guests is based on an annual trend calculation for each community. The appropriate statistical data for 1983–2004 for the study area are available for this calculation; this allows a community to be represented by the basic tourism development path. Mean monthly proportions of annual overnight visitors by community were used for a test period from 1995 to 1999. The operating status of the tourism actors, the availability of drinking water and the temperatures are also incorporated in the calculation of the demand by the tourism sector in order to be able to model the scenarios (Dingeldey 2008). This approach allows the direct and indirect quantification of environmental influencing factors on the demand by overnight guests.

In the "Water consumption model" component, the quantities of water consumed by the tourism infra- and suprastructure are calculated based on the operating status of the tourism actors (open or closed, artificial snow production or not). This is based on the surveys on the water demands completed by the tourism service providers that were carried out as a part of the project. These surveys indicated that the tourism infrastructure facilities have a direct (occurring as a consequence of the operation of the facility) water demand that is influenced to a relatively negligible extent by the number of visitors.

For ski areas with artificial snow production and golf course actors, the water consumption is calculated daily based on the simulated temperature and precipitation conditions. In order to be able to account for regional conditions, the respective sizes of the real models (number of fairways) are considered for the golf course actors, and the size of the snow-covered area is considered for snow-making machines.

The water demand by same-day visitors is obtained by multiplying the number of visitors in a community and the proportion of gastronomy guests (=0.5) as well as the average water consumption per guest (=25 l).

The water demand by overnight tourism is obtained by multiplying the number of bed nights calculated by the model with the average water demand for an overnight stay. This value is approximately 180 l per night and is largely the result of use by the overnight guest himself, and only a negligible amount is contributed by the facilities within the lodging establishment (e.g. swimming pools) (Sax 2008).

The total demand for drinking water by the tourism sector is calculated by the demand values for each existing infrastructure facility and the drinking water demand generated by the demand of tourists (overnight and day visitors). The

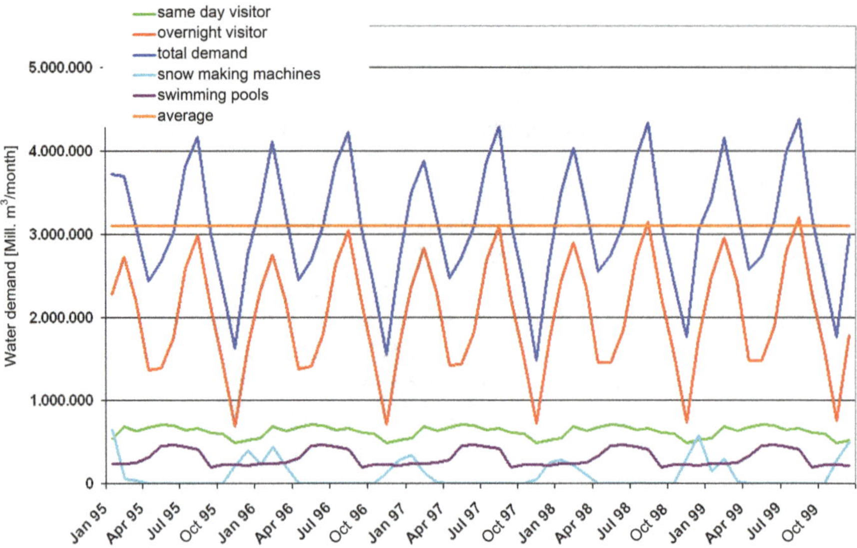

Fig. 46.2 Drinking water demand by tourism in the Upper Danube basin 1995–1999

resulting water quantities are shown for the entire year in 1998 in the map. However, the drinking water demand by the tourism sector shows a strong seasonal element (see Fig. 46.2).

46.4 Results

The demand for drinking water by the tourism sector is quite heterogeneous is space and time. Particularly high demands occur in the Alps and as a whole in the region of Munich. The seasonal variation in water demand can be portrayed by taking into account the operating times for the individual infrastructure types, the availability of drinking water (calculated in the *WaterSupply* model) and the seasonally varying numbers of overnight visitors in the study area (see Fig. 46.2). Overall, it can be concluded that there is a slight overestimate in the calculation of the demand for drinking water by the tourism sector. This is based on the fact, for example, that the values for drinking water demand by swimming pools arise not only as a result of the tourists but also from use by local residents. However, overall this error is relative minor. As a result of the heterogeneous mix of sectors in the tourism industry, there are no statistics on the water demands for it. Therefore, the demand can only be calculated independently. The values presented are thus akin to a projection. In order to assess the value of the results, the calculated values need to be compared to the magnitude of the official statistics on water supply to the end users. For example, calculated across all Bavarian communities in the Upper Danube drainage basin, the modelled drinking water demand by the tourism sector is approximately 2.8 % of the water supplied to end users. By comparison, the hospitality sector in Germany required approximately 2.4 % of the water supplied to end users (Möller 2001).

References

Dingeldey A (2008) Modellierung der touristischen Attraktivität zur Bestimmung der Übernachtungsnachfrage im Einzugsbereich der Oberen Donau unter Berücksichtigung von Umwelteinflüssen. Dr. Hut, München

Harrer B (1995) Tagesreisen der Deutschen. In: Schriftenreihe des DWIF, vol 46. DWIF, München, p 41f

Klaasen LH, Paelinck JHP, Wagenaar S (1979) Spatial systems. A general introduction. Saxon House, Teakfield Limited, Farnborough

Maschke J (2005) Tagesreisen der Deutschen. In: Schriftenreihe des DWIF, vol 50. DWIF, München

Möller A (2001) Umweltorientierung im Gastgewerbe. In: Schriftenreihe des DWIF, vol 48. DWIF, München, p 44

Sax M (2008) Entwicklung eines Konzepts zur computergestützten Modellierung der touristischen Wassernutzung im Einzugsgebiet der oberen Donau unter Berücksichtigung des Klimawandels. In: Schmude J (ed) Beiträge zur Wirtschaftsgeographie Regensburg, vol 11. Universität, Lehrstuhl für Wirtschaftsgeographie, Regensburg. Dingeldey, Schmude

Part IV
Scenarios

Chapter 47
GLOWA-Danube Scenarios

Andreas Ernst, Silke Kuhn, and Wolfram Mauser

Abstract This chapter gives an overview of the scenarios of the future used in GLOWA-Danube. Scenarios comprise quantitative and/or qualitative assumptions about possible future conditions and support the process of modelling. Based on expert group discussions and available knowledge, a GLOWA-Danube scenario describes one possible projection of the patterns of climate and social conditions into the future, thus using the best knowledge in order to limit the number of possible scenarios to those that are probable or relevant. A GLOWA-Danube scenario consists of two basic elements that are interconnected: a climate scenario (climate trend) which can take on various manifestations (climate variant) and a societal scenario, which can take place under various climatic conditions. Finally, another selection option relates to policy measures, which are targeted, point interventions (in space and/or time) to counteract or to support a trend in a selected scenario. The approach described in this section provides the basis for the definition of all scenarios of the future used within the GLOWA-Danube project.

Keywords Scenario • Climate scenario • Societal scenario • Policy intervention • GLOWA-Danube

47.1 Introduction

Scenarios are defined as hypothetical visions of the future that are formulated in words (Götze 1991). Scenarios comprise quantitative and/or qualitative assumptions about possible future conditions and support the process of preparation for decision-making in that sense that they allow to analyse consequences and

A. Ernst (✉) • S. Kuhn
Center for Environmental Systems Research (CESR),
University of Kassel, Kassel, Germany
e-mail: ernst@usf.uni-kassel.de

W. Mauser
Department of Geography, Ludwig-Maximilians-Universität München
(LMU Munich), Munich, Germany
e-mail: w.mauser@lmu.de

W. Mauser, M. Prasch (eds.), *Regional Assessment of Global Change Impacts*, DOI 10.1007/978-3-319-16751-0_47

trade-offs of different decision alternatives. Scenarios are described by a combination of different elements. The case at hand specifically involves the characteristics of key factors internal or external to the individual models.

The following additional assumptions are made regarding the scenarios:

- Scenarios are not predictions of the future. They are also not expected to be "correct".
- Scenarios must be plausible and convincing.
- Scenarios must be relevant in the sense that they highlight potential future uncertainties and conflicts.

The scenarios in GLOWA-Danube, as described below, were developed based on these guiding principles.

47.2 Formation and Logic of the GLOWA-Danube Scenarios

In its full complexity, a GLOWA-Danube scenario describes one possible projection of the patterns of climate and social conditions into the future. To make this projection, climate scenarios based on science and actor-based societal scenarios are implemented in DANUBIA.

In theory, there are an infinite number of scenarios, which a priori can all make the same claim to validity. For this reason, the first task is to make use of the available historical climatological and social knowledge in order to limit the number of GLOWA-Danube scenarios to those that are probable based on current knowledge, to the extent that the most reliable statements can be made. Accordingly, only an excerpt of the possible/conceivable and plausible scenarios is covered here, which hypothetically can be expanded as needed at any point, with adequate input.

The approach presented below for preparing the GLOWA-Danube scenarios is the result of several project workshops in which the available knowledge of the future regional trends in climate and society was discussed in detail and evaluated. The logic of the GLOWA-Danube scenarios presented here was discussed over the course of several stakeholder workshops with the policymakers and stakeholders participating in the GLOWA-Danube project and was approved by them.

A GLOWA-Danube scenario consists of two elements that are interconnected: a climate scenario, which can take on various manifestations, and a societal scenario, which can take place under various climatic conditions, since climate extrinsically affects society.

The combination of a regional climate trend and a climate variant yields a climate scenario (see Chaps. 48, 50, 51). To date, a total of four climate trends and six climate variants that are available for the preparation of a GLOWA-Danube scenario have been defined using a variety of methods. Four climate variants were statistically generated from historical climate using a climate generator (see Chap. 49), and two climate variants are based on the downscaled and bias-corrected results of regional climate models (see Chap. 51).

In addition, there is a social driver, which in particular affects the actor models. These societal scenarios exist as three manifestations. They are described in detail in Chap. 52.

No statements are made with respect to the probability of occurrence of the scenarios. The climate variants formed using the statistical climate generator (see Chap. 49) are an exception to this rule.

47.3 Scenario Period

GLOWA-Danube scenarios cover the period from 2011 to 2060, and a reference period from 1971 to 2000 is defined, to which change is referenced. The scenario period that was selected thus comprises 50 years. Although most climate models simulate climate for the whole century until 2100, the earlier 2060 endpoint was selected for GLOWA-Danube to match the social science models and their associated scenarios. The reason is that social science scenarios are estimated to show greater uncertainty regarding the estimation of future developments.

The selection of a GLOWA-Danube scenario that specifically targets an issue allows the effects of climate change on a wide range of sectors to be analysed; this analysis permits the identification and the simulation of various interventions for adapting to or avoiding the consequences of climate and ultimately to assess their effectiveness.

47.4 Scenario Selection

Figure 47.1 schematically illustrates the structure and development of a GLOWA-Danube scenario. One option is selected from each column, thus allowing the scenario to be custom arranged for each issue. In general, the selection criteria can all be combined, although the two from the bias-corrected results of the regional climate models are an exception (see the explanations under Fig. 47.1). More detail in Fig. 47.2.

The *first selection criterion* (selection 1) in compiling a GLOWA-Danube scenario involves the *climate trend*: there are four plausible regional climate trends available based on the global *IPCC* emission scenario A1B. The climate trends vary according to the magnitude of the temperature increase and the percentage change in precipitation (see Chap. 48).

However, the climate trends only provide information about the general climate trend pattern. The meteorological data for driving DANUBIA are determined with the *second selection option* (selection 2), the climate variants. These variants are in part from the statistical climate generator (see Chap. 49), which generates a series

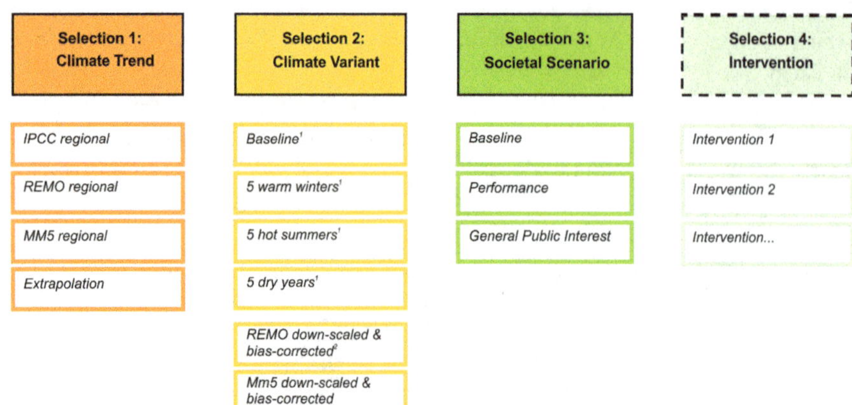

Fig. 47.1 Matrix of scenarios in GLOWA-Danube. An entire scenario consists of a path from left to right: a selection 1 + a selection 2 + a selection 3; selection 4 is optional. In principle all combinations can be joined, except the REMO downscaled and bias-corrected and MM5 downscaled and bias-corrected climate variants, which can be selected only in combination with the REMO and MM5 regional climate trends

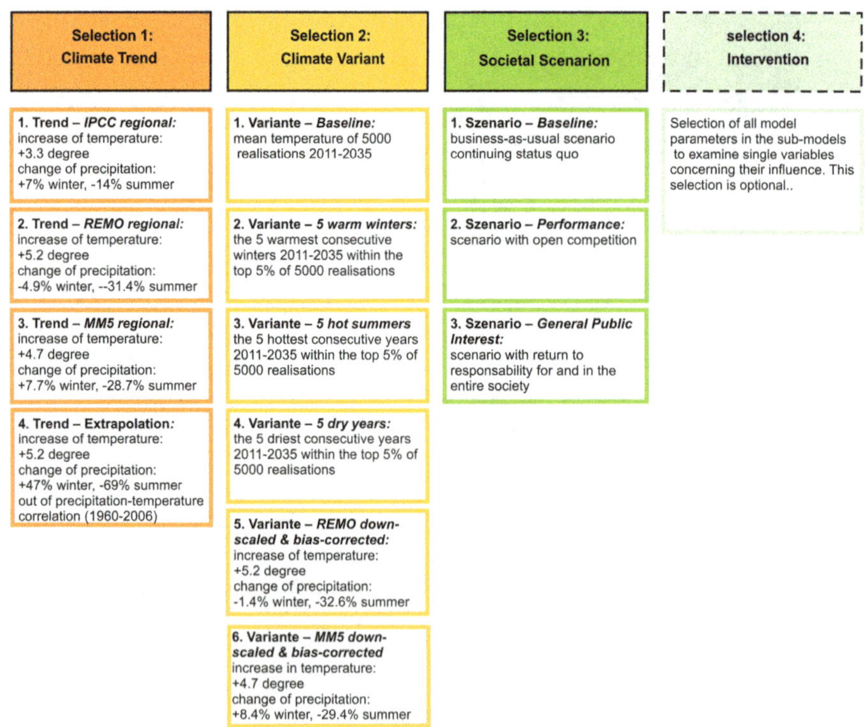

Fig. 47.2 Specification of the GLOWA-Danube scenario matrix

of different meteorological time series as climate generators, taking into account stochastic rules from historical meteorological measurement series. Those variants that create interesting scenarios (i.e. they present a particular challenge) are then selected from this time series according to predefined criteria. Alternatively, the climate variants can be based on the results of the regional climate models after bias correction and downscaling (see Chap. 51). At present, there are two such climate variants that affect the *REMO* and *MM5* models.

The climate variant options can be considered as a specification of the climate scenario. They are suited for addressing particular issues, such as the behaviour of groundwater aquifers under the condition of *5 successive dry years* (the fourth variant).

The *third selection* involves the *societal scenario* (selection 3). In addition to a *baseline* scenario in which the status quo persists, there are two opposing scenarios from which to choose. The societal scenarios are described in detail in Chap. 52. (also rf. de Vries and Perry 2007)

A *fourth selection option* relates to *interventions* (selection 4). An intervention is a targeted, point intervention (in space and/or time) to counteract or support a trend in a selected scenario; for example, this may be the use of snow canons in regions where a lack of snow threatens tourism in the *general public interest* scenario. The intervention catalogue defines the extent of the externally set treatment options. This selection option is optional.

According to the prescribed logic for the GLOWA-Danube scenarios, society can also develop on its own under various climate trends and climate variants and can pursue different courses in terms of societal change, which determines the everyday activities, the spirit of the times and the perception of diverse themes.

The approach described above is the basis for the definition of all the future scenarios used in the GLOWA-Danube project. This approach provides a simple and comprehensive method for deriving the most diverse variants for the development of climate and societal scenarios from global trends and to discuss these with the stakeholders. The application of this concept has proved to be helpful in that the results of the scenario simulations can be presented and documented in a simple, unambiguous and transparent manner as "the following results x were modelled under the scenario conditions of climate trend u, climate variant v and societal scenario w".

References

de Vries J, Perry T (2007) Der demografische Wandel und die Zukunft der Gesellschaft. Navigator. In: Newsletter SinusSociovision, 2/2007

Götze U (1991) Szenario-Technik in der strategischen Unternehmensplanung. DUV, Wiesbaden

Chapter 48
The GLOWA-Danube Climate Trends

Wolfram Mauser, Thomas Marke, Andrea Reiter, Daniela Jacob, and Swantje Preuschmann

Abstract GLOWA-Danube, in order to simulate climate change impacts, needs meteorological drivers with high spatial and temporal resolution which reflect the temporal course of the regional climate change signal. Uncertainty in the amount and course of future climate change motivates to define and analyse the impact of a range for assumed temperature and precipitation changes. The GLOWA-Danube approach to define a range of climate change trends and their temporal courses is described. It combines results from global and regional climate models with a thorough analysis of observed climate data to define a realistic range of four climate trends and their different temporal courses until 2100. The results of the trend analysis are shown and discussed.

Keywords GLOWA-Danube • Climate change • Regional climate trends • REMO • MM5

48.1 Introduction

A continuous temporal change of key elements of climate can be described as trends. This applies not only to past observations but also to possible future developments that are simulated using climate models. GLOWA-Danube climate trends are defined as average trends in the climate over the course of a long period of time. In

W. Mauser (✉)
Department of Geography, Ludwig-Maximilians-Universität München (LMU Munich), Munich, Germany
e-mail: w.mauser@lmu.de

T. Marke
Institute of Geography, University of Innsbruck, Innsbruck, Austria
e-mail: thomas.marke@uibk.ac.at

A. Reiter
Bavarian Research Alliance GmbH, Munich, Germany
e-mail: reiter@bayfor.org

D. Jacob • S. Preuschmann
Climate Service Center, Hamburg, Germany
e-mail: daniela.jacob@hzg.de; swantje.preuschmann@hzg.de

W. Mauser, M. Prasch (eds.), *Regional Assessment of Global Change Impacts*, DOI 10.1007/978-3-319-16751-0_48

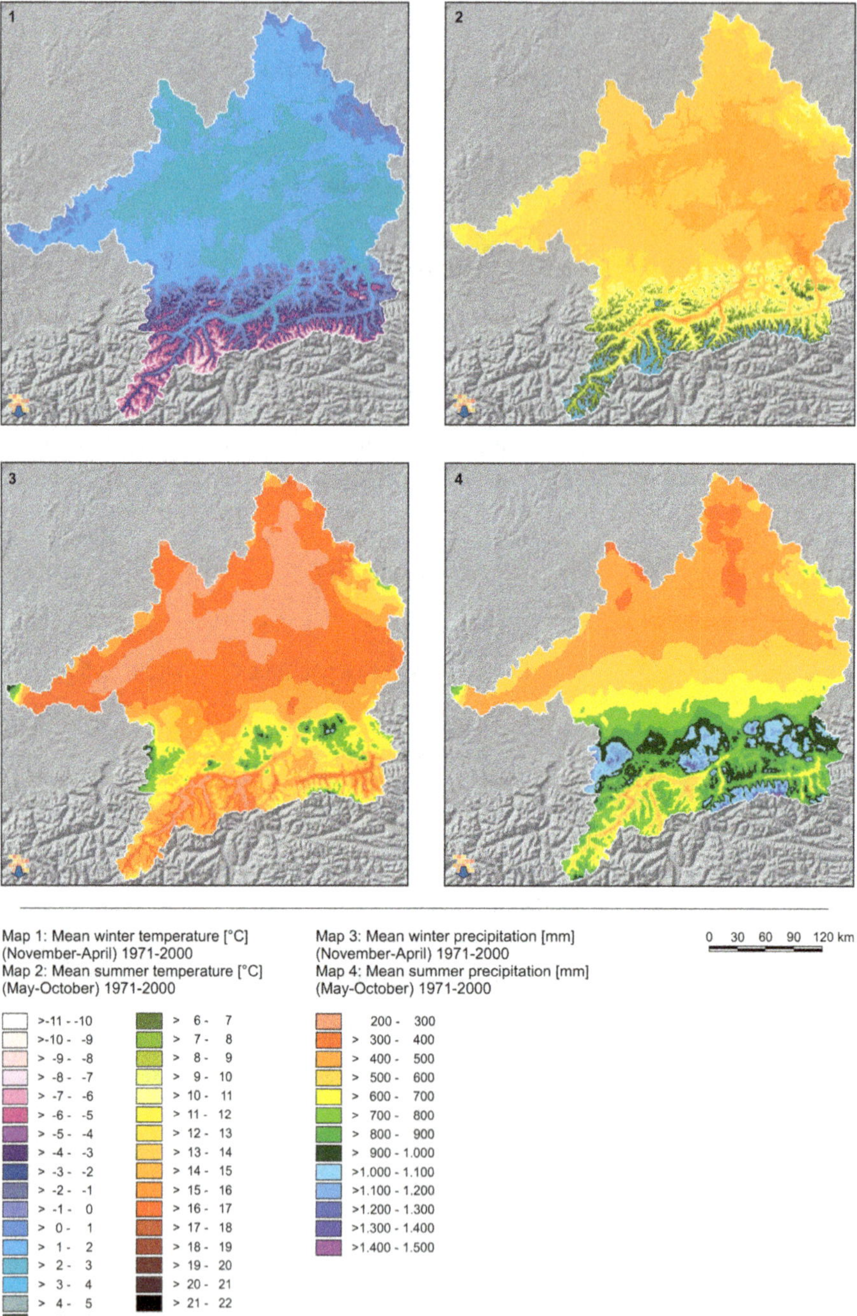

Map 48.1 Mean values of air temperature and precipitation from the past (Data source: station observations from DWD, Deutscher Wetterdienst; ZAMG, Zentralanstalt für Meteorologie und Geodynamik)

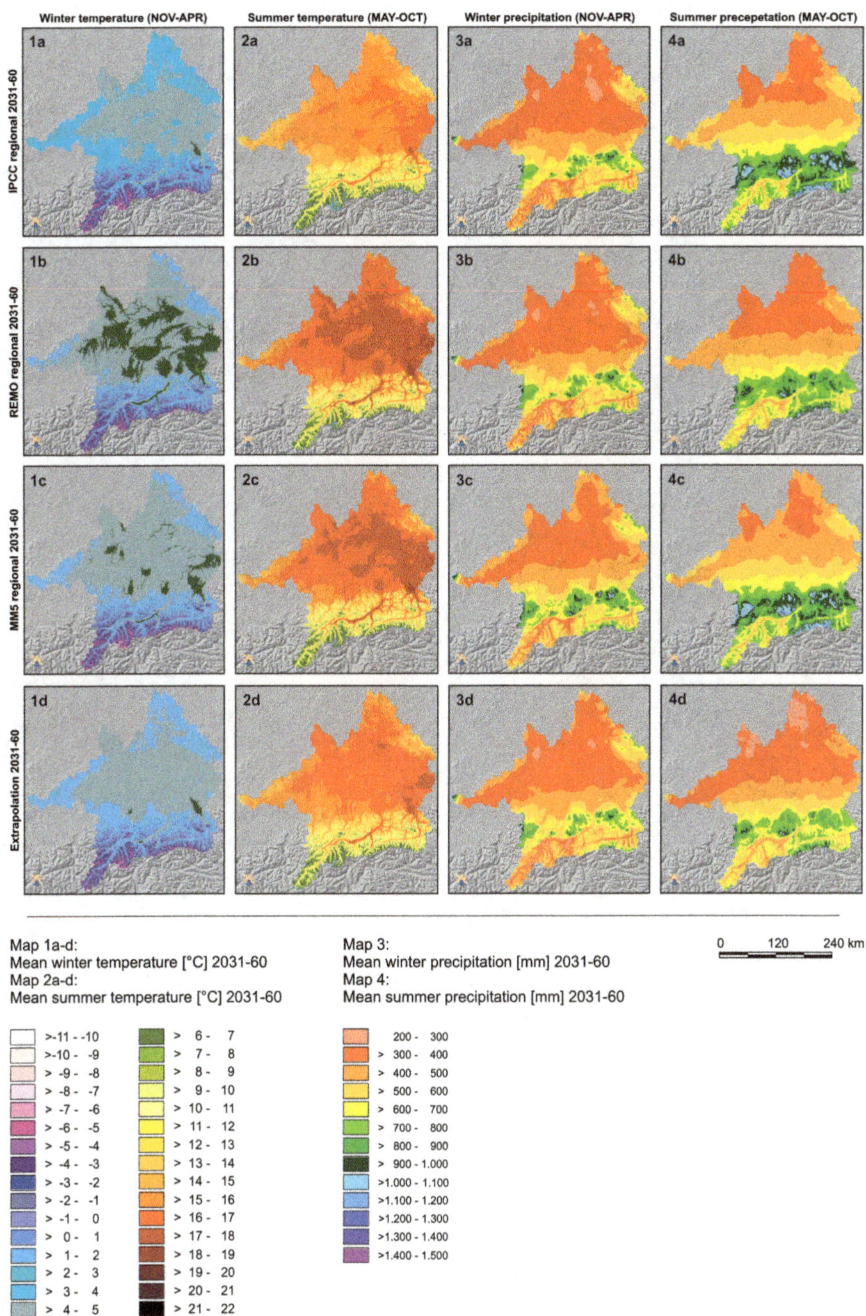

Map 48.2 Mean values of air temperature and precipitation under different climate trends and the *Baseline* climate variant

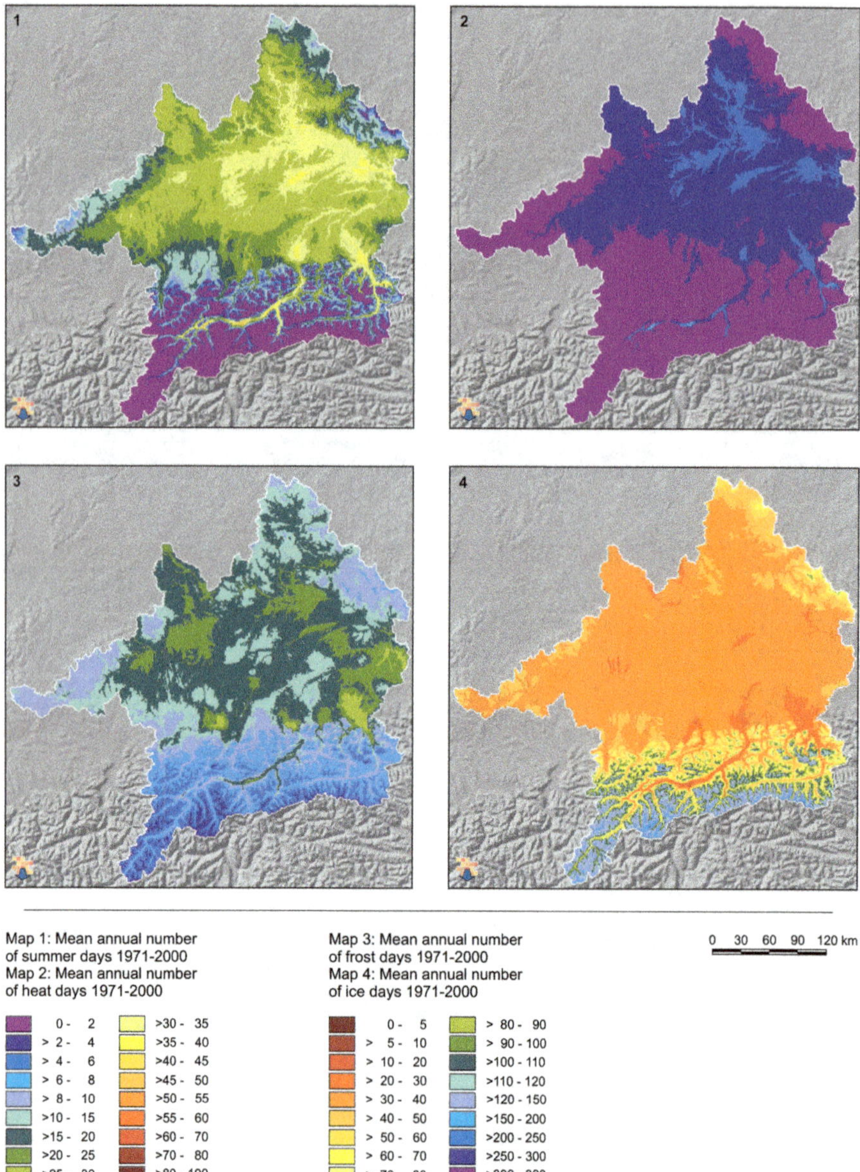

Map 48.3 Mean values of extreme days in the past (Data source: station observations from DWD, Deutscher Wetterdienst, ZAMG, Zentralanstalt für Meteorologie und Geodynamik)

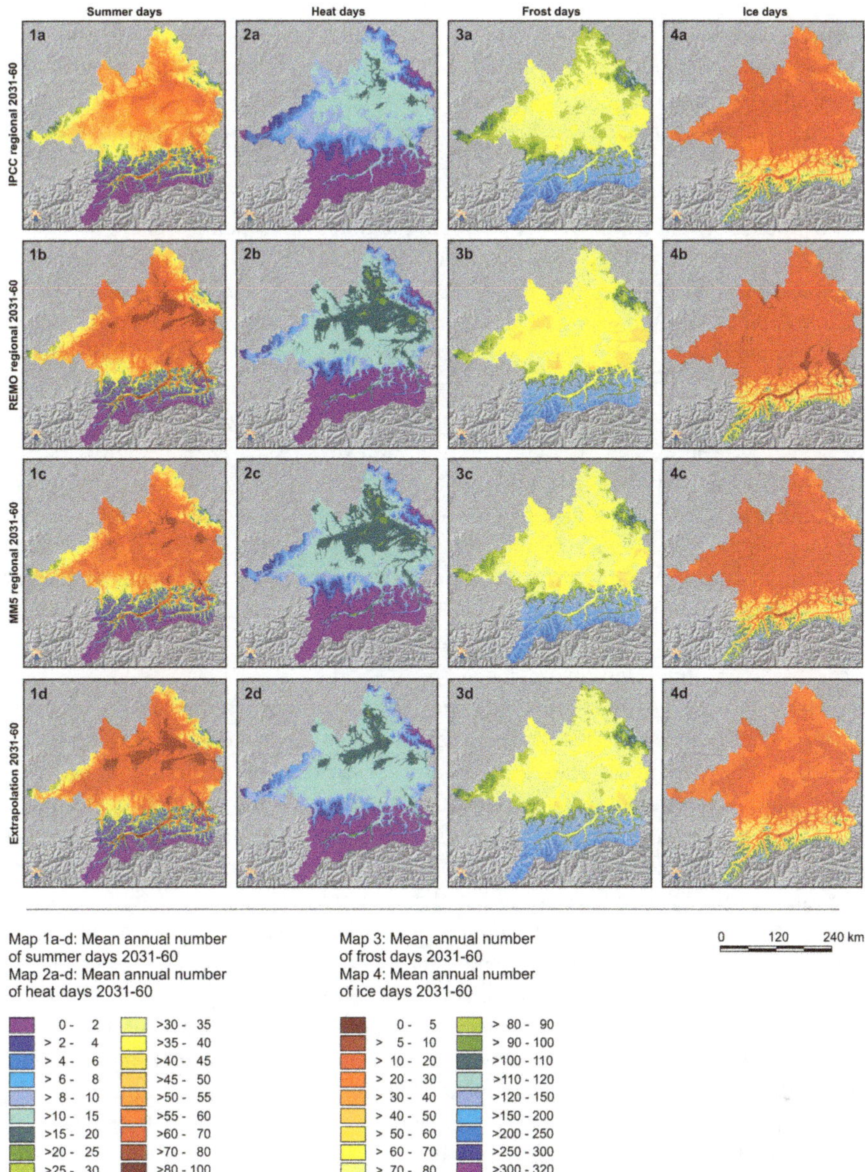

Map 48.4 Mean values of extreme days for different climate trends under the *Baseline* climate variant

contrast, in the context of GLOWA-Danube, climate variants are the meteorological drivers that are entered as high spatial and temporal resolution meteorological variables into the DANUBIA model system (see Chap. 50).

Future global climate trends have been derived for quite some time and in large numbers, based both on known historical climate changes and on the results of simulations of the future, using a variety of global climate models. These models are summarised and edited in the reports of the Intergovernmental Panel on Climate Change (IPCC) from the United Nations (IPCC 2001, 2007). These simulations of the future by the global climate models primarily cover the period from 1990 to 2100. Their results were in turn determined by global emission scenarios that are documented in the Special Report on Emission Scenarios of the IPCC (2000, abbreviated as SRES). They describe the potential future global trends in concentrations of climate-changing greenhouse gases. For the fifth IPCC assessment report (Stocker et al. 2013), Representative Concentration Pathways (RCPs) were defined on the basis of changes in radiative forcing, which substitute the SRES emission scenarios (Collins et al 2013). The logic of RCPs in terms of climate science is somewhat different from the SRES scenarios in that the former use greenhouse gas concentration trajectories, whereas the latter use greenhouse gas emission trajectories as scenarios. Both produce similar climate signals when driving climate models, which are used as basis to study future climate change impacts The GLOWA-Danube project uses the SRES emission scenarios.

The IPCC (2007) provides a good overview of the emission scenarios that underlie the model calculations and the global and continental climate developments which the global climate models, based on these emission scenarios, have simulated. Since the simulations by the global climate models have usually been carried out at a resolution lower than 100×100 km^2, by necessity, many regionally significant details cannot be considered in the calculations. For example, the topographic elevation of the grid cells south of Munich is approximately 1,200 m a.s.l. in the global climate models. This is approximately the same as the elevation of the German low mountain ranges. Consequently, it cannot be expected, especially in regions with strong topographic influence, that the global climate models can accurately reproduce and simulate the regional trends in climate change. For this reason, for all issues in regional climate research, it is important to develop methods that allow the regional climate trends to be derived from the global climate trends. The methodology developed within GLOWA-Danube is schematically depicted in Fig. 48.1; the goal is to compile the meteorological input data for DANUBIA at high spatial and temporal resolution.

As shown in Fig. 48.1, the preparation of a regional climate trend consists of three steps:

1. Selection of an emission scenario that can be converted into trajectories of a possible change of greenhouse gas concentrations in the atmosphere.
2. Translation of the selected emission scenario into a global climate trend using a global climate model.

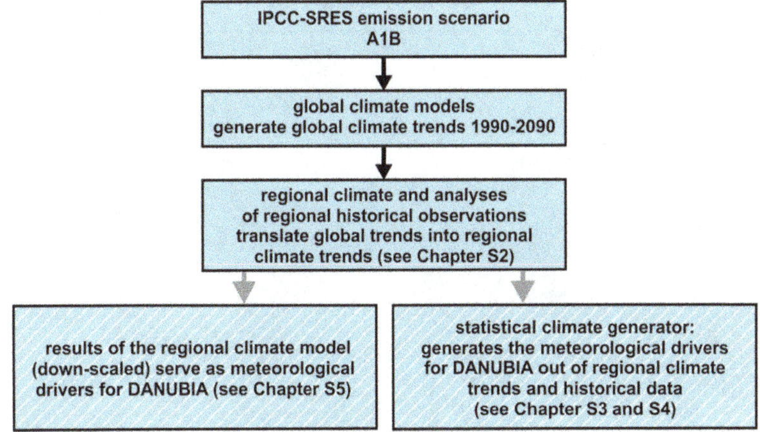

Fig. 48.1 Scheme of regional climate model development procedure, explained in this chapter. The following transfer in meteorological drivers for DANUBIA is explained in Chaps. 49, 50 and 51 (*shaded* area)

3. Implementation of the results of the global climate model into a regional climate trends for the Upper Danube drainage basin. They will form the basis of the more holistic GLOWA-Danube scenarios, which also include different socio-economic scenarios.

48.2 Selection of the Emission Scenario

A total of 12 IPCC-SRES emission scenarios were studied for this purpose. Figure 48.2 illustrates the expected increase in atmospheric CO_2 concentrations and temperatures over the period from 1990 to 2100 for the different emission scenarios; these are based on the values expected based on the assumptions of the IPCC global emission scenarios.

According to Fig. 48.2a, the future CO_2 concentrations will change as a result of various assumptions about the future CO_2 emissions sometime after 2025. In the most optimistic scenario, they stabilise by the year 2100 at around 500 ppm (scenario B 1) or, as described by the rather pessimistic A1FI scenario, increase without noticeable slowing to almost 1,000 ppm. Following extensive discussion among the project groups, it was decided that the IPCC-SRES-A1B emission scenario forms the basis for the further development of the GLOWA-Danube scenarios. The A1B emission scenario selected for GLOWA-Danube represents a moderate scenario that leads to a slowing of the increase in CO_2 concentrations and a CO_2 concentration of approximately 700 ppm by 2100. The assumptions of this scenario are described in detail in IPCC (2000).

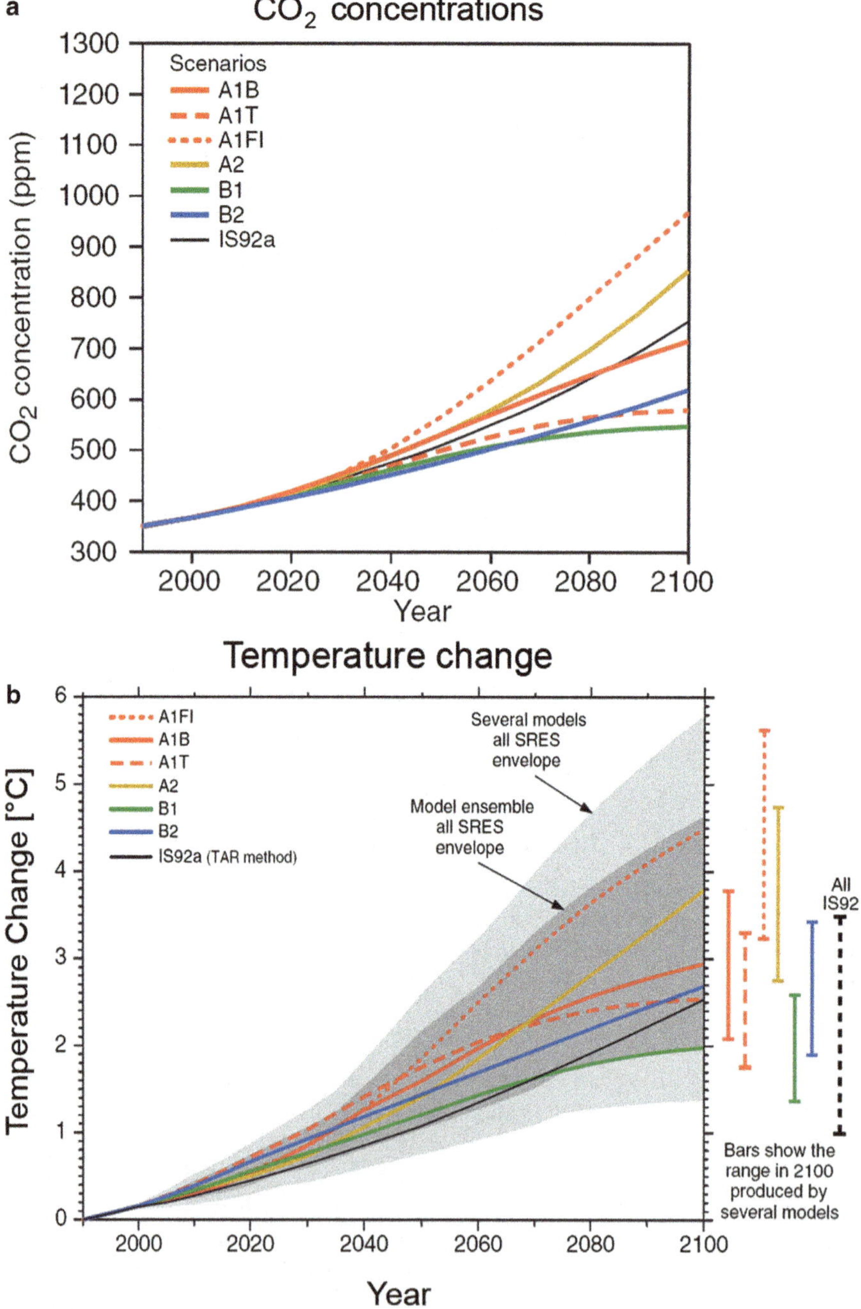

Fig. 48.2 (**a**) Projected increase of global mean CO_2 concentration in six different SRES scenarios (extract of IPCC 2001, Climate Change 2001 – Working Group I – The Scientific Basis, SPM 5). (**b**) Resulting increase of global mean temperature (extract of IPCC 2001, Summary for Policymakers, p. 14, fig. 5)

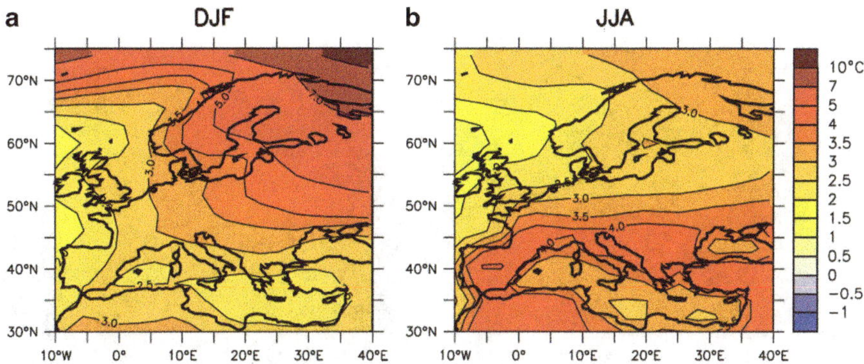

Fig. 48.3 Increase of mean temperature in Europe, 1980–1999 to the period 2080–2099 (IPCC-SRES-A1B scenario as mean value of 21 global climate models); (**a**) December–January–February; (**b**) June–July–August (extract of IPCC 2007, p. 875, fig. 11)

48.3 Translation of the Global Emission Scenario into a Global Climate Trend

48.3.1 Results of the Intergovernmental Panel on Climate Change IPCC

The published results of a comparison of 21 different global climate models conducted by the IPCC 2007 were used for this step; each global climate model was driven by the IPCC-SRES-A1B emission scenario for 1990–2100.

The results are presented in Figs. 48.3, 48.4 and 48.5.

Figure 48.3 reveals that the global climate models predict a temperature increase between the two periods compared, 1980–1999 and 2080–2099, of around 3.3 °C for the Upper Danube basin, both for the winter and summer months.

Figure 48.4 shows the percent change in precipitation in Europe between the two periods compared, 1980–1999 and 2080–2099, as the mean simulation results from 21 global climate models for the summer and winter months based on the IPCC-SRES-A1B emission scenario. A non-uniform picture emerges for the Upper Danube basin. In summer, there is a mean decrease in precipitation by approximately 14 %, whereas in winter, the models predict a slight increase on average of approximately 7 %. Figure 48.5 provides insight into the uncertainties within the statements about the climate change signal for precipitation. For each grid point, the number of models that predict an increase in precipitation is shown. In Scandinavia, almost all models consistently predict that precipitation will increase in summer and in winter. In North Africa, almost all models consistently predict that precipitation will not increase throughout the year. The Upper Danube basin is situated at the boundary between these two systems, in which statements about precipitation based on the global climate models show the largest number of contradictions. For the

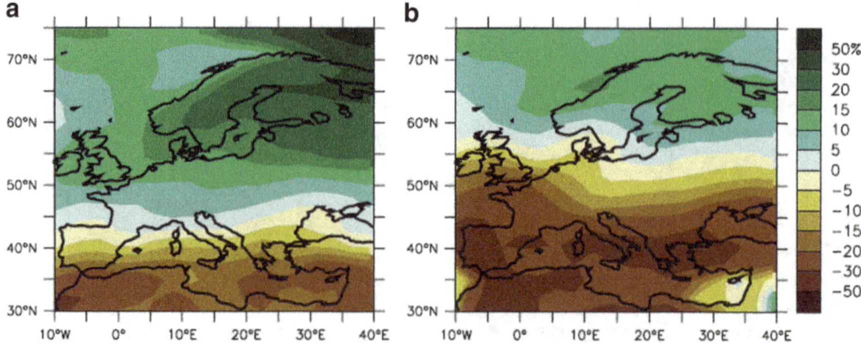

Fig. 48.4 Percental change of precipitation in Europe, 1980–1999 to the period 2080–2099 (IPCC-SRES-A1B scenario as mean value of 21 global climate models); (**a**) December–January–February; (**b**) June–July–August (extract of IPCC 2007, p. 875, fig. 11.5)

Upper Danube basin, the majority of the models agree that winters will be moister

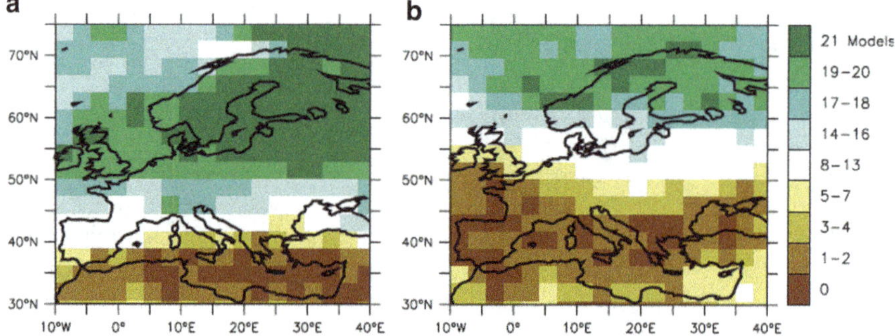

Fig. 48.5 Estimation of certainty of the statements in Fig. 48.4: number of models, which predict an increase in precipitation between the time periods 1980–1999 and 2080–2099; (**a**) December–January–February; (**b**) June–July–August (extract of IPCC 2007, p. 875, fig. 11.5)

and summers drier. However, there are contradictory results for both; this suggests that statements about the trend in precipitation compared to the Mediterranean and Scandinavia are fraught with somewhat greater uncertainty.

48.3.2 The Global Basis for the REMO and MM5 Regional Climate Models

Results from global climate models are used to drive higher-resolution regional climate models with the idea to resolve, in a nested approach, regional climate peculiarities, which global climate models with their coarser resolution are not able to resolve. In the first nesting step, both of the regional climate models used in the GLOWA-Danube project were driven with the same global dataset. This was carried out using a global simulation calculated from the coupled atmosphere-ocean circulation model *ECHAM5–MPIOM,* as the driver for the simulation of the present climate as well as possible future climates by the regional models. This simulation was prepared at the Max Planck Institute for Meteorology. While for the historical period from 1950 to 2000 actual observed greenhouse gas concentrations were used as input for the global model, the future period from 2001 to 2100 was simulated based on the changes in greenhouse gases that are postulated by the IPCC emission scenarios. Three realisations were calculated using the global model for each of the IPCC scenarios A1B, B1 and A2. Only the first realisation of the SRES scenario A1B was used for GLOWA-Danube (member 1). The model *ECHAM–MPIOM* has a spatial resolution of approximately 180 km – however, like all gridded models, it is not accurate to the grid point. A mean over several grid cells should always be selected for regional considerations. For the analyses of the ECHAM5 data presented here, the mean of 5×5 grid cells was chosen, such that the Upper Danube basin is situated in the centre of the 25 grid cells.

Figure 48.6 depicts the deviation in the annual mean temperature from the long-term average for the period 1971–2000 and the relative change in annual precipitation from the long-term average for the same period for the 900×900 km region covering the Upper Danube basin, based on the calculations by *ECHAM5–MPIOM.*

48.4 Application of the Predictions of the Global Climate Models to the Upper Danube Basin

48.4.1 The Regional Climate Trends from IPCC and ECHAM5–MPIOM

The expected climate changes for temperature and precipitation for the Upper Danube basin under the assumptions of the IPCC-SRES scenario A1B can be seen in Figs. 48.3 and 48.4. The mean changes in temperature and precipitation for the winter and summer months were determined and can be regarded as the first "regional" trends; they are listed in Table 48.4 under the *IPCC regional* climate trend.

The values of the GLOWA-Danube *IPCC regional* climate trend arise from averaging the results of 21 global models, one of which is *ECHAM5-MPIOM*. By way of example and also because *ECHAM5* is used to drive the regional models, the trend was also calculated directly from *ECHAM5* data. The *IPCC regional* climate trend was calculated as the difference of two mean conditions from the periods 2080–2099 and 1980–1999, and the trend was assumed to be linear. In contrast, for *ECHAM5*, the trend was calculated as the mean increase over the period from 1990 to 2100. This value is included in Table 48.1.

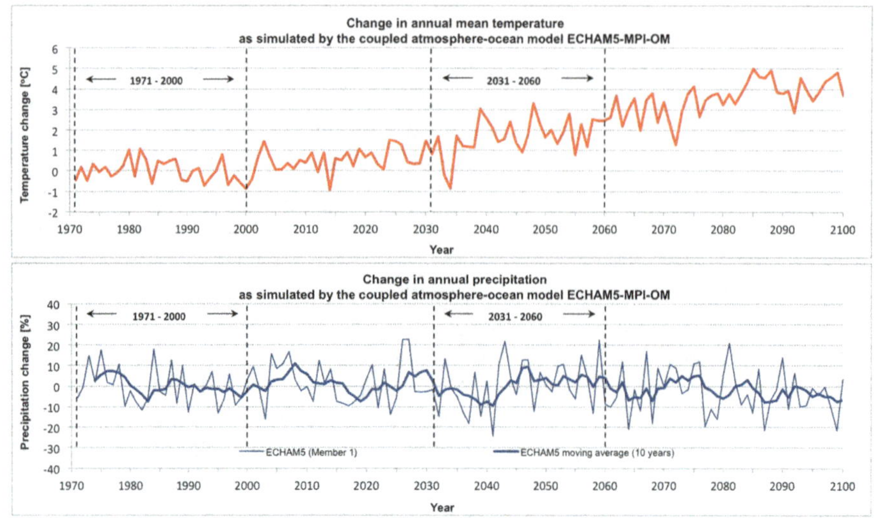

Fig. 48.6 Change in annual mean temperature and precipitation over the years 1971–2100 according to simulations of the coupled atmosphere-ocean model ECHAM5–MPIOM (Member 1, 5×5 raster elements with centres located in the Upper Danube Watershed)

48.4.2 Trends in Temperature and Precipitation Within the Upper Danube Basin Analysed Using REMO Simulations

Figure 48.5 illustrates the uncertainties of the projected changes to precipitation that were calculated using the global models. It is clear that the *IPCC regional* climate trend calculated for GLOWA-Danube should only be considered as a rough preliminary estimate, in particular because the regional climate conditions in the areas near the Alps in the Upper Danube basin are not adequately depicted in the low-resolution global models. For this reason, more accurate calculations for regional climate change have to be considered. To do so, first an analysis was carried out on the results of the regional climate simulations for Germany that were performed at the Max Planck Institute for Meteorology (MPIM) as commissioned by the German Environment Agency UBA (Umweltbundesamt = UBA), which were published in 2008 (Jacob et al. 2008). In this study, the IPCC-SRES-A1B emission scenario and the *REMO* regional climate model (Jacob 2001) were used, among others, to simulate the climate trend in Central Europe at a resolution of ~10 km. The resulting trends for temperature and precipitation in the context of this study for the Upper Danube basin are summarised in Table 48.2 for the period from 1990 to 2100.

Table 48.1 Regional climate trend of 25 grid cells around the Upper Danube basin resulting from the ECHAM5 global climate model for the time period 1990–2100

	Temperature trend ECHAM5 1990–2100 [°C]	Relative precipitation trend ECHAM5 1990–2100 [%]
DJF	+6.4	+12.8
MAM	+3.3	+3.5
JJA	+5.5	−34.1
SON	+5.0	−1.1
Annual average	+5.0	−3.6

Table 48.2 Regional climate trend in the Upper Danube basin resulting from the UBA study with the REMO regional climate model for the time period 1990–2100

	Temperature trend REMO 1990–2100 [°C]	Relative precipitation trend REMO 1990–2100 [%]
DJF	+6.8	−4.9
MAM	+3.7	+9.1
JJA	+5.3	−31.4
SON	+5.1	−14.5
Annual average	+5.2	−12.6

48.4.3 Regional Climate Trends Calculated with the MM5 Climate Model

Climate simulations similar to those for the UBA were carried out by the meteorology group within GLOWA-Danube using the community model *MM5*. In these simulations, care was taken to ensure the conditions underlying the regional climate modelling follow a similar setup to that used for the UBA study, and the same *ECHAM5* forcings were used to drive based on the same IPCC-A1B emission scenario. The use of MM5 by the meteorology group in addition to REMO, which was used by the MPIM, opened up the chance to study differences in regional climate results related to model formulations. The corresponding results from the *MM5* simulations are shown for the period from 1990 to 2100 in Table 48.3. From these results, it is clear that both models behave similarly with respect to temperature change in the basin, even though the annual mean temperature increase calculated with the *MM5* regional climate model is approximately 0.5 °C lower. However, there are differences in the maximum seasonal trends: *ECHAM5* and *REMO* calculate the maximum seasonal temperature increase for the winter months (DJF), whereas *MM5* predicts the maximum for the summer months (JJA).

The simulated precipitation changes using MM5 show even greater differences compared to *REMO*. *MM5* shows higher seasonal increases in precipitation in spring (MAM) and weaker decreases in summer and fall (JJA and SON); these results lead to a lower annual mean decrease in precipitation. The modelled trends

Table 48.3 Regional climate trend in the Upper Danube basin resulting from the regional climate modelling with the MM5 model

	Temperature trend MM5 1990–2100 [°C]	Relative precipitation trend MM5 1990–2100 [%]
DJF	+5.2	+7.7
MAM	+3.2	+13.1
JJA	+5.8	−28.7
SON	+4.8	−1.0
Annual average	+4.7	−3.5

differ marginally in the winter months (DJF). The lower temperature increase in winter simulated by *MM5* is associated with a slight increase in precipitation.

Figure 48.7 presents a comparison between the two model results for the change in the climate variables temperature and precipitation for the period from 1971 to 2100 for the Upper Danube basin. The trends for the period 1990–2100 were calculated from these results and are listed in Tables 48.2 and 48.3.

Both models show quite comparable results for the temporal course of temperature increase. The trends in precipitation also show similarities, primarily when considering the 10-year moving average for both models. There is a slight increase in precipitation in the study area up to the year 2050, whereas after this year, the precipitation decreases significantly. The high degree of consistency in the simulated temporal trends for the annual means of the two climate elements can be attributed to the use of the same global forcing (*ECHAM5*, member 1). It enters into the calculations of the regional climate model as pressure, wind speed, temperature, radiation and humidity distributions at the lateral boundaries of the simulation area and therefore strongly influences if not dominates the simulations of the regional climate trend by the regional climate models.

An interpretation of the results should also take into account that the results of the climate models always contain model uncertainties, which are around +/−5 % or more for precipitation changes, for example. When considering the trend in the relative change in precipitation compared to the long-term average for the period 1971–2000, a deviation between the models of approximately 10 % is based on the fact that both models calculate a similar trend tendency, but differ in the extremes. This is apparent from Fig. 48.7.

The simulated trends of *MM5* and *REMO* shown in Tables 48.2 and 48.3 were used as inputs for the statistical climate generator (see Chap. 49). It should be kept in mind that the climate trends describe a linear approximation of the change in temperature and relative change in precipitation over time and that thereby the specific course of temperature and precipitation change within the study period is abstracted to a single number. Nevertheless a disaggregation of this abstracted overall trend from Tables 48.2 and 48.3 into a course of annual changes of temperature and precipitation between 1990 and 2100 is required for each day of the year as input to the stochastic climate generator. This trend in mean temperature increase

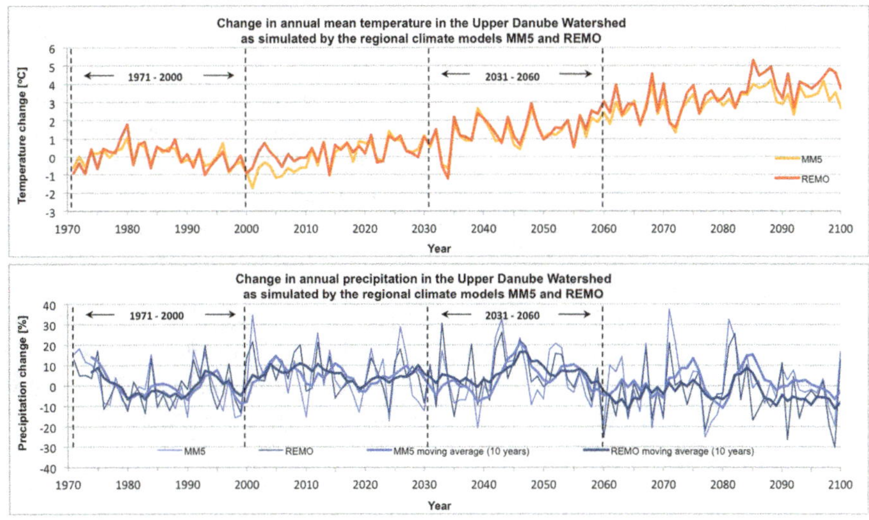

Fig. 48.7 Deviation of annual values of temperature and precipitation from the long-term mean 1971–2000 according to simulations of the regional climate models MM5 and REMO

and mean change in precipitation was assumed to behave according to the course of CO_2-concentration increase of the SRES-A1B scenario from the IPCC (see Fig. 48.2b).

48.4.4 *The Regional Climate Trend from Historical Climate Measurement Data*

The analyses of the trends in historical climate data from the DWD and the ZAMG for the Upper Danube basin for the years 1960–2006 were another important source of information for evaluating possible future regional climate change in the Upper Danube basin (Reiter et al. 2012). These data are presented in Chap. 13. They reveal a clear and marked temperature increase. The trends in precipitation that were calculated as the results of the *IPCC, REMO* and *MM5* models are also already evident in the tendencies from past observations. The winter months in the study period became moister, and the summer months tended to get drier. However, the trends in precipitation in the historical data were less obvious than the temperature trends. The continuous increase in mean annual atmospheric temperature close to the ground is shown in Fig. 48.8. It amounts on average to 0.0328 °C per year from 1960 to 2006 and the trend is approximately linear. The atmospheric CO_2 content also increases over this period, although to a lesser degree than is assumed for the IPCC-SRES-A1B scenario for 1990–2100.

In addition to the historical mean measured temperature increase, Fig. 48.8 shows the hypothetical temporal trend in temperature increase from 1990 to 2100 that is termed the *IPCC regional* trend. The latter is determined for the Upper Danube basin from Fig. 48.3. To do so, a rough approximation was first made assuming that the global and regional temperature increases by the end of the century differ in magnitude but show similar temporal trends. The trend shown is determined by first calculating the ratio of the regional temperature increase (3.3 °C) to the mean global temperature increase from the A1B scenario (2.93 °C) (see Fig. 48.2b). The resulting factor (1.13) is multiplied by the global mean trend in temperature increase for the A1B scenario indicated at the bottom of Fig. 48.2. The value of this factor (1.13) signifies that the regional temperature increase calculated from Fig. 48.3 is 13 % larger than the mean global increase from the IPCC-SRES-A1B scenario. The given curve correspondingly extends the temperature trend beyond the observed temperature from year 1990.

A second curve is presented in Fig. 48.8 and given the designation *Forward Projection*; this curve is based on the temperature increase from 1990 to 2100 extrapolated from historical measurements based on the IPCC-SRES-A1B scenario. The basic principle of the *Forward Projection* is that the future course of temperature increase should first fit in intensity to what has been observed in the past and it should at the same time consider that the SRES-A1B scenario shows sharply increasing CO_2 concentrations during the middle of the century. This implies that a simple linear extrapolation of temperature increase has to be modified to take this non-linear CO_2-concentration increase into account. The extrapolation comes from the comparison of the mean annual temperature increase of 0.018 °C per year calculated for the IPCC-SRES-A1B scenario for the period 1990–2020 with the actual temperature increase of 0.0328 °C per year over the past 47 years from measurements. The time interval from 1990 to 2020 was chosen for the comparison because in this period, the increases in temperature still follow a similar pattern as the past measurements. After 2020, a steeper increase begins in the IPCC-SRES-A1B scenario. The *Forward Projection* of the historical measurement trend was calculated, as above in the example of the *IPCC regional* curve, from the ratio of the two increases and the general trend of the A1B temperature increase from the bottom of Fig. 48.2. The result of this *Forward Projection*, as can be inferred from Fig. 48.8, is a regional increase in the annual mean temperature in the study area of 5.2 °C between 1990 and 2100. This *Forward Projection* of the warming trend derived from the past measurements is virtually identical to the results of the UBA study mentioned above, which predicted an increase in the mean annual temperature in the study area of approximately 5.2 °C between 1900 and 2100. The result is also very close to the temperature trend from the *MM5* results, which yielded a temperature increase of 4.7 °C in the study area (see Tables 48.2 and 48.3).

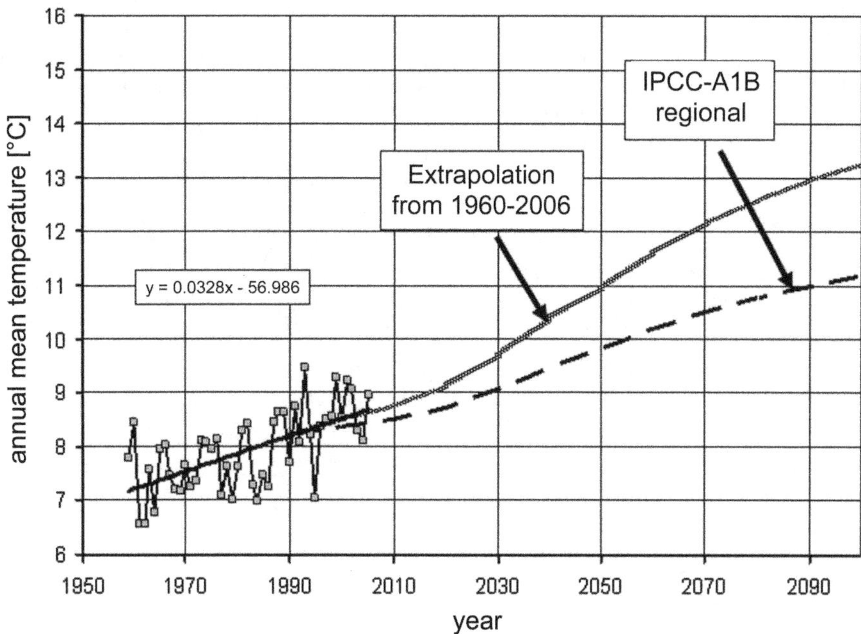

Fig. 48.8 Trend of regional temperature increase 1960–2100, temperature increase 1960–2006 observed. For IPCC regional 1990–2100, see Fig. 48.3; for Extrapolation 1990–2100, see Chap. 13

48.4.5 Synopsis of the Regional Climate Trends

Figure 48.8 and the results of the regional climate models, which are both based on the global ECHAM5 forcing (member 1), establish four regional climate trends for the temperature increase and change in precipitation in the Upper Danube basin, based on the global IPCC-SRES-A1B scenario. These are presented in Table 48.4.

The synopsis of the four GLOWA-Danube climate trends that were worked out is presented as follows:

- The regional climate trends differ considerably from the global mean temperature increase specified by the IPCC, which is listed as 2.93 °C for the year 2100 for the A1B scenario.
- From the synopsis of the available findings regarding the expected regional climate trends in the Upper Danube basin from 1990 to 2100, the following conclusions for the further studies can be made from the range of regional climate trends summarised in Table 48.4.
- The temperature trends from the *REMO* and *MM5* climate models indicate an increase in the annual mean temperature by the end of the century of between 4.7 and 5.2 °C (see Tables 48.2 and 48.3).

Table 48.4 Comparison of regional A1B climate trend and the changes in temperature and precipitation for the period 1990–2100, resulting from analyses of the Intergovernmental Panel on Climate Change (2007), the Environment Ageny Austria UBA (Jacob et al. 2008) with REMO, the project group Meteorologie München with MM5 and the analysis and Extrapolation of historical climate trends for the Upper Danube basin

A1B climate trend	Temperature trend 1990–2100	Relative precipitation trend 1990–2100	Annual precipitation trend
IPCC regional	+3.3 °C winter	+7 % winter	–4.4 %
	+3.3 °C summer	–14 % summer	
REMO regional	+6.8 °C winter	–4.9 % winter	–12.6 %
	+3.7 °C spring	+9.1 % spring	
	+5.3 °C summer	–31.4 % summer	
	+5.1 °C autumn	–14.5 % autumn	
MM5 regional	+5.2 °C winter	+7.7 % winter	–3.5 %
	+3.2 °C spring	+13.1 % spring	
	+5.8 °C summer	–28.7 % summer	
	+4.8 °C autumn	–1 % autumn	
Extrapolation	+5.2 °C winter	+47 % winter	–16.4 %
	+5.2 °C summer	–42 % spring	
		–69 % summer	
		–2 % autumn	

- The trends in the relative change in precipitation from the *REMO* and *MM5* climate models indicate a decrease in the annual precipitation by the end of the century of between –3.5 and –12.6 %.
- The annual temperature and precipitation changes calculated with *MM5* and *REMO* are of a similar magnitude as the value calculated by the global climate model *ECHAM5* (temperature change +5.0 °C, precipitation change –3.6 %).
- The changes calculated from the observations in the scenario *Forward Projection* are +5.2 °C for annual mean temperature and –16.4 % for the change in annual precipitation and are thus also in the same range as the other model results.
- All the trends that were calculated can be classified as consistent and similar. The magnitude of their differences is in the order of natural variability; i.e. the differences are similar to natural fluctuations in the non-linear climate system.
- All four approaches predict an increase in temperatures for the entire year, although *REMO* calculated the most intense temperature increase for the winter months (DJF), while *MM5* yielded the most prominent warming for the summer months (JJA).
- All four approaches consistently calculate a decrease in precipitation in the summer months, although to varying extents (–14 to –69 %). The uncertainties in the calculation of precipitation values and their changes are considerably larger than for temperature (see Chap. 13).
- The GLOWA-Danube trend for the relative change in winter precipitation (DJF) fluctuates about an interval from –4.9 to +47 %.

- Thus, overall, the variations in assumed precipitation changes calculated and used in GLOWA-Danube for both seasons are above 50 %. The significant changes in precipitation calculated with the *Forward Projection* are notable.
- Moreover, all four approaches calculate different annual variations in the trends. This result highlights the need for subsequent impact studies to use a range of possible changes in order to adequately account for the uncertainties in the calculation of regional climate change patterns.

References

Collins M et al (2013) Section 12.3.1.3: The new concentration driven RCP scenarios, and their extensions, In: Chapter 12: Long-term climate change: projections, commitments and irreversibility. Archived 16 July 2014, pp 1045–1047, In: Stocker TF et al (ed) Climate change 2013: the physical science basis. Working Group 1 (WG1) contribution to the Intergovernmental Panel on Climate Change (IPCC) 5th assessment report (AR5), Cambridge University Press, Cambridge

IPCC (2000) Special report on emissions scenarios (Nakicenovic N, Swart R, eds). Cambridge University Press, Cambridge

IPCC (2001) Climate change 2001: the scientific basis. Contribution of Working Group I to the third assessment report of the Intergovernmental Panel on Climate Change (Houghton JT, Ding Y, Griggs DJ, Noguer M, van der Linden PJ, Dai X, Maskell K, Johnson CA, eds). Cambridge University Press, Cambridge/New York

IPCC (2007) Climate change 2007: the physical science basis. Contribution of Working Group I to the fourth assessment report of the Intergovernmental Panel on Climate Change (Solomon S, Qin D, Manning M, Chen Z, Marquis M, Averyt KB, Tignor M, Miller HL, eds). Cambridge University Press, Cambridge/New York

Jacob D (2001) A note to the simulation of the annual and inter-annual variability of the water budget over the Baltic Sea drainage basin. Meteorol Atmos Phys 77:61–73

Jacob D, Göttel H, Kotlarski S, Lorenz P, Sieck K (2008) Klimaauswirkungen und Anpassung in Deutschland – phase 1: Erstellung regionaler Klimaszenarien für Deutschland. Final report of the UFOPLAN-project (Ed.: Umweltbundesamt, Dessau- Roßlau). Umweltbundesamt (Federal Environmental Protection Agency)

Reiter A, Weidinger R, Mauser W (2012) Recent climate change at the Upper Danube – a temporal and spatial analysis of temperature and precipitation time series. Clim Change 111:665–696

Stocker TF et al (ed) (2013) Climate change 2013: the physical science basis. Working Group 1 (WG1) contribution to the Intergovernmental Panel on Climate Change (IPCC) 5th assessment report (AR5), Cambridge University Press, Cambridge

Chapter 49
The Statistical Climate Generator

Wolfram Mauser

Abstract GLOWA-Danube requires high spatiotemporal resolution time series of meteorological inputs. A spatially distributed statistical climate generator was developed, which produces a network of synthetic time series of meteorological measurements. It is based on a thorough analysis of historic weather pattern, for which the basic element was chosen to be 1 week. It uses the GLOWA-Danube climate trends and a random number generator to produce time series of weekly weather pattern for the period from 2011 to 2060 including selectable climate trends. Mahalanobis distance is used as criterion to select the most similar week from historic records to build up a future dataset. The statistical climate generator is successfully validated with four realisations of past climate, for which consistent climate parameters as well as reproduction of runoff behaviour using the hydrological model ROMET were proven.

Keywords GLOWA-Danube • Climate generator • Mahalanobis distance

49.1 Introduction

Climate impact model components are driven by meteorological data to simulate and analyse the effects of climate change on different aspects of the water resources of the Upper Danube. The meteorological model inputs reproduce scenario-based climate trends. The results of the scenario simulations and the conclusions that are derived from them are generally dependent on both the assumptions underlying the future climate trends and the availability, type and quality of the meteorological data driving the impact model components.

The meteorological drivers for the climate impact model components in GLOWA-Danube consist of a time series of arrays of the meteorological parameters precipitation, radiation (incident short and long wave), atmospheric temperature

W. Mauser (✉)
Department of Geography, Ludwig-Maximilians-Universität München (LMU Munich), Munich, Germany
e-mail: w.mauser@lmu.de

W. Mauser, M. Prasch (eds.), *Regional Assessment of Global Change Impacts*, DOI 10.1007/978-3-319-16751-0_49

(2 m), atmospheric humidity (2 m) and wind speed (2 m). The impact models expect these arrays at a spatial resolution of 1×1 km^2 and a temporal resolution of 1 h. In the past, the driving meteorological input in GLOWA-Danube was generated from data from climate monitoring networks using spatial and temporal interpolation (see Chap. 51). Among other purposes, this input is used to validate the hydrological climate impact models. This approach is not possible for scenarios that involve the future. Therefore, other methods are required to prepare the meteorological drivers to be used for simulating future scenarios.

The results of regional climate models, such as *MM5*, *REMO*, can serve as the source data for the meteorological driver. Although the results of *REMO* and *MM5* do provide hourly values, the spatial resolution of the regional climate models is usually lower than 1 km, so the results need to be adapted to the scale of the climate impact models of DANUBIA using a downscaling process. To do so, also methods for the correction of biases in the results of the regional climate models are needed; these are known as bias corrections. Chapter 51 includes a detailed description of the methods developed for this as well as the associated problems and expected results.

Another source for meteorological drivers is provided by compiling the historical measurement data from the climate monitoring networks in such a manner that the climate trend underlying each scenario is reproduced. The statistical climate generator developed for this approach is described below and belongs to a family of weather generators and specifically to the group known as "nearest-neighbour resampling" (Yates et al. 2003; Buishand and Brandsma 2001; Young 1994; Orlowsky et al. 2007). In contrast to the methods that depend on weather conditions (Spekat et al. 2006), the climate generator forgoes the analysis of the given weather conditions and instead generates hourly spatial arrays for a scenario-based future weather trend.

The climate generator is based on the assumption that the measured weather patterns of the past will continue in the future in a similar fashion, although at different sequences and frequencies. A 1-week time interval was selected for characterising a weather pattern. The property of the weather pattern within 1 week in the past is characterised by weekly mean temperature and mean total precipitation averaged over all monitoring stations in the Upper Danube basin. The subsequent process of generating artificial meteorological drivers from meteorological measurements involves three steps (Fig. 49.1).

49.2 Step 1: Statistical Analysis of the Historical Measurement Data from the Climate Monitoring Network

In the first step, the available data from the DWD climate monitoring network and the Austrian Weather Service (see Chap. 11) was subjected to a statistical analysis. This consisted in determining mean weekly weather patterns by calculating the

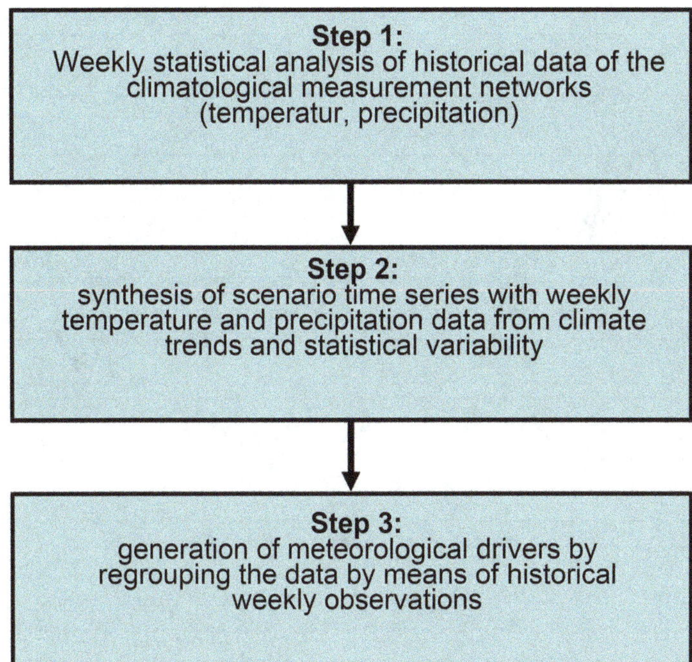

Fig. 49.1 Scheme of procedure for the generation of future meteorological input data by means of the statistical climate generator

mean temperature and mean total precipitation from the measurements made by all available stations in the study area for each weekly weather period. The week was chosen as the time interval since it corresponds well to the characteristic duration of dynamic pressure systems, which define weather conditions in the study area. Both low and high pressure systems show temporal patterns of this duration. Thus, a value for mean temperature and mean total precipitation is obtained for each week in the available 47-year historical dataset from 1960 to 2006. Subsequently, for each week of the year, the 47 pairs of values from 1960 to 2006 are analysed by linear regression. This yields three variables for each week of the year for the historical period: mean average temperature, mean average total precipitation and the covariance matrix of the two variables. The covariance matrix provides information about the co-variability of the mean temperature and mean precipitation total for all stations as well as about the correlation between the two variables over the period under consideration. Taken together, the analysis provides the average weekly course in these variables over the year and their covariance as significant characteristics of the regional climate for the period being considered. The weekly course of the correlation coefficients between mean temperature and mean total precipitation calculated from the weekly covariance matrices is shown in Fig. 49.2 for the analysis period from 1960 to 2006.

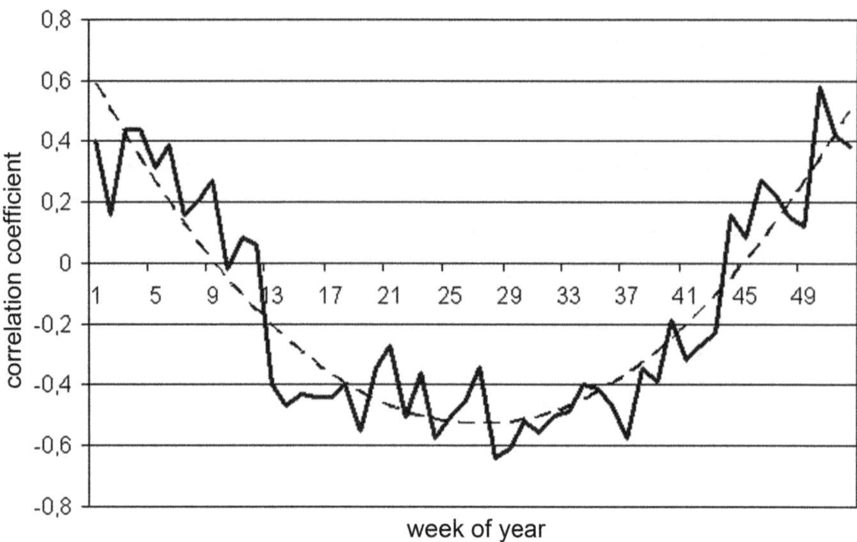

Fig. 49.2 Trend of weekly linear correlation coefficient of mean temperature and precipitation sum of the climatological stations in the Upper Danube basin; analysis period: 1960–2006

Figure 49.2 reveals a significant positive correlation with a correlation coefficient of approximately 0.4 around weeks 1 and 52. This means that there is a tendency that during these weeks, elevated mean atmospheric temperatures in the study area also correlate with the above-average precipitation. This trend is reversed for the summer weeks 25–28. Here, negative correlations are apparent, suggesting that weeks with elevated mean atmospheric temperatures show below average precipitation values. Both results are easy to explain from regional climatology. Whereas in winter an above-average temperature in the study area is normally associated with southerly inflows and elevated water vapour content and hence more precipitation, in summer, higher temperatures are usually associated with high pressure systems and hence reduced precipitation. During the transitional seasons around weeks 10 and 44, there is no correlation and hence no conclusions about tendencies can be made.

The weekly course in the statistical variables mean temperature, total precipitation and their covariance matrix thus, in the chosen approach, characterises the climate of the historical period that was analysed. Once the analysis of the historical climate is complete, the next steps serve to artificially generate possible weather patterns of a future climate trend using the statistical parameters in a reverse process. For these steps, weekly data for correlated mean temperature and mean total precipitation first have to be generated from the future climate scenario defined in Chap. 48.

49.3 Step 2: Generation of Weekly Climate Time Series from a Specified Climate Trend

In step 2, the temporal trend in weekly mean temperature and mean total precipitation are calculated for a selected regional climate trend (see Chap. 48) for the period from 2011 to 2060. Four components are temporally superimposed to obtain weekly pairs of values for mean temperature and mean total precipitation for the future:

1. The change in annual mean temperature resulting from the climate trend
2. The weekly course in mean temperature relative to the annual trend
3. The change in mean total precipitation resulting from the climate trend for each week in a year
4. The random statistical fluctuation in weekly mean temperature and total precipitation which takes into account the weekly changing correlation between the two values (see Fig. 49.2).

The interaction of the four components is shown in Fig. 49.3a–f using the example of generating an artificial weekly climate time series for the period from 2011 to 2060.

In this fictional example, it can be anticipated that the increase in temperature from 1990 to 2100 in the study area will have a value of 5 °C and should have a temporal course of the trend matching that of the A1B scenario (see Fig. 49.3a); the seasonal variation in temperature is shown in Fig. 49.3b. The anticipated percent change in weekly precipitation from 1990 to 2100 in this example is drawn from a trend analysis of the REMO-UBA scenario and is shown in Fig. 49.3c. This change proportionally follows the temperature increase for the current year. Figure 49.3d presents the calculated random statistical variation in temperature for an example year chosen from the 2011–2060 period. This variation was calculated from the weekly covariance matrices for mean temperature and precipitation using a random number generator (Visual Numerics 2006). Appropriate statistical variations can be calculated by repeatedly using the random generator for each week of the period being studied.

An artificial time series for the trend in weekly mean temperatures for the years 2011–2060 is now created, in which the temperature trend, the seasonal variations in temperature and the statistical variabilities for all weeks from 2011 to 2060 are superimposed. The trend that emerges from the superposition of weekly mean temperatures from 2011 to 2060 is shown in Fig. 49.3e. In addition to seasonal fluctuations, the trend (red line) and the statistical variability of temperature are easy to discern. The generation of the time series for weekly total precipitations takes place in the same way from Fig. 49.3a, c and a corresponding statistical variation in mean total precipitation from Fig. 49.3d, where the trend for the seasonal change is accounted for and the statistical variation in precipitation totals is simu-

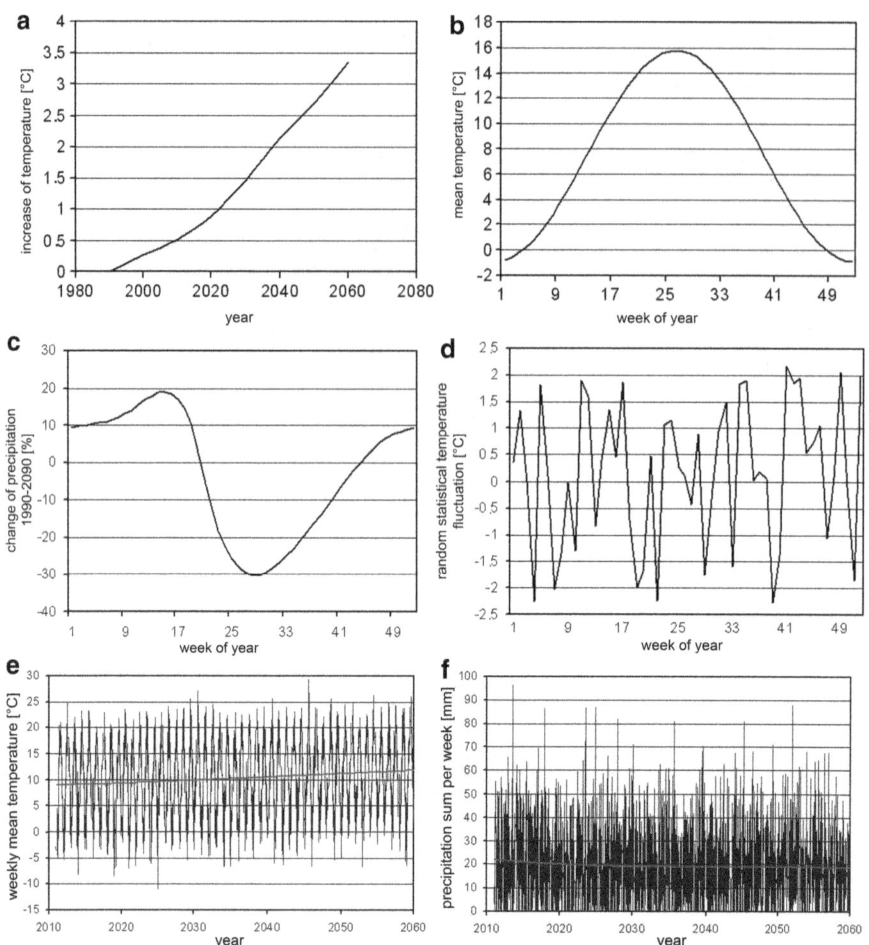

Fig. 49.3 Scheme of the generation of synthetic weekly time series of mean temperature and mean precipitation sum by means of the statistical climate generator for the Upper Danube basin: (**a**) temperature trend from IPCC-A1B, increase of 5° 1990–2100; (**b**) course of weekly mean temperatures 1960–2006; (**c**) precipitation trend as percent change of precipitation sums per week 1990–2100 after Jacob et al. (2008); (**d**) statistical variability of weekly mean temperature from the covariances of the meteorological data 1960–2006; (**e**) combination of a), b) and d) for a time series of weekly mean temperatures 2011–2060; (**f**) combination of (**a**), (**c**) and (**d**) for a time series of weekly precipitation sums 2011–2060

lated using the random number generator. A time series for the weekly precipitation totals generated this way for the period from 2011 to 2060 is shown in Fig. 49.3f.

In determining the statistical variability, it should be considered that weekly mean temperature and mean total precipitation are interdependent. As shown in

Fig. 49.2, this dependency follows a seasonal trend. The result of this is that a positive deviation from the given weekly mean in winter calculated by the random number generator for temperature results in a related random, positive deviation in total precipitation. Correspondingly, according to Fig. 49.2, for the summer months there is a negative, also random deviation. In order to take into account this interdependency between the two variables when generating the random fluctuations, the two-dimensional random number generator RNMVN from the IMSL Statistical Library (Visual Numerics 2006) was used. Each time it is invoked (i.e. for each week), this random number generator specifies two independent normally distributed random numbers for the weekly mean temperature and the weekly mean total precipitation, based on the mean values and the covariance matrices, under the assumption that both variables are normally distributed. Using the random number generator, different pairs of values for mean temperature and total precipitation were calculated for each week from 2011 to 2060, each based on different means and covariances. The temporal trend in the random portion of the variation for the time series that is generated is determined by the starting value of the random number generator. Different starting values generate reproducible, differing, statistically equivalent courses of the random fluctuation in the weekly time series.

The result of step 2 is the generation of mutually independent, artificial, future time series of weekly mean temperature and total precipitation for 2011–2060. The following assumptions were made in the generation of the time series shown in Fig 49.3e and f:

1. In the period chosen (2011–2060), the mean temperatures and precipitation totals change in accordance with the underlying climate trend. The climate trends used in GLOWA-Danube are presented in detail in Chap. 48. The functioning of the statistical climate generator is not limited to the climate trends presented in Chap. 48 and can therefore benefit directly from advances in research on regional climate change.
2. The future seasonal magnitude of the weekly mean temperatures does not significantly change, even if the temperatures increase. This assumption is supported by the statements of the IPCC (2007), according to which the Upper Danube region will experience a significant increase in the Mediterranean climate component but will not see entirely new weather trends as a result of climate change.
3. The covariances between weekly mean temperatures and mean precipitation totals and their interdependency depicted in Fig. 49.2 will not significantly change over the period from 2011 to 2060. This assumption allows the covariance matrices calculated from the data from 1990 to 2006 to be used for calculating the statistical variations in the scenario-climate time series. The assumption that the Upper Danube basin remains in the same climate zone as today despite climate change also applies in this case.

49.4 Step 3: Generating New Data Series Using Regrouping of Historical Data

The weekly time series for mean temperature and mean total precipitation generated in step 2 for a future climate change trend are not adequate as meteorological input in DANUBIA. DANUBIA requires hourly arrays of the meteorological parameters at a spatial resolution of 1×1 km^2.

In step 3, the future meteorological datasets are now generated by regrouping, based on the results from step 2 for the period from 2011 to 2060, the historical meteorological datasets from the German and Austrian Weather Services. These datasets were used in the validation of DANUBIA (see Chap. 4).

The basic idea of this process is that the weekly mean temperatures and mean precipitation totals calculated in step 1 from the historical time series characterised the weather pattern of that week for the entire study area. It can hence form a key for accessing the full set of measured values from all stations that were stored during that week in the database. They consist of 3 sets of measurements meteorological variables taken at the 7:00, 14:00 and 21:00 on each day of the week. To form meteorological datasets, which can drive DANUBIA the weekly mean temperatures and mean precipitation totals that resulted from step 2 and form the scenario, time series are used. For each future week, the historical week with the most similar combination of weekly mean temperature and mean precipitation total is now selected from the database. The complete dataset containing the meteorological measurements of the selected historical week is then adopted for the scenario week under consideration. This dataset consists of all the meteorological measurement data from all stations in the study area and forms a set of both physically and spatially consistent, though measured, meteorological station data. Once this process is carried out for all the weeks from 2011 to 2060, the result is a regrouped meteorological dataset for the entire scenario period that is based on historical measurement data. It consists of synchronous historical measurements from the stations included and at the same time portrays the selected future climate trend.

The maximum likelihood criterion was used to identify the historical week that is climatologically most similar to the selected future week. This was done using the Mahalanobis distance between all historical weeks and the future week under consideration within the feature space that is spanned by the weekly mean temperature and mean total precipitation (Mahalanobis 1936), as a measure of climatological similarity. This takes into consideration the differing variances of mean temperature and mean total precipitation, under the prerequisite of normal distribution, using the means and covariance, as well as the interdependency of the two parameters. This measure is inversely proportional to the probability that a sought-after historical week exhibits the same weather pattern as the considered future week in the scenario. The historical week that has the smallest Mahalanobis distance to the scenario week and hence has the most similar weather pattern is adopted into the new dataset describing future climate, regardless of when in the year this week occurred in the past.

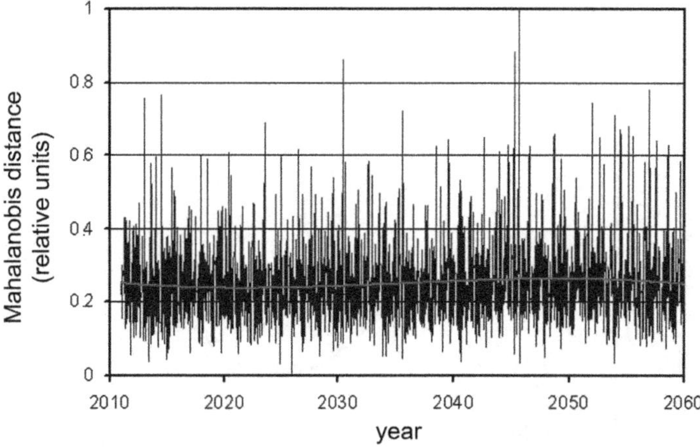

Fig. 49.4 Mahalanobis distance of selected historical and future week based on weekly mean temperature and precipitation sum

Figure 49.4 illustrates an example of the Mahalanobis distance for all weeks from 2011 to 2060 for the *Baseline* climate variant (see Chap. 50) of the *Forward Projection* climate trend (see Chap. 48). The Mahalanobis distance for this scenario period reveals no temporal trend. This means that the average similarity between the future weekly means calculated in step 2 and the selected measured weekly means from the past does not change during the scenario period from 2011 to 2060. An increase in the Mahalanobis distance over time means that the similarity of the scenario weeks from step 2 compared to the week chosen from the measured weeks becomes increasingly weaker with the increasing future effects of climate change. This indicates that in the scenario, additional weekly weather patterns arise that have no counterpart in the past and hence have not been measured in the past.

Once steps 1–3 have been completed, there is a new meteorological driver dataset, which has a defined temperature trend, follows a prescribed temporal change in precipitation and exhibits the same variabilities as the measured historical dataset. It consists of spatially and physically consistent, though measured, data in the same format, of the same quality and at the same spatial and temporal resolution as the historical data that was used to validate DANUBIA. The spatiotemporal dataset of the meteorological scenario driver is then prepared using spatial and temporal interpolation, as was the case for the validation of DANUBIA (see Chap. 4).

In addition to these advantages, this procedure also has disadvantages. Thus, extremely heavy precipitation that exceeds historical precipitation events cannot be represented.

A change in the persistence of weather conditions for a period in excess of a week can also not be accounted for, and randomly distributed gaps arise between adjacent weeks; however, the extent of these gaps does not exceed that which is observed naturally. Nonetheless, the method introduced for the statistical climate generator is definitely capable of generating new, not yet observed moist or dry

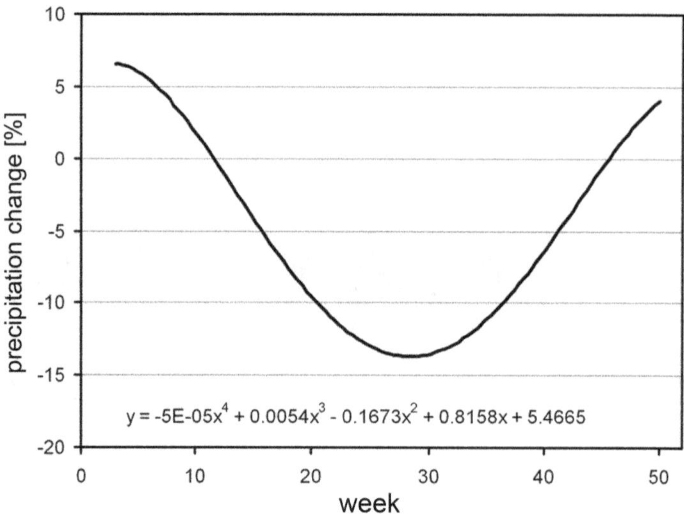

Fig. 49.5 Weekly hours of percent change of precipitation in the Upper Danube basin from the comparison of the results of global models from IPCC (2007) in the period 1990–2100

periods, while the increasing temperature trend produces more frequent and longer-lasting successions of existing dry or moist periods.

49.4.1 Weekly Patterns in the Climate Trends

By way of example, the climate generator has been presented thus far in its basic mode of functioning. It forms the basis for further implementation of the four climate trends outlined in Chap. 48. In addition to the annual temperature trend, the climate generator thus requires as input variables both the seasonal cycle of change in the weekly mean temperatures and the change in weekly precipitation totals for each climate trend. Therefore, the weekly distribution of precipitation change must be derived from the seasonal data of potential precipitation change in Table 48.4 of Chap. 48. This derivation is carried out by interpolation between the sampling points supplied by the table while ensuring the given seasonal mean changes are preserved. For the weekly trend in the change of precipitation between 1990 and 2100, the interpolation was carried out using the trend shown in Fig. 49.5 for the *IPCC regional* climate trend.

The data on the seasonal change in precipitation and temperature from the *REMO* and *MM5* regional climate trends (see Chap. 48, Table 48.4) allow an interpolation both of the weekly percent changes in precipitation and in the weekly changes in temperature for the period 1990–2100. Here too, the interpolation took place such that the seasonal means were preserved. The results are shown in Fig. 49.6.

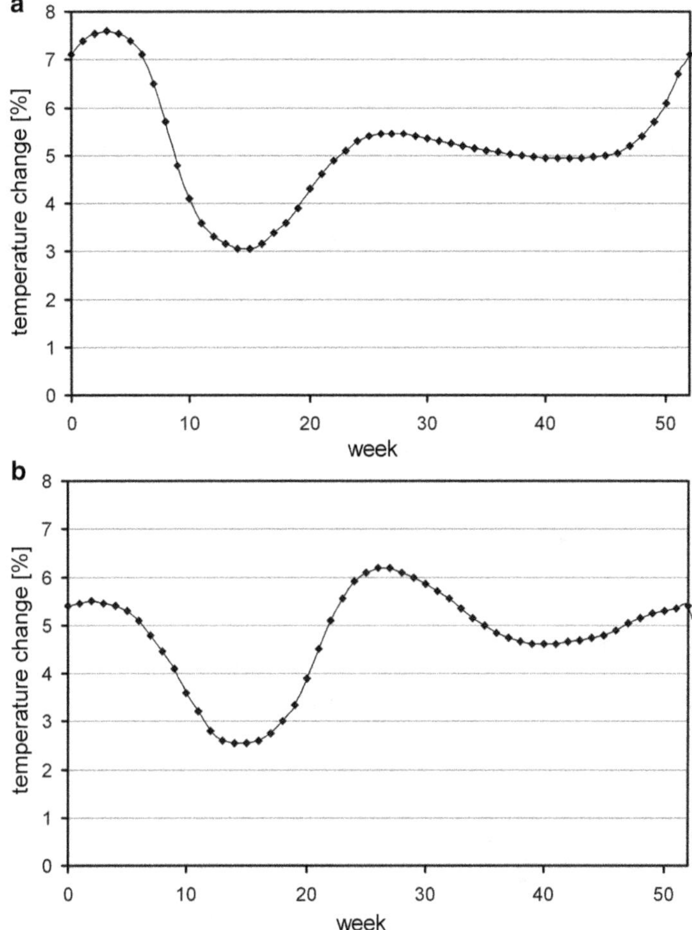

Fig. 49.6 Weekly course of temperature change in the Upper Danube basin based on seasonal results from (**a**) the REMO regional climate model after Jacob et al. (2008) and (**b**) the MM4 regional climate model 1990–2100

The weekly trend in temperature change from *REMO* in Fig. 49.6a shows a distinct maximum in late winter, with a peak at more than +7 °C. In comparison, the temperature increase in spring with a minimum of 3 °C is considerably lower. In the remaining seasons, it fluctuates around a value of 5 °C. The changes in temperatures from *MM5* fluctuate with the same magnitude but show much weaker seasonal variations. The minimum temperature increase for both models occurs in spring.

The changes in precipitation within the Upper Danube basin according to IPCC (see Fig. 49.5) and the regional climate model results of *REMO* (see Fig. 49.7a) differ in the overall trends, primarily in winter. IPCC shows an increase in precipitation, while in the *REMO* simulation, precipitation decreases for the period under consideration. *REMO*, in contrast, shows a significant increase in precipitation in the

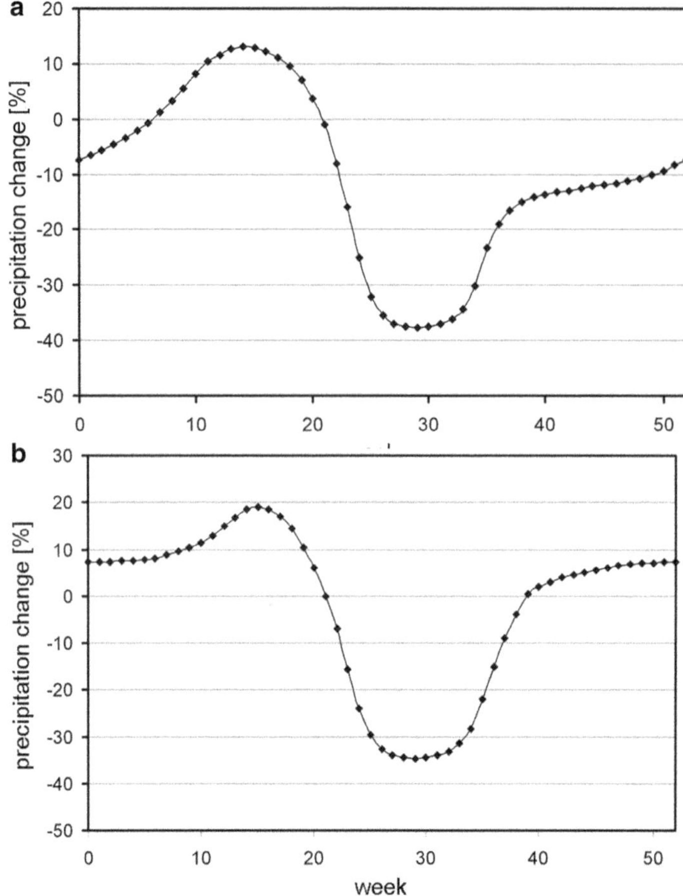

Fig. 49.7 Weekly course of percent change of precipitation in the Upper Danube basin based on seasonal results from (**a**) the REMO regional climate model after Jacob et al. (2008) and (**b**) the MM4 regional climate model 1990–2100

spring months. Overall, the seasonal changes in the results of the *REMO* regional climate model are considerably more marked than in the averaged results from the global climate model.

The same is generally true for the results of the regional climate model results of *MM5*, where the drying trend is more muted in *MM5* than for *REMO*. However, here too, precipitation increases most in spring up to 2100, though there is also an increase in winter, whereas *MM5* also shows a clear decrease in precipitation during the summer months.

For the fourth climate trend chosen, the *Forward Projection* of the temperature trend based on the measured historical temperature increases, an estimate of the weekly future changes in precipitation in the form of seasonal percent changes is not appropriate as a result of the low statistical significance of the historical precipi-

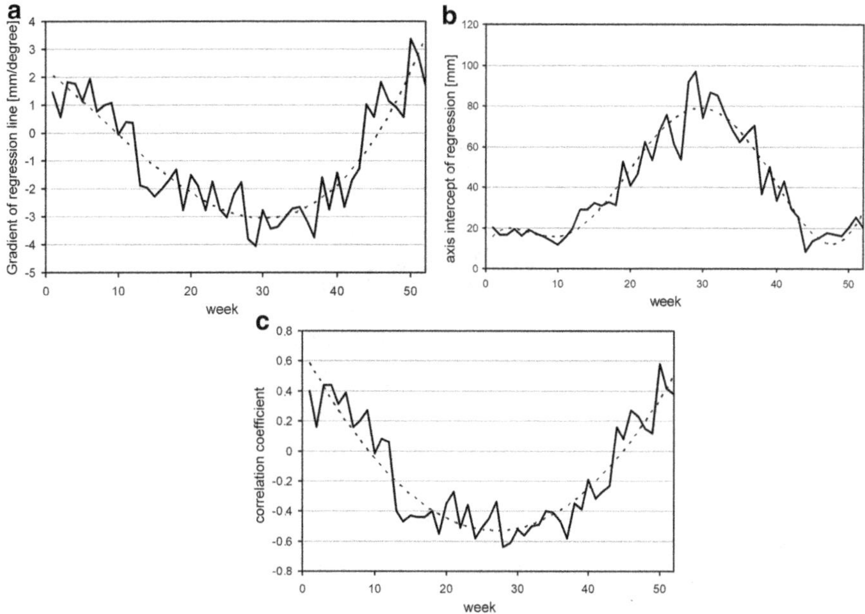

Fig. 49.8 Results of the linear weekly regression analysis of the relation between mean temperature and precipitation sum in the observed climatological data 1960–2006 ($n = 47$) in the Upper Danube basin; (**a**) gradient, (**b**) axis intercept and (**c**) correlation coefficient of the regression lines

tation trend that was indicated in Chap. 13. Therefore, another strategy must be chosen for the calculation of the possible future changes in precipitation.

This strategy consists of analysing the statistical relationship between the weekly mean temperature and the weekly total precipitation within the historical measurements. For this purpose a linear regression analysis is carried out for each week in the year for 47 years with measurements available. The changing relationship between the two variables over the course of the year is summarised in Fig. 49.8a–c. In Fig. 49.8a, the sensitivity of the mean total precipitation to changes in temperature is plotted as slope of the regression line for each week in the year. In Fig. 49.8b, the intercept of the regression line is shown, and Fig. 49.8c shows the weekly correlation coefficients of the regression lines. Figure 49.8c reveals significant positive correlation coefficients in winter and negative correlation coefficients in summer. Assuming that these relationships do not change in the future, this means that with increasing atmospheric temperatures, there will be increasing precipitation in winter and decreasing precipitation in summer. These results are consistent with those from the IPCC and, with limitation, also those from the UBA in Figs. 49.6 and 49.7. The weekly relationships shown can now be used to quantify the precipitation changes associated with the assumed future temperature increase.

Under the assumption that these statistical relationships can be applied to future meteorological conditions, the percent changes in weekly precipitation totals associated with the temperature increase of 5.2 °C determined from the *Forward*

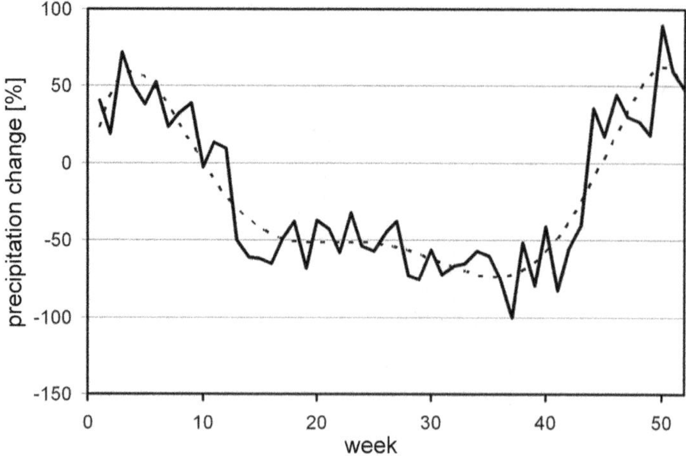

Fig. 49.9 Percent change of weekly precipitation sums 1990–2100 in the Extrapolation climate trend with an assumed temperature increase of 5.2 °C 1990–2100 working with the historical statistical relation between temperature and precipitation (see Fig. 49.8)

Projection climate trend for the period from 1990 to 2100 can now be calculated from these statistical analyses. These are presented in Fig. 49.9 and are consistent with the results in Fig. 49.5.

The values resulting from a seasonal summary of the percent changes in precipitation for the *Forward Projection* climate trend in Fig. 49.9 are given in Table 49.1.

The percent change trend in precipitation in the Upper Danube basin between 1990 and 2100 is shown in Fig. 49.9. It was calculated based on the historical statistical relationship between temperature and precipitation in the period from 1960 to 2006 and is comparable to the one from the *IPCC regional* climate trend, which was presented in Fig. 49.5; however, the trend of Fig. 49.9 shows significantly larger amplitude both in winter and in summer.

49.4.2 Validation of the Statistical Climate Generator

A prerequisite for using the statistical climate generator in the GLOWA-Danube project is the successful validation of the method. Once it has been ensured, as shown in Fig. 49.4, that the uncertainty in the selection of the most similar historical week for the meteorological scenario dataset does not accrue with time during the scenario, the suitability of the climate generator in reproducing the past climate and its statistical properties can be examined. To do this, four artificial meteorological drivers for the years 1960–2006 were generated with the climate generator following the method described above. These drivers differ from each other only in the selected starting value for the random number generator. The starting values were themselves randomly selected to avoid a systematic bias in the results. The four

Table 49.1 Seasonal change of precipitation in the Upper Danube basin based on the Extrapolation climate trend

Season	Precipitation change 1990–2100 (%)
Winter DJF	+47.0
Spring MAM	−41.8
Summer JJA	−68.8
Autumn SON	−2.1
Annual mean	−16.4

datasets are statistically equivalent to the measured dataset, which, for the case under consideration, means that they:

1. Are similar in value and seasonal trend in the most important mean climatologic parameters (temperature and precipitation) to the measured climate dataset (i.e. they belong to the same statistical population). The comparison of the measured and synthesised datasets considered the weekly mean temperature and its distribution, the mean weekly precipitation total and its distribution and the intensity of precipitation and the number of precipitation events per week. The use of the random number generator based on the means and covariance matrices calculated from the measured data ensures this assumption is met.
2. Provide values of the derived characteristic meteorological statistical parameters for the historical climate that are comparable to the measured values. In our case, the parameters selected were annual mean temperature, mean annual precipitation total, number of summer days and number of frost days.
3. Provide values of simulated hydrological parameter that are comparable to the measured hydrological parameters by driving DANUBIA with the artificial meteorological datasets. In this case, the values selected were mean annual discharge MQ, 1-year flood peak discharge HQ_1, 100-year flood peak discharge HQ_{100} and the mean duration curve at gauge Achleiten (outlet of the Upper Danube drainage basin, $A = 77{,}000\ km^2$) as well as Oberaudorf on the Inn (inflow of the Inn to Germany, $A = 9000\ km^2$).

In Fig. 49.10a, the course of weekly mean temperatures in the basin for 1960–2006 from the measured dataset is compared to the synthetically generated data. In general, the courses of all five datasets are consistent. Figure 49.10b shows the standard deviations in the weekly mean temperatures. Here too, there is good consistency between the measured and artificially generated data.

Figure 49.11 illustrates the corresponding courses of the mean weekly precipitation totals averaged across all stations. Again, there is overall consistency between the measured and synthetically generated courses, although the synthetically generated datasets show significant deviations from the measured mean for some weeks. The distribution of weekly precipitation totals about the mean shown in Fig. 49.11b is generally identical in all datasets and decreases substantially in summer with convective precipitation events.

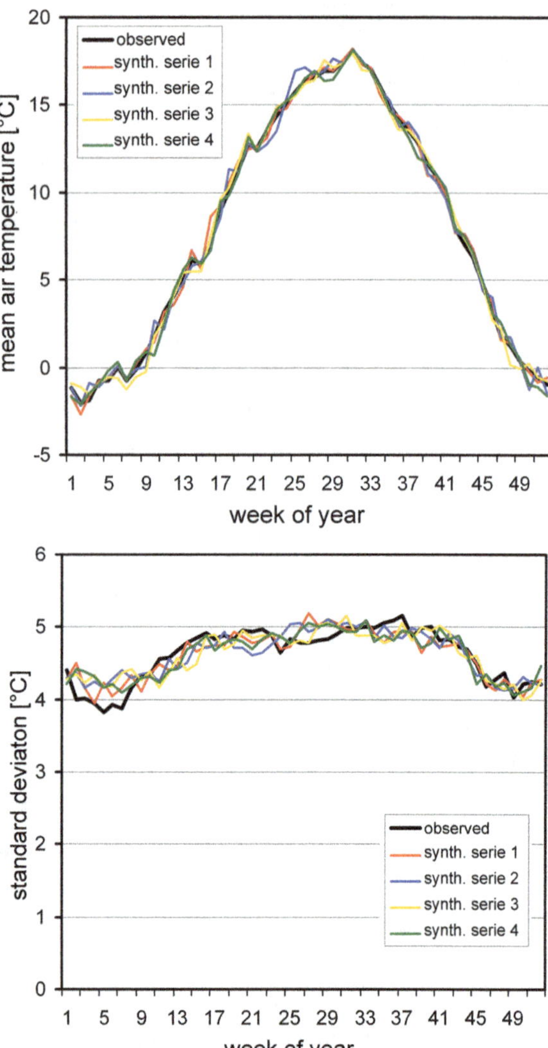

Fig. 49.10 Comparison of courses of averaged weekly mean temperatures in the study area 1960–2006 (*top*) and comparison of standard deviation of weekly mean temperatures 1960–2006 (*bottom*), each for one observed and four synthetic climatological data series

Additional criteria for the comparison between the measured and synthetically generated climate data were the mean intensity of precipitation (mm precipitation per measurement point) and the mean number of precipitation events per day. These parameters should indicate whether the average dynamics of the precipitation events have changed in the synthetic generation of data series. The results of the analysis, shown in Fig. 49.12, confirm the good consistency between the measured and synthetically generated climate data series.

The results of the analysis of the four climate parameters (mean temperature, annual precipitation, number of summer days and number of frost days) are summarised in Table 49.2, both for the measurements and for the four synthetically

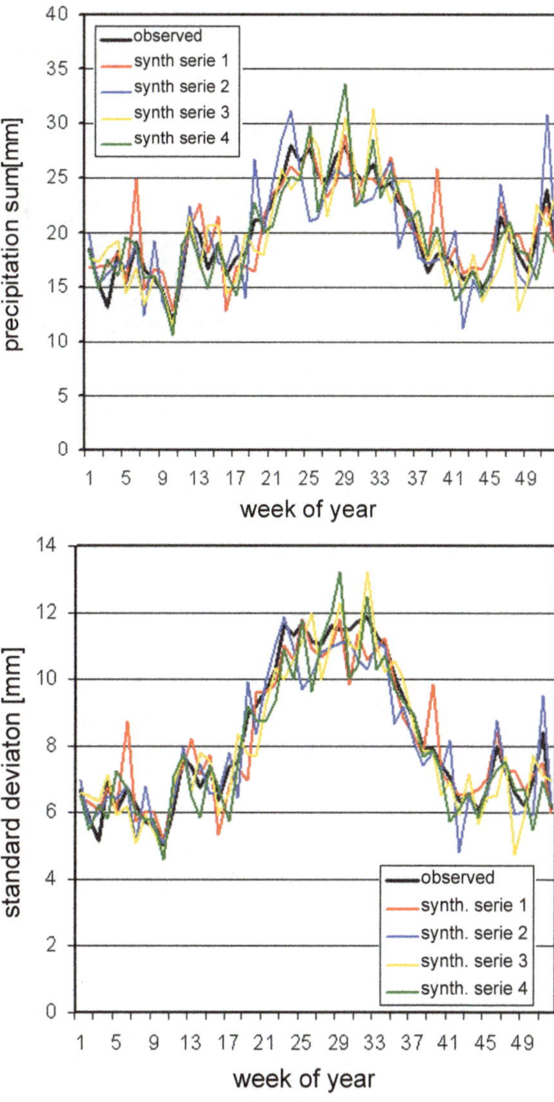

Fig. 49.11 Comparison of annual courses of averaged weekly mean precipitation sums in the study area 1960–2006 (*top*) and comparison of standard deviation of weekly mean precipitation sums 1960–2006 (*bottom*), each for one observed and four synthetic climatological data series

generated data series for the period 1960–2006. The values from all grid cells in the Upper Danube basin were averaged for the calculation of climate parameters. Table 49.2 indicates that the selected climate parameters for the measured and synthetically generated meteorological drivers are also generally consistent. As a second step in validating the statistical climate generator, the measured dataset and the four synthetically generated datasets were used as input data for DANUBIA to calculate the hourly discharge for the period 1960–2006 for each proxel in the Upper Danube basin and to aggregate these values to daily values. The discharge data, which were obtained using the measured dataset, correspond to those in Map 25.1,

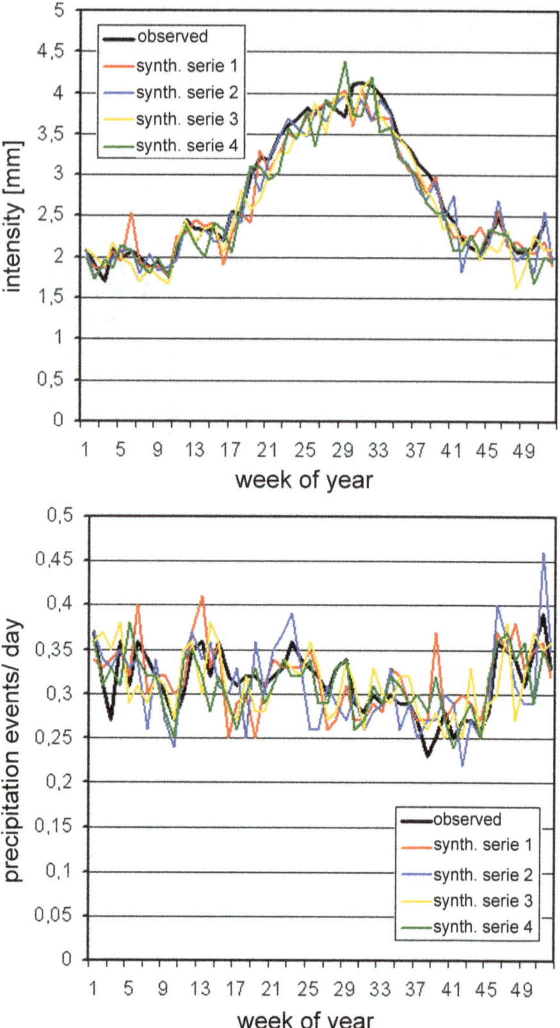

Fig. 49.12 Comparison of courses of weekly mean precipitation intensity in the study area 1960–2006 (*top*) and comparison of mean daily number of precipitation events 1960–2006 (*bottom*), each for one observed and four synthetic climatological data series

in which the validation of the hydrological part of DANUBIA is presented. The variables presented in Table 49.3 were determined for the discharge time series of the five simulations.

Although the mean discharges that result from the measured and synthetically generated drivers yield similar values, a significant overestimate of HQ_{100} in column 3 is immediately noticeable; this value was calculated based on measured climate data. The overestimate of HQ_{100} in the model is largely attributable to the fact that the model does not include human interventions on the flood development through targeted control of the available storage reservoir and the overflowing of the rivers banks.

Table 49.2 Comparison of climate parameters of observed data with the four synthetic time series 1960–2006

Time series 1960–2006	Observed values	Synthetic series 1	Synthetic series 2	Synthetic series 3	Synthetic series 4	Mean value column 3–6
Mean temperature (°C)	7.95	7.97	7.92	8.14	8.00	8.00
Annual precipitation (mm)	1,039.9	1,034.6	1,031.1	1,060.0	1,037.2	1,040.7
Number of summer days	19.5	19.7	20.2	19.3	19.6	19.7
Number of frost days	130.6	129.2	131.5	133.2	131.7	131.4

Table 49.3 Comparison of hydrological parameters (mean discharge MQ, 1-year HQ1 and 100-year flood discharge HQ100 and 100-year mean 7-day low flow discharge NMQ7) of the four synthetic time series, the observed discharges at gauge Achleiten and the modelled discharges (with observed climatological data) in the time period 1960–2006

Time series 1960–2006	From gauge data 1960–2006	Modelled with observed climatological data	Modelled with synthetic series 1	Modelled with synthetic series 2	Modelled with synthetic series 3	Modelled with synthetic series 4	Mean value column 4–7
MQ	1,450.0	1,498.1	1,463.7	1,456.5	1,560.1	1,454.0	1,483.6
HQ_1	3,827	4,206.4	3,841.6	3,900.3	3,928.3	3,506.1	3,794.1
HQ_{100}	6,755	8,187.4	6,662.6	9,987.4	6,045.5	6,828.5	7,381.0
$NM7Q_{100}$	476	418.9	350.7	345.6	291.4	296.7	321.7

Both factors led to a reduction in the measured flood peaks for the highest flooding events that occurred, especially between 1997 and 2005. In addition, there is wide variation in the HQ_{100} values among the results from the four synthetically generated times series that were used. This indicates that HQ_{100} is very sensitive to extreme precipitation events that occurred in the period under consideration (1960–2006). Also notable is the slight underestimate of the $NM7Q_{100}$ in the model results from the measured data as well as moderate to gross underestimate of the $NM7Q_{100}$ in the results from the synthetically generated times series. The slight underestimate in column 3 can be attributed to the fact that the measures in place over the past years for raising low water levels by using the storage in the drainage basin are not reproduced by the model. The cause for the moderate to gross underestimate of the extreme low water in the results from the synthetically generated times series currently cannot be explained.

As a third step, the duration curves generated with the measured and artificial data series for gauge Achleiten at the outlet of the study area and for gauge Oberaudorf at the outlet of the Inn valley in the Alpine foothills were compared. The

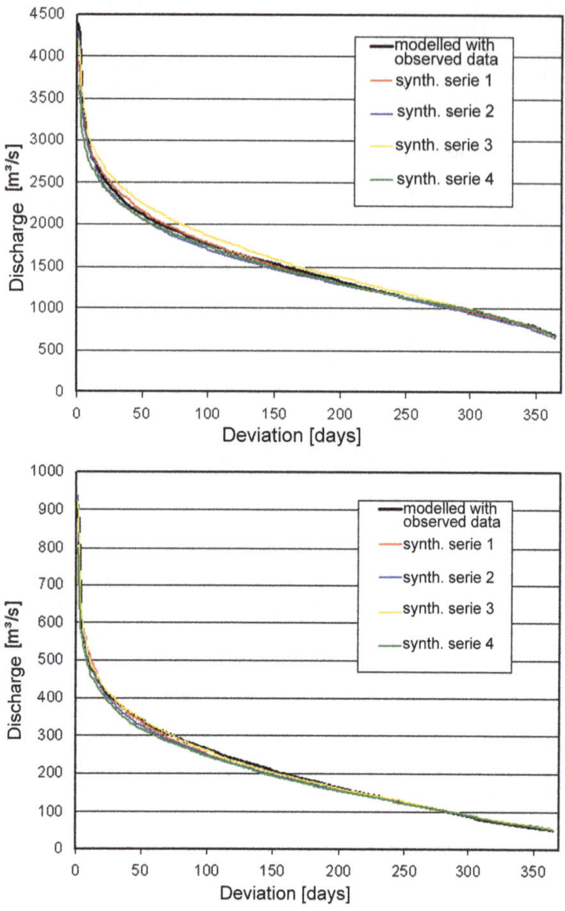

Fig. 49.13 Duration curves for the time period 1960–2006 for the modelled discharges with climatological data and for the four synthetic data series, gauge Achleiten (Danube) (*top*), gauge Oberaudorf (Inn) (*bottom*)

duration curves yield information about the number of days on which a specified discharge is over- or undershot and thus represents the dynamics of the discharge in the study area. The curves form an important basis for water management and especially for the use of hydroelectric power. Figure 49.13a and b shows the trend in the mean duration curve at gauge Achleiten near Passau and at Oberaudorf for the years 1960–2006. Overall, the results indicated good to very good consistency of the shapes of the duration curves modelled using the measured data series (black lines) compared to the duration curves modelled using the synthetically generated data series (coloured lines). However, it is evident in Fig. 49.13a that the synthetically generated duration curves are at times above the curves modelled from the measured data; this result can be attributed to the overall slight overestimate of precipitation by the statistical climate generator.

49.5 Conclusion

A statistical climate generator was developed for the generation of artificial meteorological drivers that can portray possible regional future trends in climate and at the same time have the temporal and spatial resolution that is required for regional studies. This climate generator bridges the gap in scale between coarse statements about further developments of global climate and the high-resolution, physically consistent meteorological inputs that are required for regional climate impact studies. The results from the climate generator are comparable to those from the regional climate models, which have similar objectives.

It was demonstrated that the statistical climate generator in the historical period from 1960 to 2006 accurately reproduces the prevailing climate in this period in the study area and the resulting responses of the Upper Danube drainage basin. The statistical climate generator is thus a flexible tool for preparing the meteorological drivers for the scenario simulations based on the climate trends developed within the project. As was demonstrated in the procedure for validating the selected approach to realise the climate generator, it is feasible to generate a variety of statistically equivalent datasets of meteorological drivers simply by changing the starting values of the random number generator that was used. The fact that it is also possible to specify their probability of occurrence is also of great advantage. This specific feature is used in Chap. 50, which introduces the GLOWA-Danube climate variants.

References

Buishand TA, Brandsma T (2001) Multisite simulation of daily precipitation and temperature in the Rhine Basin by nearest–neighbour resampling. Water Resour Res 37(11):2761–2776

IPCC (2007) Climate change 2007: the physical science basis. Contribution of Working Group I to the Fourth Assessment Report of the Intergovernmental Panel on Climate Change (Solomon S, Qin D, Manning M, Chen Z, Marquis M, Averyt KB, Tignor M, Miller HL, eds). Cambridge University Press, Cambridge/New York, 996 pp

Jacob D, Göttel H, Kotlarski S, Lorenz P, Sieck K (2008) Klimaauswirkungen und Anpassung in Deutschland Phase 1: Erstellung regionaler Klimaszenarien für Deutschland. In: Umweltbundesamt (ed) (2008) Climate Change 11/08

Mahalanobis PC (1936) On the generalised distance in statistics. Proc Natl Inst Sci India 12:49–55

Orlowsky B, Gerstengarbe FW, Werner PC (2007) A resampling scheme for regional climate simulations and its performance compared to a dynamical RCM. Theor Appl Climatol 92(3–4):209–223

Spekat A, Enke W, Kreienkamp F (2006) Neuentwicklung von regional hoch aufgelösten Wetterlagen für Deutschland und Bereitstellung regionaler Klimaszenarien mit dem Regionalisierungsmodell WETTREG 2005 auf der Basis von globalen Klimasimulationen mit ECHAM5/MPI – OM T63L31 2010 bis 2100 für die SRES – Szenarien B1, A1B und A2. Projektbericht im Rahmen des F+EVorhabens 204 41 138 „Klimaauswirkungen und Anpassung in Deutschland – Phase 1: Erstellung regionaler Klimaszenarien für Deutschland", p 94

Visual Numerics (2006) IMSL FORTRAN Numerical Library User's Guide Version 6. Visual Numerics Ltd, Houston

Yates D, Gangopadhyay S, Rajagopalan B, Strzepek K (2003) A technique for generating regional climate scenarios using a nearest neighbour algorithm. Water Resour Res 39(7):1–15

Young KC (1994) A multivariate chain model for simulating climatic parameters from daily data. J Appl Meteorol 33:661–671

Chapter 50
The GLOWA-Danube Climate Variants from the Statistical Climate Generator

Wolfram Mauser

Abstract GLOWA-Danube, in order to study the hydrological impact of climate change in the Upper Danube basin, relies on a variety of meteorological driver datasets, which represent uncertainty in climate change. This chapter demonstrates the application of the statistical climate generator (Chap. 49) to produce climate variants from the given climate trends (Chap. 48). Climate variants in GLOWA-Danube are statistically equivalent realisations of a climate trend with defined properties and probabilities. Stakeholder discussion resulted in the selection of four climate variants for each of the four climate trends. They represent average warming, five consecutive dry years, warm winters or hot summers. The variants are analysed for their individual behaviour and usefulness for analysing adaptation options.

Keywords GLOWA-Danube • Climate variants • Climate generator • Climate change

In GLOWA-Danube, a meteorological driver for DANUBIA that follows a specified climate trend is termed a climate variant. The meteorological driver consists of a time series of arrays of the meteorological parameters precipitation, incident radiation (short- and long-wave), atmospheric temperature, atmospheric humidity and wind speed. In general, these parameter fields are expected by DANUBIA with a spatial resolution of 1×1 km and at a temporal resolution of 1 h. There are two options for the generation of different climate variants:

1. The statistical climate generator described in Chap. 49
2. The results of the regional climate models after bias correction and downscaling, described in Chap. 51

The following chapter primarily deals with option 1, since at present the option for generating the different climate variants from the regional climate models is still

W. Mauser (✉)
Department of Geography, Ludwig-Maximilians-Universität München (LMU Munich), Munich, Germany
e-mail: w.mauser@lmu.de

W. Mauser, M. Prasch (eds.), *Regional Assessment of Global Change Impacts*, DOI 10.1007/978-3-319-16751-0_50

limited by the high computational expense. Only one climate variant from each of *REMO* and *MM5* is available for the project.

The statistical climate generator provides a method, which can be used to generate synthetic meteorological data series based on historical time series of meteorological measurements. The course of the weather events in a future synthetically generated driver follows a given regional climate trend and also in general depends on the random numbers generated by the random number generator within the statistical climate generator. The random number generator of the climate generator produces a preset quantity of random numbers based on a specific initialisation number supplied at the beginning of the software program, also known as a "seed". These random numbers follow a 2-dimensional normal distribution with prescribed means and covariances. Each individual value of the seed produces a reproducible and unique series of random numbers whenever the program is executed. Different seed values generate series of random numbers, but these numbers always represent a sample from the same basic population and are thus statistically equivalent with respect to a prescribed distribution, in our case the normal distribution. Thus, in this way the selection of different seed values generates a range of different statistical series that are all statistically equivalent (i.e. belong to the same basic population). Accordingly, different series of meteorological data can be generated by the climate generator with these different series of random numbers. With the same choice of climate trend, they too are thus statistically equivalent and belong to the same basic population. Each statistically equivalent series of meteorological data from the climate generator is therefore a climate variant of a specified climate trend.

The statistical climate generator provides the opportunity to easily generate a huge number of different climate variants for each specified climate trend. This opportunity to consider the statistical distribution in the weather patterns of specific future climate trends is used in the project primarily for the following three aspects of climate impact research:

1. Identification of a mean future weather trend from a large number of climate variants based on a specific criterion, respectively, climate trend. This addresses the problem that to date no statements about the probabilities of occurrence can be made for the results of the *REMO* and *MM5* regional climate models, since there is only one realisation based on each run of ECHAM. Hence, it is also not possible to make statements about the probabilities at which the results of the climate impact modelling might occur. However, in order to be able to make practical and relevant statements, it is necessary to infer the probability of occurrence at which the given climate variant can occur.
2. Studies to estimate the uncertainty in the statements about the changes in hydrological variables as the result of scenario simulations based on the statistical variability of the meteorological driver for the specified climate trend. These estimates involve an ensemble of climate variants, i.e. statistically equivalent future meteorological data series used as inputs in DANUBIA for each assumed climate trend, and the distribution of the simulated hydrological parameters, such as discharge, evaporation, snow cover etc., is examined.

3. Studies on the effects of extreme situations that arise in the synthetically generated meteorological data series at an identifiable probability of occurrence. The following interesting extreme situations were formulated in the context of GLOWA-Danube by the project partners and stakeholders:
   - Low precipitation totals over longer periods of several years
   - High temperatures in different seasons over a longer period of several consecutive years

In an open-discussion process with the subproject partners and the stakeholders, several selection criteria for future climate variants have thus been worked out and implemented within the statistical climate generator for the GLOWA-Danube scenario period from 2011 to 2060.

The climate variants listed in Table 50.1 and the criteria underlying each were defined on the basis of the considerations noted in Points 1–3 above. This applies

Table 50.1 GLOWA-Danube climate variants, implemented and agreed upon within the subprojects and in the stakeholder process

Climate variant	Description	Criteria
1	Baseline	Mean annual temperature between 2011 and 2035
2	Long period of warm winters between 2011 and 2035	5-year temperature maximum DJF between 2011 and 2035
3	Long period of hot summers between 2011 and 2035	5-year temperature maximum JJA between 2011 and 2035
4	Long dry period between 2011 and 2035	5-year precipitation minimum between 2011 and 2035
5	Long dry period between 2036 and 2060	5-year precipitation minimum between 2036 and 2060
6	Medium dry period between 2011 and 2035	3-year precipitation minimum between 2011 and 2035
7	Medium dry period between 2036 and 2060	3-year precipitation minimum between 2036 and 2060
8	One dry year between 2011 and 2035	Driest year between 2011 and 2035
9	One dry year between 2036 and 2060	Driest year between 2036 and 2060
10	Long period of hot summers between 2036 and 2060	5-year temperature maximum JJA between 2036 and 2060
11	Medium period of hot summers between 2011 and 2035	3-year temperature maximum JJA between 2011 and 2035
12	Medium period of hot summers between 2036 and 2060	3-year temperature maximum JJA between 2036 and 2060
13	One hot summer between 2011 and 2035	Temperature maximum JJA between 2011 and 2035
14	One hot summer between 2036 and 2060	Temperature maximum JJA between 2036 and 2060

(continued)

Table 50.1 (continued)

Climate variant	Description	Criteria
15	Medium precipitation sums between 2011 and 2035	Mean annual precipitation sum between 2011 and 2035
16	Medium precipitation sums between 2036 and 2060	Mean annual precipitation sum between 2036 and 2060
17	Baseline II	Mean annual temperature between 2036 and 2060
18	Long period of warm winters between 2036 and 2060	5-year temperature maximum DJF between 2036 and 2060
19	Long period of dry summers between 2011 and 2035	5-year precipitation minimum JJA between 2011 and 2035
20	Long period of dry summers between 2036 and 2060	5-year precipitation minimum JJA between 2036 and 2060

DJF December, January, February; *JJA* June, July, August

especially to climate variants 1, 8–9 and 13–17. These involve annual extremes and means as the key aspects. Climate variants 2–7, 10–12 and 18–20 were defined largely on the basis of the considerations of the actor subprojects for the purpose of investigating the adjustment processes undertaken by the actors in response to longer-lasting extreme conditions. Hence, in general, these actors are not responding to a single extreme year.

50.1 Generating the Climate Variants

The methodological approach to generate a climate variant that complies with one of the 20 criteria listed in Table 50.1 or other additional criteria is schematically illustrated in Fig. 50.1. A climate variant is created by generating and analysing a collection of 5,000 randomly distributed, statistically equivalent meteorological datasets for the next 50 years, each in a three-step cycle. Each of the cycles consists of a double random process. First, a starting value for the random number generator of the climate generator is selected for each of the 5,000 climate variants, using a random number generator with uniform distributions. The uniform distribution of this random process ensures that all random number series of the climate generator have the same a priori probability.

A climate variant is calculated using the statistical climate generator with the randomly chosen starting value. The procedure for calculating the climate variant is presented in detail in Chap. 49. The meteorological data that are produced are then analysed based on the criteria selected from Table 50.1 for the chosen climate variant.

This process is explained below using the example of climate variant 4, "long-lasting dry periods between 2011 and 2035". To calculate a meteorological scenario data set that exhibits a long-lasting dry period of 5 years between 2011 and 2035, a

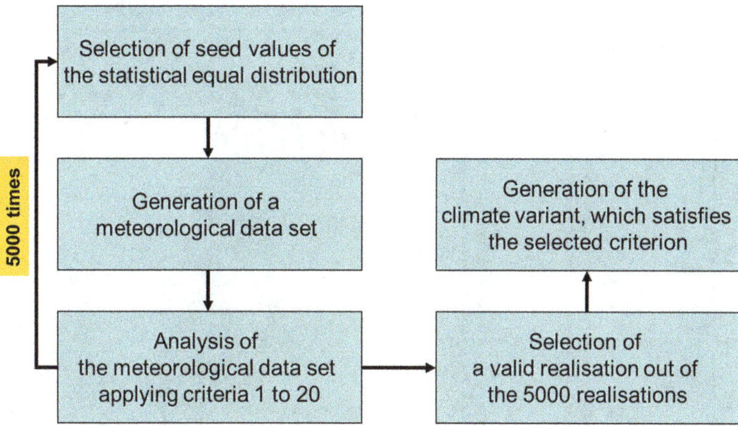

Fig. 50.1 Scheme of generation of climate variants with the statistical climate generator

5-year window for this period was slid over the synthetically generated precipitation datasets and for each 5-year period, the mean precipitation total for all meteorological stations in the dataset was calculated. Then the smallest calculated precipitation total over five successive years was saved together with the respective seed value for the random number generator of the statistical climate generator. This process was repeated 5,000 times for each climate trend. The frequency distribution for the 5,000 minimum precipitation totals over five successive years that were generated is shown in Fig. 50.2 for the *IPCC regional*, *REMO regional*, *MM5 regional* and *Forward Projection* climate trends.

A normal distribution was then fit to the frequency distribution for the precipitation totals. It is not surprising that this normal distribution was always confirmed as highly statistically significant (above the 99 % significance level) in a Student's *t*-test, since the random number generator on which the climate generator is based assumes that the variability in precipitation and temperature is normally distributed. The lower 5th percentile of the normal distribution was chosen to select a specific realisation from among the 5,000 realisations. This criterion specifies the precipitation total that is undershot in 5 % of the calculations out of the 5,000 realisations (i.e. every 100th time). The choice of the 5th percentile was decided through a process internal to GLOWA-Danube, in which a generalised approach towards defining acceptable and communicable risks in scenarios of the future development was discussed. This means that the identified minimum precipitation value occurs once in every 20th sample. This consequently represents an event that occurs at a frequency that is uncommon but not extreme. It is quite possible to change this limit and hence the extreme nature of the climate variants in the generation of other climate variants.

Depending on the climate trend being considered, the 5th percentile for climate variant 4 "long-lasting dry periods between 2011 and 2035" is 4,632 mm average

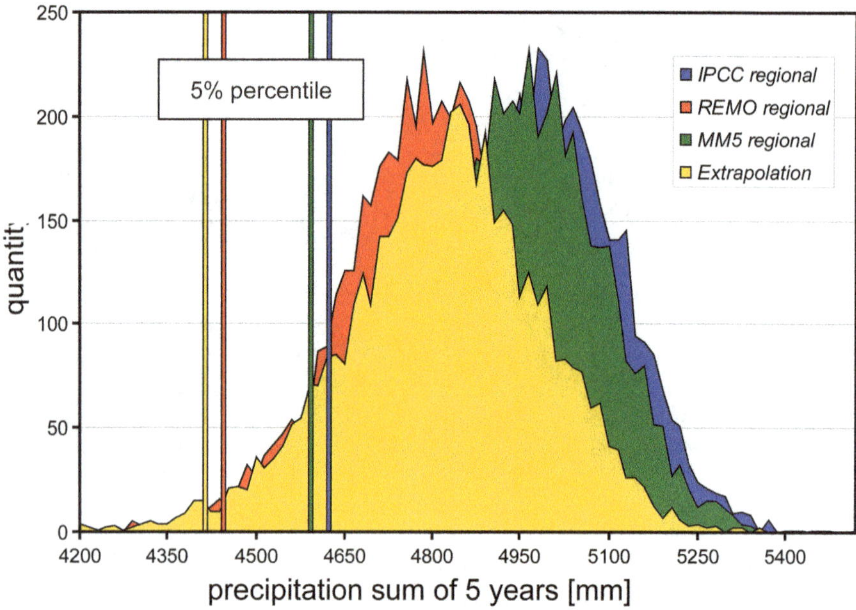

Fig. 50.2 Comparison of frequency distribution of 5,000 calculated realisations for the generation of climate variant 4 for the four climate trends *IPCC regional*, *REMO regional*, *MM5 regional* and *Extrapolation*

precipitation in the Upper Danube basin in 5 years, or 926 mm per year (*IPCC regional* climate trend), 4,450 mm in 5 years or 890 mm per year (*REMO regional* climate trend), 4,605 mm in 5 years or 921 mm per year (*MM5 regional* climate trend) and only 4,426 mm precipitation in 5 years or 885 mm per year for the *Forward Projection* climate trend. This can be compared to a mean annual measured precipitation of 1,040 mm/a with a standard deviation of 130 mm/a for the period from 1960 to 2006.

Figure 50.3 shows the development in annual precipitation totals in the selected climate variants in the period from 2011 to 2060 for the four climate trends. The periods that are particularly dry are highlighted with coloured time windows. The four different 5-year dry periods with their respective starts in the years 2013 (*IPCC regional*), 2021 (*REMO regional*), 2029 (*MM5 regional*) and 2031 (*Extrapolation*) are easy to detect, even without the colouring.

The described method was also used for generating other climate variants. Using the 5th percentile, the relevant climate variant was selected from the frequency distributions of each of the 5,000 realisations of the different climate variants (as in Fig. 50.2).

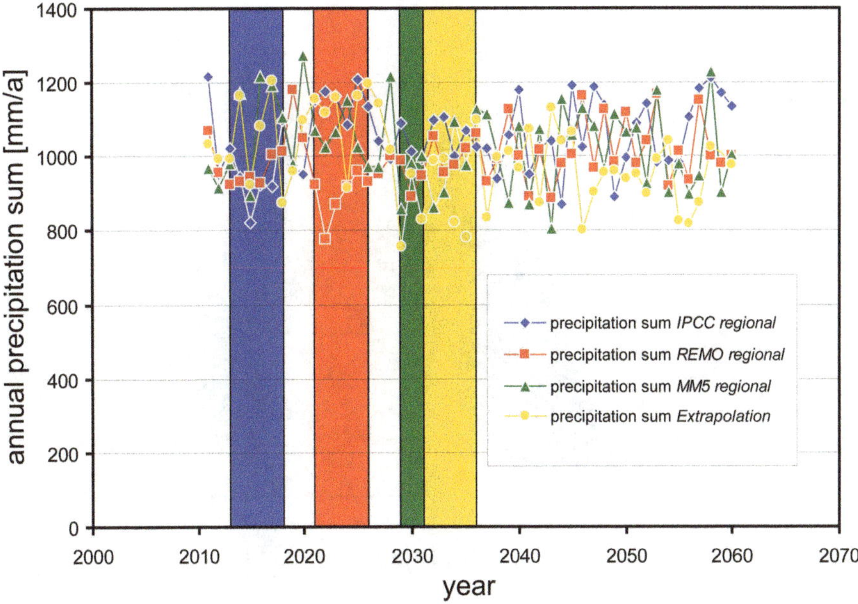

Fig. 50.3 Course of annual distribution of precipitation sums of the selected realisations of climate variant 4 for the four climate trends *IPCC regional*, *REMO regional*, *MM5 regional* and *Extrapolation*

50.2 Climate Variants in GLOWA-Danube

Following intensive discussions with the actor subproject partners involved, four climate variants were chosen with the highest priority from the list of 20 climate variants in Table 50.1. They are used in DANUBIA, under consideration of the four chosen climate trends (*IPCC regional*, *REMO regional*, *MM5 regional* and *Forward Projection* (see Chap. 48)), to study the effects of different climate variants on the water supply in the Upper Danube. These four climate variants for each climate trend are:

Climate variant 1 Average temperature increase (abbreviated: *Baseline*). This climate variant represents the average temperature increase from 2011 to 2035 from 5,000 realisations and therefore establishes the opportunity to examine a very probable weather course based on the selected climate trend and the assumptions made in the statistical climate generator. For climate variant 1, the frequency distributions for mean annual temperatures between 2011 and 2035 of the 5,000 realisations for the four different GLOWA-Danube climate trends are compared in Fig. 50.4. The figure immediately reveals the effect of the climate trend chosen on the mean annual temperatures from 2011 to 2035. They are 9.4 °C for the *IPCC regional* climate trend, 9.6 °C for the *REMO regional* climate trend, 9.5 °C for the *MM5 regional*

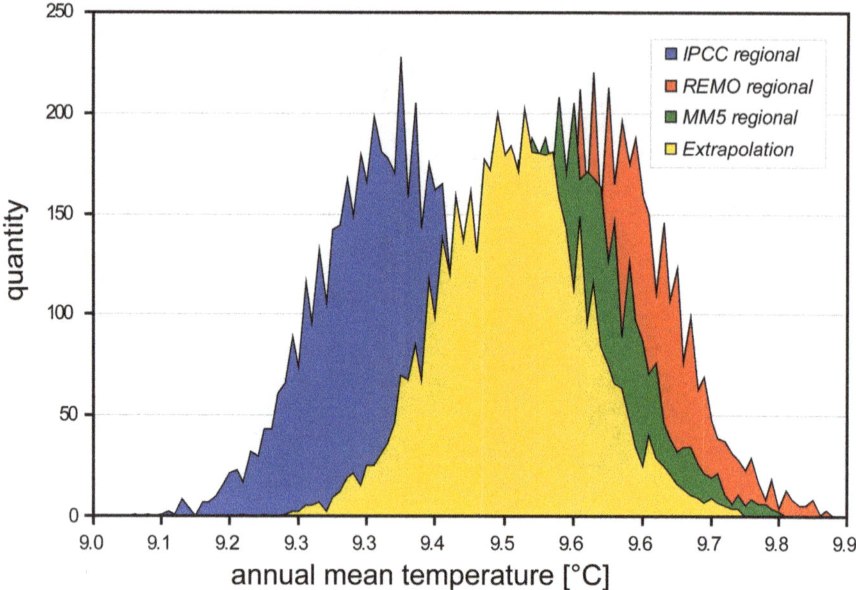

Fig. 50.4 Comparison of frequency distribution of 5,000 calculated realisations for the generation of climate variant 1 for the four climate trends *IPCC regional*, *REMO regional*, *MM5 regional* and *Extrapolation*

climate trend and 9.5 °C for the *Forward Projection* climate trend. By way of comparison, the mean annual temperature from 1960 to 2006 was 6.2 °C.

Figure 50.5 shows the trend in mean annual temperatures in the Upper Danube basin for the period 2011–2060 for the four chosen realisations that represent the mean of the histogram for each of the climate trends (*Baseline*). In addition to the overall increase in temperature from 2011 to 2060, Fig. 50.5 also reveals the different slopes and courses in the increase for the four climate variants that were selected.

Maps 50.1(1a–1d) show the mean temperature for climate variant 1 over the period from 2011 to 2035 in the four different climate trends. All four maps present similar values, but these are all significantly higher than the temperatures of the past (see Chap. 11).

Climate variant 2 Long-lasting periods of warm winters between 2011 and 2035 (abbreviated: *5 warm winters*). This climate variant establishes the opportunity to study the effects of successive warm winters that occur at a probability of 5 %, for example, on the tourism infrastructure or the supply of energy. For climate variant 2, the frequency distributions of the maximum 5-year mean temperatures in the winter months (DJF: December, January and February) in the period from 2011 to 2035 of the 5,000 realisations for the four different climate trends are compared in Fig. 50.6 along with the 5th percentile resulting from the adjusted normal distributions. Figure 50.6 immediately reveals the effect of the climate trend that is selected on the

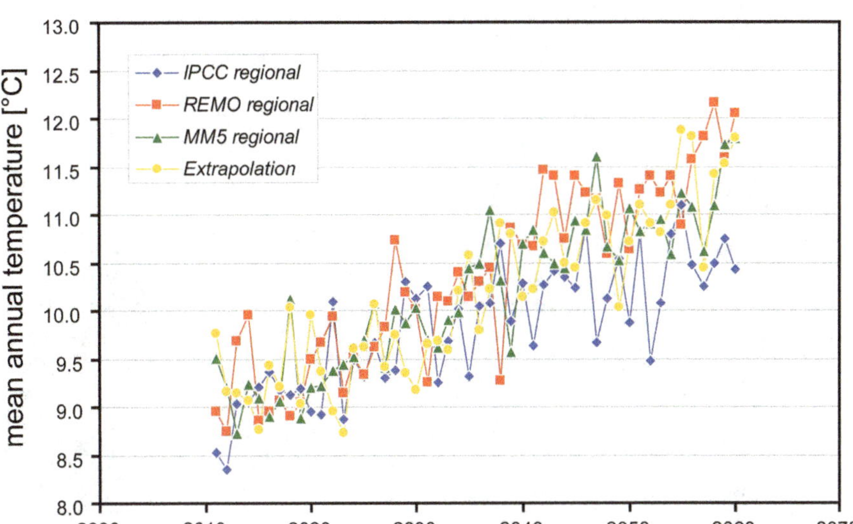

Fig. 50.5 Course of annual mean temperature of the selected realisations of climate variant 1 for the four climate trends *IPCC regional*, *REMO regional*, *MM5 regional* and *Extrapolation*

extremes of the maximum 5-year-DJF mean temperatures from the period 2011–2035. In Fig. 50.7, the trend in the mean winter temperatures in the Upper Danube basin is shown for the four selected realisations that comply with the criterion for the four climate trends selected at the 5th percentile. The periods of the warmest winters are shown in colour.

Maps 50.1(2a–2d) illustrate the mean winter temperatures from the 5 years in which the highest temperatures occurred. If these especially warm winters are compared to the mean historical winter temperature (see Map 32.1(4) in Chap. 32), there is a distinct warming, particularly for the *REMO regional* and *Forward Projection* climate trends.

Climate variant 3 Long-lasting period of persistently hot summers between 2011 and 2035 (abbreviated: *5 hot summers*). This climate variant establishes the opportunity to study the effects of successive hot summers that occur at a probability of 5 %, for example, on the generation of power, developments in glaciers and the risk of forest fires as well as for agriculture and the water supply.

For climate variant 3, the frequency distributions of the maximum 5-year mean temperatures in the summer months (JJA: June, July and August) in the period from 2011 to 2035 of the 5,000 realisations for the four different climate trends are compared in Fig. 50.8 along with the 5th percentile resulting from the adjusted normal distributions. Here too, it is immediately apparent how the selected climate trend affects the extremes in the maximum 5-year-JJA mean temperatures for the period 2011–2035.

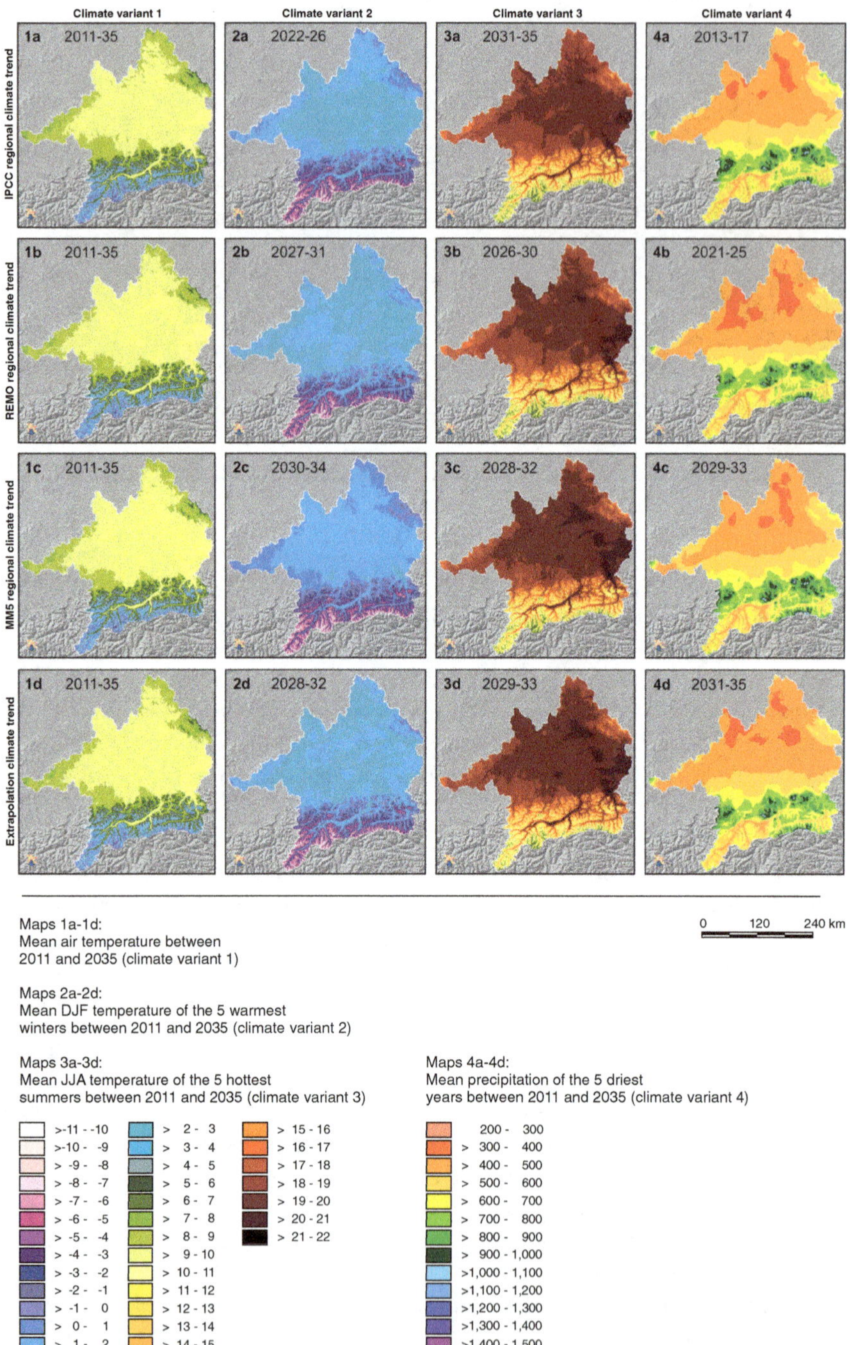

Map 50.1 The GLOWA-Danube climate variants of the statistical climate generator

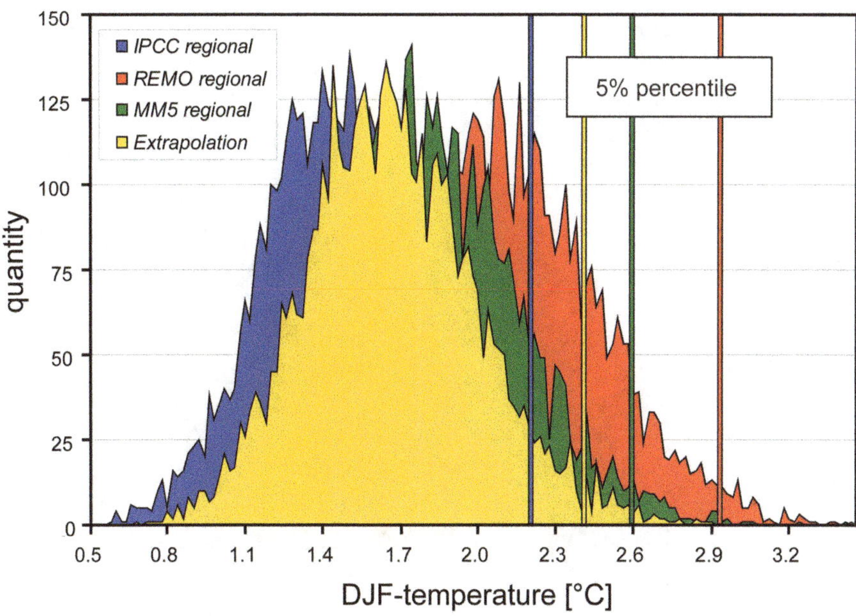

Fig. 50.6 Comparison of frequency distribution of 5,000 calculated realisations for the generation of climate variant 2 for the four climate trends *IPCC regional*, *REMO regional*, *MM5 regional* and *Extrapolation*

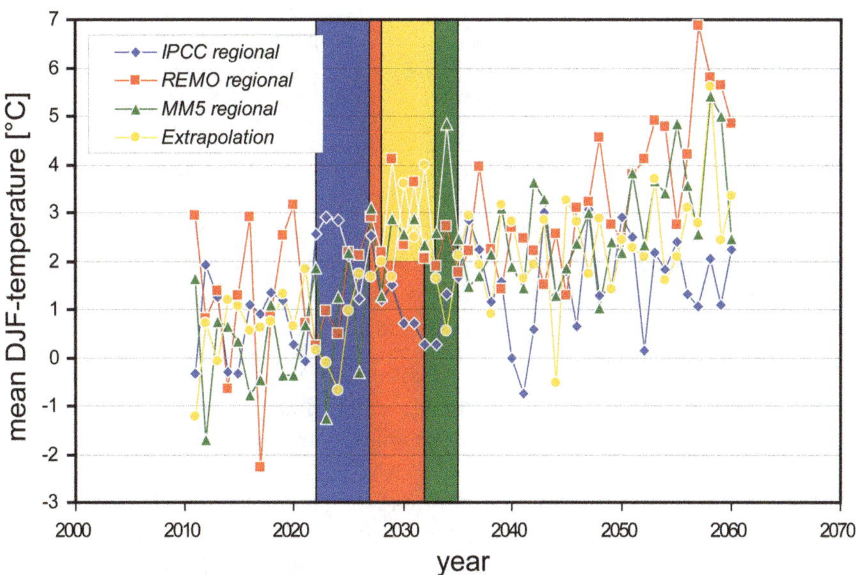

Fig. 50.7 Course of annual mean winter temperature (DJF) of the selected realisations of climate variant 2 for the four climate trends *IPCC regional*, *REMO regional*, *MM5 regional* and *Extrapolation*

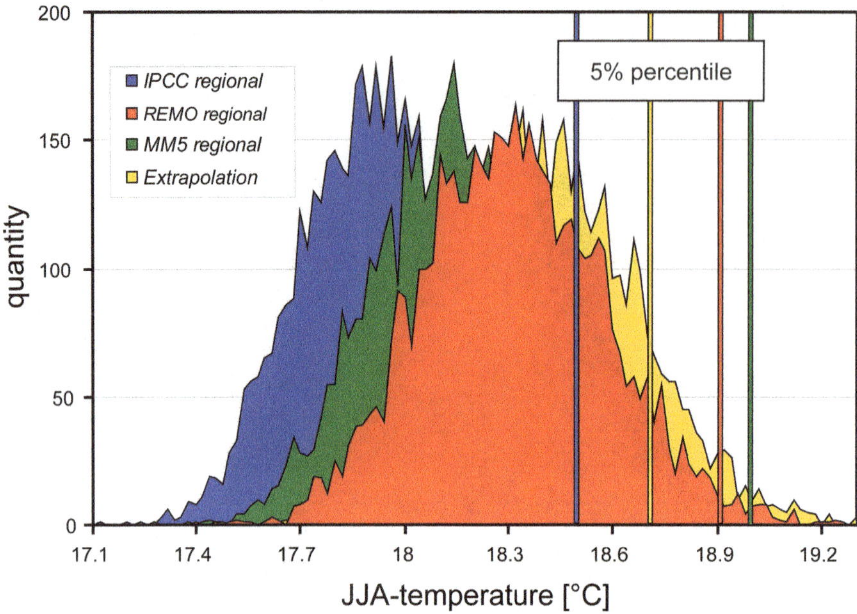

Fig. 50.8 Comparison of frequency distribution of 5,000 calculated realisations for the generation of climate variant 3 for the four climate trends *IPCC regional, REMO regional, MM5 regional* and *Extrapolation*

In Fig. 50.9, the trend in the mean summer temperatures in the Upper Danube basin is shown for the four selected realisations that comply with the criterion for the four climate trends selected at the 5th percentile. The periods of the hot summers are quite close together, with their respective starts in the years 2031 (*IPCC regional* climate trend), 2026 (*REMO regional* climate trend), 2028 (*MM5 regional* climate trend) and 2029 (*Forward Projection* climate trend).

The mean temperatures of the 5 warmest summers between 2011 and 2035 for the different climate trends are presented in Map 50.1(3a–3d). They differ only marginally from each other, but reveal a noticeable warming compared to the historical mean summer temperatures (see Map 32.1(2) in Chap. 32).

Climate variant 4 Long-lasting persistent dry period between 2011 and 2035. This climate variant establishes the opportunity to study the effects of dry periods that occur at a probability of 5 %, for example, on the generation of power, water stress in agriculture and water supply.

The comparison of the frequency distributions of the 5,000 realisations for preparing climate variant 4 for the four different climate trends and the trend in annual precipitation totals of the selected realisations from climate variant 4 are shown in Figs. 50.2 and 50.3. Map 50.1(4a–d) each show the mean annual precipitations from the 5 years in which the minimum precipitations occurred. If these especially dry

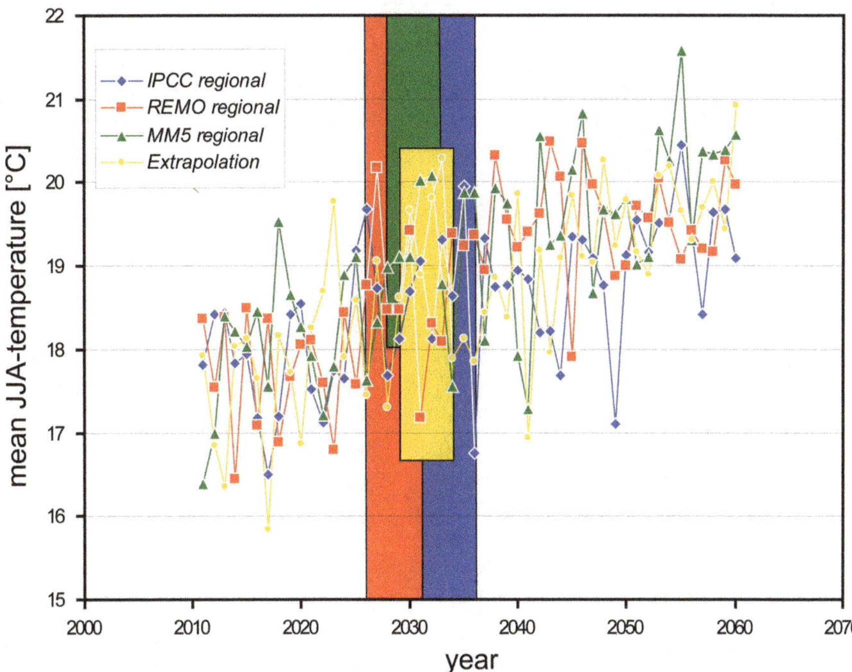

Fig. 50.9 Course of annual mean summer temperature (JJA) of the selected realisations of climate variant 3 for the four climate trends *IPCC regional, REMO regional, MM5 regional* and *Extrapolation*

years are compared with the mean historical precipitation (see Map 11.1(4) in Chap. 11), a significant reduction in precipitation over the entire drainage basin is apparent.

Table 50.2 summarises the results of the calculation of the four climate variants that were selected from Table 50.1 with highest priority. It presents the values for the 5th percentile, the time periods in which the criterion from Table 50.1 arose which defined the different climate variants and the historical means for comparison.

The table reveals that, in the selection of the climate variants that represent temporally limited, extreme conditions and which can occur at a probability of 5 %, there are some notable changes that can be expected within the next 25 years compared to the historical mean. This is especially relevant for the "5 dry years" climate variant, in which, in a quite usual scenario, the mean precipitation of the 5-year period under consideration is reduced by up to 20 % compared to the long-term mean. This, for example, transforms water availability for agriculture towards a situation, which is commonly found in Northern Italy.

Table 50.2 Values of 5 % percentiles in the different climate variants and climate trends, and time periods, which fulfil the criteria leading to the climate variants

	Climate variant 1: Baseline (mean temperature 2011–2035)	Climate variant 2: 5 warm winters	Climate variant 3: 5 hot summers	Climate variant 4: 5 dry years (minimum precipitation in 5 years/year)
Climate trend 1: *IPCC regional*	9.4 °C	2.2 °C	18.5 °C	4,632 mm/926 mm
	Period: –	Period: 2022–2026	Period: 2031–2035	Period: 2013–2017
Climate trend 2: *REMO regional*	9.6 °C	2.9 °C	18.9 °C	4,450 mm/890 mm
	Period: –	Period: 2027–2031	Period: 2026–2030	Period: 2021–2025
Climate trend 3: *MM5 regional*	9.5 °C	2.6 °C	19.0 °C	4,605 mm/921 mm
	Period: –	Period: 2030–2034	Period: 2028–2032	Period: 2029–2033
Climate trend 4: *Extrapolation*	9.5 °C	2.4 °C	18.8 °C	4,426 mm/885 mm
	Period: –	Period: 2028–2032	Period: 2029–2033	Period: 2031–2035
	Mean annual temperature	Mean winter temperature (DJF)	Mean summer temperature (JJA)	Mean annual precipitation
Past 1960–2006	6.2 °C	−2.3 °C	14.5 °C	1,040 mm

50.3 Conclusion and Outlook

This chapter presented criteria and methods of the development of the GLOWA-Danube climate variants based on the four selected climate trends and their implementation with the statistical climate generator. The climate variants, which were selected based on extensive discussions of climate scenarios both among the project groups and with the stakeholders, were described and documented. Therefore, at present, four meteorological climate variants were selected with highest priority for each climate trend, all of which have a specific statistical probability of occurrence based on a huge ensemble of 5,000 statistically equivalent meteorological driver data sets which comply with the substantive criteria.

The illustrations have revealed that in the context of the defined climate trends, which should cover the range of expected regional development in temperature and precipitation due to global climate change in the period from 1990 to 2100, there is a wide variation of statistically equivalent synthetic weather realisations for the years 2011–2060. For each climate variant, a statement can be made about their probability of occurrence based on the assumptions made within the climate generator. The whole palette of the 16 different (four climate trends times four variants) resulting meteorological drivers are input to DANUBIA. They generate a range of results from the coupled hydrological-socioeconomic actor modelling, for all of which the probability of occurrence is known.

With the criteria listed in Table 50.1 for the climate variants, the opportunities for the methods introduced for the climate variants are nowhere near exhausted. Issues that cannot be addressed by any of the four climate variants that were selected will require additional criteria, which, depending on the complexity of the requirements, are feasible at varying expenses.

Chapter 51
Climate Variants of the *MM5* and *REMO* Regional Climate Models

Thomas Marke, Wolfram Mauser, Andreas Pfeiffer, Günther Zängl, Daniela Jacob, and Swantje Preuschmann

Abstract As global-scale climate scenarios at high spatial detail are not yet available due to existing limitations in computational resources, regional climate models are often applied to allow fine-scale consideration of climate change at the regional scale. This chapter describes the derivation of climate variants from the simulations of the two regional climate models *REMO* and *MM5*, supplementing the climate variants generated by applying predefined climate trends in combination with a statistical climate generator (Chaps. 49 and 50). To overcome limitations arising from the fact that (1) the small-scale climatic variability, primarily in the southern Alpine regions of the study area, cannot be reproduced despite the comparatively high spatial resolution of the regional climate models and (2) systematic deviations between meteorological simulations and observations for the past exist (often referred to as "biases"), a method for refinement (downscaling) and bias correction of RCM data is described that is applied to the RCM data prior to its application as input for DANUBIA. Biases in the applied RCM data are shown exemplarily for simulated precipitation. Moreover the effect of bias correction on simulated discharge is

T. Marke (✉)
Institute of Geography, University of Innsbruck, Innsbruck, Austria
e-mail: thomas.marke@uibk.ac.at

W. Mauser
Department of Geography, Ludwig-Maximilians-Universität München (LMU Munich), Munich, Germany
e-mail: w.mauser@lmu.de

A. Pfeiffer
Institute of Atmospheric Physics, DLR, Oberpfaffenhofen, Germany
e-mail: a.pfeiffer@dlr.de

G. Zängl
Deutscher Wetterdienst (DWD), Offenbach, Germany
e-mail: guenther.zaengl@dwd.de

D. Jacob • S. Preuschmann
Climate Service Center, Hamburg, Germany
e-mail: daniela.jacob@hzg.de; swantje.preuschmann@hzg.de

W. Mauser, M. Prasch (eds.), *Regional Assessment of Global Change Impacts*, DOI 10.1007/978-3-319-16751-0_51

illustrated by comparing the discharge duration curve simulated by DANUBIA with and without correction of biases in the RCM data. The climate change signal characterising the climate variants based on scaled and bias-corrected *REMO* and *MM5* data is analysed by considering changes in annual and monthly mean temperature and precipitation as well as spatial patterns and seasonal changes in temperature and precipitation change in the Upper Danube watershed.

Keywords Climate variants • Regional climate models • Downscaling • Bias correction • Climate change signal • GLOWA-Danube

51.1 Regional Climate Models

Global-scale climate scenarios are being developed at several institutions throughout the world. The climate of the Earth can thereby be calculated using computer modelling on the basis of physical equations. However, despite huge advances in computer technology in recent years, the spatial resolution of the global models is still limited by the existing computational capacities. A direct consequence of this limitation is that many interactive effects that occur between the topography and the atmosphere are still not captured at the resolution of the models; these interactions have particularly great significance in topographically complex terrain like that in the Upper Danube watershed. Therefore, regional climate models are an important tool for more fine-scale consideration of climate changes at the regional scale. With approximately 10–50 km, their spatial resolution is considerably finer than that of global models, such that small-scale topographic features and meteorological processes can be better accounted for.

A model for calculating climate at the regional scale was developed at the Max Planck Institute for Meteorology in Hamburg. The *REMO* regional climate model (Jacob 2001; Jacob et al. 2007) utilises the principle of double nesting to deal with the mismatch in scale from the global to the regional level. This approach first simulates climate with a global model at a coarse spatial resolution of ~180 km and then with the regional model in two additional steps, each at a finer resolution (~40 km, ~10 km); thus, the resulting data from each coarser dataset is entered into a finer calculation as a lateral boundary forcing. *REMO* is used by the Environment Agency Austria (Umweltbundesamt = UBA), among other groups, to calculate possible regional climate changes by the year 2100 for Germany, Austria and Switzerland, at a resolution of 10 × 10 km, based on the emission scenarios worked out by the *IPCC*. These data are particularly significant as drivers for DANUBIA, since the spatial resolution is unparalleled to date.

A second model is the *MM5* model, which has already been in place since the beginning of the GLOWA-Danube project (see also Chap. 32). *MM5* also makes use of a multiple-nesting approach. However, only a single nesting step (45 km) is used to translate from the global to the regional scale. The justification for this is that, when using reanalysis data as a lateral driver, quite realistic simulations can be achieved already with a single nesting in a configuration of *MM5* that is optimised especially for precipitation in the Alpine region for present-day climate (Pfeiffer and Zängl 2010). Moreover, unlike *REMO*, there is also an "online" integration of *MM5* via two-way coupling in the simulation runs of DANUBIA; this would be virtually impossible at finer spatial resolution because of the high computational expense of the meteorological model. The combination of the physical and dynamic refinement of the global simulations through the *MM5* regional climate model and the subsequent statistically based downscaling to 1 km (see paragraph 2) is the best compromise between the technical capabilities and the requirements for high-resolution meteorological model results for DANUBIA.

The means by which regional climate trends can be derived from the results of the *REMO* and *MM5* regional climate models have already been described in Chap. 48. Climate variants can be generated by applying the trends in the development of temperature and precipitation that are inherent in the results of the regional climate models in combination with a statistical climate generator; these variants can be used as input for DANUBIA (see Chaps. 49 and 50).

In addition to the trends derived from the results, the output from the *REMO* and *MM5* regional climate models also contains hourly arrays of the meteorological variables radiation, wind, temperature, atmospheric humidity and precipitation. At least in theory, this output is suitable for use as meteorological drivers for DANUBIA. Two factors limit the direct use of this output from the regional climate models as a driver for DANUBIA:

(a) The small-scale climatic variability, primarily in the southern Alpine region of the study area, cannot be reproduced despite the comparatively high spatial resolution of the regional climate models. While the model scale of DANUBIA is 1 km, *REMO* and *MM5* operate at 10 km and 45 km resolution, respectively. As a result, key details for hydrology, such as valleys and glaciers, are lost in the regional climate models (see Fig. 51.1). This difference in scale must be bridged by a refinement of the spatial resolution (downscaling) of the regional climate simulations.

(b) The accurate quantitative modelling of precipitation, especially in the highly structured topography of the Alpine region, is a huge challenge for climate models; over- or underestimations can arise at the scale of the regional models as a result of a horizontal misalignment of the simulated precipitation events, for example.

Deviations from observed values in precipitation are not necessarily the result of deficiencies in the intrinsic processes described by the regional climate model, but to a large extent can be attributed to the global forcing that is used as input at the boundaries of the model domain (see Chap. 48, Pfeiffer and Zängl 2011). Thus, the comparison of the modelled historical climatology (using the ECHAM5 driver) from *REMO* and *MM5* with a dataset of observational measurements prepared in the context of the GLOWA-Danube project reveals systematic deviations by both models from the measurements. Especially in the winter half of the year, the monthly precipitation in the basin is significantly overestimated by both regional climate models (see Fig. 51.2).

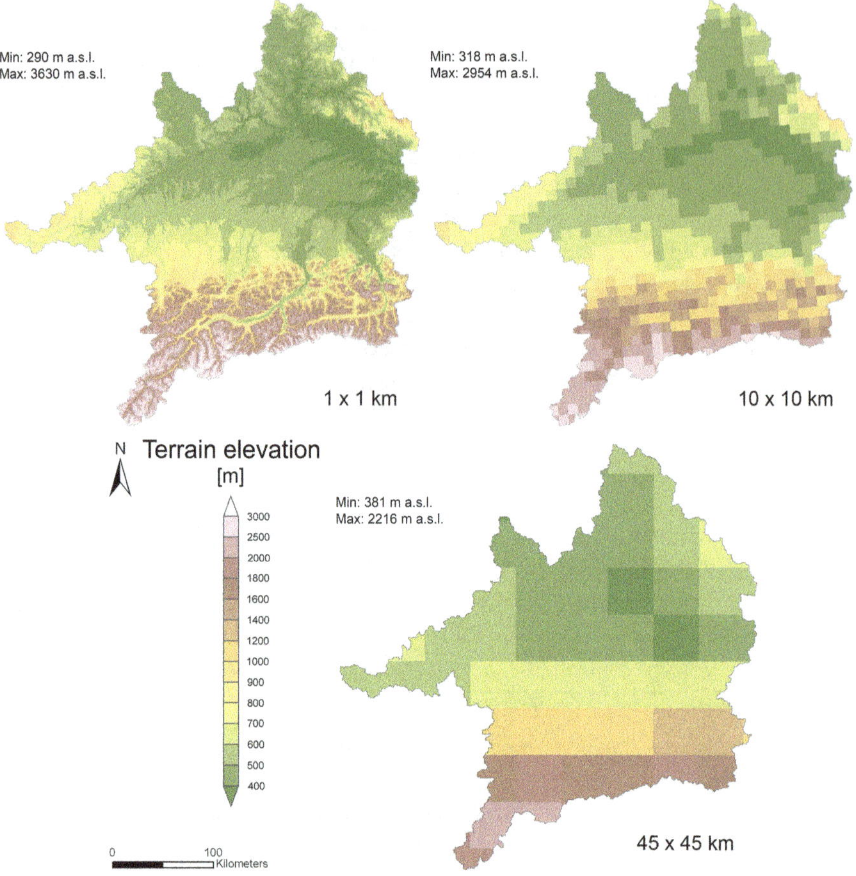

Fig. 51.1 Terrain elevation in the Upper Danube watershed at the different spatial resolutions of the models DANUBIA (1 km), *REMO* (10 km) and *MM5* (45 km)

If the model results are used directly as inputs for DANUBIA, it would not be possible to realistically reproduce the statistical properties of the measured hydrology. The systematic deviations must be removed by a bias correction, under the assumption that they will continue in the future in the same manner.

The significance of the bias correction for hydrology is clarified in Fig. 51.3. This figure presents the duration curve for discharge at the outlet of the drainage basin at Achleiten that was modelled using the downscaled *MM5* and *REMO* data. As shown in Fig. 51.3, the correction of the subgrid-scale variability in the results of both regional models (see Fig. 51.2) is not sufficient to reproduce the duration curve observed at the outlet of the Upper Danube with DANUBIA.

The overestimation in simulated precipitation in case of both models leads to a significant overestimation of discharge when a bias correction is not performed. The actual trend in the duration curve at the outlet can only be realistically modelled by adding the bias correction in the downscaling process.

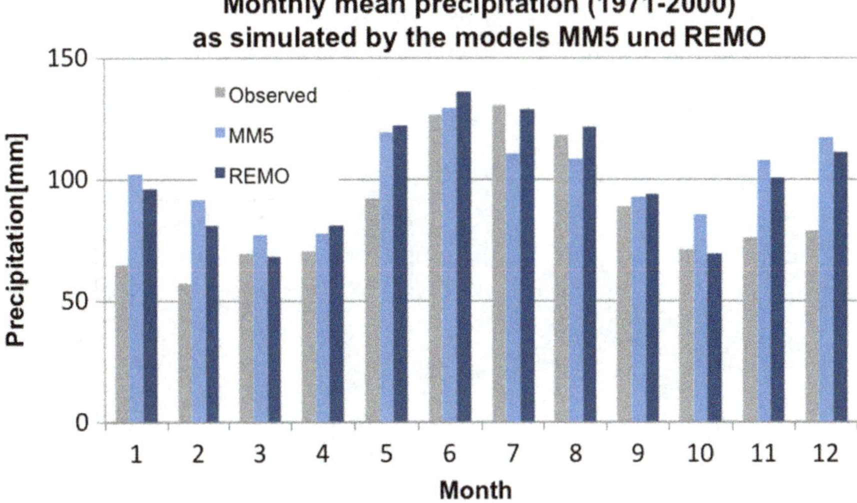

Fig. 51.2 Mean monthly precipitation in the Upper Danube watershed (1971–2000) according to uncorrected simulations of the models *MM5* and *REMO*

51.2 Downscaling and Bias Correction of the Model Results

The downscaling and bias correction of the hourly calculations of *MM5* and *REMO* takes place via the scaling interface SCALMET (Marke 2008; Marke et al. 2011). SCALMET combines various methods for up- and downscaling the meteorological parameter arrays that are used as meteorological drivers for DANUBIA. A statistical scaling method was used for the results presented here. The method is based on Früh et al. (2006) and has been extended by Marke et al. (2011). It utilises monthly scaling functions for downscaling the regional climate simulations to the 1 × 1 km resolution required for calculations within DANUBIA (see Fig. 51.4).

In a first step for this method, the subgrid-scale variability of the observed climate within each grid square for each regional climate model is derived from a high-resolution climatology (see Chap. 32). This high-resolution climatology (1 km) is aggregated to the spatial resolution of each respective regional climate model, while preserving the energy and mass, such that the result of the process is an observed climatology at the resolution of *MM5* and *REMO*. The subgrid-scale variability within each climate model pixel is calculated by comparing the dataset that results from the aggregation with the original observational data (1 km) (i.e. for each 1 × 1 km pixel within a climate model pixel, the extent to which the high-resolution monthly temperatures are above the mean for a pixel at the resolution of the climate model under consideration is noted).

The first step in the processing corrects only the subgrid-scale variability, but not the deviations between the modelled and observed climatologies; therefore, in a second step, model-specific functions for bias correction are derived from the

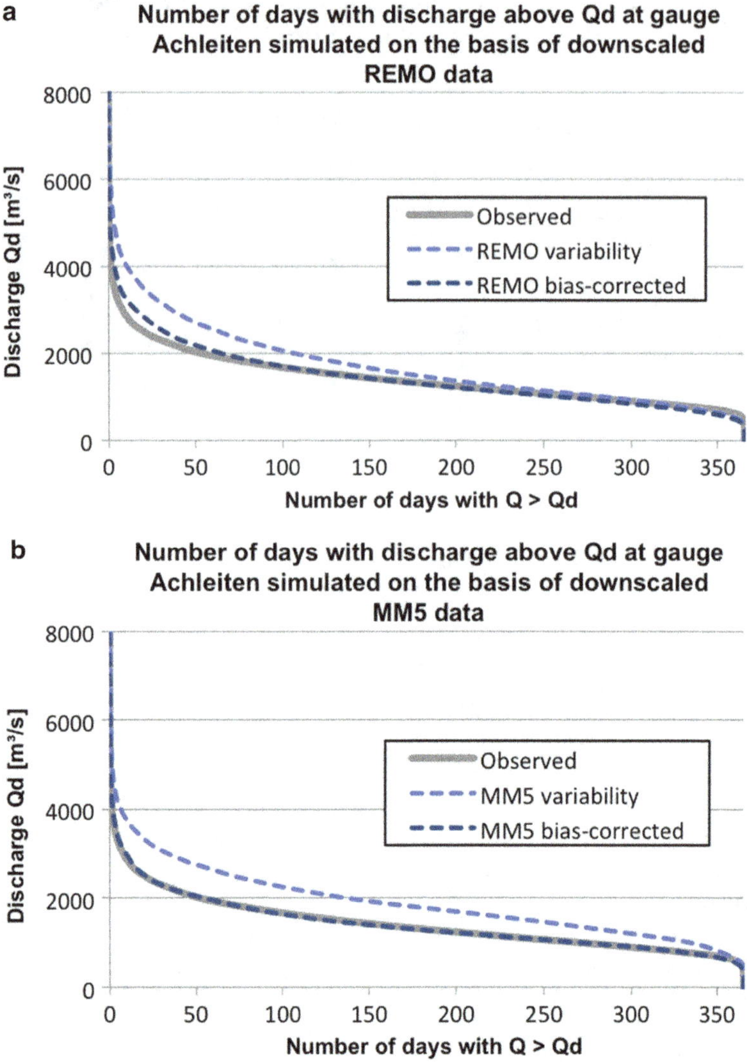

Fig. 51.3 Number of days with discharge above discharge Qd as simulated with DANUBIA for 1972–2000 on the basis of downscaled *MM5* and *REMO* data. Downscaling is carried out (**a**) and (**b**) here stand for the REMO (**a**) and MM5 (**b**) model, not for the different scaling techniques (these are represented by the lines)

observed climatology and the modelled climatology for both models. These functions are derived from the deviations of the coarse climate simulations from the monthly mean of the aggregated observations. The following overall correction (f_{total}) derives from the functions for correcting the subgrid-scale variability ($f_{\text{variability}}$) and the functions for bias correction (f_{bias}):

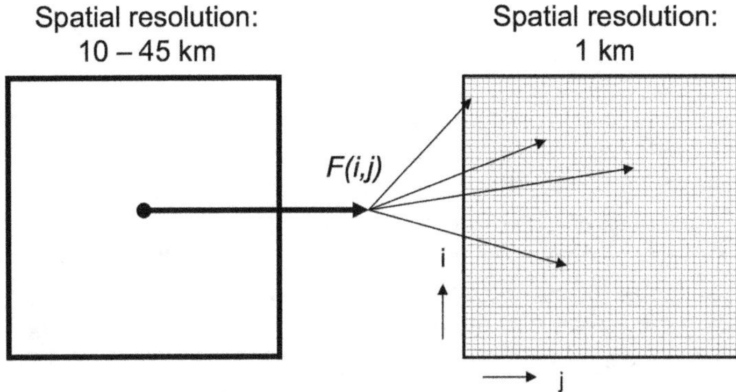

Fig. 51.4 Schematic illustration of the downscaling process on the basis of statistical downscaling functions

$$f_{\text{total}} = f_{\text{variability}} * f_{\text{bias}}$$

By comparing the observed and modelled climatologies according to this method, there are correction functions for all meteorological parameters for each month in the year. These functions account for both the subgrid-scale variability and the bias at the level of the respective model grid.

In the case of precipitation, the results are spatially distributed correction factors at a spatial resolution of 1 × 1 km for each month in the year, which are incorporated into the hourly precipitation arrays of the regional climate models. For other meteorological parameters, such as temperature, the correction can also be done using additive correction values that represent the offset against the simulated values. This method allows the comparison of the modelled data for the present climate with the observed data, such that the downscaled and bias-corrected model climatologies of the past are then identical to the observed climatologies per construct of the scaling procedure.

The respective model-specific scaling functions are then applied to the model results from the future simulations, whereby the desired effect of an observed and statistically based resolution refinement to 1 × 1 km is achieved. The maps for this chapter therefore present the results of a combination of dynamic (regional climate models) and statistical (SCALMET) scaling methods. As a result of the correction method used, the data shown should be distinguished from the original climate simulations by the *MM5* and *REMO* models, which were used to derive the regional climate trends in Chap. 48. Hereafter, the model data resulting from the scaling and bias correction are termed climate variants according to the logic of the GLOWA-Danube scenarios presented in Chap. 47, where the terms *MM5 downscaled and bias-corrected* and *REMO downscaled and bias-corrected* are used, depending on the underlying model.

51.3 Results

In the following, the simulated climate change in the Upper Danube watershed will be shown for the meteorological variables temperature and precipitation. In Fig. 51.5 (top), the temporal trend in the change in annual mean temperature for the period 1970–2100 is compared to the annual mean temperature for the reference period (1971–2000) after downscaling and bias correction. Since the scaling for temperature takes place using an additive correction term, the temperature change remains unaffected by the scaling. The trend in temperature change shown in Fig. 51.5 (top) therefore corresponds to the temperature change for the unscaled climate model data as shown in Fig. 48.7 in Chap. 48. The *MM5 downscaled and bias-corrected* and *REMO downscaled and bias-corrected* climate variants both contain a significant increase in the annual mean temperature. In addition, both climate variants show quite similar trends in temperature change over time, and only at the beginning and the last third of the twenty-first century do they exhibit differing temperature changes, with a difference of approximately 1 °C (see Fig. 51.5 top). The warming calculated for the Upper Danube watershed shows an increase in the annual mean temperature of up to 3 °C by the year 2060. Although there are still years in this period with an annual mean temperature below the average value for the reference period, the annual mean temperature by the end of the twenty-first century for both model simulations is without exception above the mean (up to +5 °C) for the reference period.

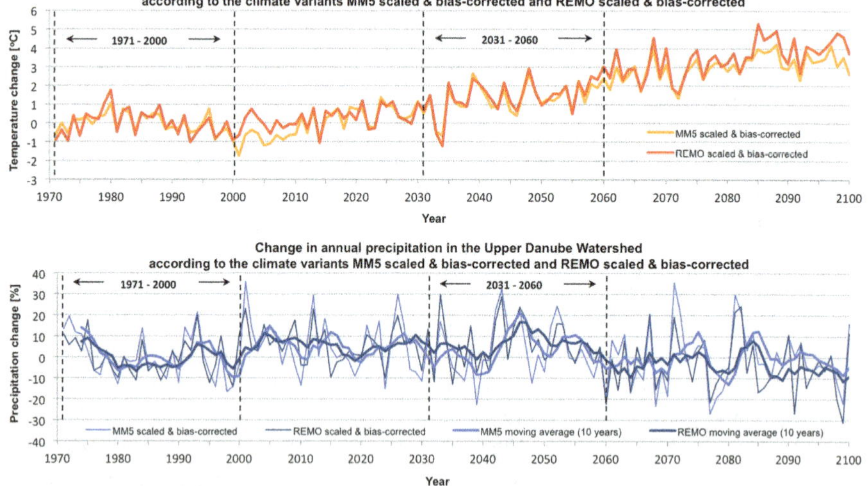

Fig. 51.5 Climate change signal in simulated temperature and precipitation for the period 1971–2100 according to downscaled and bias-corrected *MM5* (*top*) and *REMO* (*bottom*) simulations

The temporal trend in the relative change in precipitation is shown in Fig. 51.5 (bottom). For the purpose of the visualisation, each 10-year moving average (thick lines) is shown behind the highly variable annual change (thin lines). In contrast to the scaling of temperature, a multiplicative correction term is used in the case of precipitation. The result is that in addition to spatial patterns and absolute values, the change in precipitation is also affected by the scaling. Nonetheless, a comparison between Fig. 51.5 (bottom) and the trend in precipitation in Fig. 48.7 from Chap. 48 shows that the impact of the scaling on the change in precipitation is quite minor and can be ignored compared to the uncertainties in the simulated precipitation trend.

If the *MM5 downscaled and bias-corrected* and *REMO downscaled and bias-corrected* climate variants are compared, both reveal a trend in the same direction for the calculated precipitation change. However, it should be noted that the high degree of correspondence can largely be attributed to the common global driver used for the *MM5* and *REMO* model (ECHAM5; see Chap. 48, Fig. 48.6). Even so, the likelihood of the simulated climate change actually occurring is not higher, since the probability of occurrence of the results of the underlying ECHAM run is not known.

Although there is no significant increase or decrease in the annual precipitation for the first half of the twenty-first century, the data from both climate variants deviate more strongly from each other beyond 2060. Although *MM5 downscaled and bias corrected* shows considerable increases of almost 40 %, *REMO downscaled and bias corrected* reveals more pronounced decreases (up to −30 %). The trend in the moving average for the second half of the twenty-first century thus tends to be characterised by more precipitation increases in the case of the *MM5* climate variant, whereas in the case of *REMO* there tend to be more precipitation decreases.

If the linear trend in climate change is calculated across the years 1990–2100, the results are the changes in temperature and precipitation presented in Tables 51.1 and 51.2. In addition to the annual means and linear trends, the spatial pattern and seasonal changes in temperature and precipitation can be important for addressing some issues in climate change research. The climate change signal presented on

Table 51.1 Linear trend of temperature change in the Upper Danube watershed (1990–2100) in relation to the mean of 1971–2000

Temperature change 1990–2100 [°C]		
Time	*MM5 downscaled and bias corrected*	*REMO downscaled and bias corrected*
Winter	+5.2	+6.7
Spring	+3.2	+3.7
Summer	+5.8	+5.3
Autumn	+4.8	+5.1
Year	+4.7	+5.2

Table 51.2 Linear trend of precipitation change in the Upper Danube watershed (1990–2100) in relation to the mean of 1971–2000

Precipitation change 1990–2100 [%]		
Time	*MM5 downscaled and bias corrected*	*REMO downscaled and bias corrected*
Winter	+8.4	−1.4
Spring	+14.5	+10.7
Summer	−29.4	−32.6
Autumn	−2.7	−12.6
Year	−6.2	−12.4

Maps 51.3 and 51.4 were calculated as the difference between the downscaled and bias-corrected model calculations for the scenario period 2031–2060 and the historical scaled model results from 1971 to 2000.

51.3.1 Temperature

As shown in Table 51.1, the data from the *REMO downscaled and bias-corrected* climate variant show a stronger increase in temperature for the annual mean compared to the *MM5* climate variant.

An examination of the linear trends reveals that the seasonal temperature changes for the climate variants show significant differences from each other, especially in winter. Hence, in the last two decades of the twenty-first century, the temperature change in winter derived from the *MM5* data (+5.2 °C) is 1.5 °C less than the increase calculated from the downscaled and bias-corrected *REMO* data.

The annual mean temperatures averaged over the period from 2031 to 2060 for the Upper Danube watershed calculated according to the *MM5 downscaled and bias-corrected* and *REMO downscaled and bias-corrected* climate variants are similar, at approximately 1.5 °C higher than the mean for the reference period (see Maps 51.1 and 51.2). However, if only the higher mountainous regions are considered, the most marked warming for the climate variant derived from *MM5* data is approximately 1.6 °C, but for the variant calculated from *REMO* simulations, it is approximately 2.3 °C.

The spatially distributed change signal derived from the *MM5 downscaled and bias-corrected* climate variant is characterised by a rather uniform increase in the temperature change from north to south, but the changes from *REMO downscaled and bias corrected* reveal smaller-scale patterns. This significantly smaller-scale variability in the temperature change compared to the data based on *MM5* can be explained on the basis of the higher spatial resolution of the underlying *REMO* model. Indeed, the scaling of the model results in the case of both models produces subgrid-scale patterns at a spatial resolution of 1 × 1 km (see Maps 51.1 and 51.2),

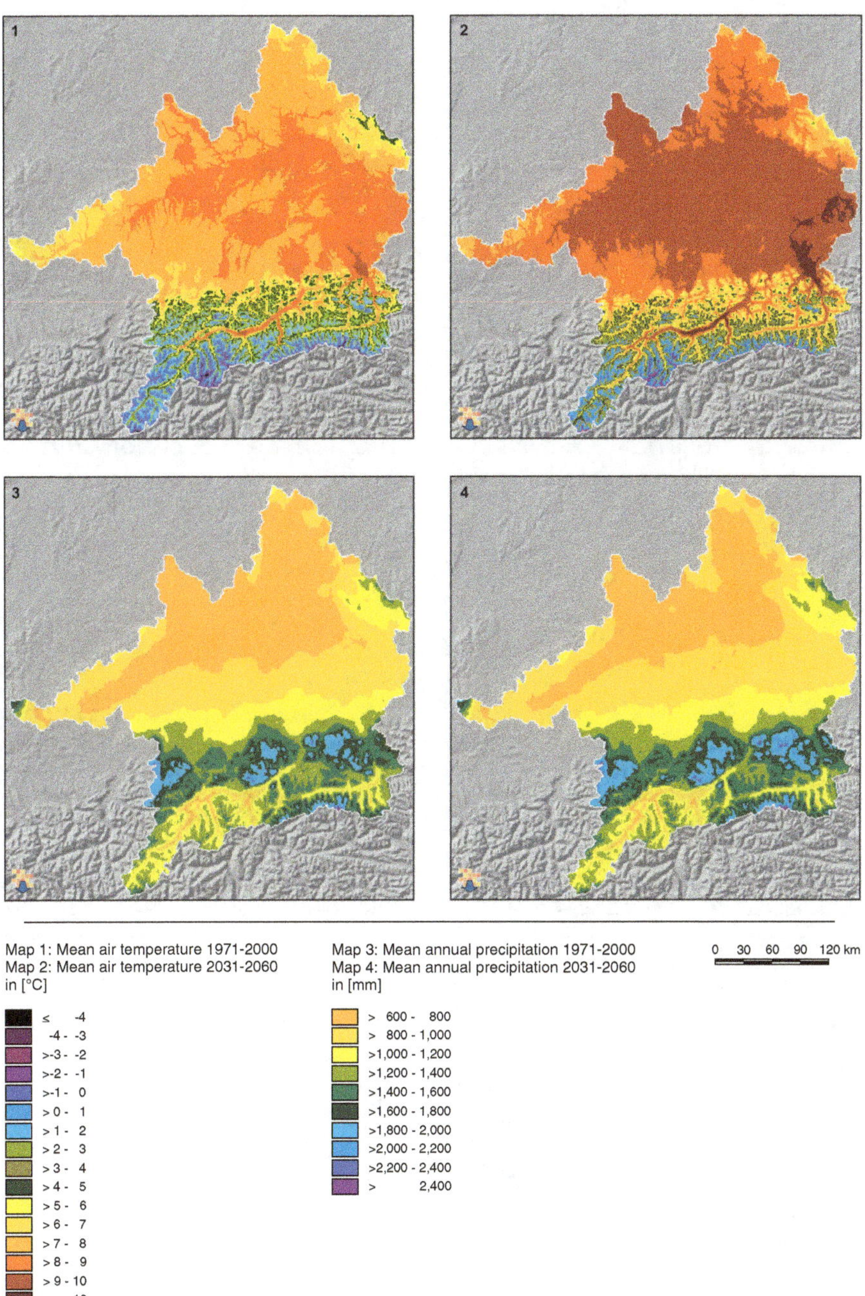

Map 51.1 Simulation of air temperature and precipitation with *MM5 downscaled and bias corrected* (global forcing: ECHAM5-MPIOM coupled atmosphere-ocean circulation model; dynamic regionalization: *MM5* regional climate model, calculations by Meteorological Institute, LMU Munich; data for the statistical scaling: DWD, Deutscher Wetterdienst; ZAMG, Zentralanstalt für Meteorologie und Geodynamik)

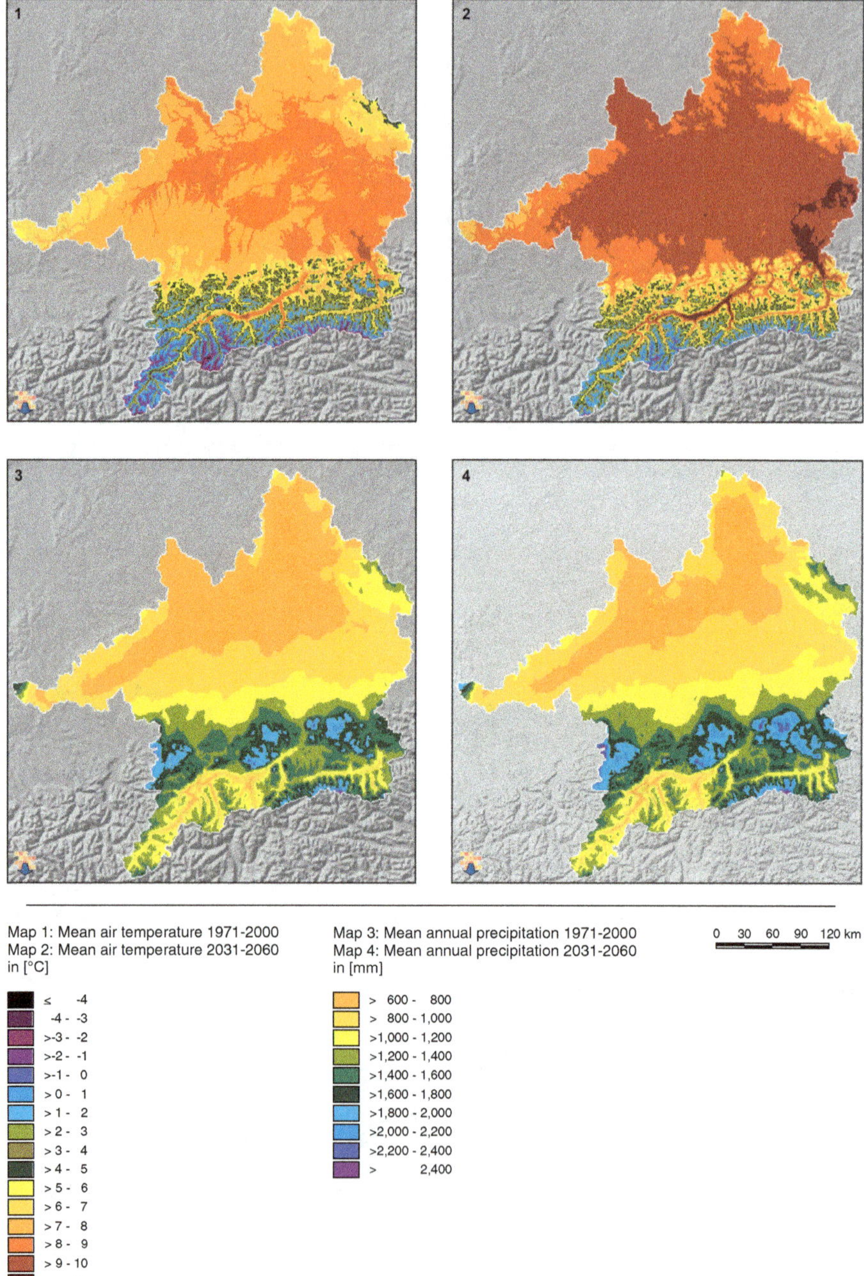

Map 51.2 Simulation of air temperature and precipitation with *REMO downscaled and bias corrected* (global forcing: ECHAM5-MPIOM coupled atmosphere-ocean circulation model; dynamic regionalization: *REMO* regional climate model, calculations by Max Planck Institute for Meteorology (MPI-M), financed by the Federal Environment Agency (UBA); data for the statistical scaling: DWD, Deutscher Wetterdienst; ZAMG, Zentralanstalt für Meteorologie und Geodynamik)

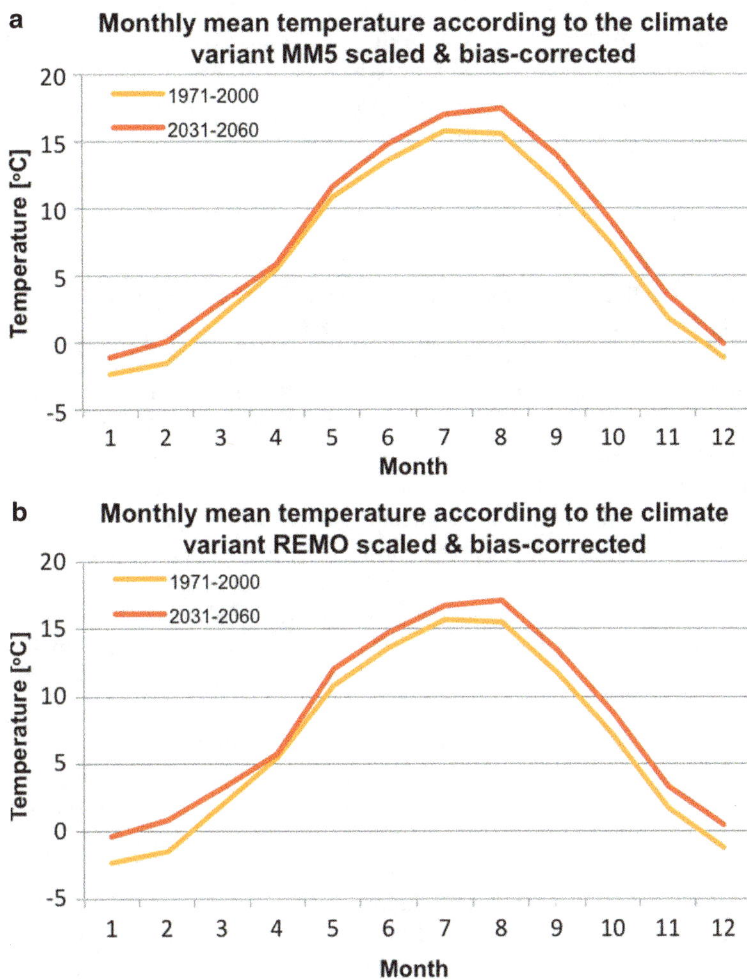

Fig. 51.6 Mean monthly temperature in the Upper Danube watershed according to downscaled and bias-corrected *MM5* (**a**) and *REMO* (**b**) simulations for the reference (1971–2000) and scenario period (2031–2060)

although the scales of the temperature change signals are dominated by the horizontal resolution for each regional climate model.

Like the trend analysis, there are significant differences in the comparative images for the winter season DJF for both climate variants. While the winter according to *REMO downscaled and bias corrected* represents the season with the highest temperature increase (+2 °C), the increase in winter calculated from the *MM5* data is significantly smaller (+1.3 °C) and is in fact less than the increases in summer (+1.5 °C) and fall (+1.9 °C). These differences can be attributed to the various trends in the individual months (see Fig. 51.6).

As Map 51.4 shows, the data from the *REMO downscaled and bias-corrected* variant reveal a maximum warming in winter and fall, whereas *MM5 downscaled and bias corrected* show the highest temperature increases in summer and fall.

51.3.2 Precipitation

In the case of both climate variants, the linear trends for the changes in precipitation in the Upper Danube watershed show a decrease in the annual precipitation by the end on the twenty-first century; however, in the case of the *MM5 downscaled and bias-corrected* climate variant, this decrease is less marked (see Table 51.2).

The trends for the seasonal changes in precipitation derived from the climate variants are generally consistent, especially in summer and spring.

The linear trends based on both climate variants show a considerable decrease in precipitation in summer and a significant increase in precipitation in spring. The future patterns in winter precipitation differ based on the linear trends in both the *MM5 downscaled and bias-corrected* and *REMO downscaled and bias-corrected* climate variants. Whereas the linear trend for the *MM5 downscaled and bias-corrected* variant is characterised by an increase in winter precipitation by 8.4 % by the end of the twenty-first century, the winter precipitation in the *REMO downscaled and bias-corrected* variant decreases by 1.4 %.

If the precipitation trends derived from the climate variants are compared to the results from the comparison of the reference and scenario periods, the differences are obvious: while the linear trend in the precipitation change (1990–2100) describes a decrease in the annual precipitation in both models, the comparison of the mean annual precipitation in the scenario period (2031–2060) with the precipitation in the reference period (1971–2000) shows an increase in precipitation in both models (see Maps 51.3 and 51.4). These results highlight the dependency of the climate change signals on the length and position of the period being considered within the modelled time series, as well as the dependency of the results on the method used to study climate change.

Although the increase in the annual precipitation in the data from the *REMO downscaled and bias-corrected* climate variant, calculated by comparing the periods 1971–2000 and 2031–2060, is somewhat higher in some regions compared to the *MM5 downscaled and bias-corrected* climate variant, the horizontal distribution of the precipitation change signal in both downscaled model results is generally consistent. This distribution is characterised by an increase in precipitation in the region of the Bavarian Forest, but primarily in the north-eastern Alpine region. In contrast, the south-western part of the basin is characterised by a decrease in the annual precipitation. In addition, the seasonal precipitation changes in the Upper Danube watershed are very similar in both climate variants (see Map 51.4(9–16)). Both downscaled model results indicate an increase in the mean regional precipitation for spring (*MM5 downscaled and bias corrected* +27 mm, *REMO downscaled and bias corrected* +32 mm) and fall (*MM5 downscaled and bias corrected* +19 mm, *REMO downscaled and bias corrected* +38 mm), as well as a decrease in precipitation in summer (*MM5 downscaled and bias corrected* −6 mm, *REMO downscaled and bias corrected* −13 mm).

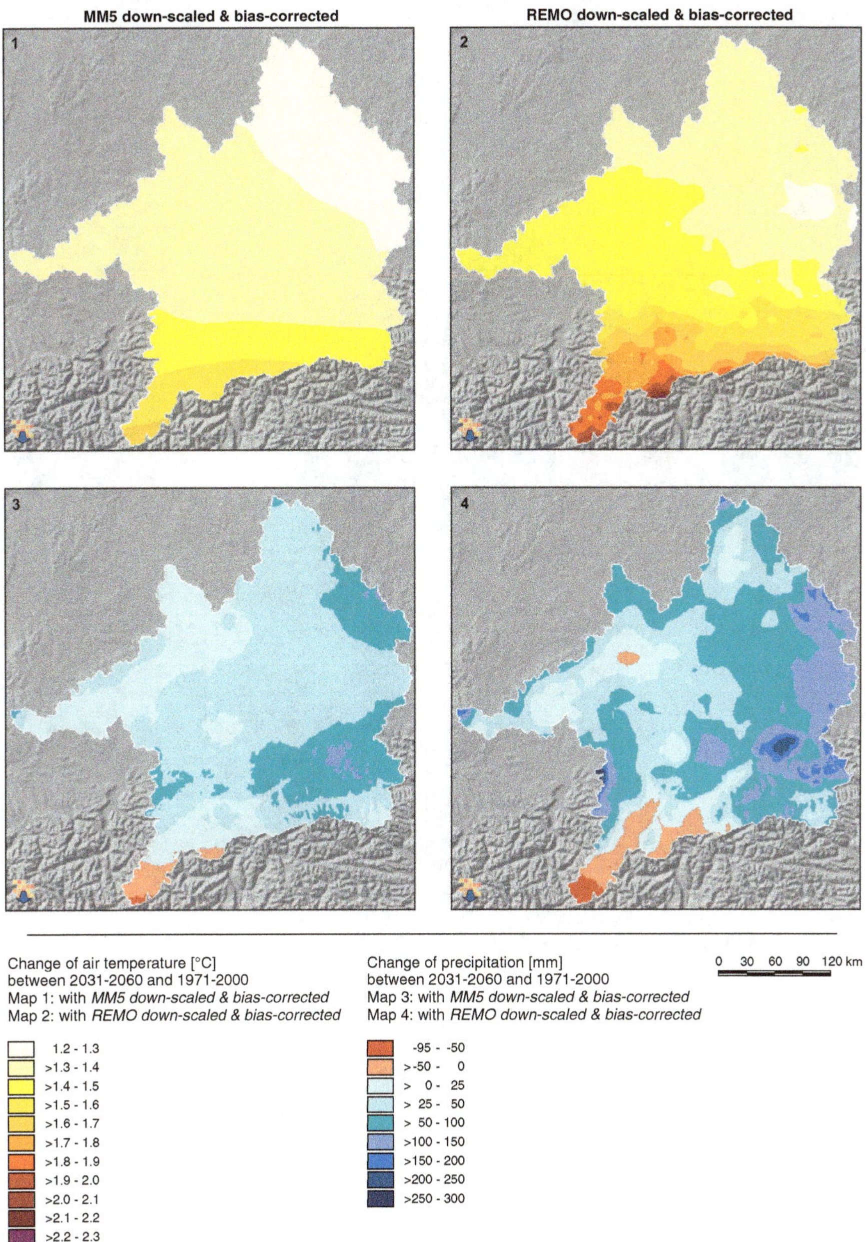

Map 51.3 Change of air temperature and precipitation between 2031–2060 and 1971–2000 with *MM5* and *REMO*, both downscaled and bias corrected (global forcing: ECHAM5-MPIOM coupled atmosphere-ocean circulation model; dynamic regionalization: *REMO* regional climate model, calculations by Max Planck Institute for Meteorology (MPI-M), financed by the Federal Environment Agency (UBA); *MM5* regional climate model, calculations by Meteorological Institute, LMU Munich; data for the statistical scaling: DWD, Deutscher Wetterdienst; ZAMG, Zentralanstalt für Meteorologie und Geodynamik)

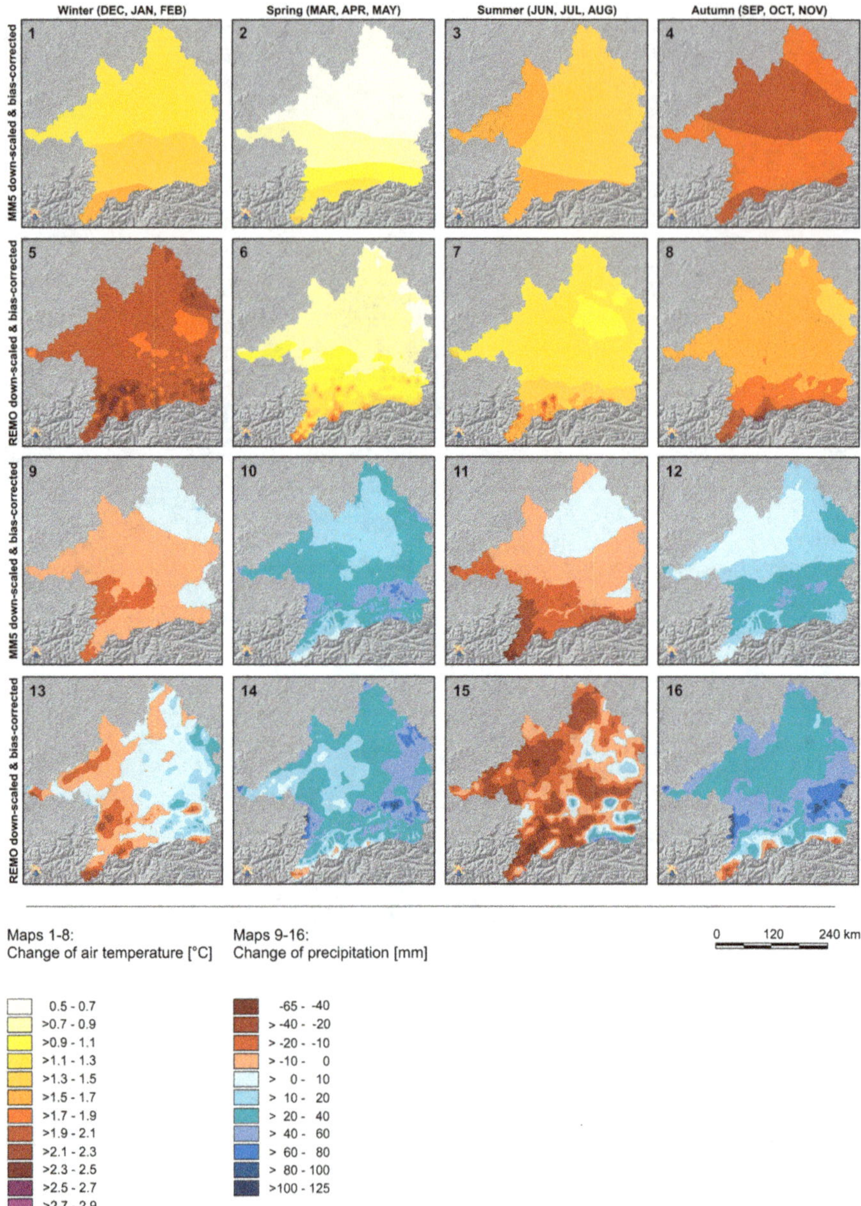

Map 51.4 Seasonal change of air temperature and precipitation between 2031–2060 and 1971–2000 with *MM5* and *REMO*, both downscaled and bias corrected (global forcing: ECHAM5-MPIOM coupled atmosphere-ocean circulation model; dynamic regionalization: *REMO* regional climate model, calculations by Max Planck Institute for Meteorology (MPI-M), financed by the Federal Environment Agency (UBA); *MM5* regional climate model, calculations by Meteorological Institute, LMU Munich; data for the statistical scaling: DWD, Deutscher Wetterdienst; ZAMG, Zentralanstalt für Meteorologie und Geodynamik)

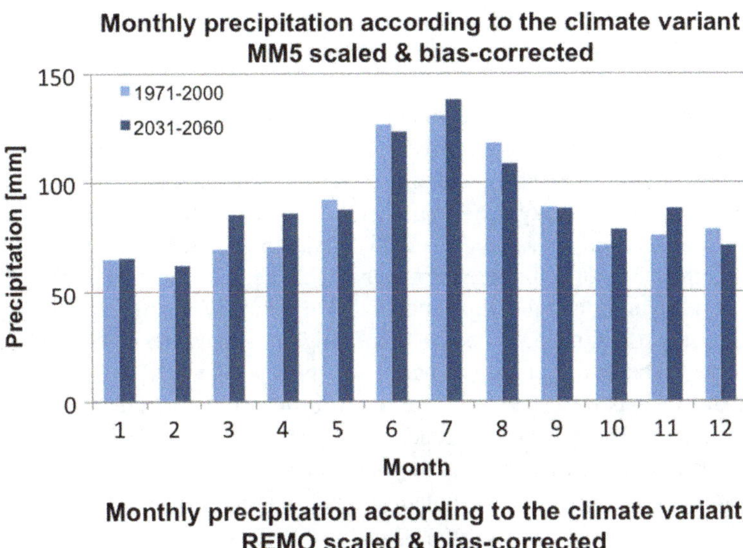

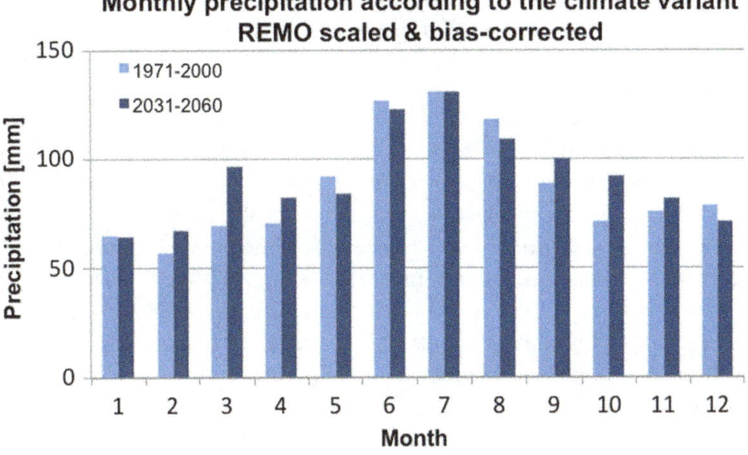

Fig. 51.7 Mean monthly precipitation in the Upper Danube watershed according to downscaled and bias-corrected *MM5* (*top*) and *REMO* (*bottom*) simulations for the reference (1971–2000) and scenario period (2031–2060)

In a comparison of the periods 1971–2000 and 2031–2060, the results for the winter show overall a less-marked difference, but opposite trends for both climate models. While the data based on *REMO* reveal an absolute increase in winter precipitation by 2 mm, the results derived from the *MM5* model calculations indicate a decrease in winter of approximately 3 mm.

If the monthly precipitation in the reference and scenario periods for the two models are compared (see Fig. 51.7), it is obvious that the differing precipitation change in winter in general can be attributed to the differing precipitation for the month of February (2031–2060). In addition, Fig. 51.7 reveals that the maximum increase and maximum decrease in both climate variants occur in the same months (March: increase and August: decrease).

51.4 Summary

As presented in this chapter, the climate variants, as downscaled and bias-corrected model results of the *MM5* and *REMO* regional climate models, both describe rather similar changes of temperature and precipitation for the periods in question. This is primarily due to the use of the same lateral driver for both models, which comes from the results from the same ECHAM5 simulation (A1B, Member 1).

The comparison of different analysis methods (comparison of the reference and scenario simulations and the trend analyses) and the analysis time periods indicated that the climate change signal is sometimes highly dependent on the method used and the period being considered. Especially in the case of annual and winter precipitation, the various data analyses resulted in different, sometimes even opposite, change signals. There were differences in the simulated changes to precipitation between the models; this fact, in combination with the fact that a difference of approximately 10 % for the precipitation changes in fall and winter arose for both climate variants when different analysis methods were used and when different time periods were considered, indicates the challenges and uncertainties in calculating future precipitation totals.

In the analysis and interpretation of the results, it is important to bear in mind that only a few realisations among a few predefined global climate trends have been considered, and these have then been further refined with *MM5* and *REMO*. As a result, it is not possible to make statements about the range of expected climate changes and the probabilities of occurrence of specific climate patterns based on the present results from the regional climate modelling. Different realisations within the scenarios must be considered in order to more accurately estimate the range of possible climate changes and their effects. The required range of model realisations is unfortunately not yet available to date, not least because of the high computational expense of such simulations.

Nevertheless, the results of the *REMO* and *MM5* simulations do reveal the expected variability of the simulated climate trends. Thus, the linear trends in precipitation from the *MM5 downscaled and bias-corrected* and *REMO downscaled and bias-corrected* climate variants for fall and winter indicate a difference of almost 10 percentage points. However, for spring and summer, the trends in precipitation from the model calculations are generally similar, with a difference of only 3 percentage points. The comparison of the trends shows that *MM5* and *REMO* simulate strong deviations in precipitation only for the last 15–30 years, but show quite similar trends for the period from 1970 to around 2070.

The increase in temperature in the Upper Danube watershed is very similarly represented by the *MM5 downscaled and bias-corrected* and *REMO downscaled and bias-corrected* climate variants; the temperature increase in the variant based on *REMO* calculations is generally somewhat larger. In this case as well, the differences in the change in temperature between the two climate variants result from the different calculations of the regional climate models for the last two decades of the twenty-first century.

References

Früh B, Schipper JW, Pfeiffer A, Wirth V (2006) A pragmatic approach for downscaling precipitation in alpine-scale complex terrain. Meteorol Z 15(6):631–646

Jacob D (2001) A note to the simulation of the annual and inter-annual variability of the water budget over the Baltic Sea drainage basin. Meteorol Atmos Phys 77:61–73

Jacob D, Bärring L, Christensen O, Christensen J, de Castro M, Déqué M, Giorgi F, Hagemann S, Hirschi M, Jones R, Kjellström E, Lenderink G, Rockel B, Sánchez E, Schär C, Seneviratne S, Somot S, van Ulden A, van den Hurk B (2007) An inter-comparison of regional climate models for Europe: model performance in present-day climate. Clim Chang 81:31–52

Marke T (2008) Development and application of a model interface to couple regional climate models with land surface models for climate change risk assessment in the Upper Danube watershed. Dissertation, Ludwig-Maximilians Universität München (LMU)

Marke T, Mauser W, Pfeiffer A, Zängl G (2011) A pragmatic approach for the downscaling and bias correction of regional climate simulations: evaluation in hydrological modelling. Geosci Model Dev 4:759–770

Pfeiffer A, Zängl G (2010) Validation of climate mode MM5-simulations for the European Alpine region. Theor Appl Climatol 101:93–108

Pfeiffer A, Zängl G (2011) Regional climate simulations for the European Alpine region – sensitivity of precipitation to large scale flow conditions of driving input data. Theor Appl Climatol 105:325–340

Chapter 52
Societal Scenarios in GLOWA-Danube

Andreas Ernst and Silke Kuhn

Abstract This chapter deals with the societal scenarios used in GLOWA Danube. The so-called societal megatrends serve as the basis for the generation of the societal scenarios within the sub-projects Economy, Farming, Household, Tourism and WaterSupply. They are based on the experience and knowledge from SinusSociovision, a German marketing company, and extrapolate potential trends based on currently observable tendencies in society. Three scenarios are described: First, a "Baseline" scenario depicts the continuation of society as it is, in the form of a business as usual scenario. The "Performance" scenario describes a world where free competition has been generally accepted, and society has become more hedonistic, market-oriented and materialistic. The scenario "Public Interest" finally paints a society where a general awareness of responsibility sets in. To apply these scenarios to the domain of the different sub-projects and to implement them in the running models, a specification was carried through to provide parameter values on key factors. Going along with that, storylines, i.e. small narratives, were developed that ensure plausibility and relevance in each sub-project as well as consistency across the whole model.

Keywords Societal scenario • Megatrend • Baseline scenario • Performance scenario • Public interest scenario • GLOWA-Danube

52.1 Introduction

By definition, a GLOWA-Danube scenario consists of both a climate scenario and a societal scenario. The precise structure and composition of the GLOWA-Danube scenarios is described in Chap. 47. While Chaps. 48, 49, 50 and 51 are devoted to the development and logic of the climate trends and climate variants, this chapter deals with the societal scenarios used in GLOWA-Danube.

A. Ernst (✉) • S. Kuhn
Center for Environmental Systems Research (CESR), University of Kassel, Kassel, Germany
e-mail: ernst@usf.uni-kassel.de

W. Mauser, M. Prasch (eds.), *Regional Assessment of Global Change Impacts*, DOI 10.1007/978-3-319-16751-0_52

A working group consisting of the following social science sub-projects using DeepActor models (see also Chap. 3) was formed for the development of the societal scenarios:

- *Economy*
- *Farming*
- *Household*
- *Tourism*
- *WaterSupply*

52.2 The Scenarios

The use of the Sinus milieus in one sub-project (e.g. see Chap. 42) makes it logical also to utilise the experience and knowledge from SinusSociovision (see de Vries and Perry (2007)) for the development of the societal scenarios, so that a unified, consistent model design is preserved.

For this reason, the so-called societal megatrends developed by SinusSociovision serve as the basis for the generation of the societal scenarios within the sub-projects listed above.

52.3 The Sinus Scenarios

The basic assumption of the SinusSociovision scenarios is that society is subject to constant change, which influences everyday activities, the spirit of the times and the perception and treatment of various themes. Within the last 20–30 years, there have been vital developments in working environments, communication, media and several other fields; these developments have radically changed our values, our attitudes and our everyday activities. One example of this is the significance of environmental awareness, which has taken on ever greater importance in society since the 1980s. Another example is communication technology, which allows us to easily be in contact with people far away at any time (by telephone, email, Skype, etc.).

The initial issue for the development of the social scenarios was which prototypical mindsets, which mentalities and which paradigms might become established? Subsequently, these societal trends and their effects could be evaluated and implemented in the sub-projects.

Figure 52.1 illustrates the scenario logic underlying the development process: potential trends were designed based on currently existing tendencies in society. Depending on the scenario, other mindsets and paradigms gain priority and shape the scenario. Other mindsets and paradigms continue to persist, but lose significance.

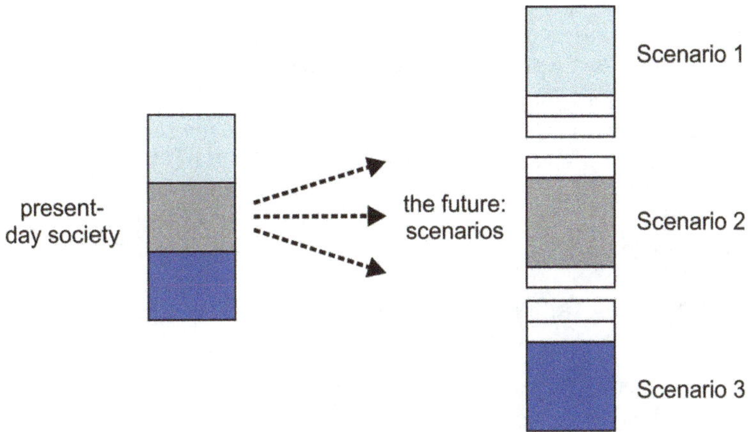

Fig. 52.1 Scenario logic used in the development of the societal scenarios (Modified after de Vries and Perry 2007)

In addition to new technologies and globalisation, dimensions that were important in the formation of the scenarios included demographic, economic and political developments.

The Sinus scenarios are presented in brief below.

52.3.1 Free is Fair *Scenario*

In this scenario, free competition has been generally accepted; society has become more hedonistic, market-oriented and materialistic. Things are judged by whether they function. According to the logic of the free market, that which is good is accepted. This also applies to people, because the individual is responsible for making the best of his or her options. Accordingly, collective responsibility and duty have lost priority. Bureaucracy is considered repressive. Public subsidies are only available for those who truly cannot help themselves. The gap widens between the rich and poor and between power and helplessness. Education and culture and spirituality and religion are private affairs.

52.3.2 Shared Destiny *Scenario*

This scenario is a counter draft to the *Free is Fair* scenario. An awareness of the responsibility of society as a whole sets in, based on the knowledge that the harmful side effects of unrestricted growth will continue to accumulate. People recognise that society has values and goals other than economic success alone.

The political and economic systems are required to assume responsibility, orientation towards the general welfare and the interests of the country. Political issues become more important, political involvement by the public increases. Accordingly, there is a return to binding values and a greater focus on non-economic dimensions. Trust, harmony and balance are considered the leading indicators for the quality of life.

52.4 The Societal Scenarios in GLOWA-Danube

The social scenarios developed and described by SinusSociovision relate to the overall social situation and have a correspondingly high level of abstraction. In order to render the scenarios useful for the GLOWA-Danube sub-projects, a specification was carried through in the sense of a certain contextual manifestation specific to each sub-project. The methodology for this approach is described in the paragraphs that follow.

52.5 Methodology

52.5.1 Step 1: Assessing Suitability

In the first step, for each sub-project and the respective submodels, the scenarios developed by SinusSociovision were evaluated for their suitability as a basis for the GLOWA-Danube scenarios. A prerequisite for this assessment was that in each model at least one so-called key factor was available that could be assigned plausible, varying manifestations, depending on the scenario. Since this was the case in all the relevant sub-projects, both Sinus scenarios were selected for the GLOWA-Danube scenarios.

New labels were chosen for those scenarios to distinguish them from the original scenarios from SinusSociovision and at the same time ensure alignment clear semantics. The *Free is Fair* scenario was renamed *Performance*; the *Shared Destiny* scenario was termed *General Public Interest* in the GLOWA-Danube context.

A decision was made to also include a *Baseline* scenario. A *Baseline* scenario is a *business as usual* scenario. Depending on the sub-project, it forms the current status quo or the established behaviour pattern and extrapolates this into the future. This scenario is rather important, especially in light of the interpretation of the two other scenarios and their deviations from the baseline scenario. Hence, there are three societal scenarios:

- Scenario 1: *Baseline*
- Scenario 2: *Performance*
- Scenario 3: *General Public Interest*

52.5.2 Step 2: Scenario Description

To implement the scenarios within each sub-project, a precise description of the societal scenarios of SinusSociovision was essential. For this reason, the second step involved the detailed description of the two chosen antagonistic scenarios in their fundamental socio-political aspects, which was then made available for use in the sub-projects.

52.5.3 Step 3: Implementation Specific to the Sub-projects

The specification of the scenarios took place in the third step. The overall societal aspects that had until now been described only at a highly abstract level were implemented into the sub-projects' specific domains. This took place in the form of small narratives, so-called storylines.

Example of a Storyline
Tourism in scenario 2 ***Performance*** (excerpt)

The touristic demand for overnight stays in the study area increases on the one hand as a result of globalisation and hence new guest types and on the other hand as a result of increased hedonist motivations in a thrill-seeking society. The same is true for the ski areas of the drainage basin: there are signs of growing tendencies. Because of the expansion of the technical infrastructure in the skiing areas, it is required to raise the economic viability threshold to earn back the investment costs. This is supported in part by shifting the use of snow cannons towards zero degrees Celsius, since all technical possibilities have to be used.

A corresponding description of the *Baseline* scenario and the domain-specific formulation of the other two scenarios can be found in the chapters of the respective sub-projects.

52.5.4 Step 4: Scenario Drivers

In the preceding step, the aspects of a scenario that are specific to a sub-project were described verbally. In the final step, the model parameters relevant for implementing a scenario in each sub-project were identified. The model parameters that are selected implement the established key factors for the societal scenarios from the point of view of each sub-project.

The result is a list of the driving factors contained within each submodel and their attribute values (Table 52.1).

Table 52.1 Driving factors and their attribute values in the three scenarios for the submodel *Tourism* in Chap. 46 (excerpt)

Factor	Attribute value in scenario 1 – Baseline	Attribute value in scenario 2 – Performance	Attribute value in scenario 3 – General public interest
Day tourism activities (relating to the catering industry)	Unchanged	Increases	Unchanged
Water use in catering industry	Unchanged	Increases	Unchanged
Water use in hotels	Unchanged	Increases	Unchanged
Expansion of snow cannon infrastructure	Yes, moderate expansion	Yes, strong expansion	No expansion
Economic viability threshold of skiing areas (minimum days per year open)	Unchanged	Increases	Unchanged

Reference

de Vries J, Perry T (2007) Der demografische Wandel und die Zukunft der Gesellschaft. In: Navigator. Der Newsletter von SinusSociovision, vol 2/2007, SINUS Markt- und Sozialforschung GmbH, Heidelberg. www.sinus-institut.de/en

Part V
Integrative Results

Chapter 53
Impact of Climate Change on Water Availability

Florian Zabel

Abstract Climate change affects all elements of the water balance: precipitation, evapotranspiration and runoff. Decreasing annual precipitation by 55 mm and increasing annual evaporation by 47 mm leads to a decrease in annual water availability by 102 mm in the Upper Danube river basin when comparing the periods 2036–2060 with 1971–2000 when choosing the *REMO Regional Baseline* Scenario. The highest impact was found at the northern edge of the Alps, where water availability decreases by up to 400 mm. This article analyses the impact of the changed water balance on the discharge regime. We conclude a reduction of discharge between 60 and 370 m^3/s at the catchment outlet in Achleiten until 2060. A serious change in monthly discharge was found during summer, when discharge in Achleiten decreases by almost 40 % until 2060, while the annual maximum shifts from summer to spring.

Keywords Climate Change • Water balance • Water availability • Precipitation • Evapotranspiration • Runoff • Discharge • Scenarios • GLOWA-Danube

53.1 Introduction

The water availability within a basin is composed of the elements precipitation (*P*), discharge (*Q*) and evapotranspiration (*ET*). The water balance equation describes the relationship among these factors as:

$$Q = P - ET + \Delta S$$

F. Zabel (✉)
Department of Geography, Ludwig-Maximilians-Universität München (LMU Munich), Munich, Germany
e-mail: f.zabel@lmu.de

W. Mauser, M. Prasch (eds.), *Regional Assessment of Global Change Impacts*, DOI 10.1007/978-3-319-16751-0_53

For long periods of time and stable climate conditions, it can be assumed that the change of stored water (ΔS) remains constant and only a redistribution takes place among the links of the water balance. However, over the course of climate change, a change in the volume of stored water is expected, although this is not considered within this article.

The precipitation that is not returned to the atmosphere via evapotranspiration flows into the sea during a perpetual "natural water cycle" following the terrain. Only this portion of the water in the "natural water cycle" is available for human use (e.g. for agriculture and food production, irrigation, energy production, industrial use, public supply and drinking water). In this way, the water availability characterises the hydrological properties of a drainage basin and provides an overview of the available water resources. The water availability is a result of the given climate conditions. Thus, if these conditions change, the basin's water availability changes.

The goal of this section is therefore to present the effects of climate change in the Upper Danube on the availability of water. The results have great significance for long-term planning and water management.

53.2 Water Availability

In the reference period from 1971 to 2000, an average annual precipitation of 1039 mm was observed for the area of the Upper Danube (see Map 53.1(1a)). The highest annual precipitations occur in the northern borders of the Alps, in the Central Alps and in the Bavarian Forest in the northeast as well as in the Black Forest in the western part of the basin.

From the fallen precipitation evaporate 388 mm/a via transpiration by plants, evaporation and interception (see Map 53.1(1b)). Evaporation is spatially heterogeneous, since it is highly dependent on the heterogeneous land use/cover and soil conditions. In Map 53.1(1b), an elevation gradient of evaporation is apparent, since evaporation is controlled to a large extent by temperature. Forested regions at lower terrain elevation (e.g. Hofoldinger Forest, south of Munich) show the highest annual values of evaporation rates. On average over the entire basin, this yields a water availability for the past of 651 mm/a (see Map 53.1(1c)). A spatially-differentiated presentation of the available water resources from 1971 to 2000 shows a very uneven distribution. While the southern part of the basin (south of 48° N) and the Bavarian Forest are areas with great water excesses, the northern part of the basin (north of 48° N) clearly received less precipitation and at the same time shows higher rates of evapotranspiration. This leads to less water transformed to runoff.

In the future, the interaction of the elements of the water availability will change. Climate change has multiple impacts on the availability of water. The *REMO regional* climate trend with the *Baseline* climate variant (see Chaps. 47, 48, 49, 50) was chosen for a detailed scenario analysis. In this GLOWA-Danube climate scenario, a total of 984 mm of precipitation still falls on average in the Upper Danube basin for the second half of the scenario period from 2036 to 2060.

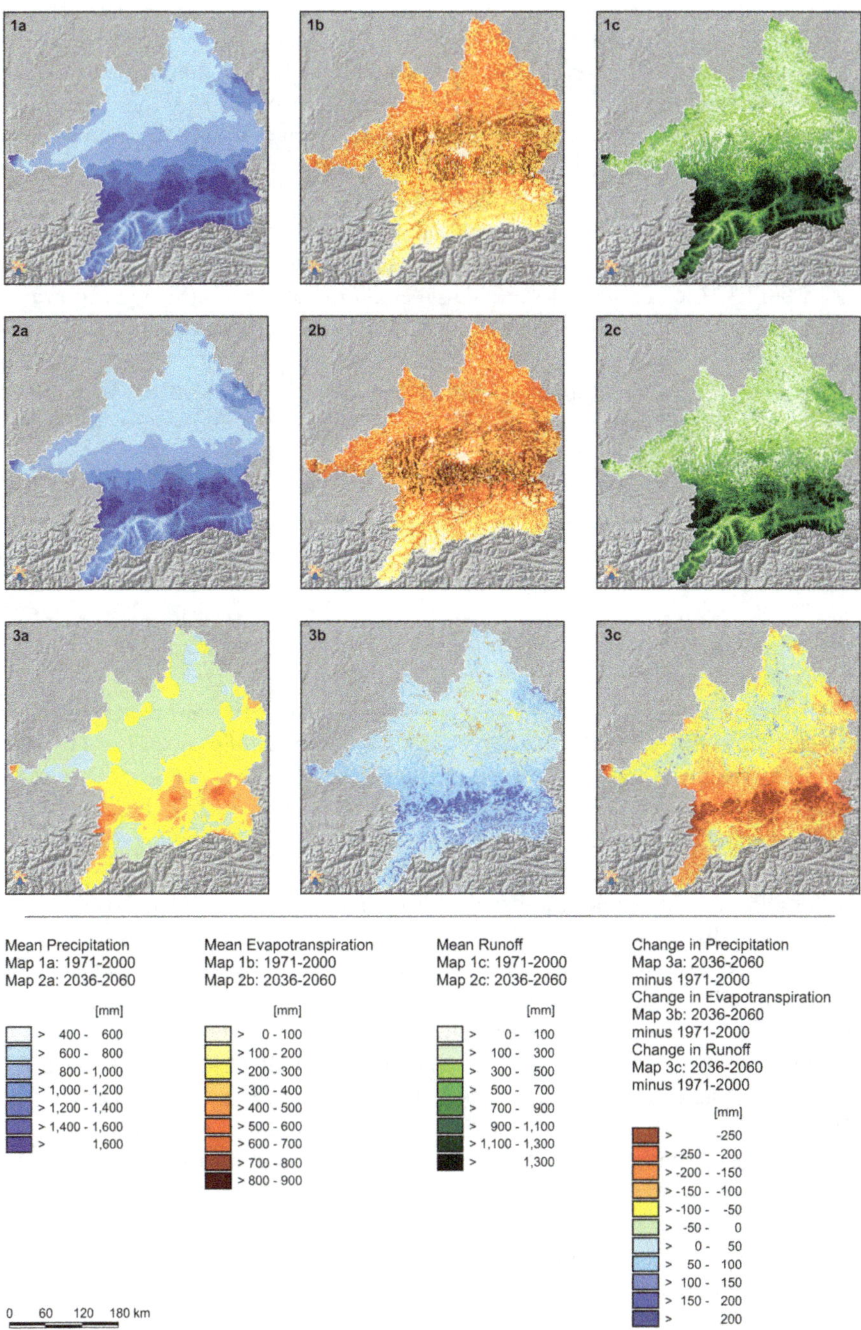

Map 53.1 Impact of climate change on water availability (*REMO regional* climate trend and *Baseline* climate variant)

Of this, 435 mm evaporate. Thus, 548 mm per year still contribute to runoff (see Map 53.1 – 2a–c). For a better representation of the spatial changes of water availability, Maps 53.1(3a–c) illustrate the change in water balance from 2036–2060 compared to 1971–2000 as a difference. The water supply over the entire Upper Danube basin is reduced on average by 102 mm, caused by a reduction of precipitation by 55 mm and an increase of evapotranspiration by 47 mm.

The greatest change in water cycle takes place at the northern Alpine fringe. Here, a significant decrease in the water supply by up to 400 mm is evident (see Map 53.1(3c)), due to decreasing precipitation along with increasing evapotranspiration rates that are triggerd by warmer temperatures and atmospheric CO_2 increase.

On the other side, evapotranspiration decreases in the northern part of the catchment, that results in a damped decrease in water availability here, while precipitation also slightly decreases.

53.3 Discharge

The runoff that concentrates in rivers, finally leaves the Upper Danube basin at the drainage outlet in Achleiten.

On the left in Fig. 53.1, the mean annual discharge (MQ) measured at the gauge in Achleiten is shown for 1971–2000. The MQ for the preceding period is 1,418 m^3/s. On the right in the figure, the temporal course of discharge for all four climate trends is shown, each depicted as the mean of the four associated climate variants from the statistical climate generator. The range between the minima and maxima for all 16 GLOWA-Danube climate scenarios appears in grey.

For all averaged climate variants within a climate trend, Fig. 53.1 shows a negative trend in discharge at gauge Achleiten during the scenario period from 2011 to 2060. The decrease in the mean climate variants from the linear trend in this period is between 60 m^3/s (*IPCC regional* climate trend) and 370 m^3/s (*Extrapolation* climate trend). The *REMO regional* climate trend decreases by 189 m^3/s.

Fig. 53.2 illustrates the trends in mean discharge at Achleiten based on the down-scaled and bias corrected results from the *REMO* and *MM5* models (see Chaps. 47 and 51). For the preceding period 1971–2000, these show a mean discharge of a similar range (1,450 m^3/s), when compared to the measurements from the gauge in Achleiten.

For the scenario period until 2060, the simulated trend in mean annual discharge calculated with the results of the down-scaled and bias corrected regional climate model also shows a decrease both for *REMO down-scaled and bias-corrected* and for *MM5 down-scaled and bias-corrected.* Since the results of the regional climate model are available for up to 2100, the mean discharges were continuously calculated up to 2100. There is a significant decrease in annual discharge after 2060, as a result of driving the regional climate models at the edges with data from the global climate model ECHAM5 (Simulation A1B, Member 1). Between 2011 and 2100,

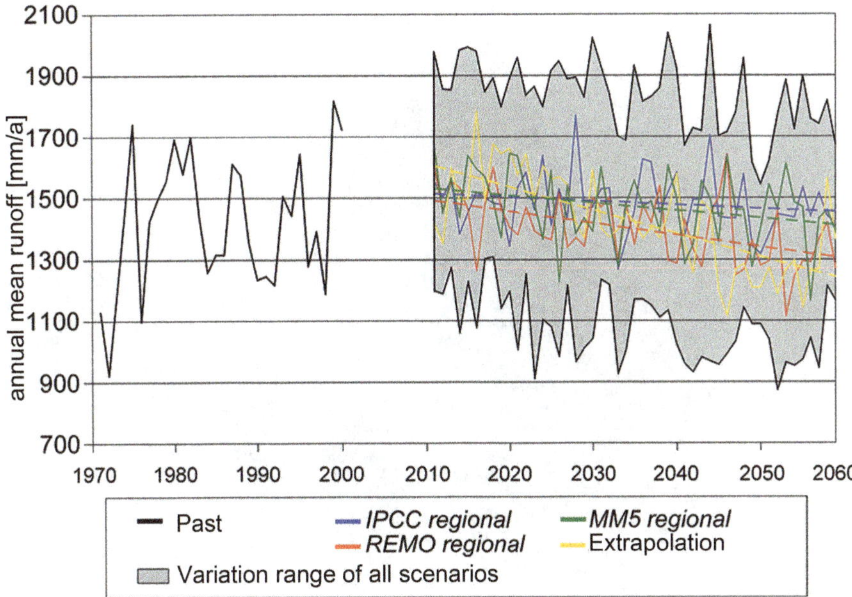

Fig. 53.1 Temporal evolution of discharge for the reference runoff (1971–2000) and the different GLOWA-Danube Scenarios (2011–2060) aggregated over the four statistical climate variants of each climate trend; the *dashed lines* represent linear trend *lines*

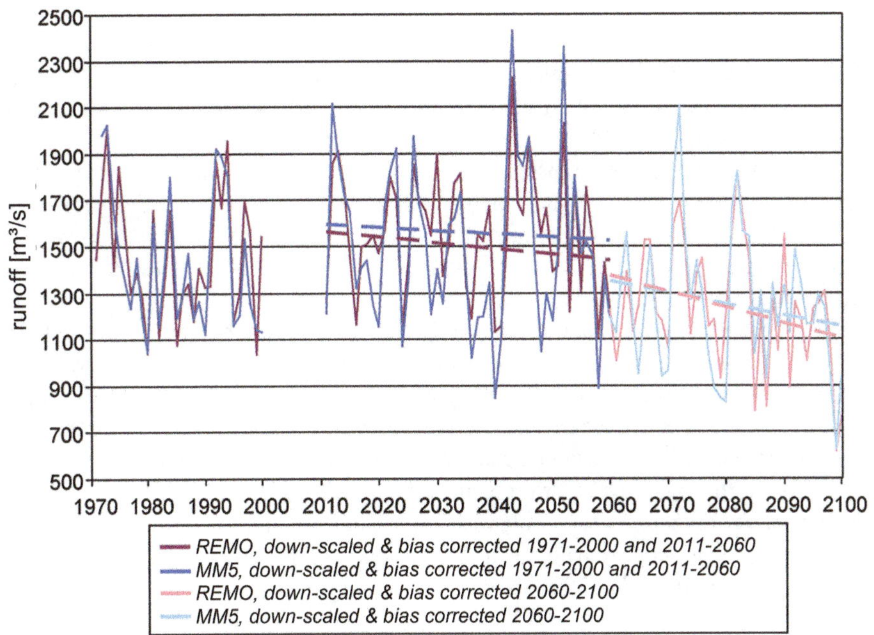

Fig. 53.2 Annual mean runoff at the gauge in Achleiten

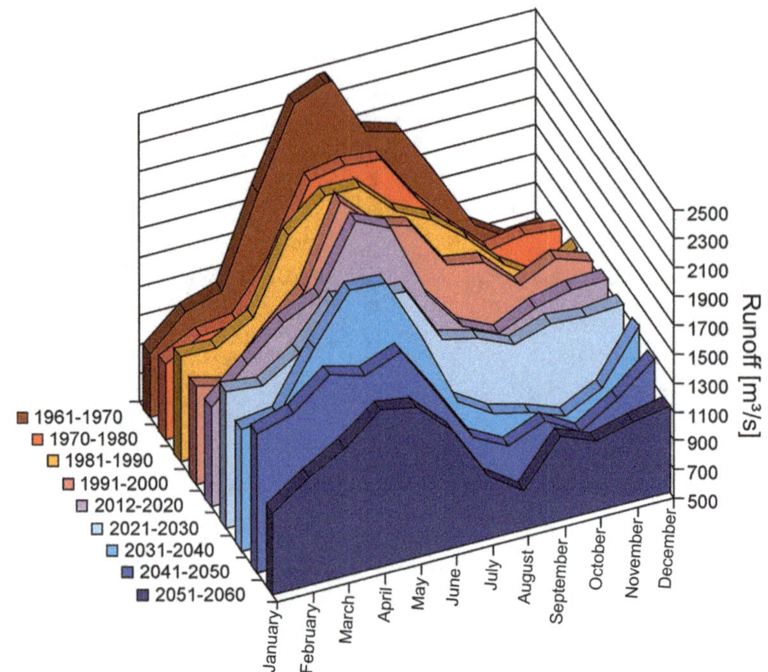

Fig. 53.3 Monthly mean runoff values at the outlet gauge in Achleiten averaged over the shown decades. From 2012, the GLOWA–Danube *REMO regional – Baseline* scenario is shown

the discharge at Achleiten reduces by 560 m^3/s for the *REMO down-scaled and bias corrected* climate variant and by 450 m^3/s for the *MM5 down-scaled and bias corrected* climate variant.

In addition to the annual discharge, the analysis of the discharge regimes yields important statements about the trend in water availability. Figure 53.3 shows the trend in mean monthly discharge for each decade between 1961 and 2000 as well as between 2012 and 2060 for the *REMO regional* climate trend with the *Baseline* climate variant.

A significant and gradual change in the discharge regime can be observed. To date, the mean monthly discharges at gauge Achleiten reaches a maximum in the summer months as a result of summer precipitation and snow melt in the Alps. The lowest mean discharges are recorded in the winter months. Measurements over the past three decades already detected a shift in the annual maximum of the mean monthly discharge towards spring. Until 2060, this trend continues with a simultaneous, steady reduction of monthly discharge. The annual minimum shifts from the winter months to August. The mean discharge in August reduces from 2,023 m^3/s (1971–1980) to 767 m^3/s (2051–2060).

53.4 Conclusion

The availability of water in the Upper Danube basin will be significantly reduced in the future, depending on the climate scenario chosen. The southern part of the basin can expect a drastic decrease in the in water availability, while water availability will only decrease slightly in the northern part.

The results of the down-scaled and bias corrected regional climate models *REMO* and *MM5* are suitable for only general statements concerning the simulation period until 2060. Nonetheless, they indicate a decline in the mean discharge by 2060 followed by an abrupt decrease until 2100.

The climate trend for the statistical climate generator averaged over all climate variants shows a reduction range between 60 and 370 m^3/s at the outlet gauge in Achleiten from 2011 to 2060. A significant change in the discharge regime at gauge Achleiten accompanies the general decline in the water availability. Over the course of the scenario period, the highest monthly gauge discharges shift from the summer months to the spring.

Chapter 54
Scenarios for the Development of Low Flow in the Upper Danube Basin

Wolfram Mauser, Florian Zabel, Thomas Marke, and Andrea Reiter

Abstract The projected future impact of climate change on low flows in the Upper Danube basin is analysed for a broad range of climate change scenarios based in a stochastic climate generator or on results of regional climate simulations. The analysis was carried out from 2011 to 2060 both for selected gauges and specifically the outlet gauge at Achleiten as well as of the spatial distribution of the 50-year return period weekly low flow. All selected regional climate scenarios result in a decrease of annual low flows at the outlet of the Upper Danube until 2060 between 15 and 50 %. The picture is very differentiated when looking at the spatial distribution of the low flow change as a result of climate change. Whereas low flow can be expected to increase in Alpine values, it may decrease in the Alpine forelands. The results are summarised in four key statements.

Keywords GLOWA-Danube • Low flow • Uncertainty

54.1 Introduction

Low flow is a significant limiting factor in the utilisation of water resources. Low flow is typically characterised as a lasting, lower than average discharge at a gauge within a river system. There are a variety of causes for low water levels. These

W. Mauser (✉) • F. Zabel
Department of Geography, Ludwig-Maximilians-Universität München (LMU Munich), Munich, Germany
e-mail: w.mauser@lmu.de; f.zabel@lmu.de

T. Marke
Institute of Geography, University of Innsbruck, Innsbruck, Austria
e-mail: thomas.marke@uibk.ac.at

A. Reiter
Bavarian Research Alliance GmbH, Munich, Germany
e-mail: reiter@bayfor.org

W. Mauser, M. Prasch (eds.), *Regional Assessment of Global Change Impacts*, DOI 10.1007/978-3-319-16751-0_54

range from reduced precipitation or increased evapotranspiration to diminished water stored naturally as snow and ice in the mountains. Usually only a combination of these factors causes abnormally low water levels that lead to restrictions for hydropower plants, for example, navigation or cooling in connection with thermal energy production. Furthermore, low water in conjunction with elevated temperatures can negatively impact water quality. Before appropriate measures can be taken to adjust for climate change in the field of low flow, it is therefore necessary to understand the impact of the expected climate change on the low flow situation.

In order to make statements about the expected extent of the changes in low flow caused by climate change, the water supply of the Upper Danube was modelled and analysed using 18 GLOWA-Danube climate scenarios for the period 2011–2060 at gauge Achleiten as well as for the entire region of the Upper Danube drainage basin. The scenarios chosen are described in Chaps. 47, 48, 49, 50, 51 and comprise all the combinations of climate trends and variants based on the statistical climate generator introduced there as well as on down-scaled and bias-corrected results from the *REMO* and *MM5* regional climate models. In addition, the low flow development of a null scenario was examined, in which no temperature increase takes place from 2011 to 2060.

54.2 The Future Development in Low Flow at Gauge Achleiten

The discharges of the Upper Danube were calculated hourly for all climate variants for the period from 2011 to 2060 for the entire Upper Danube drainage basin, and these values were aggregated and stored as daily values. Discharges at the drainage outlet of the Upper Danube (gauge Achleiten, near Passau) were then analysed together with the historical reference period 1971–2000 for the minimum annual average 7-day discharge (NM7Q). The NM7Q was selected as the parameter, because together with the NM30Q, it has been most commonly used in ongoing studies of low flow situations. This parameter is approximately of the same order of magnitude as the lowest daily average, but is less susceptible to measurement errors or short-term anthropogenic factors, since these are levelled off (Schiller 1978; Helbling et al. 2006) (Fig. 54.1).

The null scenario was then examined to ensure that there are no model instabilities that might simulate a future change in low water discharge without climate change. The historical measurements and future simulations were assumed to belong to the same data population. Trend analysis of the linear regression over the entire period indicates that despite a slight decrease in low flow, there is no significant temporal trend at the 99 % probability level.

The temporal trend in the resulting NM7Q for the ensemble of 18 GLOWA-Danube climate scenarios, which consists of combinations of four climate trends and four climate variants from the climate generator as well as the two climate variants

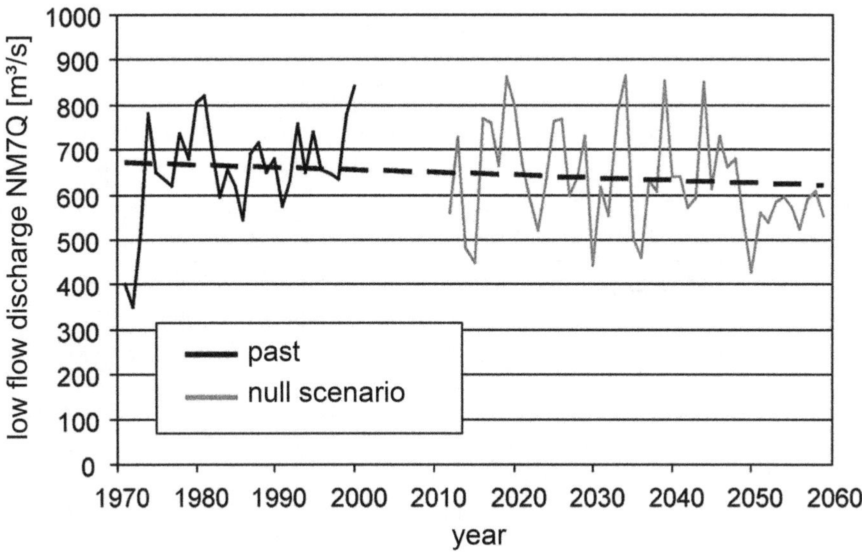

Fig. 54.1 Measured (1971–2000) and simulated (2011–2060) annual low flow discharge (NM7Q, minimum average 7-day discharge) at gauge Achleiten, based on the null scenario (gauge data: © Bayerisches Landesamt für Umwelt, www.lfu.bayern.de)

from the *REMO* and *MM5* regional models, is shown in Figs. 54.2 and 54.3 along with the historical low flow discharges. The left sides of Figs. 54.2a and b indicate the NM7Q calculated using measured discharges at gauge Achleiten for the historical period. In contrast, on the left side of Fig. 54.3, the past is modelled using *REMO* and *MM5* driven by the global ECHAM5 simulator (see Chap. 51). The right side of Fig. 54.2a shows the low flow discharges simulated as modelling results for all 16 climate variants from the statistical climate generator. There is notable statistical spread in the trend that is caused by random variation in the climate generator and which reproduces natural variability. For improved clarity, Fig. 54.2b depicts the mean of each of the four climate variants chosen (*five warm winters, five dry years…*) for each of the four climate trends. In comparison to the null scenario, there is a significant reduction in the low flow discharge at gauge Achleiten for each of the averaged climate variants of the four climate trends over the study period. This reduction is weakest for the *IPCC regional* climate trend and strongest for the *Extrapolation* (forward projection, WM: we should clarify whether we should use this term throughout the text) climate trend. The spread caused by the various assumptions of each climate trend ranges from a reduction of low water discharge by 15–50 % of the current value by the year 2060.

Figure 54.3 illustrates the simulation of annual low flow discharge using the down-scaled and bias-corrected results of the *REMO* and *MM5* regional climate models (see Chaps. 47 and 51). The mean simulated low flow discharges in the past are consistent with the means from the gauge measurements shown in Fig. 54.1.

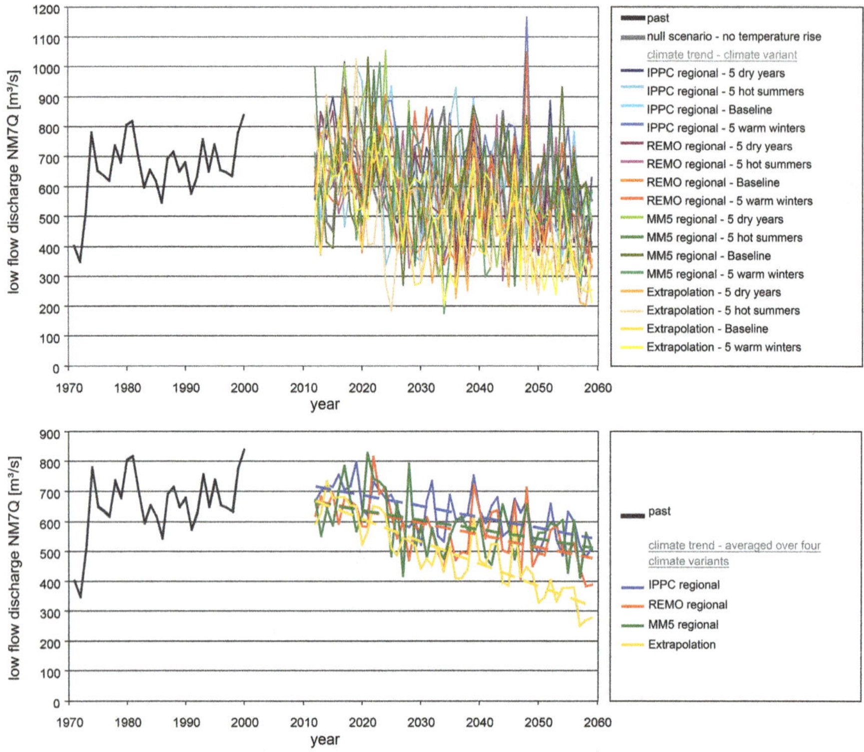

Fig. 54.2 Annual low flow discharge (NM7Q: minimum average 7-day discharge) at gauge Achleiten 1971–2000 (measured) and 2011–2060: (**a**) 16 statistical GLOWA-Danube climate variants (**b**) each of the four climate trends averaged over the four climate variants; *dashed lines* represent linear trend lines (gauge data: © Bayerisches Landesamt für Umwelt, www.lfu.bayern.de)

Both *REMO down-scaled* and *bias-corrected* and *MM5 down-scaled* and *bias-corrected* involve a simulation of historical climate conditions that are expected to mirror reality only in statistical properties and not in actual weather patterns. Nevertheless, the mean measured NM7Q value (658 m^3/s) is clearly consistent with the simulated value based on the down-scaled and bias-corrected climate model (639 m^3/s), as in the case of the statistical climate generator. In contrast to the results of the low flow analysis using the climate generator, the simulated trend in low flow discharge for 2011–2060 from the results of the regional climate models is largely consistent (see Fig. 54.3). This results from the fact that both models are driven by the global model ECHAM5. With the exception of the decade 2030–2039, precipitation amounts in these particular model runs significantly decline only by 2060. In order to demonstrate this, in Fig. 54.3, the time frame was extended by 40 years to 2100. This reveals a strong decrease in NM7Q after the year 2060. The increase of an imaginary linear trend from 2011 to 2100 (–4.45 m^3/year) corresponds approximately

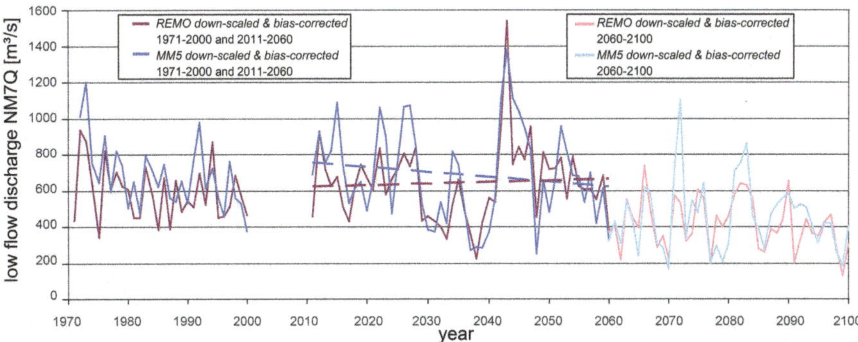

Fig. 54.3 Annual low flow discharge (NM7Q: minimum average 7-day discharge) for 1971–2000 and for 2011–2100 at gauge Achleiten with the down-scaled and bias-corrected data of *REMO* and *MM5* regional climate models; the *dashed lines* represent linear trend lines; averaged low flow discharge 1971–2000: 639 m^3/s

to the average from the four climate variants of the *REMO regional* climate trend (–4.05 m^3/year). This consistency can be explained by the fact that in the case of the *REMO regional* climate model, the climate generator derives the long range trend in precipitation based on the trend from 1990 to 2100 directly from the results from *REMO* and reconstructs the change in precipitation to 2060 from these results (see Chap. 48).

In addition to studying the changes in annual low flow discharge, scenario statements on possible changes of return periods of extreme low flow events are of key concern. Such statements involve the annual low flow discharge that occurs at a specific statistical probability of occurrence. The annual probabilities of occurrence are generally chosen between 10 % (every 10 years), 2 % (every 50 years) and 1 % (every 100 years). The low flow discharge occurring every 50 years was used to investigate the changes in extreme low flow discharges. This value was determined from each 25-year time period of annual low flow discharges (NM7Q), (DVWK 1992), with the assumption that the annual low flow discharges follow a log-normal distribution (DVWK 1983, 1992). To investigate the trend over time of the 50-year low flow discharge, the 25-year time period that was used was shifted year-by-year over the available period of annual low flow discharges. In this way, six values of 50-year low flow discharge were obtained for the historical period 1971–2000, which represent the periods from 1971–1995 to 1976–2000. In the same way, 26 different 50-year low water discharge values were obtained for the period 2011–2060, representing the periods 2011–2035 to 2036–2060.

The result of the analysis of the temporal trend in the 50-year low flow discharge NM7Q50 (which denotes a 50-year return period NM7Q) for different periods in the past and future is shown in Fig. 54.4. The standard deviation of the mean

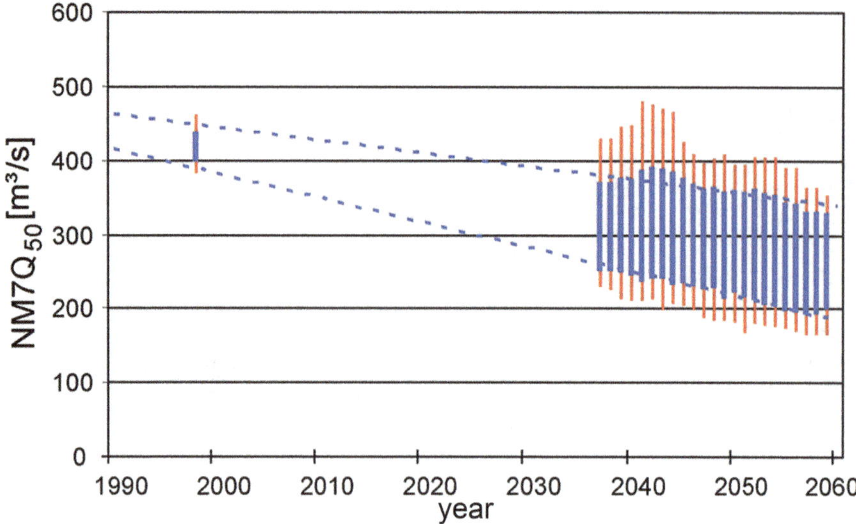

Fig. 54.4 50-year low flow discharge (NM7Q50) at gauge Achleiten; *blue bars*: standard deviation of the mean value; *red bars*: minimum/maximum; *dashed lines*: regression line of minimum and maximum level of the standard deviation of the years 2036–2060, extrapolated until 1990

NM7Q50 low flow discharge for the six 25-year periods of the years 1971–2000 is shown as a single blue bar in the left of the figure. The red lines above and below the blue bars indicate the NM7Q50 maxima and minima calculated for the period. Because the simulations of future trend begin in 2011, values appear only after 25 years, in the hydrological year 2037. Here, the blue bars represent the standard deviations of the mean NM7Q50 low flow discharge for all 16 statistical climate variants.

It is clear that the variation in low flow values was low in the past and continues to increase over time into the future. One explanation for this result is that in the past it consisted of the realisation of a single reality, which consisted of measured values; another reason is that the effects of the different climate scenarios on low flow become more distinguishable from year to year and hence the variation of the 50-year low discharges simulated with different climate scenarios increases with time.

Two regression lines are drawn through the upper and lower limits of the future standard deviation (blue) bars and extrapolated into the past. This extrapolation clearly encompasses the low flow values based on the measurements. Whereas the 50-year low flow discharges in the past were between 450 and 500 m^3/s, by the year 2060, these values are, based on the selected scenarios, only between 200 and 350 m3/s.

54.3 The Regional Development of Low Flows in the Upper Danube Basin

As the mean response in the drainage basin, the results of the analyses of the discharges at gauge Achleiten shown in Fig. 54.4 reveal a reduction in the 50-year low discharges to 2060 by up to two thirds of today's value. Gauge Achleiten at the outlet of the Upper Danube integrates various trends in the different sub-basins within the Upper Danube basin into a mean change for the overall basin. Thus, the regional differences and changes are often averaged out. Hence, the subsequent analyses sought to clarify the trends in low flow in the different regions of the Upper Danube. The method described above for calculating $NM7Q_{50}$ at a specific level of significance was thus extended, applied to each 1×1 km^2 model proxel, and the resulting spatial distribution and temporal trend of $NM7Q_{50}$ were analysed. The coefficient of determination R^2 from the regression equation of the logarithmic low flow discharges sorted according to their magnitudes with their rank was used as the significance criterion. R^2 is 1 for a perfect log-normal distribution and 0 for a random distribution. In the subsequent analyses, all proxels that have $R^2 < 0.8$ resulting from the calculation of the 50-year low flow discharge were excluded from analysis; this value indicates that less than 80 % of the variation in the data is explained by the regression equation. From Fig. 54.2b, where non-significant proxels were whitened, it is clear that this threshold means that the small receiving water courses in the basin are not sampled in the analysis and it is predominantly larger rivers that are considered. Using the method described, the values for the 50-year low water discharges were calculated for two periods (2011–2035 and 2036–2060) for all significant proxels in the Upper Danube basin for the *REMO regional* climate trend and the *Baseline* climate variant. This climate scenario was used as an example since it takes on a middle position in the future change of low flow discharges at gauge Achleiten (see Fig. 54.2b). The values calculated for the $NM7Q_{50}$ are shown in Maps 54.1 (1a and b) for all proxels with a coefficient of determination of $R^2 > 0.8$. The main river network with the Danube and Inn rivers is easy to discern. In a subsequent step, the quotient of the future (2011–2035 and 2036–2060) and past (1971–2000) 50-year annual low flow discharge values was calculated to determine the regional changes in low flow discharge that are caused by climate change. A quotient >1 signifies a future increase in low flow discharge, whereas a quotient <1 means a reduction in low flow discharge. The quotients are presented in Maps 54.1 (2a and b) for the *REMO regional* climate trend and the *Baseline* climate variant for the proxels with a coefficient of determination of $R^2 > 0.8$.

In the maps, a separation in the response of the low flow discharge to climate change in the basin is apparent both in the quotients between the values for 2011–2035 and 1971–2000 (Map 54.1 (2a)) and even more clearly in the quotients between the values for the period 2036–2060 and 1971–2000 (see Map 54.1 (2b)). Although the low flow discharge slightly decreases in the first period (2011–2035) in the Alpine foothills and north of the Danube, it increases slightly in the Alps. This tendency is even clearer in the period from 2036 to 2060. In this period, the low

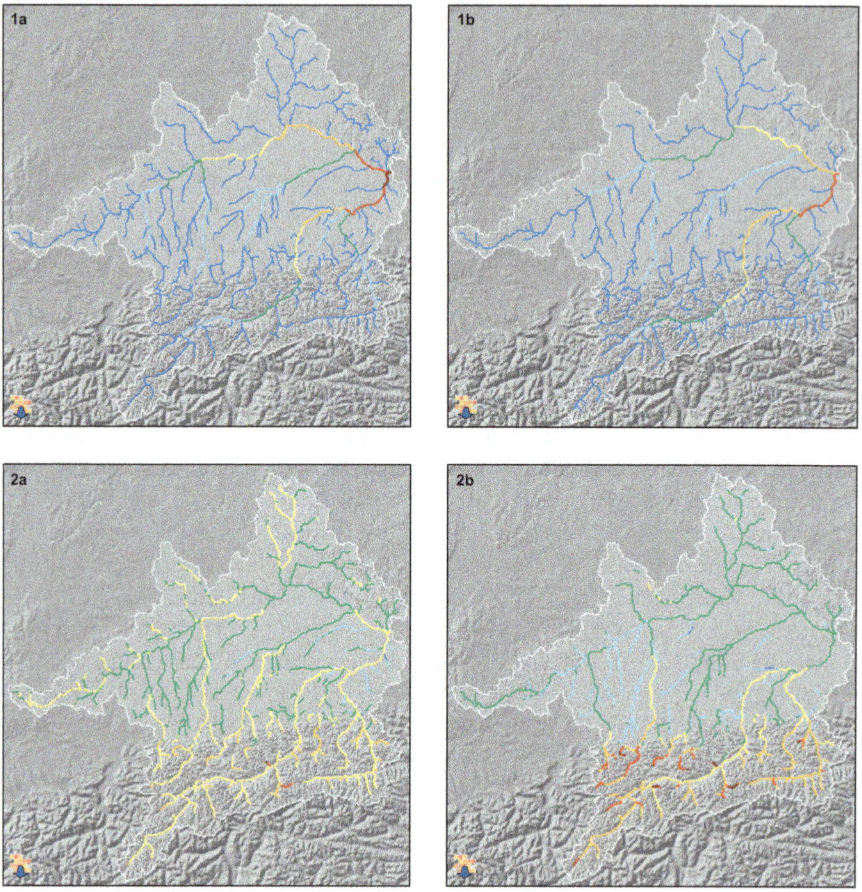

Map 1a: 50-year low flow discharge (NM7Q_{50}) in m³/s for the time period 2011-2035 for the main rivers in the Upper Danube Basin under the *REMO regional* climate trend and the *Baseline* climate variant

Map 1b: 50-year low flow discharge (NM7Q_{50}) in m³/s for the time period 2036-2060 for the main rivers in the Upper Danube Basin under the *REMO regional* climate trend and the *Baseline* climate variant

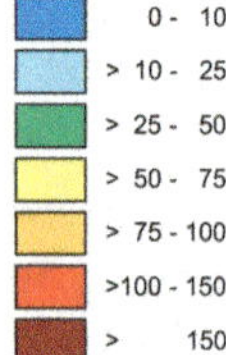

Map 2a: Quotient of future 50-year low flow discharge for low flow discharge in the past (2011-2035)/(1971-2000) under the *REMO regional* climate trend and the *Baseline* climate variant

Map 2b: Quotient of future 50-year low flow discharge for low flow discharge in the past (2036-2060)/(1971-2000) under the *REMO regional* climate trend and the *Baseline* climate variant

0 30 60 90 120 km

0 - 10
> 10 - 25
> 25 - 50
> 50 - 75
> 75 - 100
>100 - 150
> 150

0 - 0.25
>0.25 - 0.50
>0.50 - 1.00
>1.00 - 1.50
>1.50 - 2.00
>2.00 - 2.50
> 2.50

Values of the quotient between 0 and 1 represent a decrease, values greater than 1 respresent an increase of low flow discharge.

Only proxels with a coefficient of determination $R^2 > 0{,}8$ are shown.

Map 54.1 Scenarios for the development of low flow in the Upper Danube Basin

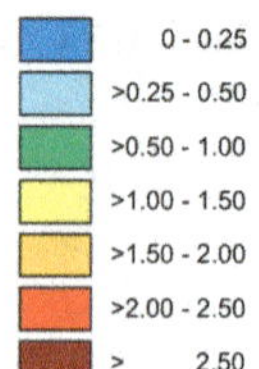

flow discharge in the Alpine foothills and north of the Danube sometimes drops down to 30 % of the mean value in the past, while the low flow discharge in the Alps almost even doubles in many sections of the rivers. One exception from this general trend is the Lech River in the Alpine foothills, where the low flow discharge increases in the future. This basic statement is evident for all the climate scenarios that were studied, although with stronger or weaker tendencies in each case. The scenario illustrated confirms its middle position.

The causes for the dichotomous trend are based on a complex interaction of elevated precipitation amounts and reduced snowfall in winter precipitation and enhanced evapotranspiration with reduced precipitation in the summer months. In the Alps, where the low water discharge arises today in winter as a result of lower temperatures, the elevated precipitation amounts and snow melt in winter will increase the low flow discharges during that season in the future. In contrast, the decreasing amounts of precipitation (see Chap. 48) and increasing evaporation in summer will significantly reduce the low flow discharges in the Alpine foothills and north of the Danube and thereby lead to an overall future reduction of discharge in summer.

54.4 Summary

The results of the studies on low water can be summarised as follows:

The low flow situation at gauge Achleiten will significantly worsen in the future as a result of climate change. Based on the ensemble analysis with the statistical climate generator, a reduction in low flow discharge MN7Q at Achleiten is expected by the year 2060; depending on the climate scenario selected, this may be a reduction by 15–50 %.

The evaluation of the down-scaled and bias-corrected results of the *MM5* and *REMO* regional climate models provides for similar conclusions as the statistical climate variants beyond 2060. This is based on the fact that the climate model was driven by ECHAM5, and this permitted significant changes in precipitation only after 2060. Thus, the result achieved up to 2060 should not lead to the conclusion that the low flow discharges up to 2060 remain constant and will only change after this date. At the time of the study completion (2009), no additional statements about the development in low flow discharge can be made with the regional climate models as a result of the lack of a statistical basis.

The future development in low flow discharge that arises as a result of climate change shows a regional split; this split shows an approximate doubling of the low flow discharges on average in the Alps by 2060, whereas the simulations indicate that there could be a significant reduction to up to 30 % of today's value of low flow discharge in the Alpine foothills and north of the Danube.

References

Deutscher Verband für Wasserwirtschaft und Kulturbau e.V. (DVWK) (1983) Regeln zur Wasserwirtschaft. Niedrigwasseranalyse Teil I: Statistische Untersuchung des Niedrigwasser Abflusses, vol 120

Deutscher Verband für Wasserwirtschaft und Kulturbau e.V. (DVWK) (1992) Regeln zur Wasserwirtschaft. Niedrigwasseranalyse Teil II: Statistische Untersuchung der Unterschreitungsdauer und des Abflussdefizits, vol 121

Helbling A, Kan C, Marti P (2006) Niedrigwasser kleinste Mehrtagesmittel des Abflusses. Hydrologischer Atlas der Schweiz

Schiller H (1978) Die Trockenperiode 1976 – Eine hydrologische Monographie und eine Niedrigwasseranalyse. In: Schriftenreihe des Bayerischen Landesamts für Wasserwirtschaft vol 12, München

Chapter 55
Scenarios for the Development of Floods in the Upper Danube Basin

Wolfram Mauser

Abstract The projected future impact of climate change on flood peak discharge and return period in the Upper Danube basin is analysed for a broad range of climate change scenarios based in a stochastic climate generator or on results of regional climate simulations. The analysis was carried out from 2011 to 2060 for the outlet gauge at Achleiten and for the two adjacent gauges in the main tributaries. In addition, the spatial distribution of the 100-year return period annual peak discharge was analysed. The result for the outlet gauge of the Upper Danube shows no significant changes of annual peak discharges with climate change until 2060. This is the result from a complex interaction of a slight decrease in the Alpine forelands and an increase in the Inn, which originates in the Alps. Spatially large increases of flood peaks in the Alpine headwaters, which are caused by rainfall-induced snowmelts, contrast with decreases in flood peaks in the forelands. The results are summarised in three key statements.

Keywords GLOWA-Danube • Flood frequencies

55.1 Introduction

Floods pose a significant natural risk. Although extreme flooding events are rare, they cause considerable personal and economic damage within the Upper Danube basin. Substantial floods have occurred over the past 10 years, most notably in the years 1999 (the "Whitsun flood"), 2002 and 2005. The drainage basins at the Alpine fringe were predominantly affected by these floods, but severe flooding occurred even in the Danube. Floods in the Upper Danube are predominantly triggered by largely intensive and long-lasting precipitation in the context of an Omega weather situation. In an Omega low-pressure system, moist air from the Mediterranean is brought towards the Alpine fringe via the Balkans. At the Alpine fringe, moist air

W. Mauser (✉)
Department of Geography, Ludwig-Maximilians-Universität München (LMU Munich), Munich, Germany
e-mail: w.mauser@lmu.de

W. Mauser, M. Prasch (eds.), *Regional Assessment of Global Change Impacts*, DOI 10.1007/978-3-319-16751-0_55

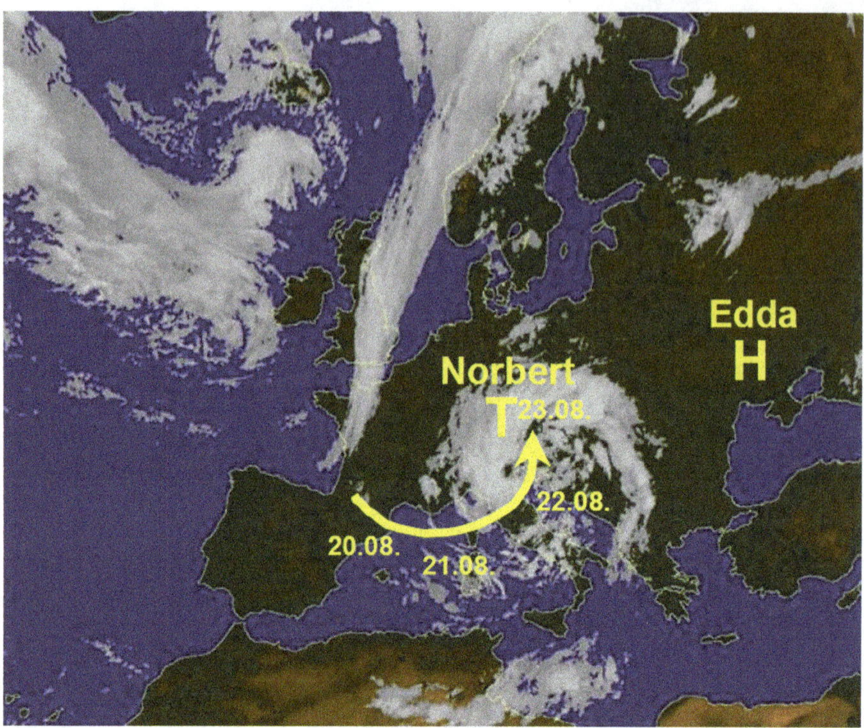

Fig. 55.1 V(5)-track cyclone, during the Danube flood in august 2005 (LfU 2007, p. 10)

masses lead to intense, long-lasting orographic precipitation events. Figure 55.1 shows exemplarily the weather condition that leads to flooding in August 2005, during an Omega weather situation. The route of the moist air masses within the "Norbert" depression is obvious.

In order to make statements about the future development in flood frequency and peak discharges that may arise through climate change, the 100-year peak discharge was used as the basis for assessment. The 100-year peak discharge at a specified location along a river, abbreviated as HQ_{100}, is calculated by fitting a statistical distribution to observed or simulated annual peak discharges. The 99th percentile of the selected distribution represents the peak discharge reached or exceeded at the selected location on statistical average in 100 years (DVWK 1999). To perform this calculation, the water balance and the river discharges in the Upper Danube basin were modelled with 18 GLOWA-Danube climate scenarios (four climate trends with each of four climate variants and two climate variants from the results from the *REMO* and *MM5* regional climate models) and analysed for the period from 2011 to 2060, both at gauge Achleiten and for all proxels that represent larger water bodies in the Upper Danube basin. The climate scenarios that were used are described in Chaps. 47, 48, 49, 50, and 51 and include all possible combinations of the climate trends and climate variants that are used by the statistical climate generator and the down-scaled and bias-corrected results from the *REMO* and *MM5* regional climate models.

Particular care is required to study floods using climate scenarios, since they are short-term events that last only a few days. The scenarios must be capable of accurately reproducing both the short-term precipitation dynamics and the long-term trends in soil moisture and reservoir accumulation and hence be able to correctly render the initial condition of the basin prior to the heavy precipitation; this is vital for accurately mapping the change in the probabilities of the occurrence of floods under future changed climate conditions. Under the restrictions for the meteorological drivers generated by the statistical climate generator (see Chap. 49) and the key assumption that in principle no new weather patterns will arise in the Upper Danube basin until 2060, the climate variants created with the climate generator are used for the further analysis of changes in flood statistics resulting from an assumed change in climate. The down-scaled and bias-corrected results of the *REMO* and *MM5* regional climate models were utilised for comparison, in order to be able to assess the quality of the statements from current regional climate models.

55.2 The Future Development in Flood Status at Select Gauges in the Upper Danube Basin

The results of the simulated discharges from 16 statistical climate variants (see Chaps. 49 and 50) were subjected to a statistical flood analysis based on the annual peak discharges with the aim to determine the 100-year return period flood. Based on the guidelines developed by the former German Association for Water Resources and Land Improvement (Deutscher Verband für Wasserwirtschaft und Kulturbau = DVWK 1999) for calculating the annuality of flood peak discharges, a log-normal distribution of annual flood peaks was assumed following the analysis of the coefficients of variation. With this assumption, the 100-year discharges at gauge Achleiten were determined for each 25-year measured discharge series from 1971 to 2000. Since there is a total of 30 years between 1971 and 2000, five values for the historical 100-year discharge are obtained by shifting the selected 25-year window one year at a time over this period. The mean, standard deviation, maximum and minimum were determined for these values. The same procedure was used for the 16 time series of annual peak discharges for the statistical climate variants for the period from 2011 to 2060.

Thus, for the 100-year discharge peaks for the first 25 years (2011–2035), this resulted in one value that was set for 2036. For 2037, the value is derived from the period 2012–2036 and so on. With this approach, there are 25 intervals for 2036–2060 for the variance of HQ_{100} for the different climate variants. The result is summarised in Fig. 55.2a–c for gauge Achleiten (beyond the junction of the Danube and Inn rivers), Ingling (on the Inn River, immediately before the junction with the Danube) and Hofkirchen (on the Danube, first gauge upstream of the junction with the Inn).

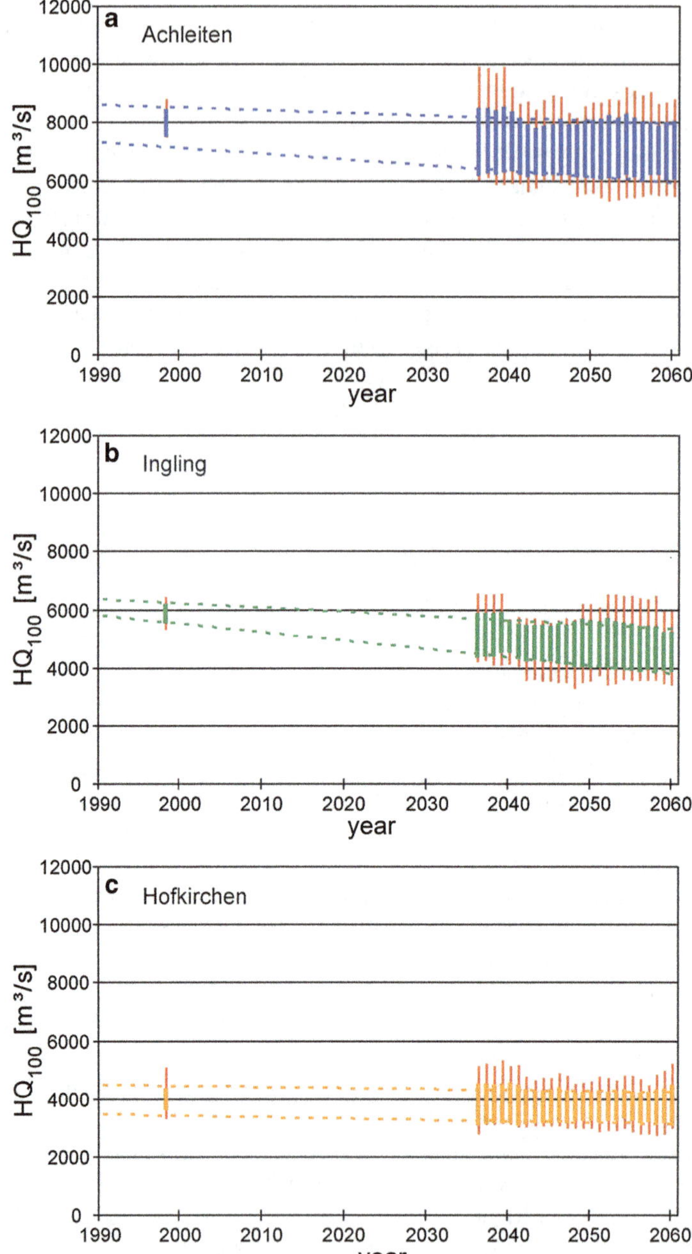

Fig. 55.2 (**a–c**) HQ100 at the gauges (**a**) Achleiten, (**b**) Ingling and (**c**) Hofkirchen, result of the statistical analysis of annual peak discharges of four climate variants of the four climate trends. *Blue* (**a**), *green* (**b**) and *orange* (**c**) *bars* standard deviation of the mean value, *red bars* maximum/minimum, *dashed lines* regression lines of minimum and maximum level of the standard deviation of the years 2036–2060, extrapolated until 1990

It is evident from Fig. 55.2a that under the climate variants from the climate generator, the peak discharge from the 100-year flood at gauge Achleiten is reduced on average from 8,000 m^3/s to approximately 7,000 m^3/s. Since processes are also conceivable within the area enclosed by the two dashed lines that yield an increase in HQ_{100}, a definitive statement about the trend is not possible.

Both the gauge at Ingling on the Inn River and the gauge at Hofkirchen on the Danube were similarly studied using the same method in order to carry out preliminary studies on the origin of this reduction. Gauge Ingling is situated immediately above the junction of the Inn and the Danube at Passau and allows the examination of the flood peaks on the Inn River, which are triggered by water that originates in the Alps. Gauge Hofkirchen is approximately 20 km upriver of Passau on the Danube and is the last Danube gauge before the junction of the Danube and the Inn at Passau.

Figure 55.2b describes the change in HQ100 at gauge Ingling based on the 16 statistical climate variants used in the study. This figure clearly reveals a reduction in HQ100 from today's approximate value of 6,000 m^3/s to an average future value of 4,800 m^3/s. Figure 55.2c depicts the trend in HQ100 at gauge Hofkirchen. This figure reveals no change in HQ100, and the peak discharges in the future continue to fluctuate around 4,000 m^3/s.

As a second step, the results of the discharge simulations were examined on the basis of the down-scaled and bias-corrected results from *REMO* and *MM5*. The results were processed using the same method that was used for the results shown in Fig. 55.2. However, since there are only two climate variants, no statistical analysis was carried out; instead, the individual results are simply presented. Figure 55.3 depicts the results for gauge Achleiten (beyond the junction of the Inn and Danube) as well as for gauge Ingling on the Inn at Passau and the gauge Hofkirchen on the Danube. Both are situated upriver from gauge Achleiten and allow the trends in HQ_{100} for the Inn and Danube to be examined separately.

Figure 55.3 provides an inconsistent picture of the trend in HQ_{100} at gauge Achleiten. First, it can be noted that the HQ_{100} values for the past at gauge Achleiten that were derived from the regional climate models are approximately 23 % (*MM5 down-scaled and bias corrected*) to 45 % (*REMO down-scaled and bias corrected*) higher than the observed values (see Fig. 55.2a, on the left). In addition, the two results for Achleiten indicate opposing trends for the future. Although HQ_{100} at Achleiten significantly increases in the case of *REMO down-scaled and bias corrected*, in the case of *MM5 down-scaled and bias corrected*, it decreases slightly. If the Inn at gauge Ingling is considered, the values for the past there are consistent. However, they diverge for the future and reveal no clear trend. The future trend at gauge Achleiten differs in these two models in the Danube basin (without the Inn), which can be seen at gauge Hofkirchen. There, the inferences from both models diverge in the same way as for Achleiten.

It is clear from Fig. 55.3 that there are even more critical uncertainties regarding the statements about the development of flood discharges based on regional climate models. These uncertainties originate from varying patterns related to modelling of

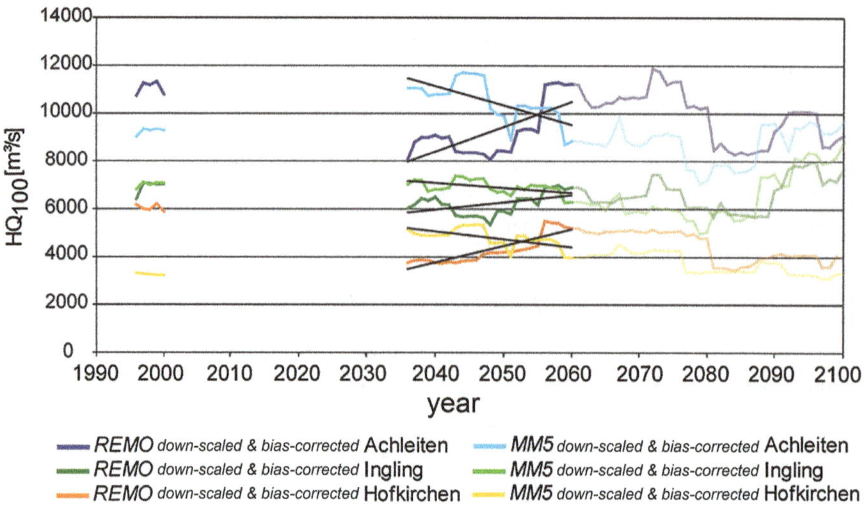

Fig. 55.3 HQ100, each for 25-year time periods between 1971–2000 (*left*) and 2011–2060 (*right*) for the climate variant *REMO down-scaled and bias-corrected* and *MM5 down-scaled and bias-corrected* for the gauges Achleiten, Ingling and Hofkirchen; the *black lines* represent linear regression lines

extreme values in the two regional climate models, *REMO* and *MM5*. This is initially surprising, since they were rather consistent in terms of the statements about mean temperature and precipitation trends, as shown in Chap. 51.

55.3 Spatial Trends in the 100-Year Peak Discharge

The next step involved an examination of the spatial distribution of the response of HQ_{100} as simulated using the stochastic climate generator to climate changes assumed to take place under the climate scenarios that are used. The goal of this analysis was to identify spatial patterns in the trend of HQ_{100}. This involves determining HQ_{100} for the three time periods 1971–2000, 2011–2035 and 2036–2060 for each of the 16 statistical climate variants and for every proxel in the drainage basin from the results of the discharge simulation, according to the method described in the DVWK (1999). The method includes the calculation of a regression line for the annual peak discharges that has been linearised with the appropriate statistical distribution (log-normal or Weibull) and sorted. The associated coefficient of determination (R^2) was used as a measure of the validity of the statements regarding HQ_{100} for each grid element. It was assumed that no significant assertions about HQ_{100} are achievable for values of $R^2 < 0.8$.

Map 55.1 – 1a and 1b show the distribution of HQ_{100} from the simulated discharge data for the past (1971–2000) and the distribution of mean HQ_{100} for the 16

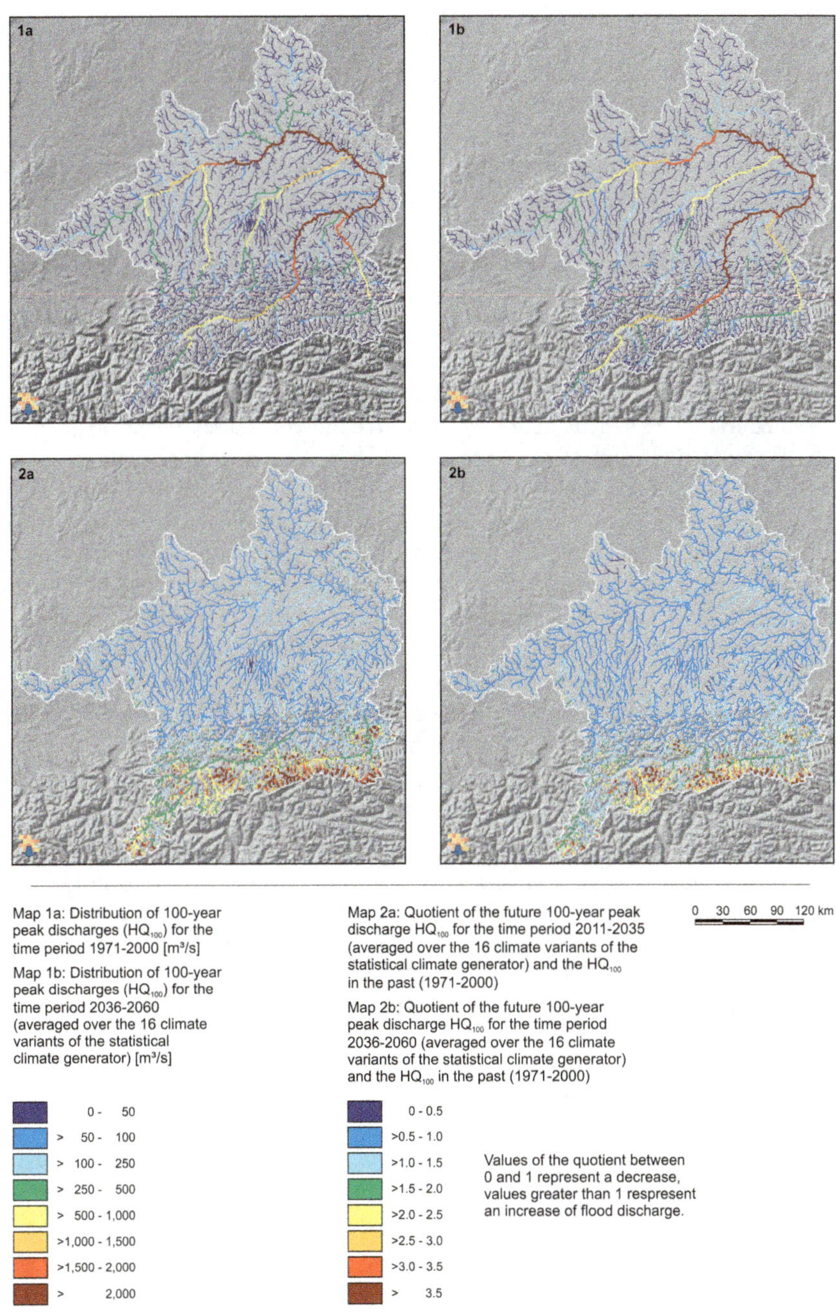

Map 55.1 Scenarios for the development of floods in the Upper Danube basin

statistical climate variants (see Chaps. 49 and 50) for the period 2036–2060 with values of $R^2>0.8$. All proxels for which $R^2<0.8$ in the calculation of HQ_{100} are not shown. It follows from Map 55.1 – 1a and 1b that significant assertions about HQ_{100} are achievable primarily for the main river channels. As expected, HQ_{100} increases with the increase in area of the drainage basin; this fact is easy to see if the path of the Inn River is followed at the outlet of the Alps. The two time periods that were examined show distinct differences in this respect. The comparison of Map 55.1 – 1a and 1b indicates that on average across all the climate variants, the flood peaks in the central sections of the Inn will increase in the future. A change of regime in the Inn from an alpine to a subalpine river results in massive reductions in the flood peaks at gauge Ingling in the simulations, as shown in Fig. 55.2b.

The change in regime and the associated response of HQ_{100} to climate change are presented in Fig. 55.4. This figure illustrates the ratio of future to past HQ_{100} for the periods 2036–2060 and 1971–2000 for select proxels along the courses of the Salzach and Inn rivers. Analysis of the Inn begins at its headwaters at a basin area of approximately 500 km² and ends at gauge Ingling with an area of 26,000 km². At the headwaters of the Inn in the Alps, the analysis shows a future increase in the 100-year flood peaks by a factor of 1.4. In the course of the increase in upstream area, the pattern of a future increase in flood peaks changes by the factor 1.0 at the junction with the Salzach, at an upstream area of approximately 12,000 km². Up to the outlet into the Danube, the 100-year flood peaks then reduce by a factor of 0.8. A similar pattern is evident for the Salzach. In that case, there are extreme increases in flood peaks by a factor of 1.8 at the headwaters, with approximately 200 km² upstream area. Again, the pattern changes as the Salzach exits the Alps, and there is a reduction by a factor of 0.9 at the junction with the Inn. This increase in flood

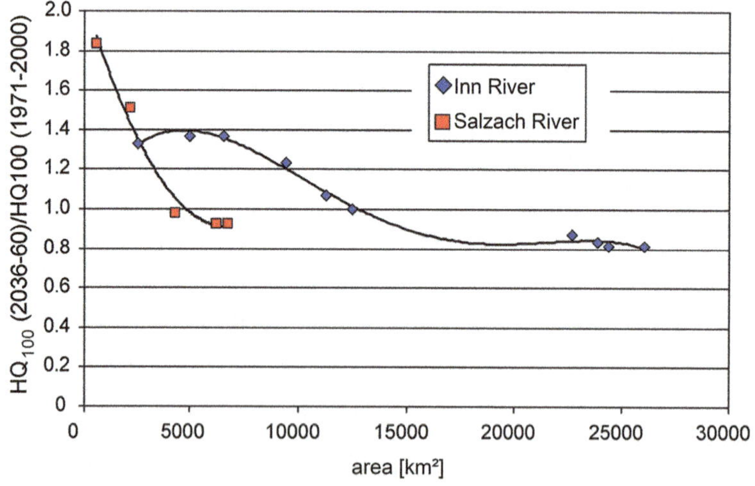

Fig. 55.4 Relation of future HQ100 of the time period 2036–2060 to past HQ100 of Inn and Salzach rivers at selected spots along the river course as a function of upstream area

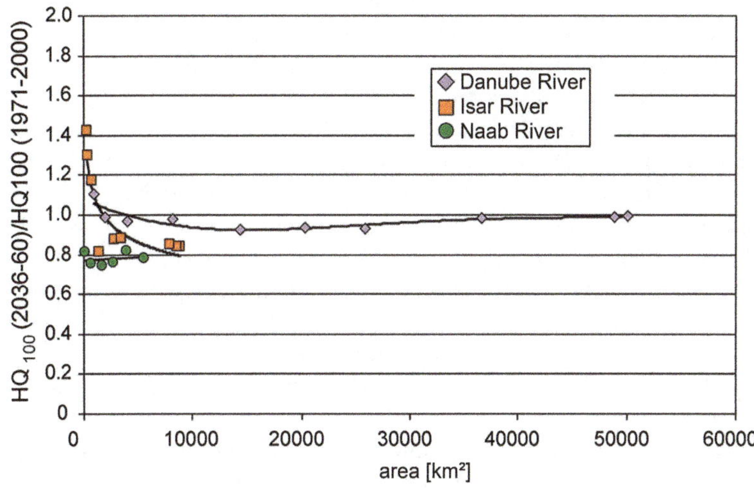

Fig. 55.5 Relation of future HQ100 of the time period 2036–2060 to past HQ100 of Danube, Isar and Naab rivers at selected spots along the river course as a function of upstream area

peaks can be explained with an increasing effect of warm rainfall on Alpine snowpacks, which accelerates their melt and in their combined impact leads to an overproportional increase in flood peak runoff in the Alpine headwaters.

The Danube shows a different pattern at gauge Hofkirchen compared to the Inn at Ingling. At gauge Hofkirchen, on average, the flood peaks are the same for the periods 1971–2000 and 2035–2060. This is also evident in Fig. 55.5, which, like Fig. 55.4, presents the change in the 100-year flood peaks along the Danube, Isar and Naab in relation to the upstream area. There is no change at all in HQ_{100} for the Danube, until the junction with the Inn (see Fig. 55.2a, c). In contrast, like the Inn and the Salzach in Fig. 55.4, the Isar, which is influenced by the Alps, shows a noticeable increase in the 100-year flood peaks in the headwater regions and an equally clear decrease in the Alpine foothills. The Naab, a subbasin that is not influenced by the Alps, shows a general reduction in flood peaks, independent of upstream area. Overall for the Danube, the increase in flood peaks in the southern regions is evidently mutually balanced by a reduction in the northern regions.

Overall, the trends described for the two subbasins of the Inn and the Danube lead to a slight decrease in the future flood peaks at the outlet of the drainage basin beyond the confluence of the Inn and the Danube at Achleiten; this result is also evident in Fig. 55.2a.

The relative changes in HQ_{100} between the periods 2011–2035 and 2036–2060 and the historical reference period 1971–2000 are illustrated in Map 55.1 – 2a and 2b. Proxels for which $R^2 < 0.8$ are not shown.

A distinct separation of the change signal for the studied basins is obvious in Map 55.1 – 2a and 2b. Thus, the flood peaks are constant and up to 50 % reduced in the Alpine forelands, along the Danube and in the Swabian and Franconian Jura as result of the averaged HQ_{100} values for the 16 climate variants. In contrast, in the Alpine valleys and in isolated areas of the Bavarian Forest, the simulated flood peaks more than doubled. The changes in the flood peaks are lowest in regions near the Alpine fringe. Although the decreasing trend in the Alpine foothills and in the Swabian and Franconian Jura accelerates with time (compare Map 55.1 – 2a and 2b), in parts of the Alpine valleys, there is a deceleration in the trend during the period 2036–2060 compared to 2011–2035. This is due to a reduction in water that is stored in snowpacks which can be melted by rainfall in spring and which then lead to runoff peaks.

There is a variety of causes for the heterogeneous pattern in the trends for 100-year flood peaks. It is important to note when interpreting the results that simple superposition of the values for 100-year peak discharges is not valid, since a change in peak discharge leads to differing flow velocities and hence to varying degrees of overlap and synchronisation in the high water flows from the subbasins. This makes attributing single causes to the specific details difficult. In the Alpine foothills, snow cover is widely reduced (see Chap. 57), and thus melting snow can no longer contribute to the formation of floods to the same extent as in the past. Increased evaporation in the foothills and along the Danube also contributes to overall drier soils that can absorb more water and hence reduce the amount of direct discharge in the case of floods (see Chap. 58). The result is that flood peaks are reduced. In the past, significant flooding events on the Danube (e.g. the Whitsun flood in 1999) were generally still associated with the storage of larger quantities of precipitation in the form of snow in the Alps and in the regions bordering the mountains. This considerably contributed to easing the flood situation. However, with increasing temperatures in the future, even in flood situations, the portion of precipitation that falls as snow in the Alps will decline. The associated increase in floodwaters is particularly marked if rain falls onto snow and accelerates the melting of this stored snow water such that it too is discharged. As a result, the flood peaks are intensely elevated during the first period from 2011 to 2035. As soon as the snow storage gradually decreases in the second period from 2036 to 2060, the flood peaks accordingly continue to reduce, since the amount of the meltwater in the flood peaks originating from snow and glaciers continues to drop.

55.4 Summary

Overall, Map 55.1 – 1a–b and 2a–b provide an initially somewhat surprising, though perfectly reasonable, picture of the change in the 100-year flood peaks in the drainage basin. The result makes it clear that it is generally not feasible to make a single generalised statement about future trends based on the outlet gauge in a complex drainage basin like the Upper Danube.

The following key concluding statements can be derived from the analyses:

The results of the analyses of the climate variants based on the statistical climate generator do not generate any definitive conclusions about the changes to HQ_{100} at gauge Achleiten. Based on the range of uncertainty in the results of the 16 simulated climate variants depicted in Fig. 55.2a, a change in HQ_{100} at the outlet of the drainage basin of between +4 % and −28 % is likely.

The 100-year discharges that result from both the *MM5 down-scaled and bias-corrected* and the *REMO down-scaled and bias-corrected* climate variants for gauge Achleiten do not reproduce the values determined for the past from measurements and indicate opposite trends for the future. They are therefore not suitable for making further inferences.

The results of the analyses of the statistical climate variants suggest that the 100-year flood events in the Alpine region and in regions of the Alpine foothills, the Danube, the Swabian and the Franconian Jura will behave differently as a consequence of climate change. While there will be an increase in the 100-year flood peaks in the Alps, the other regions will remain the same or decrease.

References

Bayerisches Landesamt für Umwelt LfU (ed) (2007) Gewässerkundlicher Bericht Hochwasser August 2005, Augsburg

Deutscher Verband für Wasserwirtschaft und Kulturbau e.V. (DVWK) (1999) DVWK-Merkblatt 251/100, Statistische Analyse von Hochwasserabflüssen, Bonn

Chapter 56
Influence of the Glaciers on Runoff Regime and Its Change

Markus Weber and Monika Prasch

Abstract The Alps are often described as Europe's water towers because they influence runoff regimes along the courses of the rivers by ensuring a reliable and balanced supply of water. The water originates both from increased precipitation and from the release of the precipitation stored in the snow cover and the glaciers. In the Upper Danube basin, only 0.5 % is glaciated. As efficient sources of meltwater, the glaciers play a significant role despite their small extent.

A map display of the ratio of the calculated meltwater proportion to the mean annual discharge over the period 1991–2000 was chosen to document the spatial distribution of the significance of the glacier discharge. The temporal trend was stated using diagrams with monthly mean values for discharge by decade and the proportion of the ice melt at selected gauge locations in the basin.

Overall, the results indicate a notable impact of glaciers on the runoff regime in the Upper Danube basin. However, this effect remains limited to the inner alpine regions. In the future, the discharge regime will assume a character that is dominated by snow and rain, in which the inner alpine compensation effect of glaciers disappears, and as a result, discharge variability increases, particularly in summer. In contrast, the effects on the water supply in the Danube at gauge Achleiten due to the retreat of the glaciers can be considered negligible compared to the changes in regional precipitation, because the impact of the glacier melt is decreasing downstream the river courses.

Keywords GLOWA • Danube • Glacier • Ice melt • Discharge • Gauge • Meltwater proportion

M. Weber (✉)
Commission for Geodesy and Glaciology of the Bavarian Academy of Sciences and Humanities, Munich, Germany
e-mail: Wasti.Weber@kfg.badw.de

M. Prasch
Department of Geography, Ludwig-Maximilians-Universität München (LMU Munich), Munich, Germany
e-mail: m.prasch@lmu.de

W. Mauser, M. Prasch (eds.), *Regional Assessment of Global Change Impacts*, DOI 10.1007/978-3-319-16751-0_56

56.1 Introduction

The Alps are often described as Europe's water towers because they influence runoff regimes along the courses of the rivers by ensuring a reliable and balanced supply of water. The water originates both from increased precipitation and from the release of the precipitation stored in the snow cover and the glaciers, both short and long term. From the supply of meltwater with high energy input, the snow cover and glaciers ensure an increase in discharge during periods with low precipitation and therefore low gauge water levels. The effectiveness of this balance, termed the "compensation effect" (Roethlisberger and Lang 1987), is dependent on the proportion of the mountainous area within the whole drainage basin, as well as on the proportions of melting surfaces. In the Upper Danube basin, 29 % of the area is designated as alpine, but only 0.5 % is glaciated. As efficient sources of meltwater, the glaciers play a significant role despite their small extent. They contain a considerable source of water (see also Chap. 12), which in periods of glacier retreat contributes meltwater from glaciers to the basin discharge that is equivalent to the loss of glacier ice. However, even with static or growing glaciers, a portion of the water in the mountain rivers comes from glacier ice. At the lower reaches of the Inn, the water from glaciers is actually easy to detect as a result of its characteristic green colour, and its quantity can be at best estimated based on the amount of discharge in the headwater regions. Quantitative data on the influence of the glaciers on the supply of water downstream to date have been purely speculative and limited to the flow from glaciers described above (Baumgartner et al. 1983). DANUBIA provides new insights into the proportion of water coming from glaciers in the Upper Danube and its main Alpine tributaries, the Iller, Lech, Inn and Salzach.

On a seasonal basis, the discharge within a drainage basin does not just match the current supply of precipitation but also depends on the periodic change in the available volume stored (e.g. snow cover, glaciers, groundwater, artificial reservoirs). Each of these outflows shapes the discharge hydrograph of a gauge in a particular way. For example, the formation and depletion of the snow storage adds an asymmetrical triangular trend with a prominent maximum in spring, at the time of the most extensive snowmelt in the lower-lying areas; this is shown by the hydrograph for Kempten (see Fig. 56.2). In the mountains, the snow melts throughout the summer in the higher elevations and thus also provides runoff during dry periods in summer. However, the overall characteristic of a runoff regime shaped by snow is the spring maximum. In Figs. 56.1–56.10, the monthly means for the mean discharges during the decade 1991–2000 (light blue) illustrate this point. Hence, the entire Upper Danube basin is shaped by the release of snow storage, except in the glaciated headwater regions (e.g. Vent, Huben).

In contrast, discharge from the glaciers reaches a maximum in the midsummer months of July and August, when large areas of the ice surfaces are snow free and solar radiation input is still high. Since ice melt is efficient as a result of up to fourfold rates of absorption of the radiation in comparison to snow surfaces, the discharge maximum due to ice melt is usually higher. A discharge regime that is shaped

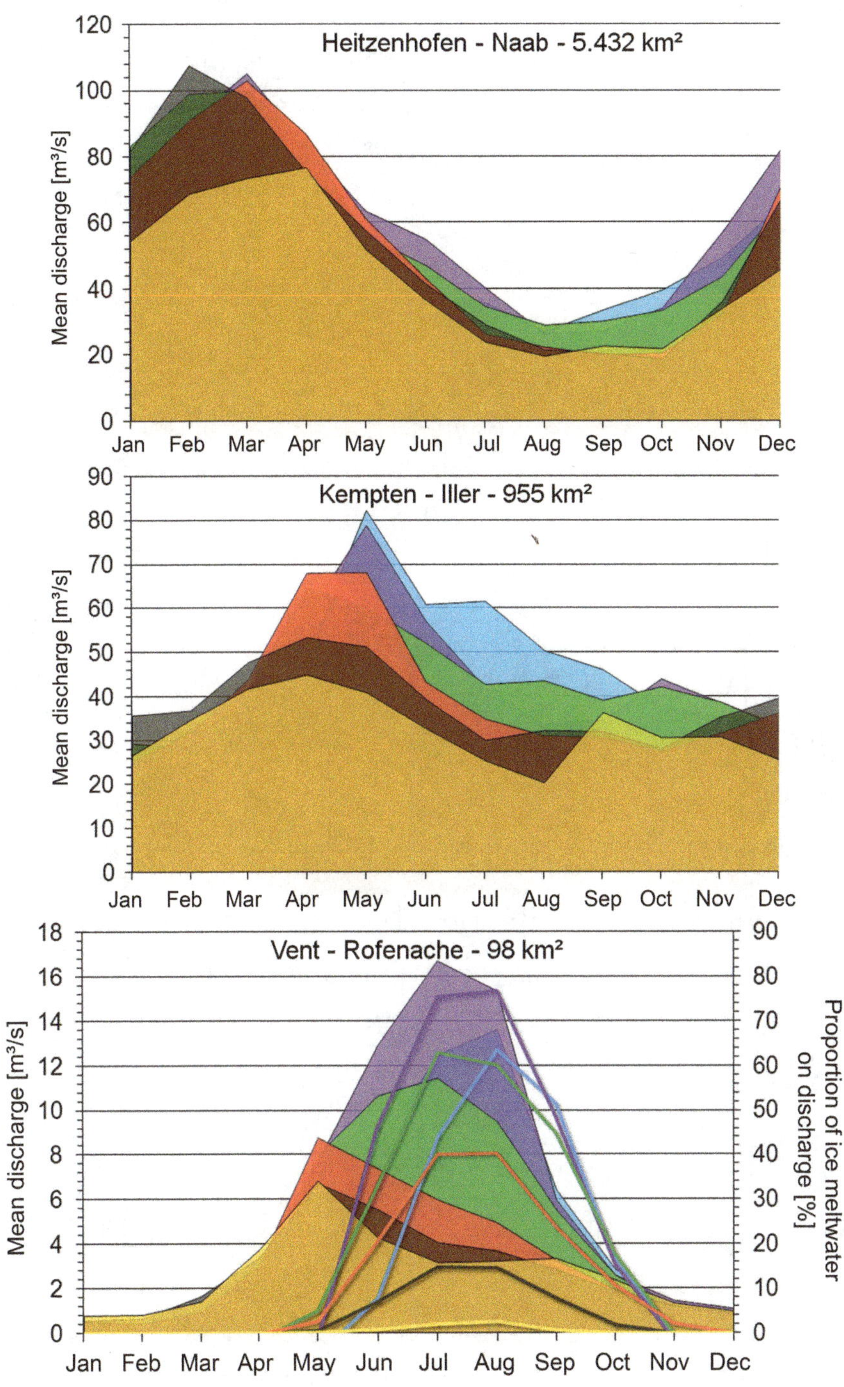

Figs. 56.1–56.10 Averaged monthly mean of discharge [m³/s] and relative proportion of ice meltwater [%] in the Upper Danube subbasins (*REMO regional climate trend* and *Baseline climate variant*)

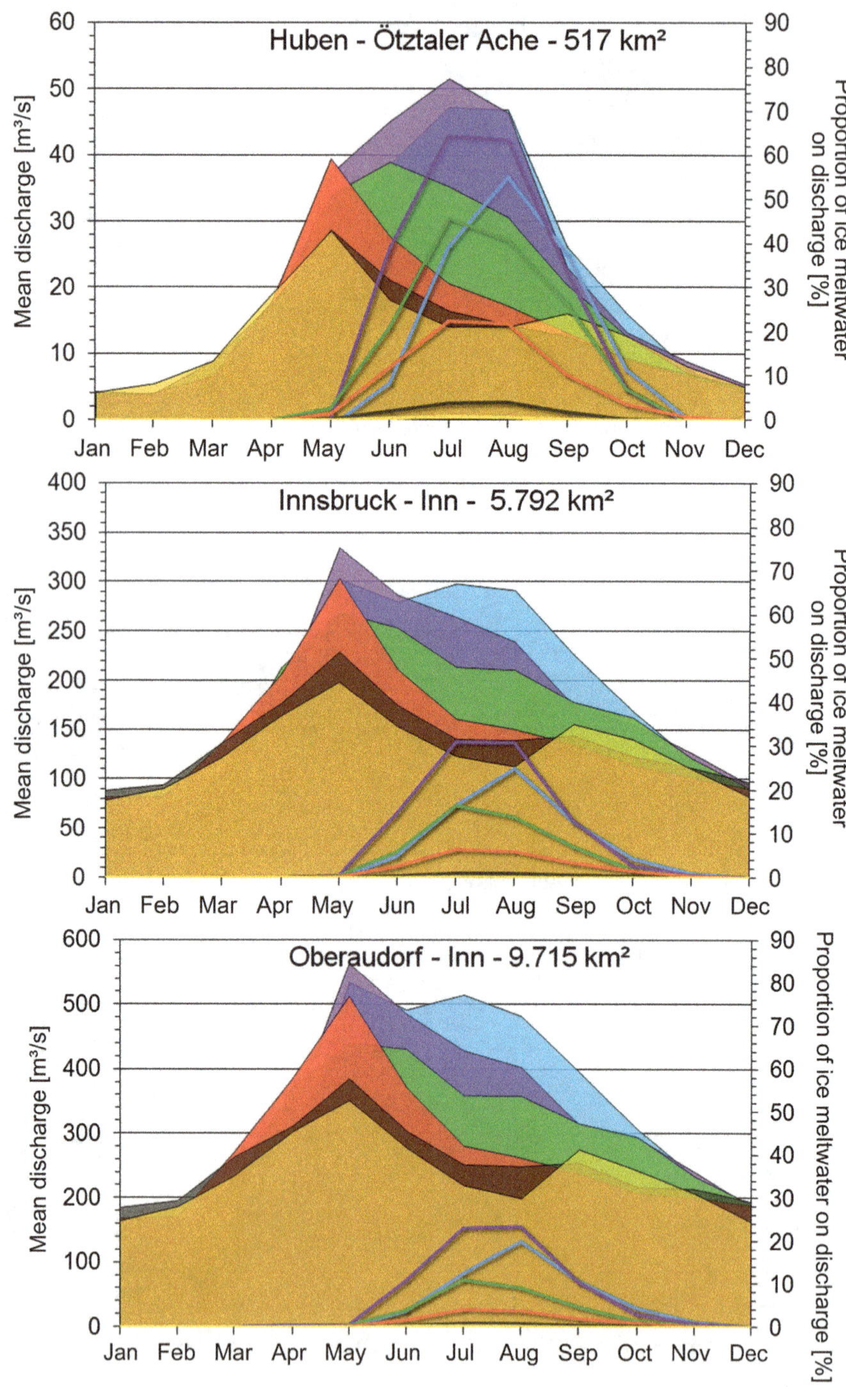

Figs. 56.1–56.10 (continued)

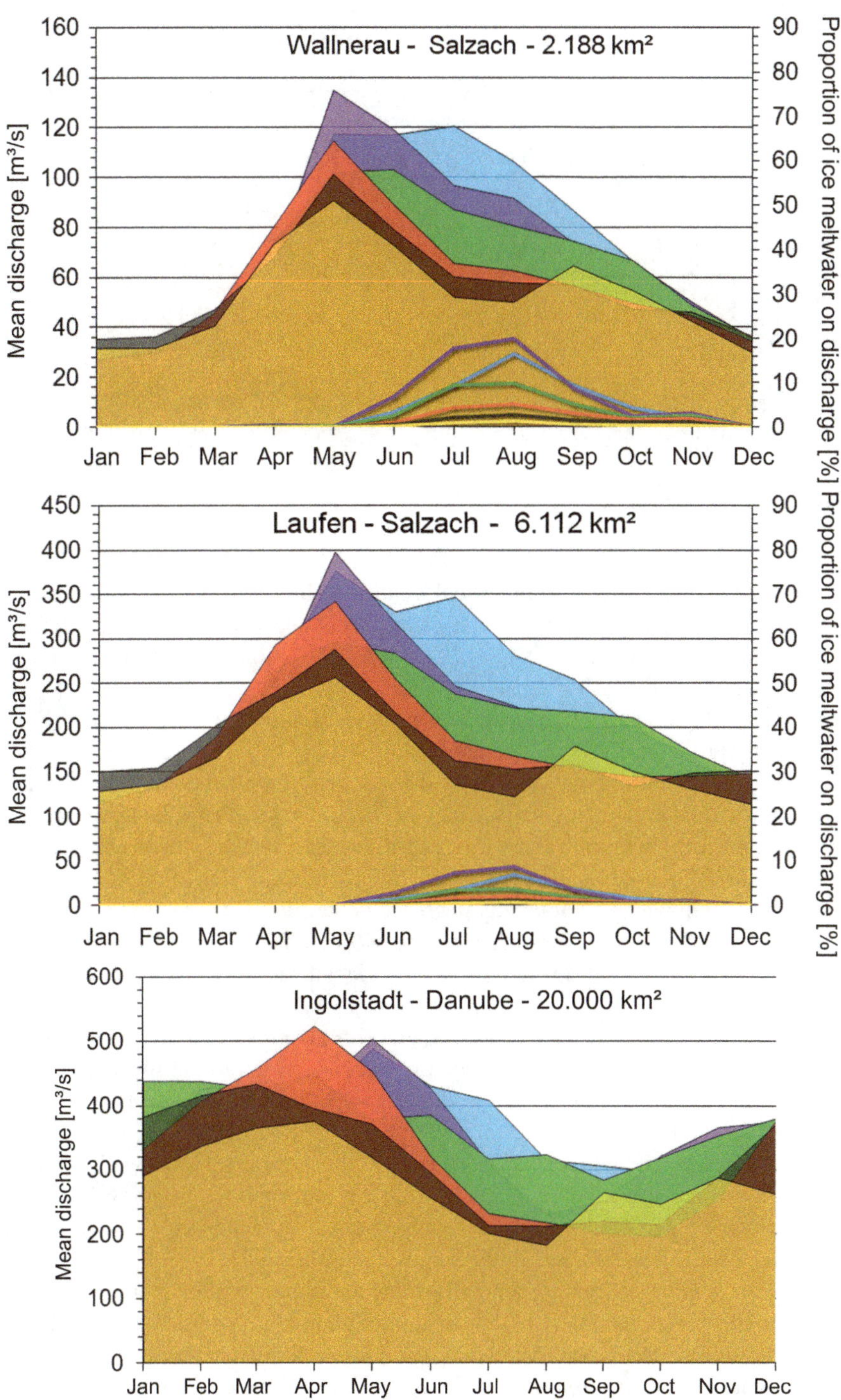

Figs. 56.1–56.10 (continued)

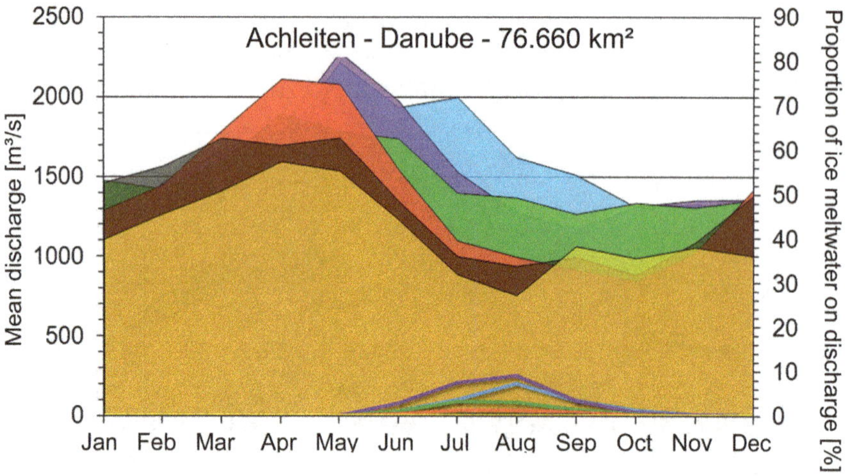

Figs. 56.1–56.10 (continued)

by glaciers thus displays a symmetrical trend with high discharges in summer and very low discharges in winter (e.g. at gauges Vent and Huben; see Figs. 56.3 and 56.4). The ratio between discharge maximum and minimum is calculated by the proportion of the ice meltwater in discharge, which decreases with increasing distance from the source and with increasing size of the drainage basin and hence decreasing proportion of glaciated area.

With the current ongoing retreat of the glaciers, the ice melt can be seen as a subsidy for areal precipitation, since it comes from the ice reservoirs that have been in existence for a long time and which supply water directly to the water channels. This additional glacier supplies is correlated with the expansion of the snow-free ice surface and will therefore gradually decrease in the future along with the shrinking ice masses and ongoing warming (see Fig. 56.11). Thus, in the highly glaciated basins, there should be a transition from the glacial runoff regime to a pluvial and nival regime.

Whether the lack of meltwater from glaciers also results in a notable decline in discharge in the subbasins depends on its contribution to the formation of the discharge. The amount of discharge decreases with increasing distance from the glaciers and first forms nivo-glacial mixtures in the runoff regime, which are characterised by a second midsummer maximum in the hydrograph as a result of the ice melt, following the maximum in spring from the snowmelt (see Figs. 56.5 to 56.8). The form and magnitude of this second maximum are essentially determined by the amount of water from glaciers in the summer months.

Gauges in larger subbasins generally show a mix between the melt of snow and glaciers and precipitation. Thus, for example, almost all gauges along the Inn or Salzach feature a second discharge maximum in the period from June to September, which can be shown to relate to both ice melt and enhanced convective summer

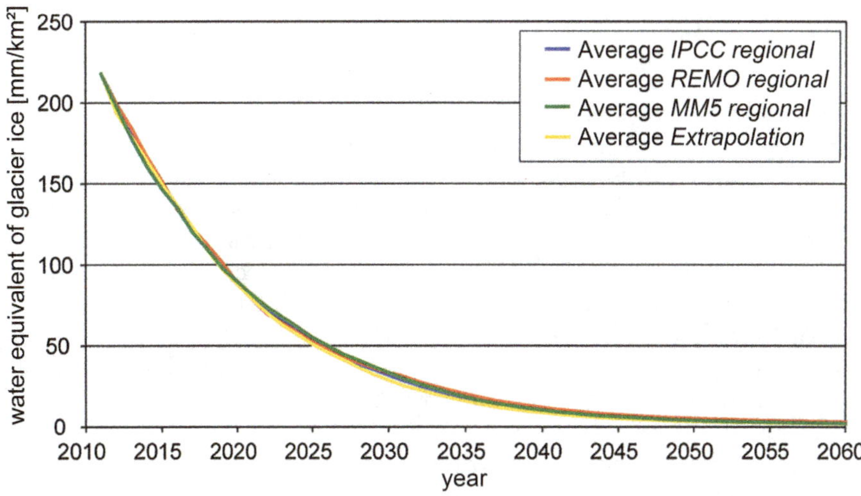

Fig. 56.11 Water equivalent of potential ice reservoir in the Upper Danube basin, gauge Achleiten. The chart displays the four averaged statistical climate variants of each climate trend

precipitation in the mountains. The details about monthly amounts of glacier melt in discharge are quite helpful for the analysis.

Great significance has been attached to the compensation effect from snow and ice melt in the mountains. This effect extends beyond the tributaries into the surrounding lowland areas and the main rivers and there dampens the variability of the summer discharge. The impact also increases with the aridity of the climate (Viviroli et al. 2003). However, the Upper Danube basin has a humid climate, so that the supply of water to the Danube up to the junction with the Inn is influenced by tributaries from the mountains only to a small extent. Nevertheless, there are considerable differences in the seasonal water supply in sections of the river that are associated with the Alpine water tower and those that have their sources exclusively in the lowlands. For example, gauge Heitzenhofen on the Naab (see Fig. 56.1–56.10) shows a significant low water status in the summer months of July and August; this low water level does not appear at gauge locations with inflow from the mountains.

This chapter uses modelling with DANUBIA and the *SURGES* glacier model (see Chap. 31) to address the question of the currently still available potential of the Alpine glaciers in the formation of discharge in relation to precipitation and snowmelt. In addition, the local seasonal differences in the discharge regimes and, by way of example, their future changes are demonstrated under the conditions of the *REMO regional* climate trend and the *Baseline* climate variant (see Map 56.2 and Figs. 56.1–56.10). The climate scenario that was chosen is characterised by both an increase in the mean temperature and a significant decrease in summer precipitation, both factors that contribute to notable effects on hydrology (see Chaps. 47, 48, 49, and 50).

56.2 Data Processing

The separation of the proportion of discharge that originates from ice melt is not possible from observational data. At best, the contributions have only been able to be crudely estimated using the scarce measurement data that are available on the losses in mass and discharges from the glaciers. However, the coupled hydrological model in DANUBIA allows a very detailed calculation of discharges in channels in basins, which are over 100 km^2 in area. The *SURGES* glacier model directly forwards the meltwater quantity calculated for the individual glaciated areas to the channel system, where it is subjected to the processes of discharge formation being considered in the model. The current quantity of ice meltwater at a specified gauge within the drainage network is the difference between the discharge calculated in a model run with glacier ice and a second run without glacier ice. The ratio of this difference to total discharge is a measure of the current significance of the glacier area for the section of the river under consideration.

To prepare the data, for every available climate variant of the GLOWA-Danube climate scenarios, a coupled modelled run with and a reference model run without glacier ice were carried out for the period 2011–2060 and analysed by decade. Only the results of the run with the *REMO regional–Baseline* climate scenario are presented in detail in this chapter. Similar results for the *IPCC regional* climate trend and *Baseline* climate variant can be found in Weber et al. 2009.

As a reference for the recent past, the modelling of the period 1991–2000 based on measurement data at the Weather Service's climate stations was used. For some subbasins of varying sizes and amounts of glaciation within the region of the Inn River, Table 56.1 provides the calculated annual means for specific discharge and the components from glaciers over the decade in addition to the regional characteristics. Both the absolute and the relative amounts of glacier water in discharge exceed by up to three times the published glacier ice melt contributions (e.g. Baumgartner et al. 1983) for other sites in the same drainage basin for the period 1930–1960. However, this does not at all document such a drastic increase in

Table 56.1 Mean yearly mass flow of glacier water and its proportion on basin discharge for subbasins of different sizes and *SURGES* model results of glacier proportion on discharge for the reference period 1991–2000

Basin	Area [km^2]	Proportion of glacier ice [%]	Discharge 1991–2000 modelled [mm]	Modelled ice melting [mm]	Proportion on discharge [%]
Vent/Rofenache	98	35	1,367	505	36.9
Huben/Ache	517	17	1,149	305	26.5
Innsbruck/Inn	5,792	4	939	79	8.4
Oberaudorf/Inn	9,715	3	1,026	67	6.5
Wasserburg/Inn	11,980	2.4	998	54	5.4
Achleiten/Danube	76,660	0.5	639	10	1.6

glacier melt towards the end of the millennium, but is instead caused by the fact that in the present study, for the first time the entire ice meltwater entering the channels from the glaciers has been calculated, while previous studies have only estimated the inflow into discharge using the observed losses of glacier mass and those extrapolated to the basin (Lambrecht and Mayer 2009).

In the case of static or growing glaciers, the methods of Baumgartner, Lambrecht and Mayer yield a contribution of zero or even a negative contribution from ice melt. However, even under these conditions, ice meltwater is generated during the summer on the glacier tongues, and this is fed into the glacier streams and thus elevates the discharge at this time compared to ice-free basins. This contribution, which is not accounted for by simply using the mass balance in the calculations, is certainly not negligible and can only be determined using a detailed glacier model. In the past, estimates of the glacier ice melt contribution for a group of mountains used a mean elevation profile for the specific mass balance, calculated using the few observational data that were available, which introduces a considerable source of error.

In contrast, *SURGES* calculates the elevation profile of the mass balance individually onto each 1×1 km^2 proxel, which quite closely approximates the individual calculation of the mass balance of each of the 550 glaciers in the study area.

The importance of considering the ice melt in the calculation of discharge is clearly illustrated in Fig. 56.12. The meltwater from snow cover is far from enough to reproduce the measured values. The discharge hydrograph can only be determined to correspond well with observations when calculated in combination with the *SURGES* glacier model.

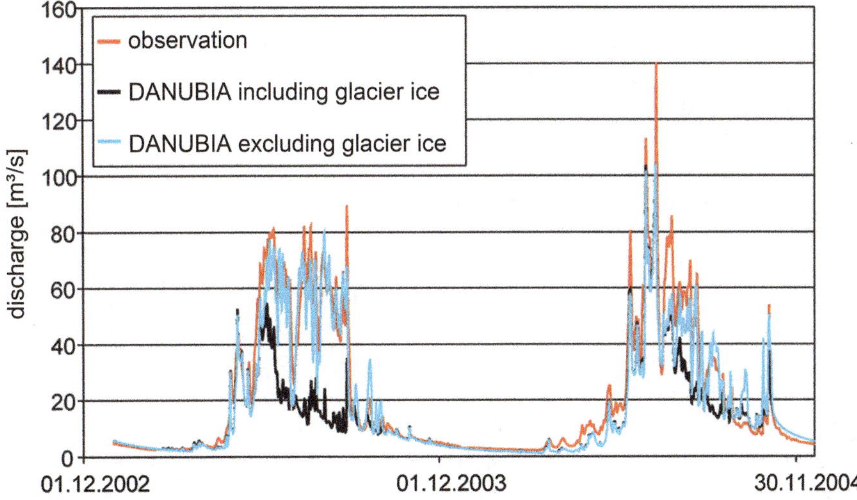

Fig. 56.12 Comparison of daily mean discharges observed at gauge Huben (*red*), modelled with (*blue*) and without glacier ice (*black*)

56.3 Results

A map display of the ratio of the meltwater proportion to the mean annual discharge over the period 1991–2000 was chosen to document the spatial distribution of the significance of the glacier discharge (see Map 56.1). The temporal trend was determined using diagrams with monthly mean values for discharge by decade and the proportion of the ice melt at selected gauge locations in the basin (see Figs. 56.1–56.10).

Map 56.1 depicts the rainbow colour-coded mean discharges and the proportion of water from glaciers in the mean annual discharge, based on the modelling of the period 1991–2000 in the Upper Danube basin river network. It is clear that only in the Inn and Salzach tributaries of the Danube an appreciable quantity of meltwater stems from the glaciers; the Danube itself, up to the junction with the Inn at Passau, is practically devoid of such meltwater. Nevertheless, owing to the quite fine-scale that was used, the water from the small Zugspitz glaciers can be detected more than 50 km downstream the flow of the Partnach and the Loisach, but with a mean proportion less than one tenth of a percent of discharge (see Map 56.1, Category 1, dark red), it has negligible hydrological significance. The same reasoning holds true for the ice relicts in the Lech catchment. Virtually all of the glacier water in the catchment is fed into the Inn, which transports the maximum amount of just 10 % of the annual volume in the section between Landeck and Innsbruck. There, the tributaries join in from the Kaunertal, the Ötztal and the Stubaital, the most highly glaciated mountain groups of the Eastern Alps (see also Chap. 12). The blue to green colour in Map 56.1 highlights the exceptionally high proportion of glacier water in the discharge from these headwaters.

The ice melt is limited to the period of ablation in which dark, snow-free ice surfaces (bare ice) are visible on the glaciers. This period generally lasts from June to October. The highest values of meltwater production are reached in July and August, when they are two to three times the annual average. To compensate for low water levels in dry periods in August, the amount of glacier water must thus be approximately 25–30 % on average. This condition is met in the sections of rivers marked in blue to green, whereas for sections marked in yellow to red in Map 56.1, the gauge levels are scarcely affected by the glaciers. This also applies to wide stretches along the Salzach.

In contrast, Figs. 56.1–56.10 show both the local seasonal trend in discharge and its gradual change, using the means across the decades for the model years 2011–2020, 2021–2030, 2031–2040, 2041–2050 and 2051–2060 under the *REMO regional–Baseline* climate scenario. The catchments of gauges Heitzenhofen (see Fig. 56.1–56.10), Kempten (see Fig. 56.2) and Ingolstadt (see Fig. 56.9) have no glaciers, and the Naab at Heitzenhofen conducts no water out of the Alps. The comparison among the figures reveals a discharge regime shaped entirely by ice melt in the reference decade (light blue) only at the Vent (see Fig. 56.3) and Huben

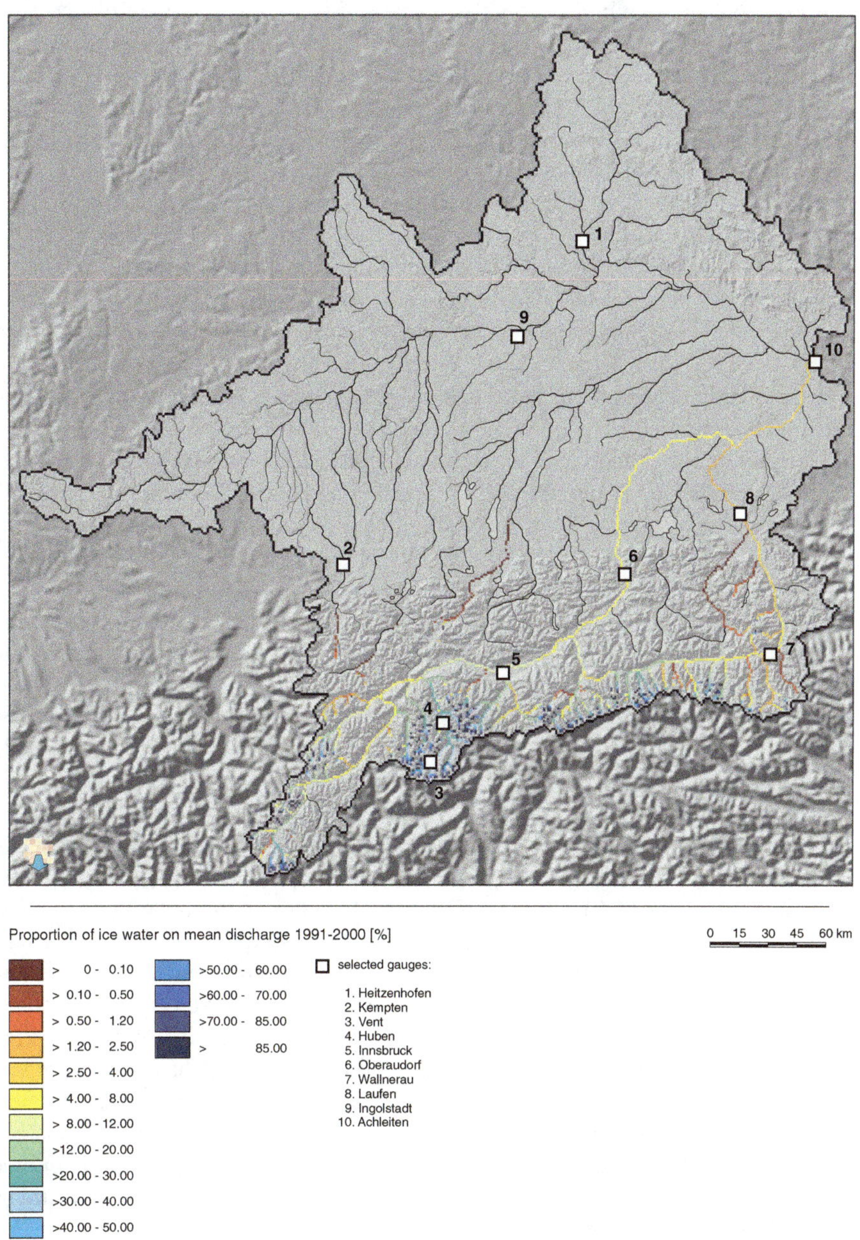

Map 56.1 Influence of the glaciers on the discharge regime in the past (DEM: Jarvis et al. 2008, DANUBIA river network)

(see Fig. 56.4) gauges in the strongly glaciated Ötztal. There, the meltwater from the glacier supplies more than 60 % of the discharge in July and August, overall approximately 1/4 to 1/3 of the annual total. On the Inn at Innsbruck (see Fig. 56.5), the regime is characterised by snow and glaciers, i.e. the snowmelt gains importance. The amount of glacier melt is reduced to less than 10 % of the annual total. This melt accumulates from May to October and reaches a maximum of 20 % of the discharge in July/August. Hence, the maximum summer discharge is comparable to the snowmelt.

As the Inn continues towards the Alpine fringe (Oberaudorf; see Fig. 56.6), the contribution of glacial water only negligibly increases with the inflows from the Zillertal, but the discharge doubles compared to Innsbruck when the increasing size of the drainage basin is considered. The significance of the glacial contribution decreases further, but remains detectable in hydrographs as a result of the high summer concentration.

The glacier regions situated on the main Alpine ridge in the eastern parts drain via the Salzach. The dominant red tones in the tributaries there are the result of lower amounts of glacial water from less glaciation in the basin (see Map 56.1). Gauge Wallnerau (see Fig. 56.7) on the Salzach also exhibits a discharge regime characterised by snow and glaciers; only there the maximum proportions of ice melt at 15 % are lower than in the Inn at Innsbruck. At the lower reaches near Laufen (see Fig. 56.8), the glacier water portion decreases to 7 %, and the hydrograph is similar to that for Kempten in that it has a regime characterised by snow and precipitation; in this regime, the second maximum is determined primarily by the summer precipitations that are more abundant in the Alps compared to the lowlands. There is indication of the distinctly elevated proportion of glacier water only during the first decade (2011–2020) of the climate scenario in the discharges for the months of July to August. Although the meltwater in the Inn in midsummer is noticeable, it has only marginal relevance for the discharge of the Danube at gauge Achleiten (see Fig. 56.10). Only 1.6 % of the total annual amount derives from the Alpine glaciers, and, even in July/August, the mean proportion in the discharge over the decade 1991–2000 is only 7 %. This is significantly below the variability in the discharge from year to year and can therefore not compensate for potential precipitation deficits that may arise. For the Inn, the contribution from precipitation and snowmelt from the mountains is significantly greater than the proportion from ice meltwater. In this way, the mountains definitely shape the discharge in the Danube beyond the confluence with the Inn. However, even as soon as Ingolstadt (see Fig. 56.9), the Danube discharge does not show the minimum in summer that is characteristic on the Naab, since the tributaries from the mountains flow in via the Lech and the Iller rivers.

How might the discharge regime of the future change as a result of the glacier retreat in and around the Alps? As can be seen in Fig. 56.11, the total potential of the glacier contribution at gauge Achleiten distributed over the entire drainage basin is somewhat more than 200 mm, which is only about 20 % of the average annual total precipitation. The depletion of these reserves occurs in the same manner in all the averaged climate variants of the four climate trends, such that the future trend in the mean annual course of discharge is presented only using the results of the *REMO regional–Baseline* scenario. This climate scenario is characterised by a significant decrease in summer precipitation (-31 % in the period 1990–2100) and an increase in spring precipitation (+9 % in the period 1990–2100) (see Chaps. 47, 48, 49, and 50). The consequence of this latter change is that both within the Alps and in the surrounding foothills, the nival character of the discharge regime remains in the future. Typically, the maximum in the snowmelt discharge is broadened, and the peaks are increasingly lower. The discharge regime experiences the most extensive transition both in the Alps and in the foothills as a result of the decrease in precipitation in the summer months from June to August. In this case, there is a minimum in discharge towards the end of the simulation period, which previously only appeared at gauges without headwaters in the Alps. The effect of the decrease in precipitation is manifested especially in the Alps and cannot be compensated for in the future either by the snow storage or the glacier melt.

The significant increase in the contribution from glaciers in the model decade 2011–2020 is primarily the result of a prolonged ablation period and the associated larger snow-free ice surfaces. As a result, the proportion of the glacier water in discharge increases slightly in the Inn and the Salzach. A significant increase in discharge is only seen in the glaciated headwater regions at Vent (see Fig. 56.3) and Huben (see Fig. 56.4). At gauge Achleiten (see Fig. 56.10), the proportion of the glacial water in the discharge only slightly increases to a maximum of 9 %, too small to compensate for the absence of precipitation.

The glacial contribution significantly decreases with declining glacier area over the following decades.

Map 56.2 illustrates this result using the proportion of discharge from glaciers in the decade 2031–2040, in the same way as Map 56.1. The proportion of the meltwater according to the *REMO regional–Baseline* scenario is significantly reduced in almost all sections of the river, which can be seen in the red and orange of the Inn and Salzach. Already above Oberaudorf, the glacier water in the Inn is almost insignificant. The contribution from glaciers disappears as a result of the complete melting of the glaciers, which has already occurred in some regions, for example, along the Lech, but also in the Saumnaun or the Zillertal Alps. Blue or green colour indicates an essential contribution on the map and is limited to the head watersheds in the Bernina range, the Ötztal Alps and the Tauern and Sonnblick ranges.

The consequence of this is that towards the end of the simulation for the Inn and Salzach, the discharge regime changes from one that is symmetrical and dominated by glaciers or with a double peak from the effect of snow and glaciers to one that is asymmetric and dominated by snow and rain (see Figs. 56.5, 56.6, 56.7, and 56.8). The loss of the ice reservoir and hence also the reduction of the flow from glaciers at all gauge sites on the Inn and Salzach lead to both a decrease in the annual total discharge and their maxima.

Between 2020 and 2060, the annual discharge decreases in headwater regions such as Vent or Huben by approximately 30–35 % and in the rest of the drainage basin by approximately 25 %. There, however, the decrease is largely the consequence of the decrease in precipitation in the *REMO regional–Baseline* scenario.

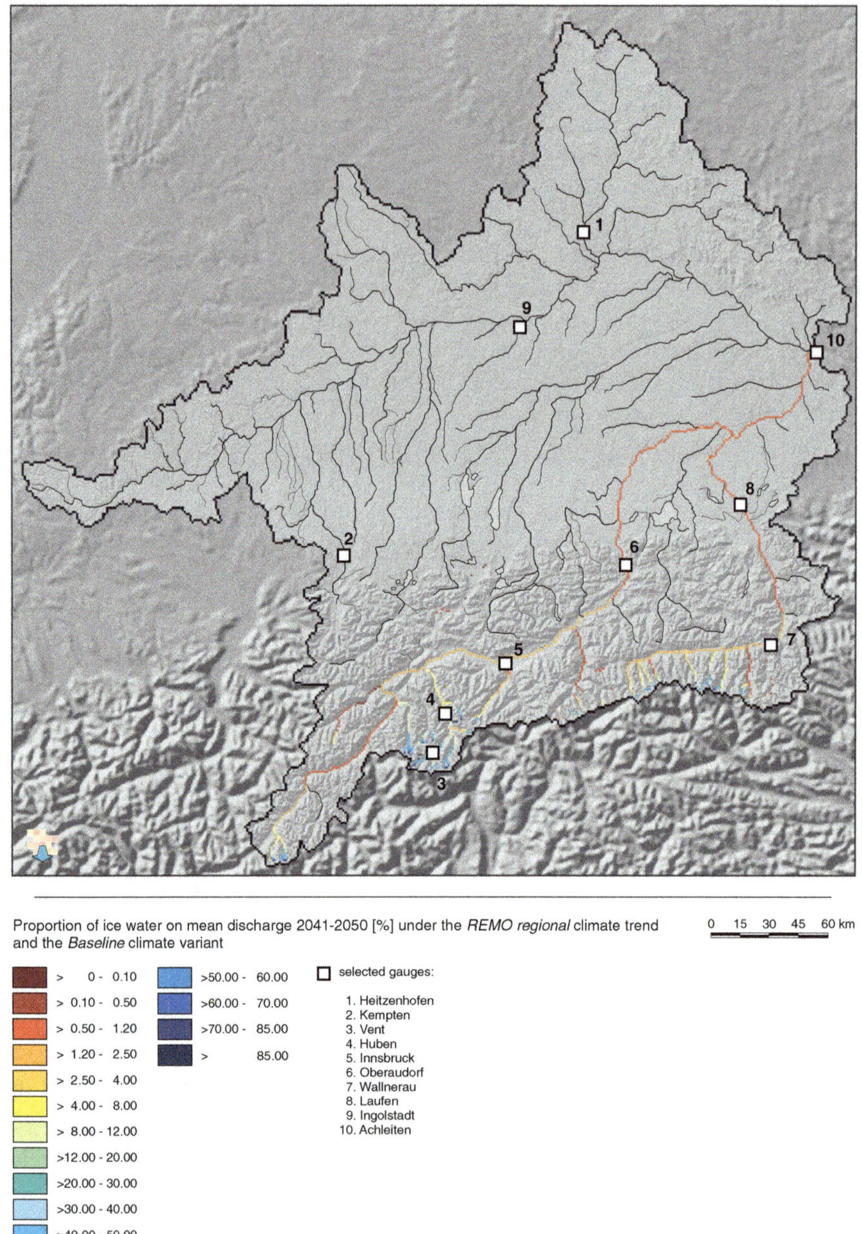

Map 56.2 Influence of the glaciers on the discharge regime in the future (DEM: Jarvis et al. 2008, DANUBIA – river network)

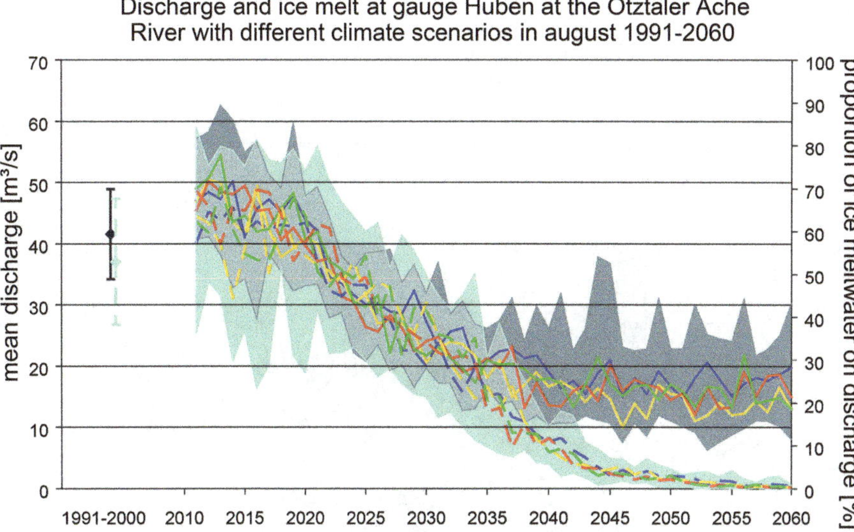

Fig. 56.13 Mean discharge in August and percental proportion of glacier melt for the gauge Huben. Blue areas represent the fluctuation margin of proportion of glacier melt from all 16 statistical climate variants. *Grey* areas represent the fluctuation margin of mean discharge in August

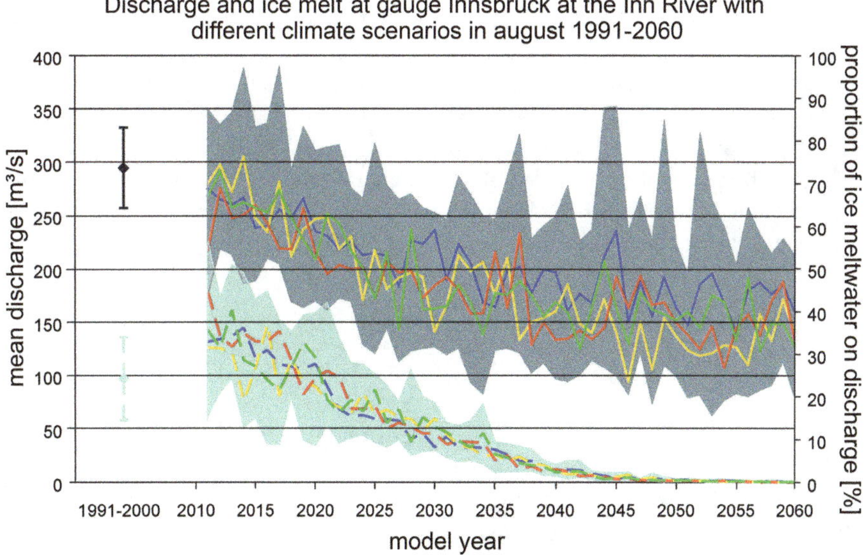

Fig. 56.14 Mean discharge in August and percental proportion of glacier melt for the gauge Innsbruck. Blue areas represent the fluctuation margin of proportion of glacier melt from all 16 statistical climate variants. *Grey* areas represent the fluctuation margin of mean discharge in August

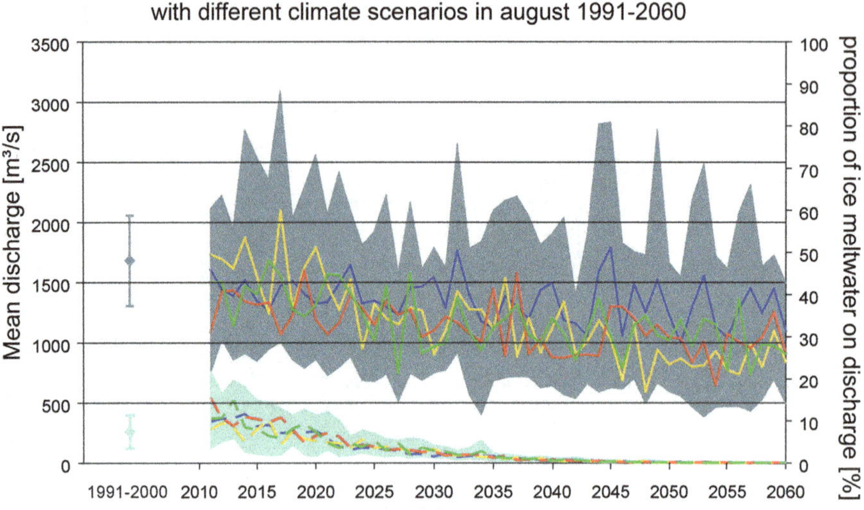

Fig. 56.15 Mean discharge in August and percental proportion of glacier melt for the gauge Achleiten. Blue areas represent the fluctuation margin of proportion of glacier melt from all 16 statistical climate variants. *Grey* areas represent the fluctuation margin of mean discharge in August

In Figs. 56.13, 56.14, and 56.15, the mean monthly average discharge and the percentage of ice melt are shown for the three gauges Huben, Innsbruck and Achleiten; also depicted is the range from all simulations for August, since in general the maximum discharge from the glaciers is in August. Whereas there is a clear decrease in discharge in the Ötztal (see Fig. 56.13) and at the Inn (see Fig. 56.14), which is the result of the reduced glaciated area, this decrease is no longer significant at Achleiten (see Fig. 56.15). Especially for the Inn, an increase in the variability can be identified that indicates the loss of the compensation effect.

56.4 Summary

Overall, the results indicate a notable impact from glaciers on the runoff regime in the Upper Danube basin, in particular for the Inn and Salzach tributary rivers. However, this effect remains limited to the inner alpine regions. In the future, the discharge regime will assume a character that is dominated by snow and rain, in which the inner alpine compensation effect of glaciers disappears, and as a result, discharge variability increases, particularly in summer. In contrast, the effects on the water supply in the Danube at gauge Achleiten due to the retreat of the glaciers can be considered negligible compared to the changes in regional precipitation. The *REMO regional–Baseline* scenario, with its significant decrease in summer precipitation even in the mountains, clearly indicates the effect of the Alpine water tower. Precipitation deficits can no longer be compensated for by meltwater from the remaining ice reservoir; in particular, the inflow from the glaciated headwater is drastically reduced. This will especially affect the existing hydropower plants, which will rank among the winners in the next 20 years as a result of the enhanced inflows, but which later may not be able to fill their reservoirs through the summer (Koch et al. 2011). Because the impact of the glacier melt is decreasing downstream the river courses, these effects are limited to the inner alpine regions, when excluding the decreased precipitation prescribed by the climate scenario.

References

Baumgartner A, Reichel E, Weber G (1983) Der Wasserhaushalt der Alpen: Niederschlag, Verdunstung, Abfluss und Gletscherspende im Gesamtgebiet der Alpen im Jahresdurchschnitt für die Normalperiode 1931–1960. Oldenbourg Wissenschaft, München

Jarvis A, Reuter HI, Nelson A, Guevara, E (2008) Hole-filled seamless SRTM data V4. International Centre for Tropical Agriculture (CIAT). Available via DIALOG. http://srtm.csi.cgiar.org. Accessed 19 Sept 2014

Koch F, Prasch M, Bach H, Mauser W, Appel F, Weber M (2011) How will hydroelectric power generation develop under climate change scenarios? A case study in the Upper Danube Basin. Energies 4(10):1508–1541. doi:10.3390/en4101508

Lambrecht A, Mayer C (2009) Temporal variability of the non-steady contribution from glaciers to water discharge in western Austria. J Hydrol 376:353–361

Roethlisberger H, Lang H (1987) Glacial hydrology. In: Gurnell AM, Clark MJ (eds) Glacio-fluvial sediment transfer: an alpine perspective. Wiley, Chichester, pp 207–284

Viviroli D, Weingartner R, Messerli B (2003) Assessing the hydrological significance of the world's mountains. Mt Res Dev 23(1):32–40

Weber M, Prasch M, Braun L (2009) Die Bedeutung der Gletscherschmelze für den Abfluss der Donau gegenwärtig und in der Zukunft. In: Mitteilungsblatt des Hydrographischen Dienstes in Österreich, vol 86, Lebensministerium, BMLFUW, Wien

Chapter 57
Mean Snow Cover Duration from November to June Under the *REMO Regional* Climate Trend and the *Baseline* Climate Variant

Markus Weber and Monika Prasch

Abstract The change in snow cover under the conditions of climate change is of great importance for the hydrological conditions of the Upper Danube basin. In addition to snow depth, which is quite heterogeneous in space, a particularly meaningful parameter is the duration of snow cover, SD. This variable is defined by the number of days with a snow cover greater than 1 mm water equivalent within the winter, which in the Alps is generally from November to June of the following year (see also Chap. 30). SD is significant for both ecological and economic reasons.

The map for this chapter illustrates the change in the snow cover duration in the Danube drainage basin as 30-year means for the model years 2011–2040, 2021–2050 and 2031–2060; these changes are based on the results of modelling with DANUBIA and the *Snow* snow model presented in Chap. 30 under the conditions of the *REMO regional* climate trend and the *Baseline* climate variant (see Chaps. 48, 49, 50).

The simulation results indicate that in a warmer climate, a snow cover in the lowlands would become even rarer than before, although absolutely no snow is not expected, even in the lowest areas. Otherwise in the future SD will be shortened at any altitude. The most far-reaching effect will be manifested in the glaciated regions, where the snow-free areas of the glaciers will continue to enlarge and the ablation period will continue to be extended. This mainly contributes to an accelerated retreat of glaciers (see Chap. 56).

M. Weber (✉)
Commission for Geodesy and Glaciology of the Bavarian Academy of Sciences and Humanities, Munich, Germany
e-mail: Wasti.Weber@kfg.badw.de

M. Prasch
Department of Geography, Ludwig-Maximilians-Universität München (LMU Munich), Munich, Germany
e-mail: m.prasch@lmu.de

W. Mauser, M. Prasch (eds.), *Regional Assessment of Global Change Impacts*, DOI 10.1007/978-3-319-16751-0_57

Keywords GLOWA-Danube • Snow • Snow cover • Snow cover duration • Lowlands • Deglaciation • *REMO*

57.1 Introduction

The change in snow cover under the conditions of climate change is of great importance for the hydrological conditions of the Upper Danube basin. In addition to snow depth, which is quite heterogeneous in space, a particularly meaningful parameter is the duration of snow cover, SD. This variable is defined by the number of days with a snow cover greater than 1 mm water equivalent within the winter, which in the Alps is generally from November to June of the following year (see also Chap. 30).

In general, SD increases as a result of the decrease in atmospheric temperature at higher elevations; above the climatic firn line, the snow may stay year-round, and hence, the SD is then from the period of November to June or up to 242 days.

The snow cover duration is significant for both ecological and economic reasons. The key ecological factors that also impact economic concerns include the glacial mass balance (see Chap. 31), the runoff regime and water storage and the vegetation cover. Specific economic factors include the costs to maintain operating capacity of transportation routes or the direct impact on winter tourism (see Chap. 62).

Using the *Snow* snow model presented in Chap. 30, the future changes to the snow cover duration are presented in the form of a map; these changes are based on the results of modelling with DANUBIA under the conditions of the *REMO regional* climate trend and the *Baseline* climate variant (see Chaps. 48, 49, 50).

57.2 Data Processing

Snow cover duration was determined for each proxel using the analysis of snow depths calculated at hourly time steps by the snow model. Although with this method the model calculation generates a continuous series over 50 years from the defined starting point, this should not be considered as a prediction for the future course of weather patterns. Thus, specific individual events, such as an episode with extremely low precipitation or a particularly small amount of snow in a model year, do not need to be tied to the time in which they occurred. They can be considered as random parts of the process.

Meaningful is the statistical information of the results for modelling the water equivalent of the snow cover in the context of a scenario where a climate trend was overlaid. Like other climatological variables, the averaging interval is a period of 30 years. These mean values are determined at the interval of a decade, beginning in the model year 2011, in order to pool the data. These three overlapping intervals each represent different climatic conditions with a temperature level that increases according to specifications as the model years pass.

Table 57.1 Mean snow cover duration (*SD*) in days [*d*] between November and July for the specified altitude ranges and time periods with the conditions of *REMO regional* climate trend and *Baseline* climate variant

SD [d]	Snow cover duration [d]				Precipitation [mm]			
Altitude range [m a.s.l.]	1971–2000	2011–2040	2021–2050	2031–2060	1971–2000	2011–2040	2021–2050	2031–2060
290–500	58	40	30	22	797	799	794	760
500–1,000	87	63	52	40	1,037	1,038	1,031	984
1,000–1,500	170	149	139	125	1,460	1,438	1,428	1,361
1,500–2,000	190	170	163	153	1,410	1,394	1,380	1,325
2,000–2,500	204	184	178	168	1,365	1,348	1,334	1,287
2,500–3,000	221	201	195	185	1,339	1,317	1,303	1,262
3,000–3,500	235	221	216	207	1,380	1,367	1,347	1,315
>3,500	242	237	235	228	1,714	1,678	1,665	1,606
Entire basin	101	79	70	59	1,039	1,035	1,028	983

Table 57.2 Mean percental proportion of snow on precipitation (P:S) for the alpine part of the drainage basin, listed separately for the specified altitude ranges and model time periods. The second column shows the percental area proportions of the altitude ranges on the entire basin area; the last row shows the mean snow percentage of the entire basin for comparison

P:S [%]					
Altitude range [m a.s.l.]	Area proportion [%]	1971–2000	2012–2040	2021–2050	2031–2060
290–1000	5.89	21	17	15	13
1,000–2,000	14.70	38	34	33	31
2,000–3,000	7.72	62	56	55	52
>3,000	0.38	79	73	71	69
Alps	28.70	41	37	36	33
Basin	100.00	25	21	20	18

Additional statistics for the mean annual precipitation totals, divided into individual elevations, and for the proportion of the total precipitation that is solid were prepared for the Alpine region; these statistics help to interpret the map and are listed in Tables 57.1 and 57.2. The results of a modelling of the past (1971–2000) are used as a reference period.

57.3 Results

Map 57.1 illustrates the change in the snow cover duration in the Danube drainage basin as 30-year means for the model years 2011–2040, 2021–2050 and 2031–2060, as expected under the conditions of the *REMO regional* climate trend and the *Baseline* climate variant. The spatial resolution of the representation of the mean

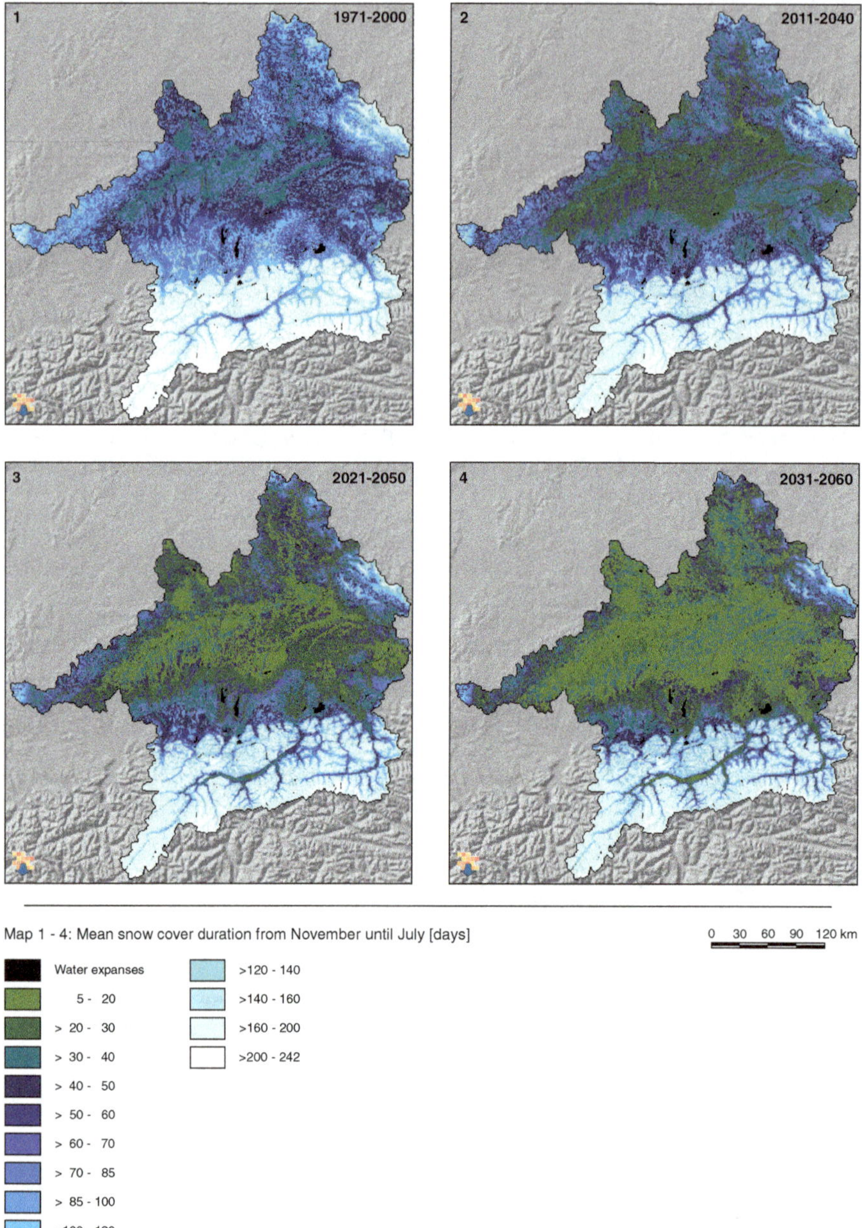

Map 57.1 Mean snow cover duration under the *REMO regional* climate trend and the *Baseline* climate variant

snow cover duration over 30 successive years is 1×1 km^2. The results of the modelling for the period 1971–2000 using observational data serve as a reference in Map 57.1(1). Even at this relatively coarse scale, the close correlation between SD and terrain elevation is evident; this is also apparent in the increase in the means over elevational belts of the basin shown in Table 57.1. Since the subscale orography within the 1×1 km^2 scale exhibits elevational differences of up to more than 100 m within the Alps, snow cover duration can vary locally on a proxel by up to 100 days compared to the mean from the scale of km. Like the example of Map 30.1 for a shorter period of averaging shown in Chap. 30, there are also regional differences that correlate with the spatial distribution of precipitation, in addition to the dominant relationship with elevation. With the overall increase in temperature and the resulting higher snow line, the number of days with snow cover decreases at all altitudes (see Table 57.1). Comparison among the maps and with the reference Map 30.1 suggests a significant decline in the snow cover duration in the lowlands. However, Table 57.1 indicates that, compared to the reference period, the mean number of snow days is gradually reduced over the course of the scenario initially by an average of 5–10 days per decade. This shortening is most apparent at the elevation belts between 1,000 and 1,500 m a.s.l., where by the end of the simulation, the decrease has reached up to 14 days per decade. Above and below these elevations, the decline in snow cover duration generally decreases; in particular, at elevations above 3,000 m a.s.l., the decrease is initially barely perceptible but drastically increases towards the end of the simulation period by about 7–9 days as well. This can be explained by the fact that with the increase in temperature over the scenario period, increasingly more of this elevation belt is situated below the firn line, above which the snow no longer completely melts. This is also the region in which glaciers can still be found. According to the second column in Table 57.2, this area is a relatively small part of the whole basin area.

The simulation results indicate that in a warmer climate, a snow cover in the lowlands would become even rarer than before, although absolutely no snow is not expected, even in the lowest-lying areas. As a result, even in these areas in the future, one needs to be prepared for winter weather. However, the duration of snow cover that is shortened by an average of 40 days towards the end of the *REMO regional* climate trend and *Baseline* climate variant at all elevations below 3,000 m a.s.l. is not correlated by a decrease in snow fall of comparable magnitude compared to the reference period 1971–2000. The decrease in precipitation totals at lower elevations is between 8 and 10 % (see Table 57.1) and at the highest elevations above 3,500 m a.s.l. declines to just around 7 %. According to Table 57.2, at least a third of the precipitation in the Alps falls as snow and still just a quarter falls as snow for the entire basin during the reference period. In the future, this will change only marginally, since the majority of snow falls in midwinter, in which even under the conditions of climate change, there are still episodes with temperatures below freezing. In areas below 2,000 m a.s.l., the snow cover will be depleted at increasing rates, tied in with the significantly earlier onset of snow melt. However, the changes for areas above 3,000 m a.s.l. are the most obvious, since there the high snowfalls in summer that occurred in the reference period are no longer expected in the future. Thus, in these areas the portion of the total precipitation that falls as snow will decrease from an average of 80 % to below 70 %. There, the annual snow accumulation along with the decrease in precipitation at all elevations will decline by approximately 250 mm of water equivalent towards the end of the simulation, compared to the reference period.

Fewer traffic hindrances will occur in the lowlands as a result of the shorter snow cover duration and the operating seasons of the ski lift facilities (see Chap. 62) will be shorter at all elevations. At the same time, there will be a longer period available for vegetation growth. Moreover, the depletion of the snow cover will occur earlier in the year and this will change the nival runoff regime of the Upper Danube basin (see Chap. 60). The most far-reaching effect will be manifested in the glaciated regions, where the snow-free areas of the glaciers will continue to enlarge and the ablation period will continue to lengthen. Together with the reduced snow accumulation, extreme negative mass balances will result and there will be an associated accelerated retreat of the glaciers (see Chap. 56).

Chapter 58
Trends in Evapotranspiration of Heterogeneous Landscapes Under Scenario Conditions

Tobias Hank

Abstract A large part of the flux of mass and energy from the land surface to the atmosphere via evapotranspiration consists of transpiration and therefore is influenced by vegetation. The assessment of evapotranspiration under scenario conditions thus requires a dynamic representation of vegetation growth in the applied model. An explicit model of photosynthesis-driven canopy development therefore is applied to simulate evapotranspiration in the heterogeneous Upper Danube basin for 16 climate scenarios based on four climate trends (*REMO*, *IPCC*, *MM5*, *Extrapolation*).

During a long-term reference period from 1960 to 2006, the annual sum of evapotranspiration in the Upper Danube basin shows a steady increase from approx. 370 mm per annum in the 1960s to about 400 mm per annum in the years after 2000. All of the respected climate scenarios agree on a continuation of this increasing trend during the scenario period from 2011 to 2060. The comparison of two 30-year periods (1971-2000 vs. 2031-2060) reveals that the forested slopes of the alpine region show the highest positive change signal. However, not only the annual amount of evapotranspiration is subject to change but also the seasonal pattern shows some distinctive shifts. While the incidence of maximum evapotranspiration tends to shift towards late July until the 2020s, this trend is reversed from 2030 onwards, where the maximum evapotranspiration again may be found in June. This most likely is due to water scarcity during times of maximum potential evapotranspiration, since both spring and autumn evapotranspiration show a steady increase until the end of the scenario period.

Keywords GLOWA-Danube • DANUBIA • Evapotranspiration • Vegetation • Gas exchange • Photosynthesis

---

T. Hank (✉)
Department of Geography, Ludwig-Maximilians-Universität München (LMU Munich), Munich, Germany
e-mail: tobias.hank@lmu.de

W. Mauser, M. Prasch (eds.), *Regional Assessment of Global Change Impacts*, DOI 10.1007/978-3-319-16751-0_58

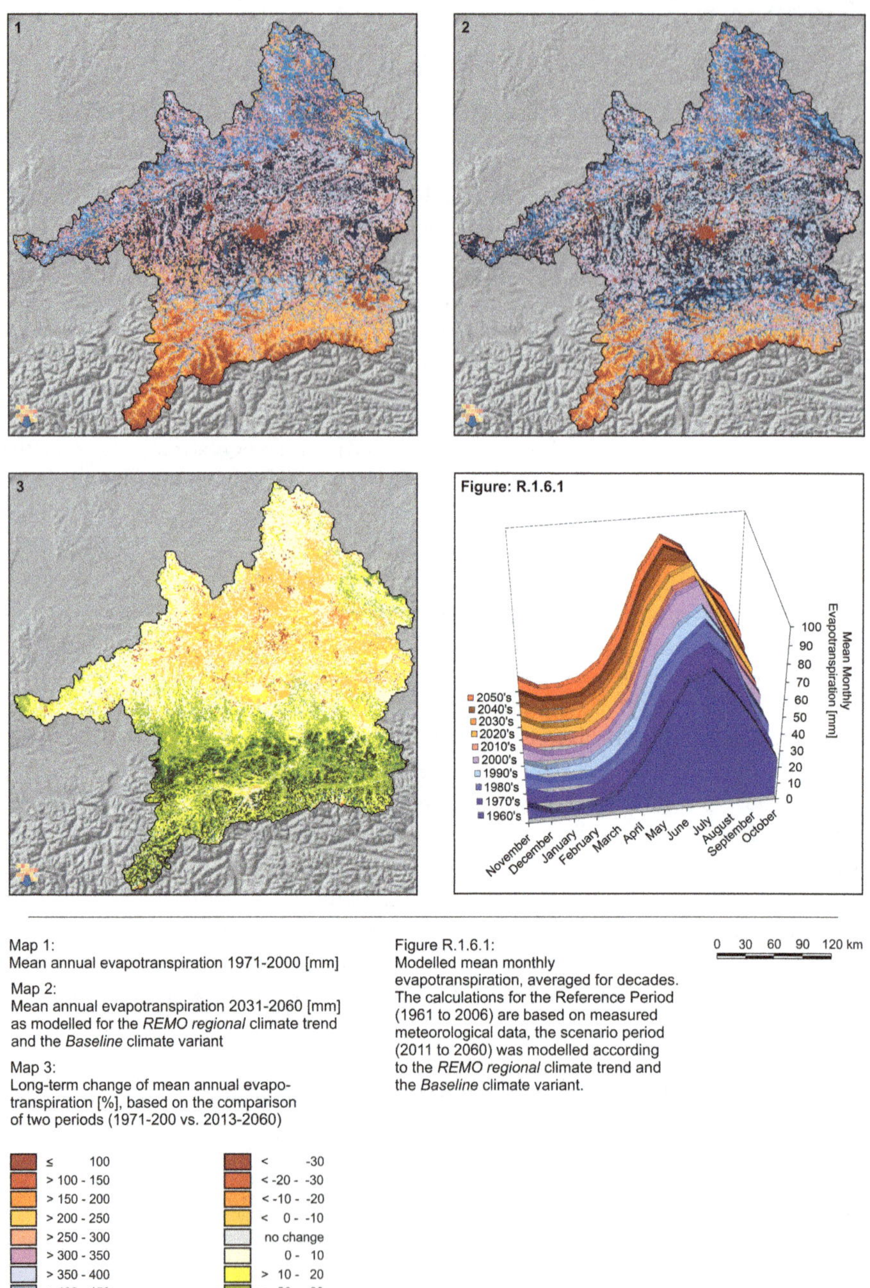

Map 1:
Mean annual evapotranspiration 1971-2000 [mm]

Map 2:
Mean annual evapotranspiration 2031-2060 [mm] as modelled for the *REMO regional* climate trend and the *Baseline* climate variant

Map 3:
Long-term change of mean annual evapotranspiration [%], based on the comparison of two periods (1971-200 vs. 2013-2060)

Figure R.1.6.1:
Modelled mean monthly evapotranspiration, averaged for decades. The calculations for the Reference Period (1961 to 2006) are based on measured meteorological data, the scenario period (2011 to 2060) was modelled according to the *REMO regional* climate trend and the *Baseline* climate variant.

Map 58.1 Trends in evapotranspiration of heterogeneous landscapes under scenario conditions (*REMO regional* climate trend and *Baseline* climate variant)

58.1 Introduction

Evapotranspiration, i.e. the flux of latent heat from the land surface, is determined by processes of active exchange between the earth's surface and the atmosphere. The composition of three different types of evaporation processes (soil evaporation, interception evaporation, transpiration) varies depending on the dominant land cover. Processes that are characteristic for non-vegetated surfaces, such as evaporation from open water surfaces or from exposed soils, where water vapour is released to the atmosphere through macro- and micropores from the groundwater or soil water reserves, are complemented by controlled transpiration mechanisms of vegetated land surfaces. By this transpiration process, water is removed from the groundwater or soil water reserves via roots and is released to the atmosphere via the above-ground organs of the actively growing vegetation. The rate of this stream of water vapour is regulated through the stomata of the plant in dependence of the growth status, which again is highly dependent on environmental variables. The process of transpiration thus gains particular importance for the determination of land-surface evapotranspiration. In addition, interception evaporation takes place on both vegetated and non-vegetated land surfaces. For example, interception occurs when puddles form on sealed ground surfaces (residential areas, bedrock, etc.) following precipitation events or when precipitation is stored long term on the land surface in the form of ice and snow (glaciers, snow cover) and also if the above-ground parts of the canopy retain a certain portion of the precipitation. Intercepted precipitation is directly returned to the atmosphere in the form of water vapour and does not contribute to growth processes. Depending on the dominant type of land cover and on the ecological intensity of land use, evapotranspiration is thus composed of differing ratios of evaporation, transpiration and interception evaporation. These dynamics are explicitly accounted for in the DANUBIA model runs.

58.2 Data Processing

For the reference period (1960–2006), a model run based on measured data from meteorological stations was carried out, while the scenario period (2011–2060) was successively modelled for four climate trends and the corresponding climate variants (see Chaps. 48 and 50) that were generated using the statistical climate generator (see Chap. 49). Thus, a range of potential future trends in land-surface evapotranspiration could be generated. The results of the different variants for each climate trend were then averaged to illustrate an average development for the different climate trends (see Fig. 58.2).

The *Baseline* climate variant characterised by the *REMO regional* climate trend represents a future scenario of intermediate values. In order to demonstrate the mean temporal dynamics of the trends in evapotranspiration under the impact of the *REMO regional–Baseline* climate scenario, the calculated daily sums of evapotranspiration

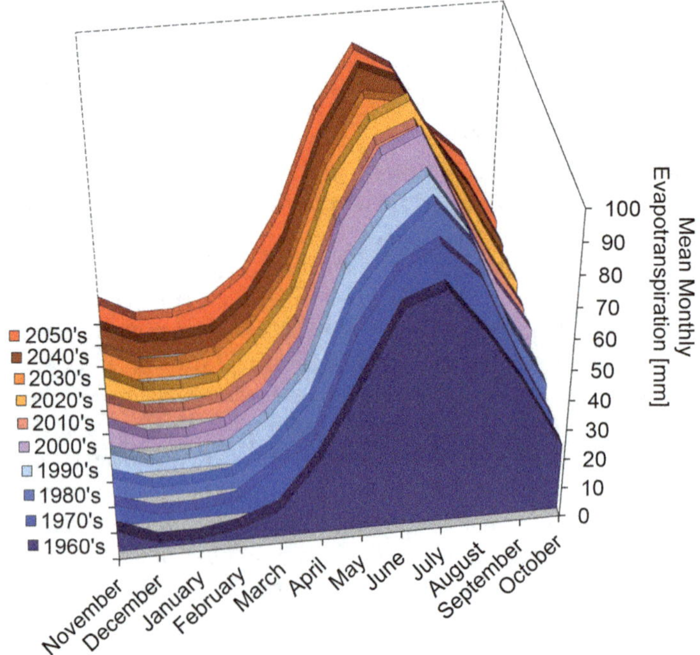

Fig. 58.1 Modelled mean monthly evapotranspiration, averaged for decades. The calculations for the reference period (1961–2006) are based on measured meteorological data, while the scenario period (2011 to 2060) was modelled according to the *REMO regional* climate trend and the *Baseline* climate variant

were first summarised as monthly totals. The monthly evapotranspiration sums were then averaged by decade (see Fig. 58.1) to also illustrate the seasonal shift of evapotranspiration patterns.

58.3 Model Documentation

In the Upper Danube basin, the evapotranspiration capacity of the land surface is principally influenced by the growth and activity of the vegetation cover. Mechanistically describing the determinative processes of exchange between plants and the atmosphere through a detailed and dynamic plant growth model thus is of major importance. The core of the vegetation module is an established model for calculating photosynthesis of C_3 plants (Farquhar et al. 1980), which is combined with a method for describing the conductivity between leaves and atmosphere (Ball et al. 1986). The model is enhanced by methods for the description also of the C_4 metabolism (Chen et al. 1994) as well as by a parameterisation for modelling forest

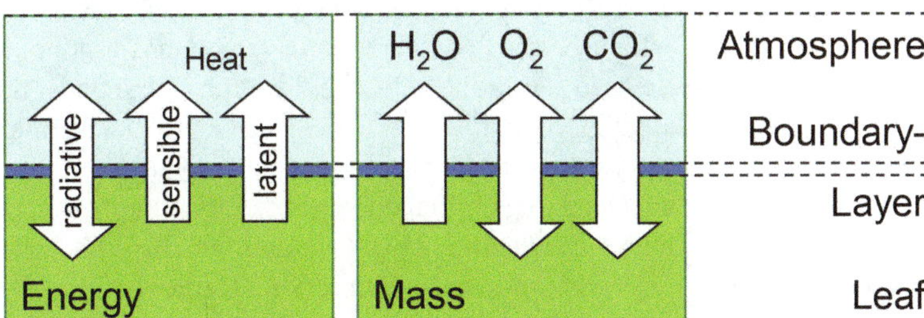

Fig. 58.2 Exchange processes of mass and energy between leaf and atmosphere as they are accounted for in the DANUBIA model

trees (Falge 1997). The model provides the energy required for photosynthesis as the remainder of an explicitly modelled leaf energy balance, where the input of energy in the form of absorbed direct and diffuse radiation is counterbalanced by three paths of energy loss via sensible heat, long-wave emission and latent heat loss through the transpiration flux (Hank 2008; see Fig. 58.2).

To account for the seasonal dynamics of vegetation development, the allocation of assimilates into the different parts of the canopy is regulated with the help of a model of phenological progress according to Yin and van Laar (2005). The phenological pattern of forest trees acquires particular importance with respect to the regional water supply. It too is dynamically implemented using a phenological model approach based on Menzel (1997). The nutrient supply for plants is assumed not to be limiting for the present calculations, taking into account that the nutrient supply in the area of interest normally is kept on an optimum level through regular fertilisations.

58.4 Results

The change signal for annual evapotranspiration is depicted in Map 58.1 – 3; this signal is the result of the comparison between two long-term periods (each 30 years). The comparison involved the mean annual total evapotranspiration of the years 1971 to 2000 (see Map 58.1 – 1, based on model calculations driven by measured meteorological data) and the mean evapotranspiration of the years 2031–2060 (see Map 58.1 – 2). The evapotranspiration totals presented in Map 58.1 – 2 result from model calculations that were driven by the statistically generated *Baseline* climate variant that is shaped by the temperature and precipitation trends from the *REMO regional* climate scenario. It is notable that the majority of the area in the Upper Danube basin is characterised by increasing evapotranspiration totals. The model calculations suggest particularly high positive rates of change for the Alpine region, where primarily the forested slopes of the northern foothills are notable for increases in evaporation totals by up to 50 %. In the face of increasing annual mean temperatures that are a part of the scenario, an increase in evapotranspiration can be explained mostly by the elevated saturation deficit in the atmosphere under the conditions of the scenario. In addition, high temperatures accelerate the chemical processes that make up photosynthesis and hence expedite gas exchange by plants. Increasing temperatures also lead to a prolonged vegetation period, by accelerating growth in the spring months but delaying leaf fall in autumn. Nevertheless, some areas of the basin signal a decline in evapotranspiration. The lower-lying areas of the Danube valley primarily indicate a decrease in the long-term evapotranspiration by up to 20 %. Also the gravel plains north of the city of Munich and the region of the Inn valley around the city of Innsbruck (see overview Map 1.1 in Chap. 1) are characterised by a decrease of the long-term evapotranspiration by up to 10 %. This development may be determined by a variety of processes. For one, the enhanced evaporation and transpiration activity taking place under higher temperatures leads to a depletion of the soil water storage earlier in the year, which is why the supply of soil water that is available for transpiration processes is limited in the months with the prime evapotranspiration capacity (July and August; see Fig. 58.1). Nonetheless, even when the water supply is adequate, the modelled change signal could indicate a decrease in evapotranspiration, for example, if elevated temperatures in areas cultivated with cereal crops lead to an acceleration of the phenological progress. If the ripening process of these crops is initiated earlier in the year and thus the life cycle of the crop is shortened, the evapotranspiration activity is curtailed. Further, many plant types exhibit increased efficiency of water use when they are subject to increased atmospheric carbon dioxide concentrations (Polley et al. 1993), which allows for maintained rates of carbon fixation at lower transpiration rates. The opposite dynamics of mechanisms that promote and inhibit evapotranspiration is evident, if the temporal character of the trend in evapotranspiration activity is considered. It is apparent that evapotranspiration during the summer months increases into the 2020s (see Fig. 58.1). However, this trend gradually is reversed after the 2030s, while there is a simultaneous shift in the month of the maximum evapotranspiration capacity from July to June. Evapotranspiration in spring (March, April) also continues to increase, as does the evapotranspiration during the fall (September, October, November).

The ensemble of a total of 16 scenario calculations covers a range of potential trends (see Fig. 58.3). All of the climate scenarios that were modelled indicate an increase in the annual evapotranspiration totals. Whereas the man of the climate variants for the *REMO regional* shows the steepest trend, the evapotranspiration totals in the *Extrapolation* climate trend lag behind considerably. This result can be largely attributed to the significant decrease in summer precipitation that is one of the assumptions of the *Extrapolation* climate trend (see Chaps. 47 and 48). Although the relative temperature increase and the associated elevated saturation deficit is modelled in a similar manner for the *Extrapolation* climate trend as for the *REMO regional* climate trend, there are more moderate rates of evapotranspiration indicated in the case of the *Extrapolation* climate trend. Evidently, a lack of surface water during the summer months leads to a stagnation of the evapotranspiration capacity in this case.

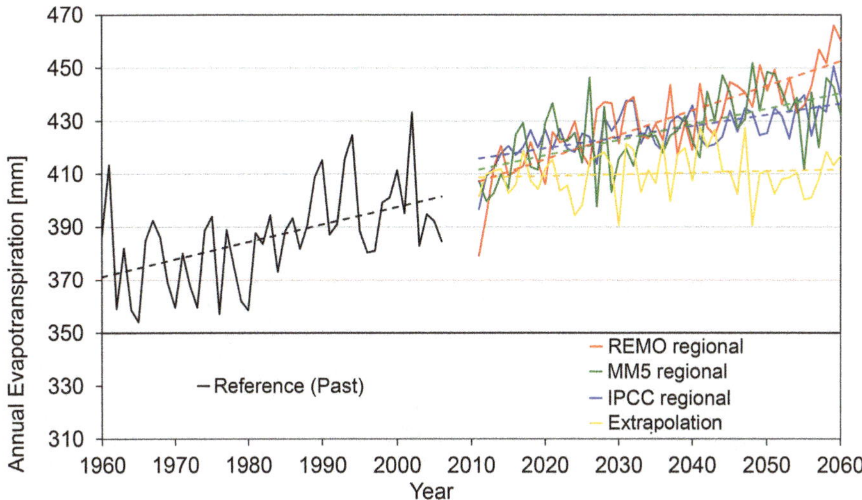

Fig. 58.3 Development of mean annual evapotranspiration during the reference period (past) and the different GLOWA-Danube climate scenarios (future). The annual sums represent averaged values of the four climate variants of each climate trend. The *dashed lines* show the respective linear regressions

References

Ball J, Woodrow I, Berry J (1986) A model predicting stomatal conductance and its contribution to the control of photosynthesis under different environmental conditions. Prog Phot Res 4:221–224

Chen D, Coughenour M, Knapp A, Owensby C (1994) Mathematical simulation of C4 grass photosynthesis in ambient and elevated CO2. Ecol Mod 73(1–2):63–80

Falge E (1997) Die Modellierung der Kronendachtranspiration von Fichtenbeständen (Picea abies (L.) Karst.). PhD, University of Bayreuth, Bayreuther Forum Ökologie, vol 48

Farquhar G, von Caemmerer S, Berry J (1980) A biochemical model of photosynthetic CO2 assimilation in leaves of C3 species. Planta 149(1):78–90

Hank T (2008) A biophysically based coupled model approach for the assessment of canopy processes under climate change conditions. Dissertation, Ludwig-Maximilians Universität München (LMU Munich)

Menzel A (1997) Phänologie von Waldbäumen unter sich ändernden Klimabedingungen: Auswertung der Beobachtungen in den internationalen phänologischen Gärten und Möglichkeiten der Modellierung von Phänodaten [Forest phenology under changing climatic conditions: analysis of observations of the international phenological gardens and approaches to modelling phenological data]. In: Forstliche Forschungsberichte München [Forestry scientific reports Munich], vol 164, Frank Munich, pp 158. ISSN 0174-1810

Polley H, Johnson H, Marino B, Mayeux H (1993) Increase in C plant water-use efficiency and biomass over 3 glacial to present CO concentrations. Nature 361:61–64

Yin X, van Laar H (2005) Crop systems dynamics. An ecophysiological simulation model for genotype-by-environment interactions. Academic, Wageningen

Chapter 59
Groundwater Recharge Under Scenario Conditions

Christoph Heinzeller

Abstract Although the Upper Danube region is currently not suffering under water scarcity and thus the availability of drinking water is not a key issue, it is important to investigate the potential impacts of climate change on groundwater recharge in the Upper Danube drainage, because more than 90 % of the drinking water in this region is sourced from groundwater. Therefore, all of the climate scenarios of the GLOWA-Danube project, including their four statistical climate variants, as well as the climate variants of the *REMO downscaled and bias-corrected* and *MM5 downscaled and bias-corrected* regional climate models for the Upper Danube basin were calculated with the DANUBIA model framework. This resulted in a total of 18 model scenario runs which all show a significant decrease in GWR particularly in the summer months due to the climate change-induced decline of rainfall and higher evapotranspiration rates in the summer. For example, the proportion of the total annual groundwater recharge in the summer months is approximately 29 % for the reference period and will decline to 17 % based on the *REMO regional–Baseline* scenario. The predicted decrease in GWR for the period 2011–2060 fluctuates between 33 and 125 mm. The results of DANUBIA show that there is a significant change in GWR due to climate change in the Upper Danube region based on the model scenarios.

Keywords GLOWA-Danube • DANUBIA • Groundwater • Groundwater recharge • Drinking water • Climate change

59.1 Introduction

In this chapter, the potential impacts of climate change on groundwater recharge (GWR) in the Upper Danube drainage basin are presented.

In DANUBIA, GWR is defined as the quantity of water that percolates through the soil pore space that is both unsaturated and penetrated by plant roots (see

C. Heinzeller (✉)
Department of Geography, Ludwig-Maximilians-Universität München (LMU Munich), Munich, Germany
e-mail: christoph.heinzeller@lmu.de

W. Mauser, M. Prasch (eds.), *Regional Assessment of Global Change Impacts*, DOI 10.1007/978-3-319-16751-0_59

Chap. 24) to reach the groundwater body. The dynamics of soil water storage is controlled primarily by the main processes of reservoir accumulation (precipitation, discharge, capillary rise) and the processes of depletion (evapotranspiration, interflow, groundwater recharge). The process of GWR or also the percolation of the soil water from the unsaturated zone into the groundwater body is realised in DANUBIA by the transfer of simulation values from the *Soil* model component to the *GroundwaterFlow* groundwater model component. The unit for GWR in the model is mm per unit time.

More than 90 % of the drinking water in the Upper Danube basin is sourced from groundwater (Emmert 1999). In order to investigate the effects of climate change on groundwater recharge and hence also on the regeneration of drinking water, all of the climate trends defined for the GLOWA-Danube project, including their four statistical climate variants, as well as the climate variants of the *REMO downscaled and bias-corrected* and *MM5 downscaled and bias-corrected* regional climate models for the Upper Danube basin were calculated (see Chaps. 47, 48, 49, 50, 51). This resulted in a total of 18 model runs. The results of these runs were quantitatively and statistically analysed (see Table 59.1).

59.2 Results

Maps 59.1-1 and 2 provide an overview of how groundwater recharge might change in the context of climate change. The maps illustrate the mean annual volume of percolating water from the unsaturated soil zone for the entire basin for the *REMO regional* climate trend with the *Baseline* climate variant as well as for the reference period. The regional differences in GWR clearly stand out here. The trend that results from the combination of the *REMO regional* climate trend with the *Baseline* climate variant (–49 mm; see Table 59.2) represents an average of all the scenarios used.

The contrast of the groundwater recharge in the summer months (June, July and August) for the reference period and for the *REMO regional–Baseline* climate scenario (see Maps (3 and 4)) should indicate which regions of the basin might see significant future declines and hence also more frequent deficits in GWR in the summer months.

59.2.1 Trends of the Scenarios

Figure 59.1 provides an overview of the ranges and trends for the scenarios. This figure plots the mean annual values for groundwater recharge for each of the four climate variants per climate trend and the special cases *MM5 downscaled and bias corrected* and *REMO downscaled and bias corrected*. Figure 59.1 and Table 59.2 both indicate a reversal of the trend in groundwater recharge for all scenarios. Thus,

Table 59.1 Statistical characteristics of groundwater recharge of all applied climate-trend climate-variant combination

	Reference run	IPCC regional*		REMO regional*		MM5 regional*		Extrapolation*		REMO downscaled and bias corrected		MM5 downscaled and bias corrected	
Period	1971–2000	2011–2035	2036–2060	2011–2035	2036–2060	2011–2035	2036–2060	2011–2035	2036–2060	2011–2035	2036–2060	2011–2035	2036–2060
Mean [mm]	554	528	513	514	492	534	527	548	478	466	453	462	435
Min [mm]	365	329	333	329	333	315	317	301	319	324	285	296	261
Max [mm]	686	729	761	757	735	728	769	716	693	625	682	706	785
STABW [mm]	88	80	88	84	80	88	86	95	81	82	97	105	141
Trend [mm]	70	−7	−60	−23	−43	−49	−27	−33	−49	2	−55	−60	−7
Change [%]	–	−4.7	−7.4	−7.2	−11.1	−3.5	−4.8	−1.1	−13.6	−15.8	−18.1	−16.6	−21.4

* Mean of all climate variants of the respective climate trend

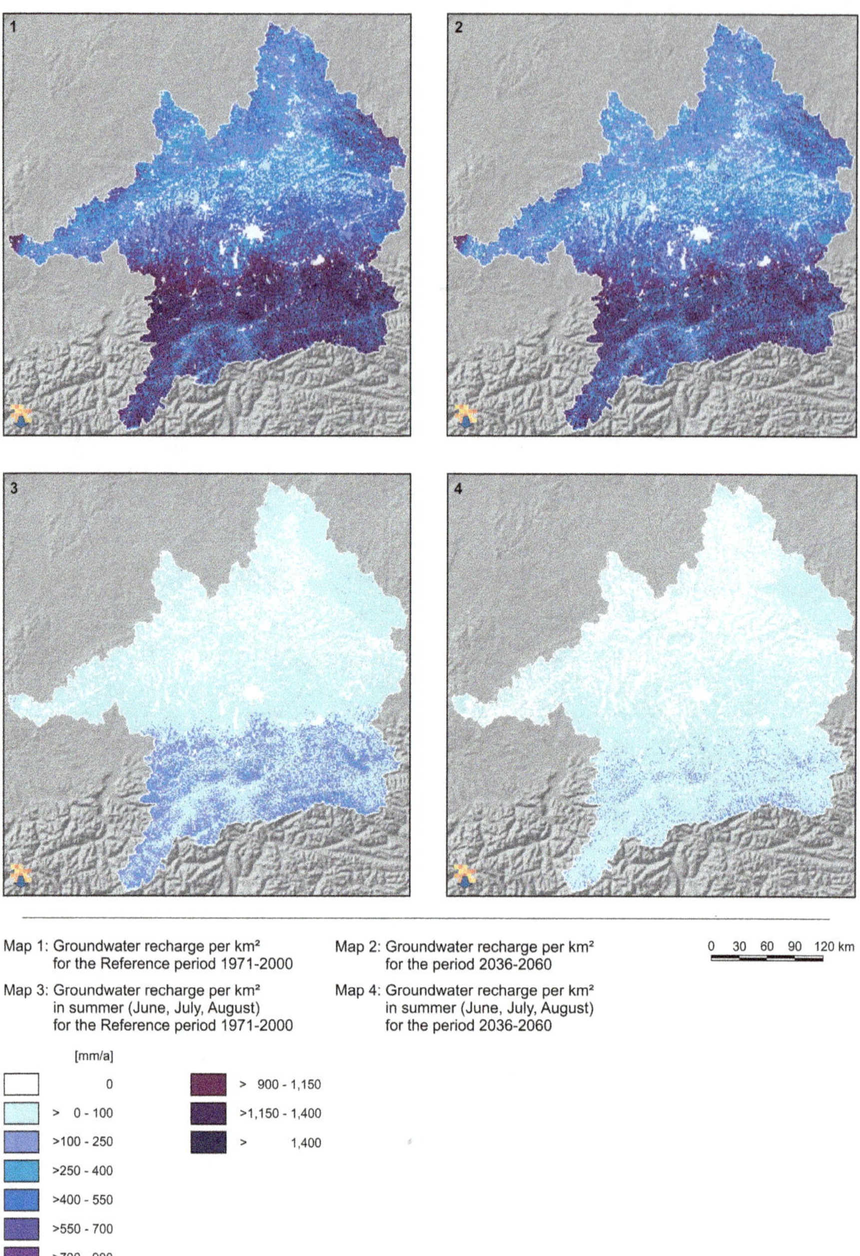

Map 59.1 Groundwater recharge under scenario conditions (*REMO regional* climate trend and *Baseline* climate variant)

Table 59.2 Mean linear trend of groundwater recharge in the Upper Danube river catchment

Run	Period	Trend [mm]
Past	1960–2006	+11
IPCC regional	2011–2060	−39
REMO regional	2011–2060	−49
MM5 regional	2011–2060	−30
Extrapolation	2011–2060	−125
MM5 downscaled and bias corrected	2011–2060	−56
REMO downscaled and bias corrected	2011–2060	−33

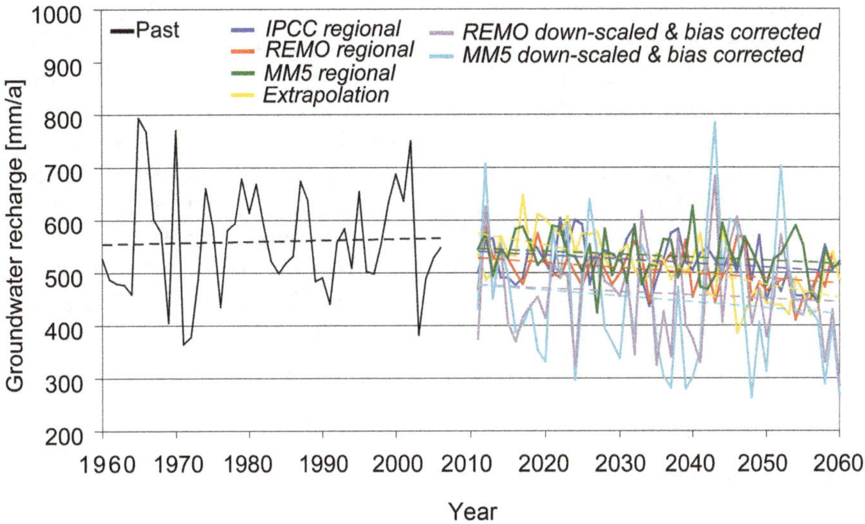

Fig. 59.1 Change of groundwater recharge for the past period and the applied GLOWA-Danube climate scenarios. Depicted are the spatial aggregated annual values, time averaged over the four statistical climate variants. The two climate variants based on the regional climate models are shown separately; the *dashed lines* are representing the trend lines

in the past, there is still a slight increase in GWR in the Upper Danube basin over the long-term average. However, there is a decrease in groundwater recharge for all scenarios. The predicted decrease in GWR for the period 2011–2060 fluctuates between 33 and 125 mm.

59.2.2 Spatial Changes

The long-term average for groundwater recharge in the Upper Danube in the reference period 1971–2000 was 554 mm/a. As shown in Maps 59.1(1–4), the GWR initially basically followed the precipitation signal (see also Chap. 11, Map 11.4).

Thus, regions with values above 550 mm/a for GWR are generally located in areas with abundant precipitation in the low mountain ranges, in the Alpine foothills and in the Alps. The GWR in these regions also benefits from the reduced evaporation, the somewhat deficient storage capacity of the soils (e.g. Limestone Alps, Pleistocene gravel) and by the alpine relief. The regions of the South German escarpment with its bedrock boundaries are characterised by areas with GWR values between 0 and 550 mm/a. The mean annual GWR for the entire basin for the *REMO regional–Baseline* climate scenario (2011–2060) is 503 mm/a.

The pattern for the spatial distribution of groundwater recharge in the scenario shares the essential features of the references period. The differences in the change compared to the past are primarily evident upon closer examination by region. Thus, it is apparent that particularly in the Alpine foothills, there are fewer areas with GWR values above 700 mm/a. The GWR for the low mountain ranges is also reduced. Likewise, there is a conspicuous decline in the Alpine region. Hence, the regions with values above 700 mm/a noticeably thin out. Using the *REMO regional–Baseline* climate scenario, there is a decrease in groundwater recharge for the entire Upper Danube basin of approximately 9 % for the period 2011–2060 (see Table 59.1). If only the change for summer is considered, the potential effects of climate change become even clearer. No groundwater recharge was calculated for the summer months for the reference period for approximately 16 % of the proxels; this means that in the *REMO regional–Baseline* climate scenario (2036–2060), the value is as much as 27 %.

59.2.3 Temporal Change

Like in many other regions in the northern hemisphere, groundwater recharge in the Upper Danube basin takes place mostly from November to April (in the hydrological winter). In this period, evapotranspiration contributes less to the decrease in groundwater recharge as a result of lower temperatures and reduced plant growth. In the summer half of the year, there is higher evapotranspiration and hence GWR is naturally lower, and this makes an additional decline caused by climate change more noticeable. Map 59.1 – 4 reveals which significant effects the expected climate change might have on groundwater recharge and hence on the availability and quality of drinking water, especially in summer.

For the Upper Danube basin, a significant decrease in precipitation particularly in the summer months is expected (see Chap. 48, Map 48.2 and Chap. 51, Map 51.4). First, this will have a direct impact on GWR. In addition, as a result of the increase in temperature, there will be an increase in evaporation and an elevated demand for water by vegetation (transpiration). The temporal change in GWR in the summer months of June, July and August is illustrated by Figs. 59.2 and 59.3.

Change of groundwater recharge for the past period and the applied GLOWA-Danube climate scenarios. Depicted are the spatial aggregated annual values, time averaged over the four statistical climate variants. The two climate variants based on

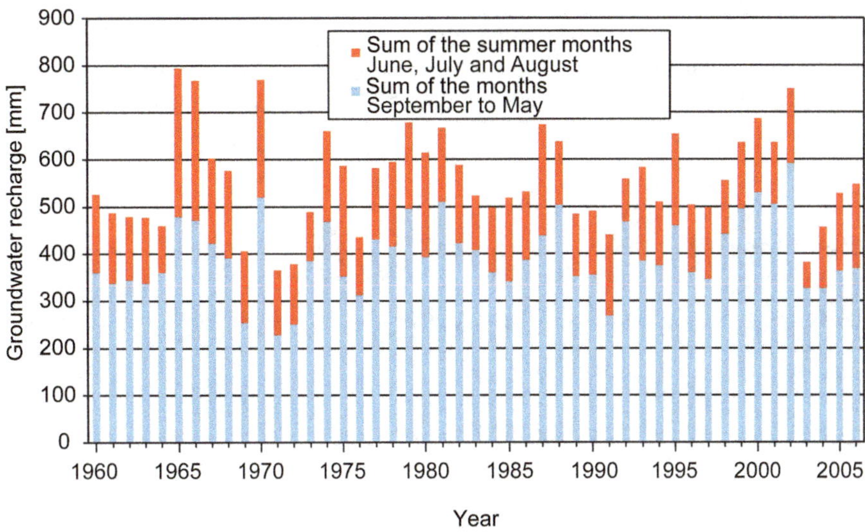

Fig. 59.2 Sum of annual and fraction of summer-groundwater recharge of the period 1960–2006 (past)

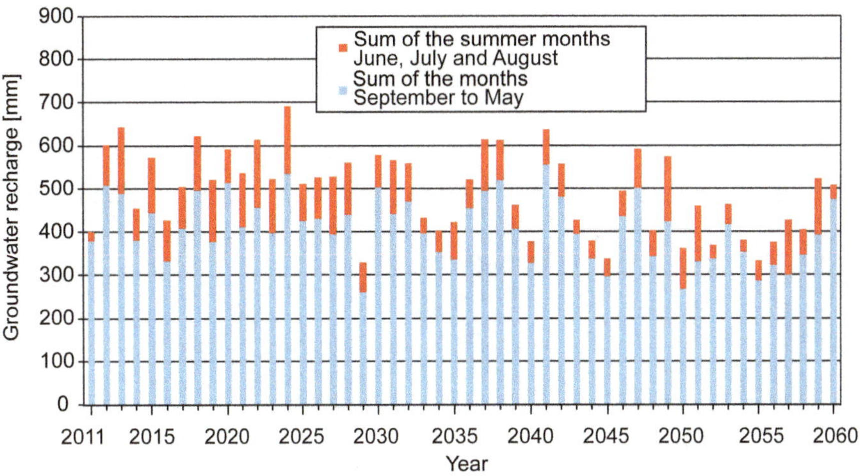

Fig. 59.3 Sum of annual and fraction of summer-groundwater recharge of the period 2011–2060 (*REMO regional–Baseline* climate scenario)

the regional climate models are shown separately; the dashed lines are representing the trend lines

The proportion of the total annual GWR that occurs as groundwater recharge in the summer months is approximately 29 % for the reference period.

For the *REMO regional–Baseline* climate scenario, this proportion of the GWR for June to August is only 17 %. In Fig. 59.3, it is also clear that the mean annual groundwater recharge only seldom exceeds 500 mm in the second half of the scenario period (2036–2060).

Reference

Emmert M (1999) Die Wasserversorgung im deutschen Einzugsgebiet der Donau. Wasserwirtschaft 89(7–8):396–403

Chapter 60
The Influence of Snow Cover on Runoff Regime and Its Change

Markus Weber and Monika Prasch

Abstract The seasonal snow cover has a special significance in shaping the local runoff regime in the Upper Danube basin, since over 8 % of the area is located at elevations above 2,000 m a.s.l., where the snow cover remains until early summer. It temporarily retains the solid portion of precipitation and successively releases it with snow melt. A causal allocation of total discharge into rain, snow melt and ice melt is needed to quantitatively analyse the influence of snow storage in subsections of the drainage network. The contribution of snow melt to discharge can only be approximated by defining an indicator variable.

The maps illustrate the indicator for the amount of snow melt in discharge for the decade 1991–2000 based on measurement data and for the decade 2031–2040 calculated with the REMO regional – *Baseline* climate scenario. The index is equivalent to the mean annual percentage of snow melt in discharge at a relative accuracy of around 10 %.

In the lowlands, the runoff regime is primarily determined by rainfall; the snow storage has maximum significance in the inner alpine valleys. The maps were enhanced with pie charts for 18 gauges, which are separated into sections according to the amount of rain, snow or ice runoff. They reveal that in high elevation glaciated basins, the discharge source is just about equally represented by snow melt, ice melt and rain. In the future, discharge will be determined more by rain than by snow, especially at the northern Alpine fringe and in the large river valleys.

Keywords GLOWA-Danube • Snow cover • Runoff • Rainfall • Snow melt • Ice melt • REMO

M. Weber (✉)
Commission for Geodesy and Glaciology of the Bavarian Academy of Sciences and Humanities, Munich, Germany
e-mail: Wasti.Weber@kfg.badw.de

M. Prasch
Department of Geography, Ludwig-Maximilians-Universität München (LMU Munich), Munich, Germany
e-mail: m.prasch@lmu.de

W. Mauser, M. Prasch (eds.), *Regional Assessment of Global Change Impacts*, DOI 10.1007/978-3-319-16751-0_60

60.1 Introduction

The seasonal snow cover has a special significance in shaping the local runoff regime in the Upper Danube basin, since over 8 % of the area is located at elevations above 2,000 m a.s.l., where the snow cover remains until early summer (see also Chap. 57). The snow cover temporarily retains the solid portion of precipitation and successively releases it with snow melt. In winter, the snow cover extends across almost the entire drainage basin and snow accumulation remains only at the highest elevations all year round. Snow melt takes place extensively and simultaneously in spring, so that the discharge hydrograph has a characteristic wide maximum for a regime dominated by snow. With increasing elevation, this maximum shifts towards the summer. In glaciated regions, the summer discharge maximum from snow melt overlaps with the maximum from ice melt, which appreciably shapes summer discharge in the headwater regions. However, the proportion from ice melt rapidly declines outside of the mountains, as was demonstrated in Chap. 56. In the lower regions of the mountain rivers, snow accumulation and snow melt determine the seasonal water supply to a greater extent than the glaciers. The addition to discharge from the stored winter precipitation is particularly important during dry summer periods and generally depends on the extent and volume of the snow cover that is still present at that time. Even over the whole year, the influence of snow storage on discharge increases with elevation, as the proportion of snowfall increases. If there are glaciers in the region, the contribution of ice melt to discharge must also be accounted for in the balance. The volume of snow storage in the higher regions of the Alps is considerably higher than in the lowlands because:

- More precipitation falls in the mountains.
- The proportion of precipitation that falls as snow increases with elevation.
- The periods with ice and snow melt are shorter at higher elevations.

Thus, the runoff regime of a drainage basin for which the majority of the area is in lowlands is primarily determined by rain and hence is known as a pluvial regime. In contrast, the supply of water in alpine rivers is mostly from snow melt, and the discharge regime in these areas shows a nival character. The discharge of river with an alpine source region can still be increased by snow meltwater from the mountains, even after the snow melt in the lowlands. In the Upper Danube basin, this influence can be observed at the tributaries (the Lech, Iller, Isar and Inn with the Salzach) and at the inflows from the Bavarian Forest.

A causal allocation of total discharge into rain, snow melt and ice melt is needed to quantitatively analyse the influence of snow storage in subsections of the drainage network. Although this can be accomplished relatively easily and accurately with DANUBIA in the case of glacial melt as a result of the small quantity and its direct input into the channel flow, the contribution of snow melt to discharge can only be approximated by defining an indicator variable. The representation of this variable on a river pixel network allows a visualisation of local significance of snow storage for the discharge regime of the past and its change in the future that is analogous to that for the glaciers (see Chap. 56). Together with the results for the glacial discharges, the distribution of the mean annual discharges from any subbasin into rain, snow and ice components permits the quantitative exploration of the function of the mountains as a water tower for the lowland areas.

60.2 Data Processing

The same approach that was used for the calculation of the portion from glacial meltwater (see Chap. 56 and Fig. 60.1) lends itself at first to calculate the contribution of snow melt to the discharge: the difference in calculated discharge compared to a model run with deactivated snow and ice storage is considered. Cutting out the snow is relatively easy to accomplish, by setting the threshold temperature to absolute 0 K to differentiate between solid and liquid precipitation in the *Snow* model (see Chap. 30). As a result, all precipitation falls as rain across the entire basin.

Figure 60.1 shows the calculated mean discharge hydrographs for gauge Achleiten at Passau as they would appear under unchanged atmospheric conditions, if various storage processes are considered or eliminated in the model configuration. The dark blue curve shows the discharge if precipitation falls exclusively as rain, and can be compared to the black curve for mean discharge calculated with both snow and ice storage (glaciers) included. The red line indicates discharge without ice melt, and the difference between the black and the red curves yields the isolated ice melt in light blue. During the period when snow cover is formed (November to March), discharge with snow is lower, while in the period with snow melt (April to October), discharge is higher with snow than without snow. The area delimited by the dark blue and the red curves is thus a good measure for the relative contribution of snow melt to total discharge.

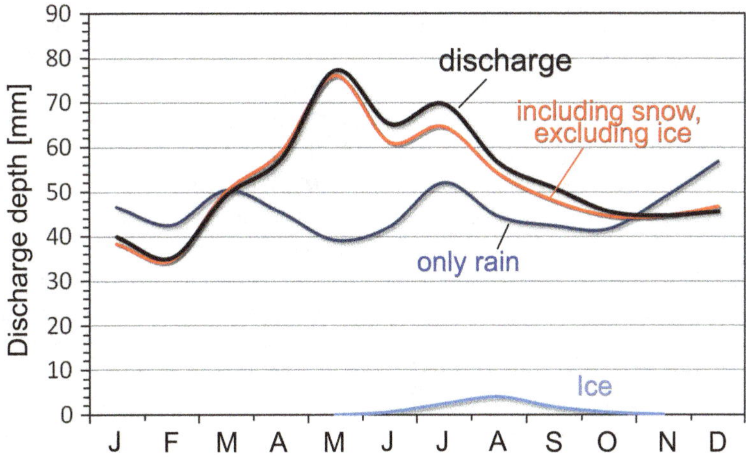

Fig. 60.1 Mean monthly depth of discharge at gauge Achleiten with differently configured model runs for the time period 1991–2000

Although the precipitation amount is identical, the annual discharge totals for the model runs with and without snow and ice storage differ slightly by up to 10 %. This is a necessary consequence of the formation of discharge, which is subject to a complex nonlinear process chain in the soil and on the surface, in addition to the inputs from precipitation and meltwater.

If the integral over the amounts for the difference from the model runs with and without snow storage are set up and the result is standardised with the discharge totals, the result is the index variable F_{snow}, which is a quantitative measure for the relative significance of snow storage in the formation of discharge. With the exception of the differences described above, it quantifies the amount of the water volume from precipitation that passes through snow storage. The calculation of F_{snow} in the entire basin area forms the basis for the representation of this contribution in the maps, interpreted as the percentage of the snow storage in the annual discharge. In addition to snow melt in the glaciated headwater regions, ice melt F_{ice} is also valid (see Chap. 56). With an efficiency that is up to four times higher, ice melt is comparable to snow melt in these areas. By neglecting the changes in storage in the soil, the variable

$$F_{rain} = 100\ F_{snow} - F_{ice}$$

can be interpreted as the contribution from liquid precipitation (rain). The largest of the three factors therefore determines the discharge regime: pluvial, nival or glacial. Moreover, the mean annual discharge formed at any gauge sites within the subbasin can be separated as to its source (rain, snow or ice storage) using these factors.

60.3 Results

Maps 60.1 and 60.2 illustrate the indicator explained above for the amount of melted snow storage in discharge for the decade 1991–2000 based on measurement data and for the decade 2031–2040 calculated with the *REMO regional – Baseline* climate scenario. This index is equivalent to the mean annual percentage of snow melt in discharge at a relative accuracy of around 10 %, based on the method.

In the lowlands, the runoff regime is primarily determined by rainfall; this is evident in the predominant red colour in the river network. The significance of snow storage and the proportion of snow melt in discharge increases from yellow to green to blue. The nival character generally increases with elevation, as shown also by the Danube tributaries flowing from the Bavarian Forest; however, this dominance is displaced by the glacial discharge in the glaciated headwater regions on the main Alpine ridge. The snow storage has maximum significance in the inner alpine valleys of the Inn and the Salzach, but especially on the northern edges of the Alps in the source regions of the Lech and the Isar, where precipitation totals are low. The remote influence on the Danube tributaries from snow melt in the Alps is typically more striking than that from the glaciers. Especially in spring, the snow melt guarantees an above average supply of water in the affected sections of the rivers.

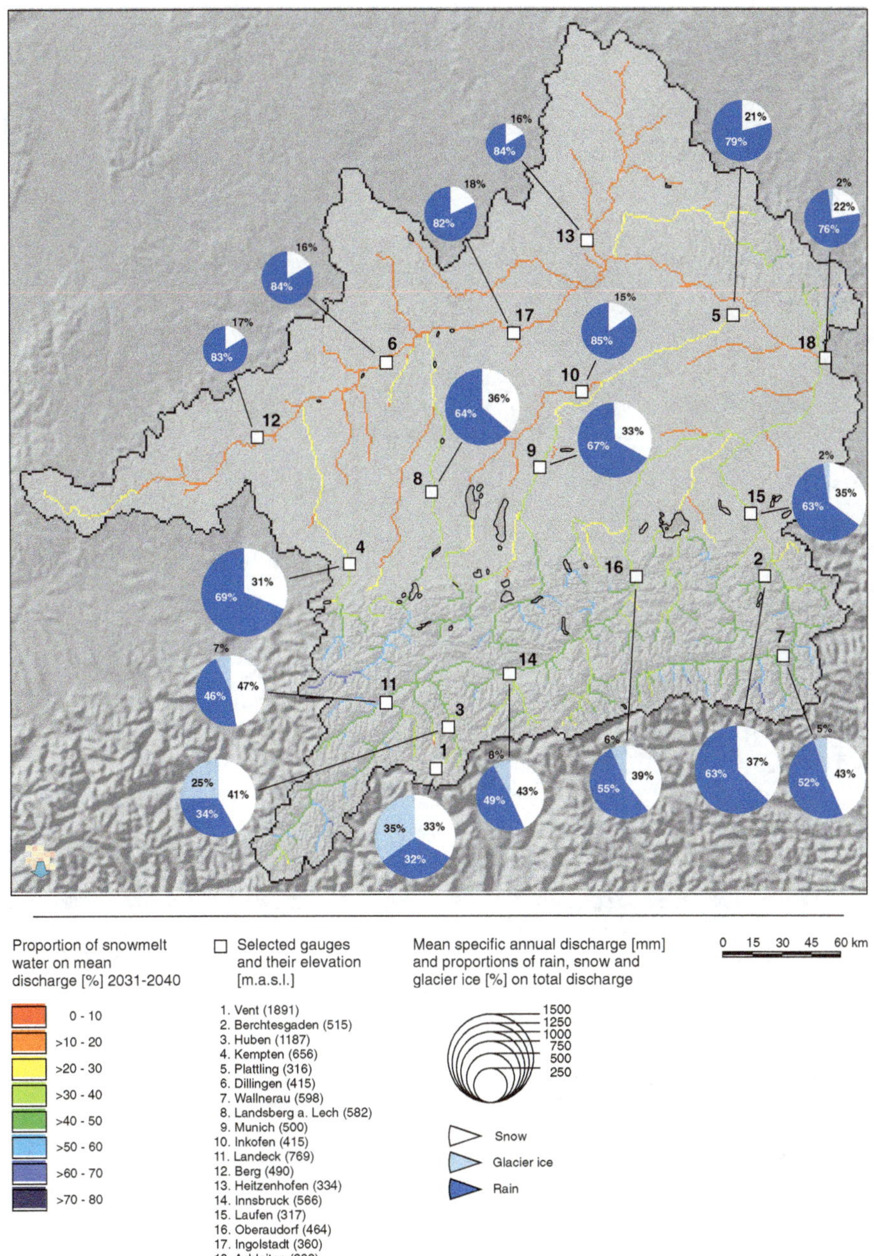

Map 60.1 Influence of the snow cover on the discharge regime in the past (DEM: Jarvis et al. 2008, DANUBIA derived river network)

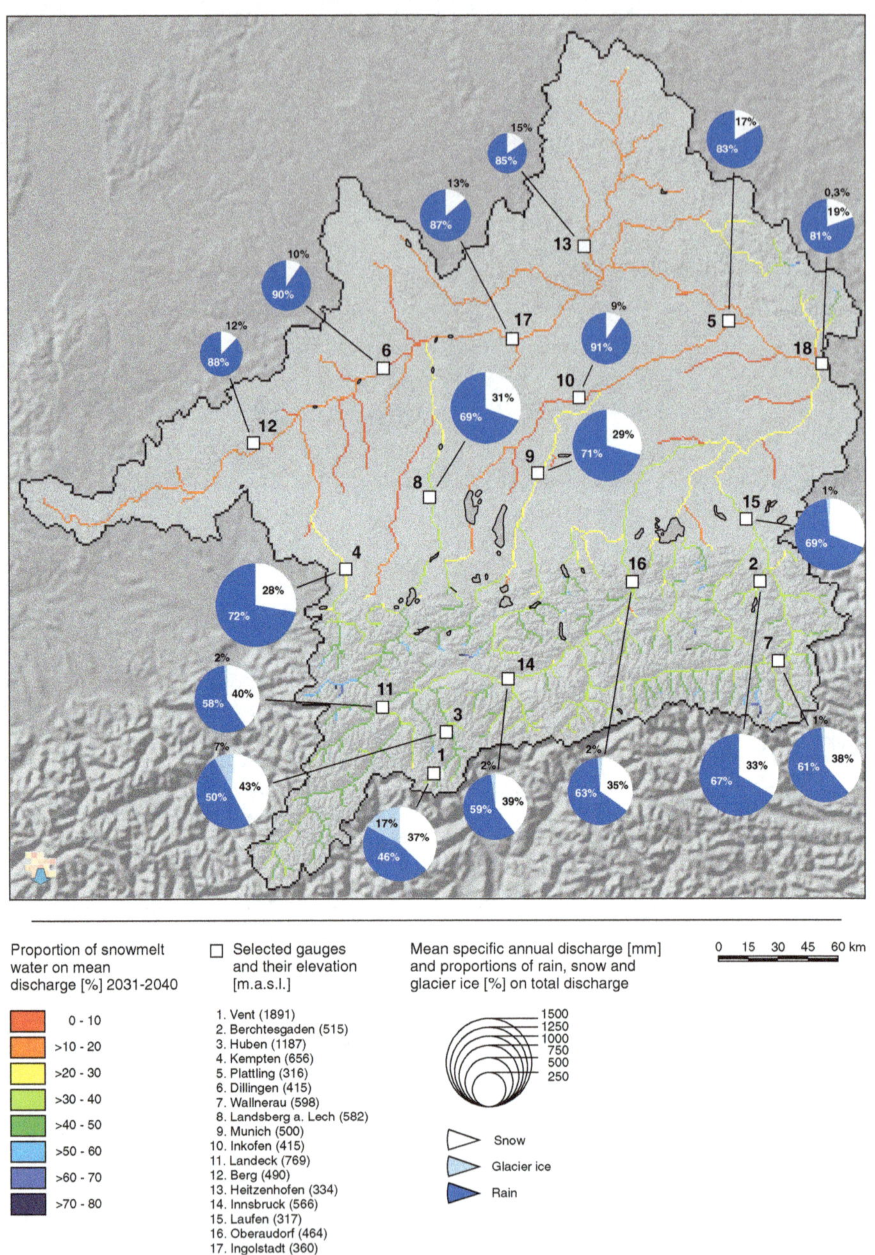

Calculated under the *REMO regional* climate trend and the *Baseline* climate variant.

Map 60.2 Influence of the snow cover on the discharge regime in the future (*REMO regional* climate trend and *Baseline* climate variant, DEM: Jarvis et al. 2008, DANUBIA derived river network)

The Maps 60.1 and 60.2 were enhanced with pie charts for 18 gauges. They illustrate their water yield as specific annual discharge of the proportional area in the associated subbasin. In general, this is highest in the inner alpine headwater regions (gauges 1–4 and 7) and lowest in the lowlands without any connection to mountains (gauges 12 and 13). The charts are separated into sections according to the amount of rain, snow or ice runoff. They reveal that in high elevation glaciated basins (e.g. Vent or Huben), the discharge source is just about equally represented by snow melt, ice melt and rain. Snow storage takes on prime importance in the lower-lying areas of the Inn valley (e.g. Innsbruck) or the Salzach valley. Outside of the Alps, for example, in Dillingen (6), the proportion of discharge from snow is relatively insignificant compared to that from rain, except where the river has its source in the mountains. A particularly striking example of this is the drainage basin of the Amper in the Alpine foothills near Munich. At gauge Inkofen (10), the proportion of discharge from snow melt is only 9 % at a specific discharge depth of 600 mm. In contrast, the adjacent gauge Munich (9) on the Isar and Landsberg (8) on the Lech, which have sources in the Alps, despite having somewhat smaller catchment areas, not only have significantly greater discharge, but the proportion that is from snow melt (over 30 %) is just as high as for catchment areas directly at the Alpine fringe. Thus, the Munich gravel plains above the Isar benefit from water from the Alps. But even at gauge Achleiten (18), the influence of the snow beyond the confluence of the Isar and the Inn is higher than that at Dillingen (6), Heitzenhofen (13) or Ingolstadt (17).

Like for the glaciers, the diagram of the snow portion of the discharge for the decade 2031–2040 shows the most significant changes within the Alps. The sharp decline in the glacial discharges is partially compensated for by snow melt only in the highest elevation headwater catchments, visible there by the change in colour of the light green sections to dark green and blue. Typically, however, even in the higher elevation catchments, there is a general decrease in the proportion from snow in favour of the proportion from rain. In the future, discharge will be determined more by rain than by snow melt, especially at the northern Alpine fringe and in the large river valleys. In this case, the proportion will decrease from an original local value over 50 % to less than 40 %. As a result, the long-distance effects of the Alpine rivers in the foothills will also be lower. In contrast, the Danube itself and its tributaries from the less mountainous regions of the basin will experience only a negligible change in its discharge regime, although a decrease in the proportion from snow melt will occur, particularly in the upper section of the rivers, comparable to what will take place in the Alpine basins. There, the discharge will be determined exclusively by rainfall and will therefore respond primarily to increases or decreases in local precipitation.

Reference

Jarvis A, Reuter HI, Nelson A, Guevara E (2008) Hole-filled seamless SRTM data V4. International Centre for Tropical Agriculture (CIAT). http://srtm.csi.cgiar.org. Accessed 19 Sept 2014

Chapter 61
Analysis of Discharge Patterns on the Danube between Kelheim and Achleiten with a Particular Focus on Navigation

Wolfram Mauser and Andrea Reiter

Abstract The Danube is an important waterway connecting Eastern with Central Europe. The role of navigation on the Danube is likely to increase in the future. The impact of climate change was analysed for five selected gauges along the Upper Danube by simulating discharges at the gauges for all GLOWA-Danube scenarios from 2011 to 2060 and converting them into water levels. Analysis of the future limitations for navigation show that more frequent low flow periods will considerably increase the number of days per year at which navigation is disturbed by low flow from currently 4 to between 30 and 100. The impact of floods on navigation is likely not to change significantly from today's levels with a changing climate.

Keywords GLOWA-Danube • Navigation

61.1 Introduction

Of the total freight traffic in Germany, at present approximately 10 % is accounted for by inland shipping (BMVBS 2010). On the Upper Danube, this percentage is considerably smaller, since the majority of Germany's domestic shipping (more than 70 %) takes place in the Rhine basin. However, inland shipping on the Danube downstream from Kelheim contributes an important part of the traffic volume and serves as an indispensible part of the German and European transportation system. As a result of the changes in society and the economy, the globalisation of the market and the eastward expansion of the European Union, the importance of inland shipping in Germany is anticipated to increase in the near future by +43 % within

W. Mauser (✉)
Department of Geography, Ludwig-Maximilians-Universität München (LMU Munich), Munich, Germany
e-mail: w.mauser@lmu.de

A. Reiter
Bavarian Research Alliance GmbH, Munich, Germany
e-mail: reiter@bayfor.org

W. Mauser, M. Prasch (eds.), *Regional Assessment of Global Change Impacts*, DOI 10.1007/978-3-319-16751-0_61

the period from 1997 to 2015 (BMVBS 2007). The potential limitations that may arise for shipping as a consequence of climate changes such as flooding or long-lasting or frequently recurring periods of low flow need to be studied in detail in the case of such developments in order to be able to evaluate the possible ramifications for shipping and to develop potential strategies for adjustment. Chapters 54 and 55 have already thoroughly documented the potential changes on the Upper Danube as a result of low flow and flood discharges under the various GLOWA-Danube climate scenarios. In this chapter, these aspects are considered with a focus on the navigability of the Danube.

61.2 Limitations for Inland Shipping

The most significant factor for safe shipping is the water level of the river being navigated. This factor determines the navigability, the achievable loading capacities and the utilisation capacity of the ships and hence the economic efficiency of the freight transport.

Five gauges in the section of the Danube between Kelheim and Achleiten were chosen to evaluate the effects of climate change on the navigability of the Upper Danube within the GLOWA-Danube project (see Fig. 61.1).

The water level is indirectly determined from discharge. The frequency at which the discharge exceeds or drops below a specific threshold value is a key measure for assessing the situation for navigation. At one extreme, shipping must be discontinued at a minimum threshold discharge, since the water level is then insufficient for navigability. At the other, at high discharges, shipping must also be ceased, since, for example, the operation of locks is then no longer possible. Regulation low water level (RNW = Regulierungsniedrigwasserstand) and the highest navigable water level (HSW = höchste schiffbare Wasserstand) were chosen as measures for these two criteria.

61.2.1 Regulation Low Water Level (RNW)

RNW is equal to the level of water in cm at which the working depth of a stream of flowing water is defined. For the Upper Danube, the currently valid RNW_{97} is the level of water for which the discharge is equal to or exceeded for 94 % of the days from the period 1961–1990. That is, only on 22 days each year there was a lower water level (WSV 2001). The currently valid RNW_{97} and the associated discharges RNQ_{97} for the Danube gauges considered in this study are given in Table 61.1. In addition, a hypothetical RNQ_{2000} was added, which was calculated in the context of the analysis of the low flow situation on the Upper Danube from gauge measurement data for the period 1971–2000.

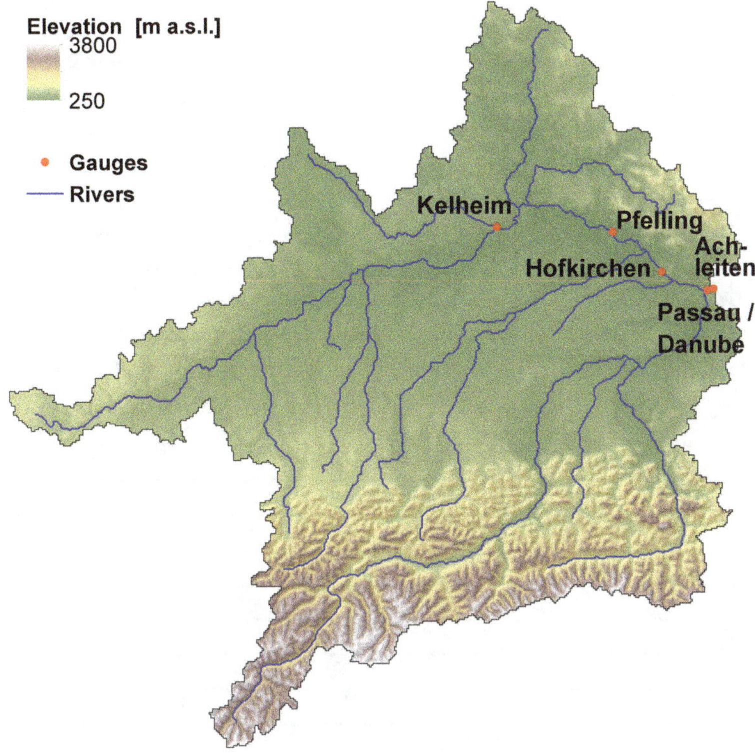

Fig. 61.1 Selected gauges at the Upper Danube (ELWIS 2010)

Table 61.1 RNW97 [cm] and RNQ97 [m³/s] at selected gauges at the Upper Danube (sources: WSV 2001; ELWIS 2010; HND 2010) and RNQ2000 [m³/s] calculated from observed gauge data from 1971 to 2000

Gauge	RNW_{97}	RNQ_{97}	RNQ_{2000}
Kelheim	250	222	216
Pfelling	290	211	260
Hofkirchen	207	324	383
Passau	415	–	399
Achleiten	260	673	795

Two climatological 30-year periods (2012–2041 and 2030–2059) were used for assessing the discharge patterns under the impact of climate change. The results from the GLOWA-Danube climate scenarios are future RNQ values for both future study periods being considered (see Fig. 61.2).

There is a significant decrease in RNW by up to 49.7 % for all GLOWA-Danube climate scenarios that were evaluated compared to the past. This decline is particularly notable in the second 30-years period.

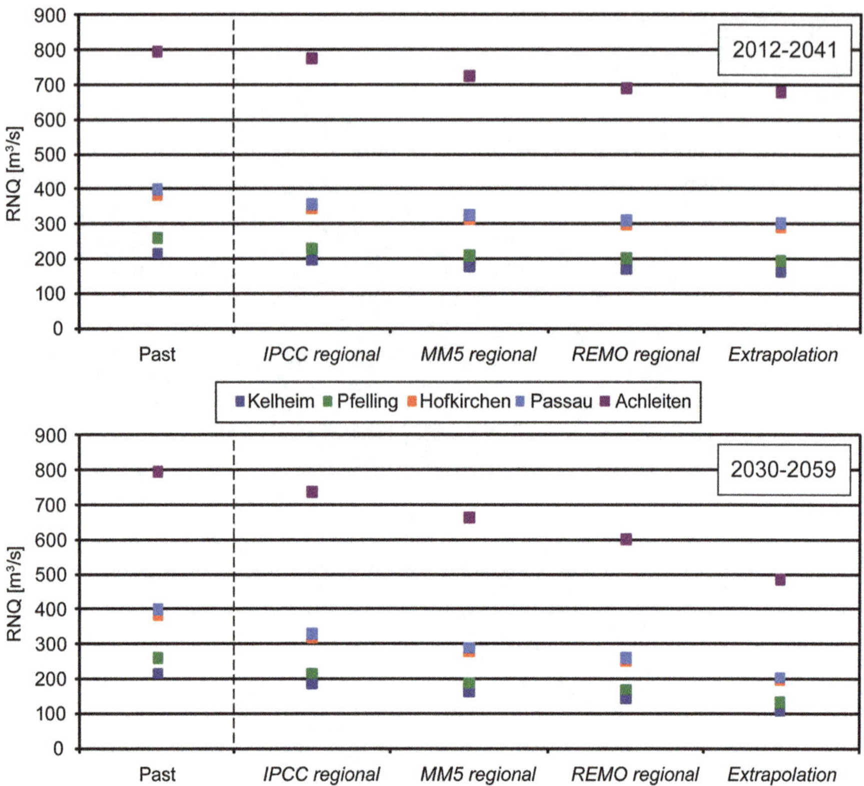

Fig. 61.2 RNQ [m³/s] at selected gauges at the Upper Danube for the past (1971–2000) and for the GLOWA-Danube climate scenarios (each averaged over four climate variants) in the time periods 2012–2041 (*top figure*) and 2030–2059 (*bottom figure*)

Table 61.2 HSW [cm] at selected gauges at the Upper Danube and respective discharges [m³/s] (sources: HND 2010, DORIS 2010)

Gauge	HSW	Discharge to HSW
Kelheim	540	1,090
Pfelling	620	1,230
Hofkirchen	480	1,570
Passau	780	–
Achleiten	502	~3,460

61.2.2 Highest Navigable Water Level (HSW)

HSW (previously termed *H*aut *N*iveau *N*avigable, HNN) identifies the highest water level in cm at which shipping is permitted (see Table 61.2).

Like RNW_{97}, HSW_{97} was specified for the Danube such that for the period 1961–1990, it was exceeded only on 1 % of days each year – this is equivalent to exceeding HSW_{97} on just four days in the year (WSV 2001). If the HSW is exceeded in the event of a flood, shipping must be discontinued. The principal reasons for this include a flow that is too powerful, danger for the nearshore marine environment, as well as the risk of collision with bridges or power lines.

The comparison of the discharges of HSW at the gauges of the Upper Danube that were examined indicates that there are no great changes for the GLOWA-Danube climate scenarios considered. In the first period (2012–2041; see Fig. 61.3), the discharges of HSW are somewhat lower than in the past. The decreases fluctuate between 18.7 % (*IPCC regional* climate trend) and 0.8 % (*Extrapolation* climate trend). The percent changes stood out most clearly at gauge Kelheim for all climate scenarios and were lower the farther downriver the gauges were located. Under the conditions of the *Extrapolation* climate trend, there were no changes in HSW, although in the other GLOWA-Danube scenarios, changes of 10 % and more are expected at all gauges.

The comparison between the two study periods revealed that the differences are quite negligible. HSW in the period 2030–2059 is again significantly lower

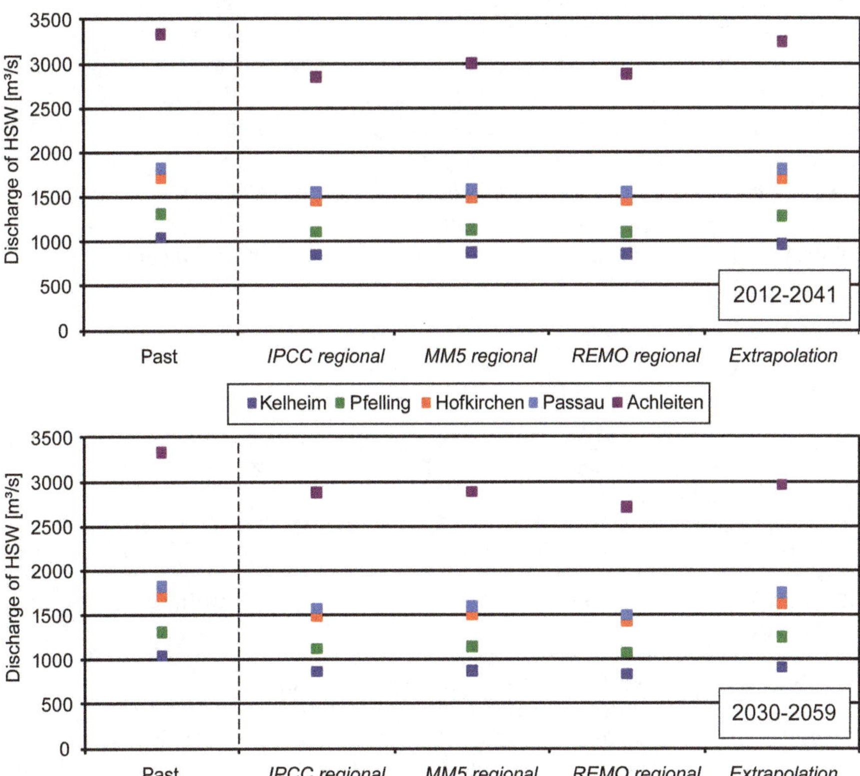

Fig. 61.3 Discharges of HSW [m³/s] at selected gauges at the Upper Danube for the past (1971–2000) and for the GLOWA-Danube climate scenarios (each averaged over four climate variants) in the time periods 2012–2041 (*top*) and 2030–2059 (*bottom*)

compared to the past and to the period 2012–2040, with a maximum decrease of 12.8 % at gauge Kelheim, but only for the *Extrapolation* climate trend.

61.3 Conclusion

The trend in RNQ will have particularly far-reaching effects for inland shipping on the Upper Danube. Hence, the RNQ will decrease at gauge Achleiten from around 800 m^3/s in the past to below 500 m^3/s in the period 2030–2059 under the *Extrapolation* climate trend. Conversely, this means that the number of days when the currently valid RNW_{97} is undershot will increase in the future to between 36 and 100. If the current navigability of the Upper Danube is to be maintained in the future, measures to adapt, such as the expansion of the depth conditions in the navigation channels (e.g. by bed load management), may be necessary. An alternative may involve the transfer of goods to small ships with lower draughts (also because of the reduced flow conditions) (BMVBS 2009). The evaluation of the negative effects on shipping that result from high water revealed a somewhat different picture. Since the HSW corresponds to the discharge that is exceeded on only 1 % of the days each year in a 30-year period, a lower HSW conversely means that the high water statuses in the studied period are not so marked or will occur less frequently. The evaluations presented in this chapter therefore indicate that in the GLOWA-Danube climate scenarios used, no negative effects are expected for shipping as a result of high waters. These conclusions are consistent with the results from Chap. 55. A stationary or perhaps slightly declining situation with respect to high water status was determined specifically for the navigable part of the Danube.

References

Bundesministerium für Verkehr, Bau und Stadtentwicklung BMVBS (2007) Schifffahrt und Wasserstraßen in Deutschland – Zukunft gestalten im Zeichen des Klimawandels. Bestandsaufnahme, Bonn

Bundesministerium für Verkehr, Bau und Stadtentwicklung BMVBS (2009) Tagungsband KLIWAS. Auswirkungen des Klimawandels auf Wasserstraßen und Schifffahrt in Deutschland, Bonn

Bundesministerium für Verkehr, Bau und Stadtentwicklung BMVBS (2010) Verkehr in Zahlen 2009/2010, Bonn

DORIS (2010) Pegelstände. http://www.doris.bmvit.gv.at

ELWIS (2010) Pegelübersicht aller Flussgebiete. http://www.elwis.de

HND (2010) Hochwassernachrichtendienst. http://www.hnd.de

Wasser und Schifffahrtsverwaltung des Bundes, WSV, Wasser und Schifffahrtsdirektion Süd (2001) Donauausbau Straubing-Vilshofen. Vertiefte Untersuchungen. Schlussbericht, Würzburg

Chapter 62
Effects of Different Scenarios on the Operating Dates of Ski Areas

Anja Berghammer, Jürgen Schmude, and Alexander Dingeldey

Abstract This chapter focuses on one part of the Tourism model, the ski area actors, their break-even points and operability. In the model, the operability is determined by the snow-making technology, the number of snow guns per hectare, the water availability and the required initial snow depth for skiing. According to the chosen societal scenario, ski areas find different preconditions concerning, e.g. the permission to expand snow-making facilities which is reflected in changed break-even points and operability. With four simulation runs, which vary by climate variant (baseline vs. five warm winters) and societal scenario (performance vs. public welfare), a corridor of ski areas' potential future developments is spanned. The maps show a small-scale strongly differentiated deviation of opening days in the 2050s compared to the 2010s in the four runs. In all runs, many ski areas will be facing a distinct lower number of opening days so that they will have to reach their break-even point faster or to shut down. As a result, today's concentration tendencies among ski areas on those being more snow reliable will even be intensified until the 2050s.

Keywords GLOWA-Danube • Ski area • Opening days • Break-even point • Snow-making facility

62.1 Societal Scenarios in the *Tourism* Model

The main function of the *Tourism* model is to calculate the water demand by the tourism infra- and suprastructure as well to analyse the operability of these facilities under the effects of climate change. Depending on the scenario, the future of tourism in the study area may take on a variety of forms.

A. Berghammer (✉) • J. Schmude
Department of Geography, Ludwig-Maximilians-Universität München (LMU Munich), Munich, Germany
e-mail: anja.berghammer@lmu.de; j.schmude@lmu.de

A. Dingeldey
BWL Reiseverkehrsmanagement, Duale Hochschule Baden-Württemberg, Ravensburg, Germany
e-mail: Dingeldey@dhbw-ravensburg.de

W. Mauser, M. Prasch (eds.), *Regional Assessment of Global Change Impacts*, DOI 10.1007/978-3-319-16751-0_62

The main determinative influencing factors for tourism include the economy, demographic changes, technical advances and environmental consciousness (Freyer 2000 and Dingeldey 2008). These factors vary depending on the societal scenario and influence the behaviour of the actors. The GLOWA-Danube climate trends and climate variants are introduced in preceding chapters and the essential features of the GLOWA-Danube scenarios are described elsewhere (see Chaps. 47, 48, 49, 50, 51, and 52); below, the effects of the various societal scenarios on the *Tourism* model are presented. The focus is the ski area actor class (Sax 2008) from the Deep-Actor approach (Barthel et al. 2008).

62.1.1 The Tourism Model in the Baseline Societal Scenario

The *Baseline* scenario, also known as the "business-as-usual" scenario, involves the assumption of an environmental consciousness that is similar to that of today. Economic development can be described as moderately expansive, such that maintenance investments are a core focus. This means that the water demand and technical facilities of the tourism infrastructure in the study area generally remain the same as the status at the beginning of the twenty-first century. Technical advances in infrastructure and mobility are moderately transformed.

The demographical trends essentially influence the tourism demands. In the coming decades, a declining population and increase in the older population are expected in Germany (Statistical Office 2006; Dingeldey 2008). However, in the *Baseline* scenario, a level of activity and health in old age that is equivalent to today is assumed, such that there are no changes to tourism demands that result from demographic changes.

Since ski areas will see no or only a moderate upgrading of the capacity for making artificial snow, the economic feasibility threshold for the operating dates will remain unchanged. The economic feasibility threshold can be defined for a number of parameters and represents the limit beyond which economic operation is possible. The operability of ski areas (which depends on snow-making technology, snow-making time windows, availability of water and amount of natural snow) also remains at today's level.

62.1.2 The Tourism Model in the Performance Societal Scenario

In the *Performance* scenario, the future is characterised by a very highly market-oriented economic system. Technical advances are implemented to satisfy demand whenever possible and financially feasible, as a result of a reduced importance of environmental consciousness. Since there is also no limitation on water use for tourism purposes in the future, the water demand by tourism facilities will increase.

The demand for overnight tourism opportunities in the study area increases, in part because increasing globalisation manages to attract new guest categories from Eastern Europe and Asia (Freyer 2000; Petermann et al. 2006). In addition, there is an assumed higher level of activity and health in old age (Opaschowski et al. 2006).

The existing ski areas produce more artificial snow and do so over larger areas. This first requires raising the economic feasibility threshold in order to match the higher investment expenses. The considerably greater water use can lead to regional water shortages and is initially managed with restrictions in the face of serious impact to the environment. However, the upgrade of the supply is based solely on the expansion of the existing ski areas and no new ski areas are established.

62.1.3 The Tourism Model in the Sustainability Societal Scenario

In the *Sustainability* scenario, environmental consciousness plays a more significant role, whereby the parameters in the *Tourism* model remain at the status quo. The values for water demand by the tourism infrastructure are to a large extent the same as for the beginning of the twenty-first century. Increased government intervention means that not all potential new technology is implemented. Despite these developments, the tourism industry within the study area enjoys good conditions, as before. Because there is an assumption that both positive and negative changes in demand balance overall, the tourism demand remains virtually unchanged (Ludwig 2007).

The capacities for artificial snow production in ski areas do not change, since there is a mindset that climate change cannot be confronted long term with technical means and therefore increased investment and expenditure of resources are not worth it. As a result, the economic feasibility threshold also remains unchanged and the operability of the ski areas is limited as a consequence of the sometimes insufficient amount of snow under the circumstances.

62.2 Implementations

Scenarios reflect possible manifestations of the future. A corridor can be generated by means of a variety of scenario runs, within which the "real future" lies with increasing probability. In order to reveal the corridor that is generated for the *Tourism* model under the assumptions of two contrasting societal scenarios (*Performance* and *General Public Interest*), two simulation runs were carried out with the *IPCC regional* climate trend and the *Baseline* climate variant. Since the focus of this chapter is skis areas and in particular their economic feasibility threshold, the *five warm winters* climate variant was also simulated, since the consequence

Table 62.1 Selected scenarios

Run	Climate trend	Climate variant	Societal scenario
1	IPCC regional	Baseline	Performance
2	IPCC regional	Baseline	General public interest
3	IPCC regional	Five warm winters	Performance
4	IPCC regional	Five warm winters	General public interest

of five years with warm winters presents a huge threat for the economic operation of lift facilities. For this chapter, the corridor boundaries are marked by the two most extreme scenarios; on one extreme is run 1 (best possible use of the given resources) and on the other extreme is run 4 (very problematic developments for winter sport tourism from expected climate change, without investment in artificial snow production). In this way, there are four complete GLOWA-Danube scenarios that are comprised as follows (Table 62.1).

The economic break-even point and the operability of the ski areas are of particular interest. If there are investments in order to remain competitive, the economic break-even point increases. That is, either more or more operating days with a high turnover are needed per season to generate the investment costs.

The operability of snow-making machines is operationalised in the simulation by the snow-making technology (maximum temperature for snow production), snow-making time window (snow-making machines per hectare), the availability of water (represented by flags, see Chap. 28) and the quantity of natural snow.

The results of the four simulations are presented and compared below.

62.3 Selected Results from the Scenario Runs

The four Maps 62.1 – 1–4 illustrate the percent variation in the simulated operating days for each ski area in the study area for the winter seasons from 2050 to 2059, compared to the period 2012–2020. In reality, the economic break-even point for operating days varies depending, among other factors, on elevation and exposure, size and investment expenses of the ski area. In the model, this fact can only be accounted for to a limited extent, and the number of required operating days calculated varies by state and region, as per discussions with experts (see Table 62.2). This represents the most meaningful differentiation that results from the model.

The four Maps 62.1 – 1–4 depict a small scale but strongly differentiated picture that once again clarifies the major advantage of DANUBIA over other more generalised approaches. Map 62.1 – 1 and Map 62.1 – 3 (*Performance* societal scenario) exhibit a significantly lower decrease in the average number of operating dates compared to the runs using the assumption of the ongoing *General Public Interest* societal scenario (run 2 and run 4). This difference can be attributed to the upgrading of snow-making capacities. However, in run 1 and run 3, the present economic break-even point for operating dates is also not reached in many ski areas. It can be con-

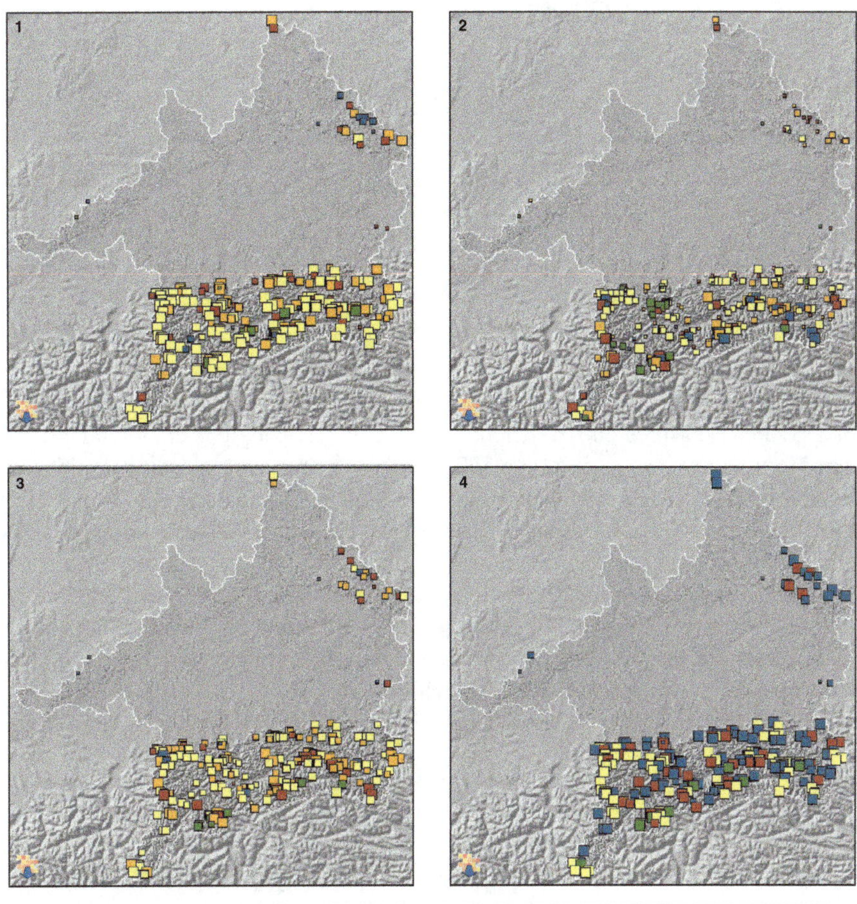

Map 1 - 4: Deviation in percentages of simulated operating days for each ski area in the study area for the winter seasons from 2050 to 2059, compared to the period 2012-2020, for the following different GLOWA-Danube scenarios.

0 30 60 90 120 km

Map 1: *IPCC regional* climate trend, *Baseline* climate variant, *Performance* societal scenario
Map 2: *IPCC regional* climate trend, *Baseline* climate variant, *General Public Interest* societal scenario
Map 3: *IPCC regional* climate trend, *5 warm winters* climate variant, *Performance* societal scenario
Map 4: *IPCC regional* climate trend, *5 warm winters* climate variant, *General Public Interest* societal scenario

Mean number of operating days 2012-2020

- 0 - 30
- 31 - 60
- 61 - 80
- > 80

Change of operating days [%] 2050-2059 compared to the period 2012-2010

- > 0
- 0 - -25
- < -25 - -50
- < -50 - -75
- < -75

Map 62.1 Effects of different scenarios on the operating dates of ski areas

Table 62.2 Economic feasibility threshold of operating days listed by regions

Region/federal state	Economic feasibility threshold of operating days
Northern Bavaria (Bavarian Forest, Fichtelgebirge)	60
Bavarian Alpine foothills/Bavarian Alps	80
Baden-Württemberg (D)	40
Upper Austria (A)	40
Tyrol (A)	80
Salzburg (A)	80
Engadin (CH)	80

Source: Sax (2008)

cluded from this that the "100-day rule" discussed in the literature (the profitability of a ski area is ensured only if there is a snow cover of at least 30 cm for at least 100 days during the ski season; Abegg 1996) will no longer have the same significance in the future as it does today. Each ski area must generate its own necessary gross margin within a shorter season, or the tourism strategy for all destinations must be reconsidered.

In some ski areas, primarily those at higher elevations where sufficient snow is ensured, a winter sport season that is virtually unlimited in the future is possible despite climate change. This will further enhance the trend toward a concentration of ski areas and winter sport sites (Dingeldey 2008).

References

Abegg B (1996) Klimaänderung und Tourismus. Klimafolgenforschung am Beispiel des Wintertourismus in den Schweizer Alpen, Zürich

Barthel R, Janisch S, Schwarz N, Trifkovic A, Nickel D, Schulz C, Mauser W (2008) An integrated modelling framework for simulating regional-scale actor responses to global change in the water domain. Environ Model Software 23:1095–1121

Dingeldey A (2008) Modellierung der touristischen Attraktivität zur Bestimmung der Übernachtungsnachfrage im Einzugsbereich der Oberen Donau unter Berücksichtigung von Umwelteinflüssen. Dr. Hut, München

Freyer W (2000) Ganzheitlicher Tourismus. FIT, Dresden

Ludwig E (2007) The future of leisure travel. In: Conrady R, Buck M (eds) Trends and issues in global tourism. Springer, Berlin/Heidelberg/New York, pp 227–235

Opaschowski HW, Pries M, Reinhardt U (2006) Freizeitwirtschaft. Die Leitökonomie der Zukunft. LIT, Hamburg

Petermann T, Revermann C, Scherz C (2006) Zukunftstrends im Tourismus. Editions Sigma, Berlin

Sax M (2008) Entwicklung eines Konzepts zur computergestützten Modellierung der touristischen Wassernutzung im Einzugsgebiet der oberen Donau unter Berücksichtigung des Klimawandels. In: Schmude J (ed) Beiträge zur Wirtschaftsgeographie Regensburg, vol 11. University of Regensburg, Regensburg

Statistisches Bundesamt (ed) (2006) 11. koordinierte Bevölkerungsvorausberechnung. Annahmen und Ergebnisse, Wiesbaden

Chapter 63
Effects of Different Scenarios on Water Consumption by Golf Courses

Anja Berghammer, Jürgen Schmude, and Alexander Dingeldey

Abstract As golf tourism has been a growth market in the last years with an increasing economic relevance and as golf courses have a high water demand, they are considered in the Tourism model as an actor class. This chapter focuses on their water consumption under different scenarios. In the model, the water demand of a golf course is determined by its area size, the number of fairways, the capacity of the reservoir and the regional climate, especially the precipitation amount during summer and local climatic peculiarities. According to the chosen societal scenario, golf courses find different preconditions concerning, e.g. the permission to irrigate artificially and the irrigation intervals. With four simulation runs, which vary by climate variant (baseline vs. five hot summers) and societal scenario (performance vs. public welfare), a corridor of golf courses' potential future water demand is spanned. The results show that the water demand depends mainly on the chosen societal scenario; the climate variant has nearly no influence. Action is required by golf course operators as in each simulation run, temperature increases while precipitation decreases during summer. Then, possible options for action are to use types of grass with an augmented drought resistance or to establish shorter lawn cuttings.

Keywords GLOWA-Danube • Golf course • Water consumption • Artificial irrigation

A. Berghammer (✉) • J. Schmude
Department of Geography, Ludwig-Maximilians-Universität München (LMU Munich), Munich, Germany
e-mail: anja.berghammer@lmu.de; j.schmude@lmu.de

A. Dingeldey
BWL Reiseverkehrsmanagement, Duale Hochschule Baden-Württemberg, Ravensburg, Germany
e-mail: Dingeldey@dhbw-ravensburg.de

W. Mauser, M. Prasch (eds.), *Regional Assessment of Global Change Impacts*, DOI 10.1007/978-3-319-16751-0_63

63.1 Introduction

The golf sport in the investigation area represents a growing tourism segment, like in the majority of source markets and destinations. Golf sport has experienced ongoing positive growth for several years (DGV 2009). Growth has been recorded both in terms of supply and demand. For example, in 2003 in all of Germany, approximately 460,000 players were associated in the German Golf Association (DGV = Deutschen Golf Verband), but this number grew by 25 % by 2007 to almost 575,000 (Sax 2008; DGV 2009).

In addition, the number of new and expanded golf courses increased over the investigation area over the past five years, by an average of 17.1 % (DGV 2009). The largest relative increase took place in the Austrian part of the GLOWA-Danube area (+28.3 %); this fact can be attributed to the low absolute starting point. In Germany, the rate of increase was correspondingly lower, since the supply was already higher 5 years ago. In all cases, the positive increase in supply can be considered an indication that golf and golf tourism are in an expansion stage.

Based on their economic importance and their high water demand, golf courses must be included as relevant factors in the *Tourism* model. Next to ski areas and swimming pools, they represent an important component of the tourism infrastructure. Therefore, the water consumption by golf courses is calculated in the model and their operability and effects on the tourism demand is investigated under the impact of climate change.

In addition to other factors (areas, number of fairways, size of the reservoir pond), the diversity of land-use is a key element for modelling the water demand of a golf course. The area requirement of a standard golf course (18 holes) is between 60 and 75 ha (Baartz 1994). Each course includes typical course elements that are planted with different grass types and that make up various surface areas (see Fig. 63.1).

The driving range, greens (the green spaces around the holes) and fairways are covered with lawn turf. They comprise about 40 % of the total area and needs sometimes be heavily irrigated, depending on the drought-resistance of the type of planted grass plated. In contrast, semi-roughs (in the boundary areas) and roughs (e.g., wooded areas) only require extensive care but do not need any irrigation (Kreyssig o. J; Sax 2008). Moreover, the regional climate also needs to be taken into consideration to model water demand by golf courses. Of particular importance is the amount of precipitation, as well as the seasonal distribution of precipitation during the summer months and specific (local) climatic conditions. As a result of elevated rates of evapotranspiration, the water demand for lawns during the summer is higher than in spring or fall. Thus, enhanced irrigation is particularly necessary during this season.

According to the surveys conducted by Sax (2008), for an 18-hole golf course, there is an annual water demand between 4,000 and 85,000 m^3, which varies according to the factors listed above and on average is around 9,000 m^3 (Schmunde and Sax 2004). The sources for this water include groundwater (wells), surface and

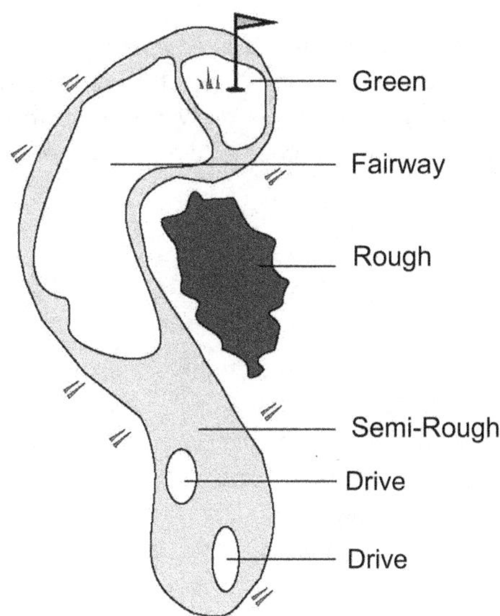

Fig. 63.1 Standard elements of a golf course (Data source: modified, after LfU 2003)

Table 63.1 Specification of factors for golf courses in the three societal scenarios

Factor	Baseline	Performance	General Public Interest
Possibility of water use	Unchanged	Unchanged	Restrictively if adverse effects for the environment occur
Watering interval	Unchanged	Possibly shortened	Stretched
Fairway watering	Partly	Area wide	No watering

drinking water (Sax 2008). In this context, there is the possibility of conflict with respect to the use of water with water management or agriculture.

63.2 Behaviour of the Golf Actors in the Three Different Societal Scenarios

In the *Tourism* model, the operability of a golf course depends, among other factors, on the option to use water resources, the watering interval and the use of fairway irrigation. Depending on the scenario being considered, the manifestations of these factors vary and the future of (golf) tourism in the study area takes on different forms (see Table 63.1).

The basic conditions of the three societal scenarios have been introduced in Chap. 62. Accordingly, the trend for golf tourism in the *Baseline* scenario proceeds in the future in the form of business-as-usual for the current situation. In the *Performance* scenario, the existing environmental protection measures are considered sufficient, so that no special measures are introduced with respect to water conservation. Fairways are basically irrigated extensively since this is desirable on the demand side. The authority to use water is essentially not limited. In contrast, in the *General Public Interest* scenario, fairways are no longer irrigated, for example, in order to avoid "unnecessary" water consumption.

63.3 Implementation

By considering two highly contrasting scenarios, a corridor is generated within which the "real future" lies with high probability. In order to reveal the corridor that is generated in the *Tourism* model under the implementation of two opposing societal scenarios (*Performance* and *General Public Interest*), two simulations were run with the *REMO regional* climate trend and the *Baseline* climate variant. Since there is a special focus in this chapter on golf courses and in particular their water consumption, the climate variant *5 hot summers* was also simulated since it can be assumed that five hot summers will significantly influence water consumption and the operability of golf courses. The result is four complete scenario runs that are comprised as follows (Table 63.2).

Select results from the four simulations are presented and compared below.

63.4 Selected Results from the Scenario Runs

Figure 63.2 provides an overview of the water consumption in m^3/a for all golf courses in the investigation area for the four scenario runs in the simulation period (2011–2060). In run 1 and run 3, the water consumption is significantly higher than in the two other runs.

It is clear that the climate variant chosen (*Baseline* or *5 hot summers*) has no significant effect on water consumption.

The main differences arise as a result of the societal scenario that is chosen. In the case of the *Performance* scenario, the golf actors are allowed to carry out irrigation of fairways, which leads to a water consumption that is 4.5 times higher on average.

The four Maps 63.1 – 1–4 illustrate the mean water consumption by golf courses at the spatial resolution of administrative districts for the period 2050–2059. In gen-

Table 63.2 Selected scenario runs

Number of model run	Climate trend	Climate variant	Societal scenario
1	REMO regional	Baseline	Performance
2	REMO regional	Baseline	General Public Interest
3	REMO regional	5 hot summers	Performance
4	REMO regional	5 hot summers	General Public Interest

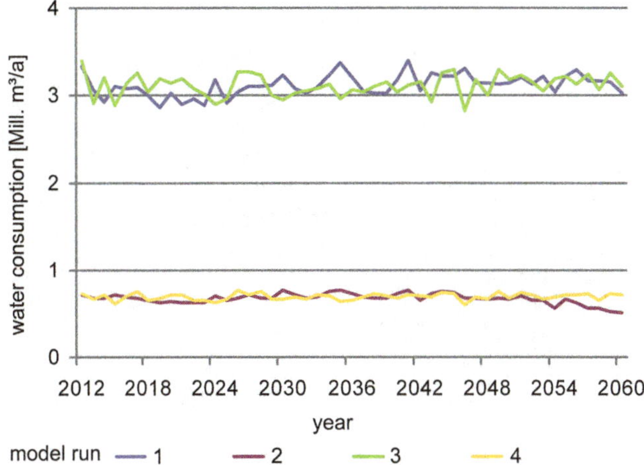

Fig. 63.2 Simulated water consumption of all golf courses in the study area in four scenario runs

eral, the majority of golf courses are located in the Bavarian portion of the investigation area, with a noticeable concentration surrounding the city of Munich. According to Fig. 63.2, water consumption varies in large part depending on the societal scenario, while the effect of climate is nearly negligible. If Maps 63.1 – 1 and 3 and Maps 63.1 – 2 and 4 are compared, initially higher water consumption is expected for Maps 63.1 – 3 and 4, since these used the *5 hot summers* climate variant. In part, this is true, but in some districts the water demand in runs 1 and 2 (*Baseline* climate variant) is higher. This can in part be attributed to the fact that climate change in the period 2050–2059 in the *Baseline* climate variant leads to a greater average temperature increase compared to in the *5 hot summers* variant. In addition, the mean annual precipitation total in the decade 2050–2059 in the *Baseline* climate scenario is lower than in the *5 hot summers* scenario.

63.5 Potential Interventions and Options

In summary, under both climate variants considered, there is a temperature increase in the simulation period as well as a decrease in summer precipitation. Thus, there is a need for action for golf course operators, if irrigation is unrestrictedly permitted, as for the *Performance* scenario.

In order to ensure the quality (condition) of golf courses, which is essential for the revenues from green fees (the fee for non-members to play on a golf course), there are a number of options for reducing the impact of climate change. These include the use of grass types with greater drought resistance to extend the necessary watering interval. Furthermore, the use of a shorter lawn cut (10 instead of 25 mm) can reduce the rate of evapotranspiration. Finally, it is sensible to invest in reservoir

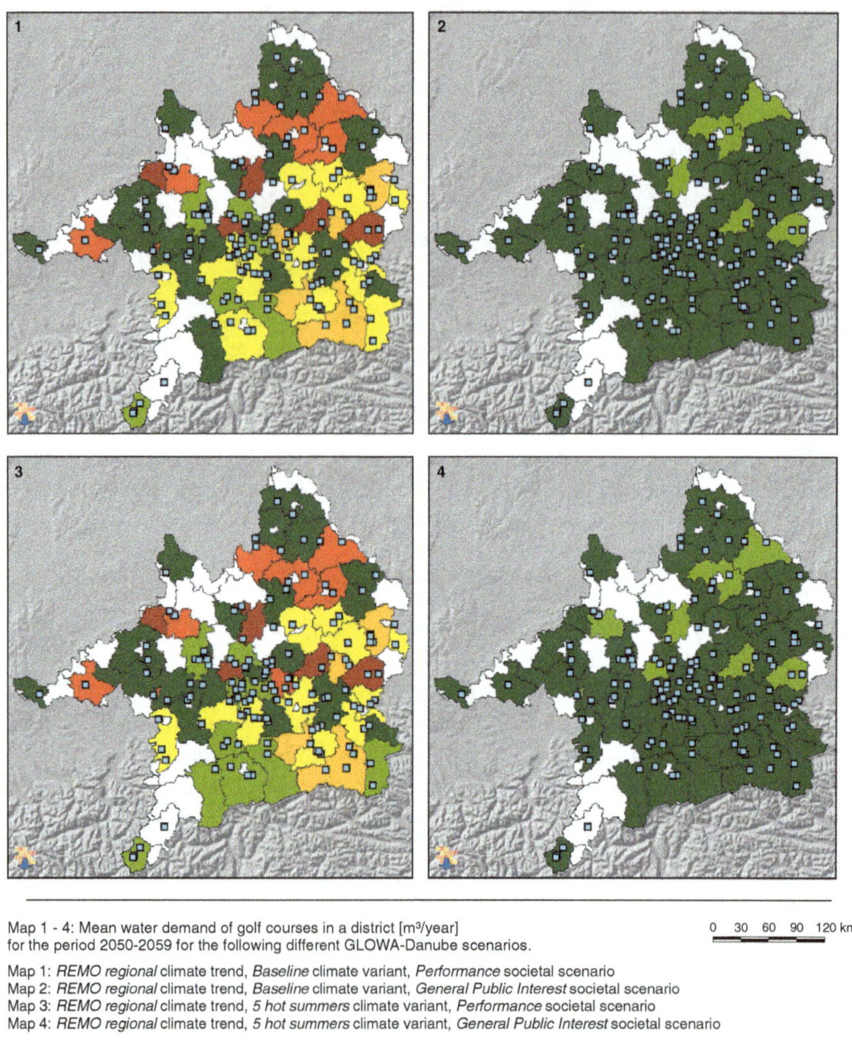

Map 63.1 Effects of different scenarios on water consumption by golf courses (With kind permission from Springer Science + Business Media: Grundwasser, Folgen des Globalen Wandels für das Grundwasser in Süddeutschland – Teil 2: Sozioökonomische Aspekte 16, 259–268, Roland Barthel, Tatjana Krimly, Micheal Elbers, Anja Soboll, Johann Wackerbauer, Rolf Hennicker, Stephan Janisch, Tim.G. Reichenau, Stephan Dabbert, Jürgen Schmude, Andreas Ernst, Wolfram Mauser, Fig. 2)

ponds, which at the same time can be integrated as obstacles on the course (LfU 2003). By the year 2007, already more than 70 % of golf courses have implemented these options.

References

Baartz R (1994) Der Konflikt zwischen Sport und Umwelt dargestellt am Beispiel der Entwicklung des Golfsports im Raum Brandenburg-Berlin. Steiner, Stuttgart

DGV (Deutscher Golf Verband e. V.) (ed) (2009) Der deutsche Golfmarkt 2008. Available via DIALOG. http://www.golf.de/dgv/details.cfm?objectid=60080400&group=109&sn=6&rc=0&mn=4&pu=6&ssn=64; Stand: 27.04.09. Accessed 27 Apr 2009

Kreyssig S (o. J.) Rechtliche und technische Grundlagen zur Wasserbeschaffung auf Golfanlagen. Available via DIALOG. http://www.ple.de/3_proj/pl_golfplatzberegnung.pdf. Accessed 27 Apr 2009

LfU (Bayerisches Landesamt für Umweltschutz) (ed) (2003) Der Golfplatz in der Landschaft, Augsburg

Sax M (2008) Entwicklung eines Konzepts zur computergestützten Modellierung der touristischen Wassernutzung im Einzugsgebiet der Oberen Donau unter Berücksichtigung des Klimawandels. In: Schmude J (ed) Beiträge zur Wirtschaftsgeographie Regensburg, vol 11. University of Regensburg, Regensburg

Schmude J, Sax M (2004) Wasser als touristische Ressource. Ein Ansatz zur Modellierung des touristischen Wasserverbrauchs. Tourismus J 8(2004):557–573

Chapter 64
Changes to the Quantitative Status of Groundwater and the Water Supply

Roland Barthel, Ralf Ziller, Anita Leinberger, and Thomas Hörhan

Abstract Global change will change the quantity and quality of groundwater resources and subsequently the prerequisites for the supply of safe drinking water. Quite often, attempts are made to simply express the status of groundwater resources by stating changes in groundwater level or concentrations. However, such results require further interpretation by experts before they can be used in decision making. This chapter presents the approach chosen in GLOWA-Danube to translate the results of model simulations of groundwater head, groundwater recharge and river discharge into an index that expresses the status of groundwater resources in one single value. This index, called a "flag", is used by the socioeconomic Actor models in DANUBIA as a basis for decision making. Here we describe the principle approach to generating flags from model results and how they are used in the decision process. Maps showing the results of flag calculations are shown for different climatic and societal scenarios.

Keywords GLOWA-Danube • DANUBIA • Groundwater • Water supply • Quantitative status • Qualitative status

64.1 Introduction

This chapter presents selected aspects of the potential future changes to the quantitative status of groundwater resources and the resulting consequences for the supply

R. Barthel (✉)
Department of Earth Sciences, University of Gothenburg, Göteborg, Sweden
e-mail: roland.barthel@gu.se

R. Ziller
Institute of Hydraulic Engineering, University of Stuttgart, Stuttgart, Germany

A. Leinberger
Albert-Ludwigs-Universtity Freiburg, Freiburg, Germany
e-mail: anita.leinberger@hydrology.uni-freiburg.de

T. Hörhan
Austrian Federal Ministry of Agriculture, Forestry, Environment and Water Management, Vienna, Austria
e-mail: thomas.hoerhan@lebensministerium.at

W. Mauser, M. Prasch (eds.), *Regional Assessment of Global Change Impacts*, DOI 10.1007/978-3-319-16751-0_64

of public drinking water in the Upper Danube basin, which is primarily based on groundwater. The results are generated primarily with the *GroundwaterFlow* model (see Chap. 26) and the *WaterSupply* model (see Chaps. 27 and 28). Additionally, these results also depend on the *LandSurface* (here primarily groundwater recharge; see Chaps. 24 and 59) and the other *Actor* (here primarily water demand; see Chaps. 39, 40, 41, 42, 43, 44, 45 and 46) components. It is also recommended to read about the climate trends and climate variants introduced in Chaps. 47, 48, 49, 50 and 51 and about the societal scenarios introduced in Chap. 52. in order to get a better idea of the input variables used.

In the Upper Danube basin, more than 90 % of the public drinking water supply comes from groundwater (Emmert 1999 and others). For this reason, investigation of the potential changes to groundwater availability under the conditions of global change is a key mandate, considering the public services and the long-term use of the natural resource. Beyond the aspect of drinking water supply, the status of the groundwater supply has decisive impacts for ecosystems that depend on groundwater. Moreover, groundwater is the only source for surface discharge in periods of drought and therefore must be considered in any investigation of low flow conditions since there are ramifications for ecology, power generation and river navigation. The quantitative changes in the supply of groundwater and especially the temporal and spatial distribution of the resources are decisive factors when it comes to supplying the population with drinking water. It can be assumed for the Upper Danube basin that even under significantly warmer and drier conditions, if water demand is constant, there is sufficient water overall for the supply of drinking water. However, local and temporary bottlenecks cannot be ruled out; these must be dealt with by an adjustment by the supply infrastructure. This chapter does not consider the potential expansion of agricultural irrigation.

64.2 Model Components and Calculation Methods Involved

The following considerations are needed in order to comprehensively evaluate the impacts of global change on groundwater resources:

- How does *groundwater recharge* change under changing climatic conditions? Groundwater recharge is calculated in the *LandSurface* component (see Chaps. 24 and 25).
- How does altered groundwater recharge affect *groundwater storage*, *groundwater levels*, *exchange between groundwater and surface water* and the spatial and temporal *dynamics of the groundwater*? The *GroundwaterFlow* provides this information.
- Where and under what conditions is the current system for *public drinking water supply* (withdrawal, supply network) no longer capable of providing sufficient drinking water for the connected end users and where and when does this lead to *ecological damage* and to a *violation of the principle of sustainability*? What

adjustments to the system might compensate for such conditions? The *WaterSupply* model provides results that are useful in this regard.

The results presented here were prepared using a model configuration in which initially groundwater recharge and water level in the channel were calculated for various climate scenarios using the soil water balance model known as PROMET (Mauser and Bach 2009). These values and the climatic data and other information from PROMET were then used to drive the *Household*, *WaterSupply*, *Economy*, *Demography* and *Tourist* actor models and the *Groundwater* components. This method avoids back coupling but allows the simulation of numerous scenarios because of a high computational speed.

64.3 Scenario Assumptions

Of the climate scenarios available (see Chaps. 47, 48, 49, 50 and 51), the *REMO regional* climate trend with the *Baseline* climate variant (see Table 50.2 in Chap. 50) was primarily used for the analyses and illustrations. Some aspects of the *climate variant "five dry years"* climate scenario are presented for comparison.

These climate scenarios were each combined with the three defined societal scenarios (see Chaps. 47 and 52), so that a total of six GLOWA-Danube scenarios are available for comparison. Only the *REMO regional-Baseline* climate scenario with both the *Performance* and *General Public Interest* societal scenarios was used for the maps.

64.4 Results

The changes in the groundwater balance and drinking water supply under the scenario conditions occur in space and time. Since not all of the aspects listed in Sect. 64.2 can be described simultaneously, the determinative variables were cumulatively presented as far as possible. The following result variables are appropriate for this purpose:

1. *GroundwaterLevel*: This was presented as the mean of 405 established groundwater zones (see Chap. 28), since otherwise local effects overlap with the general trends.
2. *GroundwaterQuantityFlag* (see Chaps. 28 and 3): This was represented as the quantitative status of the groundwater supply for 405 zones in the basin in the form of an index of 1 (very good) to 5 (very poor) and incorporates groundwater recharge, groundwater levels and basic discharge.
3. *DrinkingWaterQuantityFlag* (see Chaps. 28 and 3): This, analogous to a), depicts the quantitative status of the drinking water supply and takes into account that

suppliers of drinking water obtain water from various extraction sites (zones) and that these suppliers may be networked together.

The spatial differentiation for the trend in *flags* is depicted in Map 64.1 (see maps in Map 64.1).

The temporally differentiated trend and the comparison of different scenarios are illustrated by the time series graphs in Figs. 64.1 and 64.2. These utilized the mean of the *GroundwaterQuantityFlag* and the *DrinkingWaterQuantityFlag* for the entire Upper Danube basin. Groundwater recharge for all the climate scenarios that were included is shown in Fig. 64.1, in order to give an impression of the climatic conditions for the individual scenarios. More detailed explanations on the trends in groundwater recharge can be found in Chap. 59.

64.4.1 Map 64.1

Unfortunately, it is not possible to thoroughly present all of the spatial and temporal aspects of the complex theme of groundwater (water supply) in just a few maps. The maps presented on the left are therefore intended to provide more of an impression of the possible results than they are meant to comprehensively present the trends under the scenario conditions. The result variables that are presented include the changes in the groundwater status, *GroundwaterQuantityFlags* and *Drinking WaterQuantityFlags* (see beginning of this section), in each case for the *REMO regional-Baseline* climate scenario.

All values for the decade mean for the last decade of the simulation period (2050–2059) were calculated for the maps. The change in this decade mean relative to the decade 1990–1999 is depicted. The trends in the pattern can be easily read spatially resolved from the relative changes (difference), and at the same time absolute model uncertainties have lesser importance, since the results for the reference and scenarios are each based on the same assumptions and parameters.

The maps on the quantitative status (flags) were prepared using the available data, under specific assumptions of the scenarios and without calibration. In isolated cases, this may lead to unrealistic results arising; this situation could be improved with a better database and correction of the assumptions, which is what the stakeholder process strives for.

A comprehensive description of the approach used to evaluate the quantitative state of groundwater resources and water supply under conditions of global change is presented in Barthel et al. (2010) and Barthel (2011).

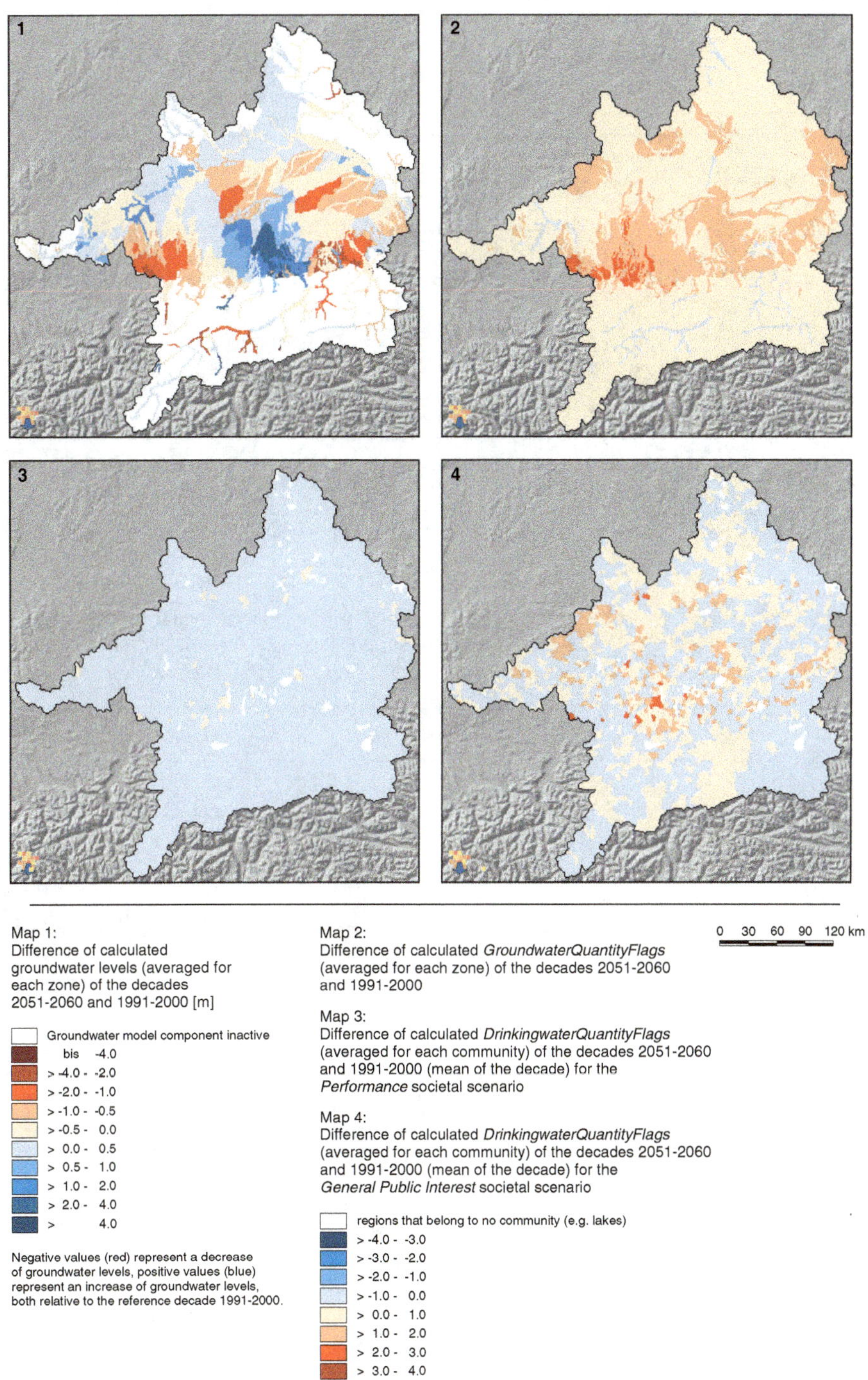

Map 64.1 Changes of the quantitative status of groundwater and water supply

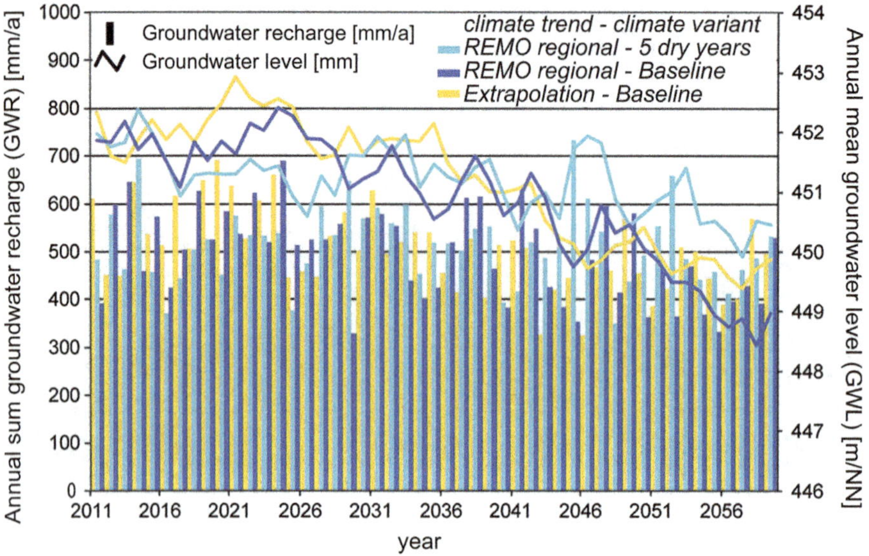

Fig. 64.1 Comparison of mean groundwater recharge (GWR) and mean annual groundwater levels (GWL) for the Upper Danube basin for three different climate scenarios: *REMO regional-Baseline* (*dark blue*), *REMO regional-five dry years* (*light blue*) and *Extrapolation-Baseline* (*yellow*)

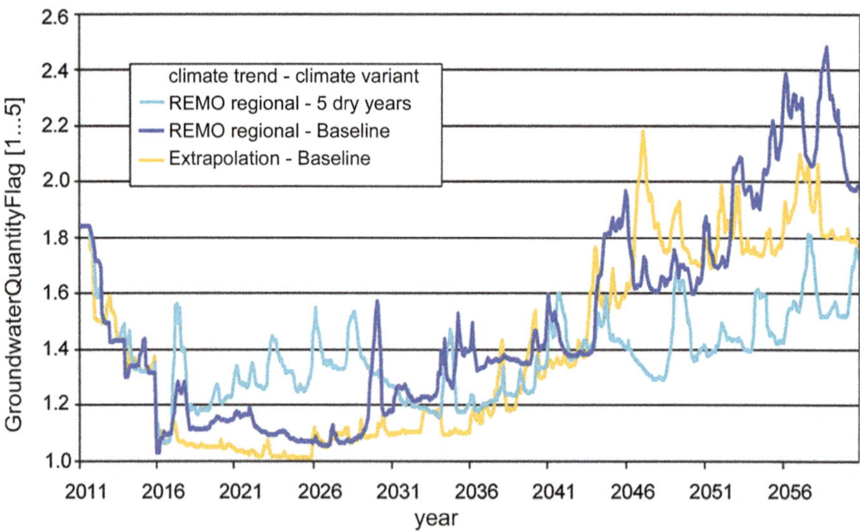

Fig. 64.2 Comparison over time of *GroundwaterQuantityFlag* (averaged over area) for the three climate scenarios *REMO regional-Baseline* (*dark blue*), *REMO regional-five dry years* (*light blue*) and *Extrapolation-Baseline* (*yellow*)

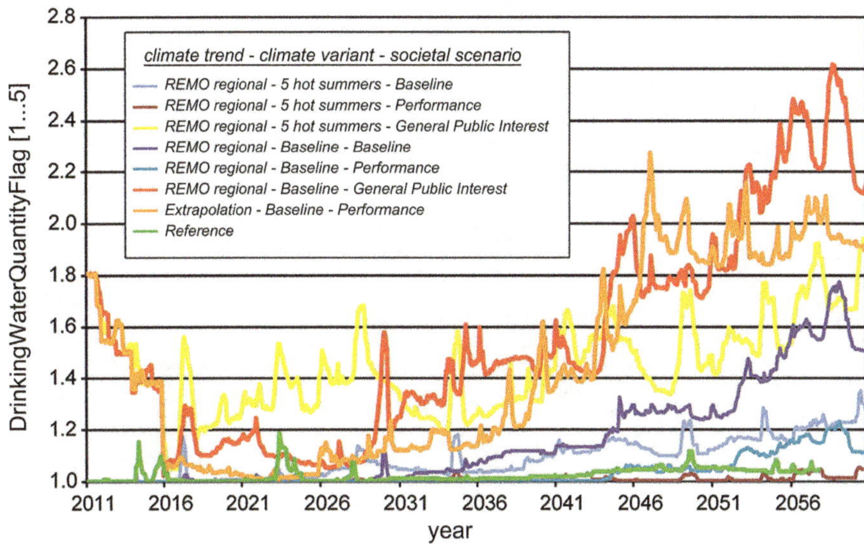

Fig. 64.3 Comparison over time of *DrinkingWaterQuantityFlag* (averaged over area) for different scenario combinations

References

Emmert M (1999) Die Wasserversorgung im deutschen Einzugsgebiet der Donau. Wasserwirtschaft 89(7–8):396–403

Mauser W, Bach H (2009) PROMET – large scale distributed hydrological modelling to study the impact of climate change on the water flows of mountain watersheds. J Hydrol 376(3):362–377

Barthel R (2011) An indicator approach to assessing and predicting the quantitative state of groundwater bodies on the regional scale with a special focus on the impacts of climate change. Hydrogeol J 19(3):525–546

Barthel R, Janisch S, Nickel D, Trifkovic A, Hörhan T (2010) Using the Multiactor-Approach in Glowa-Danube to simulate decisions for the water supply sector under conditions of global climate change. Water Resour Manage 24(2):239–275

Chapter 65
Societal Scenarios in *DeepHousehold*

Andreas Ernst, Silke Kuhn, Roman Seidl, Michael Elbers, and Daniel Klemm

Abstract In this chapter, the societal scenarios in the *DeepHousehold* model and their results are presented. The *Baseline* societal scenario describes the continuation of existing conditions with respect to the distribution of milieus and their characteristics. The values thus directly stem from the current milieu properties and from empirical studies within the project. In the *Performance* scenario, the deregulation of society has diverse direct and indirect effects on the modelled households. As water is viewed as a commodity which has to be purchased, price is becoming more important. Also, insurance, e.g. against flood risks, plays a more important role. In the "*Public Interest*" scenario however, the state is seen to have greater obligations compared to the other scenarios. This also means a focus on sustainability in terms of both the water supply and a sustainable consumption by households. The model results that are presented refer to the role of the so-called drinking water quantity flags as a mediator between natural science variables and actor variables, their impact on the activation of actors and on their behaviour (e.g. the adoption of water-saving technologies).

Keywords GLOWA-Danube • Societal scenario • *Baseline* scenario • *Performance* scenario • *Public Interest* scenario • Domestic water use • Innovation adoption

A. Ernst (✉) • S. Kuhn
Center for Environmental Systems Research (CESR),
University of Kassel, Kassel, Germany
e-mail: ernst@usf.uni-kassel.de

R. Seidl
Institute for Environmental Decisions, ETH Zürich, Zürich, Switzerland
e-mail: roman.seidl@env.ethz.ch

M. Elbers
Micromata GmbH, Kassel, Germany
e-mail: m.elbers@micromata.de

D. Klemm
wusoma GmbH, Munich, Germany
e-mail: mail@danielklemm.de

W. Mauser, M. Prasch (eds.), *Regional Assessment of Global Change Impacts*, DOI 10.1007/978-3-319-16751-0_65

65.1 Introduction

All societies are subject to constant change, and this change affects various aspects such as everyday activities or the public perception and response to different subjects. This potential for social change is accounted for in the preparation of the GLOWA-Danube scenarios (see Chap. 47). In addition to the various manifestations of the climate scenarios, the societal scenarios that might take place under different climatic conditions are also investigated.

A description of the societal scenarios that are considered and their implementation in a GLOWA-Danube scenario can be found in Chap. 52. Below, the societal scenarios in the *DeepHousehold* model and their results are presented.

65.2 Data Processing

The preparation of the data initially took place as described in Chap. 52. For each of the three societal scenarios, the effects are described in the form of narratives, also called story lines.

65.2.1 Baseline *Scenario*

The *Baseline* societal scenario in the *DeepHousehold* model describes the existing conditions with respect to the distribution of milieus and their characteristics. The values thus come from the current milieu properties and from empirical studies of our project (e.g. relating to environmental awareness, psycho-hygiene), as well as from the ratings by experts from Sinus Sociovision® (e.g. relating to price sensitivity). Thus, the *Baseline* scenario is based on the status quo and projects this without any changes into the future.

65.2.2 Performance *Scenario*

A deregulation of society has diverse direct and indirect effects on households. Water is viewed as a commodity which has to be purchased. However, it is also expected that the product is available any time. Water supply crises are generally considered as unacceptable. Since the state has largely withdrawn from this sector, households are responsible for their own backup and insurance, e.g. for flood risks. This means higher insurance rates for risk areas and lower real estate prices. Households that can keep or obtain insurance for their property remain protected, as they were in the past.

65.2.3 General Public Interest *Scenario*

In this scenario, the state is seen to have greater obligations compared to the other scenarios. This also means that the authorities are responsible to secure water supply and flood protection. The public is dissatisfied if there are problems. In contrast to the *Performance* scenario described above, water is considered an essential item and an indispensible resource and not as a commodity. This also means a focus on sustainability, in terms of both the water supply and also in a sustainable consumption by households. People operate on the principle that water should not be wasted. Sustainable management of the resource is principally endorsed.

For both the *Performance* and the *General Public Interest* scenarios, the different actor profiles (e.g. with respect to price sensitivity, orientation to the future and environmental consciousness) were determined in close collaboration with Sinus Sociovision® based on the *Baseline* societal scenario.

65.3 Results

The calculations presented here are carried out within the context of the GLOWA-Danube *REMO regional-Baseline* climate scenario (see Chaps. 47, 48, 49 and 50). The three specified societal scenarios (*Baseline*, *Performance* and *General Public Interest*) were calculated to assess their effects on the household actors.

A key driver for *DHH* is the *DrinkingwaterQuantityFlag* (DQF), an output of the *WaterSupply* model (see Chap. 28). Figure 65.1 shows the total of the *QuantityFlags* for each of the three societal scenarios. It is clear that the number of *QuantityFlags*

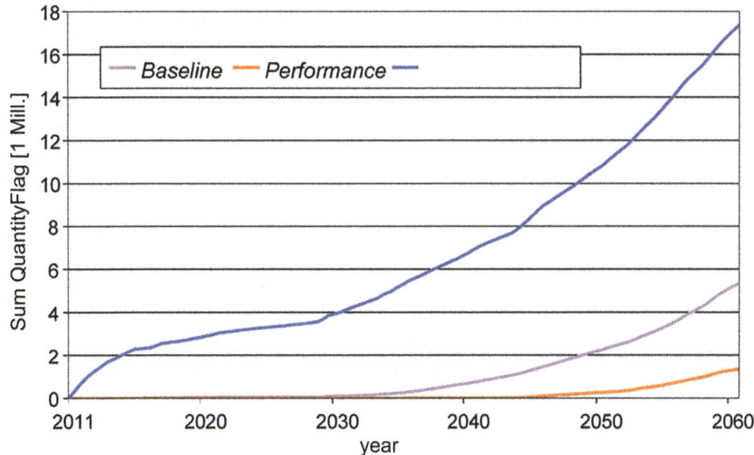

Fig. 65.1 Time courses of the sum of *QuantityFlags* for all proxels in the three societal scenarios

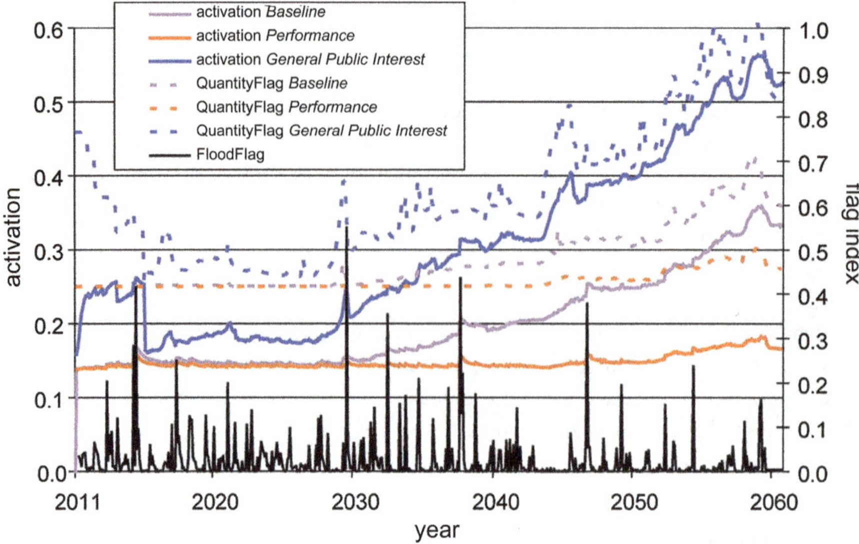

Fig. 65.2 Time courses of activation and flag indices (*FloodFlag* and *QuantityFlag*) in the three societal scenarios (aggregated values for all milieus and proxels)

in the *General Public Interest* scenario is significantly higher. The lowest number of higher level flags is present in the *Performance* scenario.

The value of the flags and their frequency of occurrence are converted to a flag index (with values from 0 to 1) in *DHH*, which reflects the severity of the drinking water or flood situation. Figure 65.2 illustrates the temporal patterns in the *QuantityFlags* (means) for each societal scenario.

The households respond to the *QuantityFlags* calculated by *WaterSupply* both with a reduced consumption of drinking water (see Map 65.1) and with an increased activation (see Chap. 44).

It is clear from Fig. 65.2 that the activation of the actors is highest in the *General Public Interest* scenario. The activation is lowest in the *Performance* scenario. In the latter scenario, no higher level flags are shown, but in the *General Public Interest* scenario, there are several of them. The *FloodFlags* shown are also unrelated to the societal scenario.

An examination of the scenarios with respect to the milieus (see Figs. 65.3 and 65.4) indicates that in all scenarios, the post-materialist actors have the highest activation values, followed by the hedonist milieus. The lowest values for activation are found in the mainstream milieus. The values for the leading milieus and the traditional milieus show intermediate values. The differences between the actor types become particularly clear in the *General Public Interest* scenario (see Fig. 65.4): although the post-materialists generally have higher activation values than most of the other actor types (the Hedonist, Leading, and Traditional milieus); the mainstream milieu actor is well below the others.

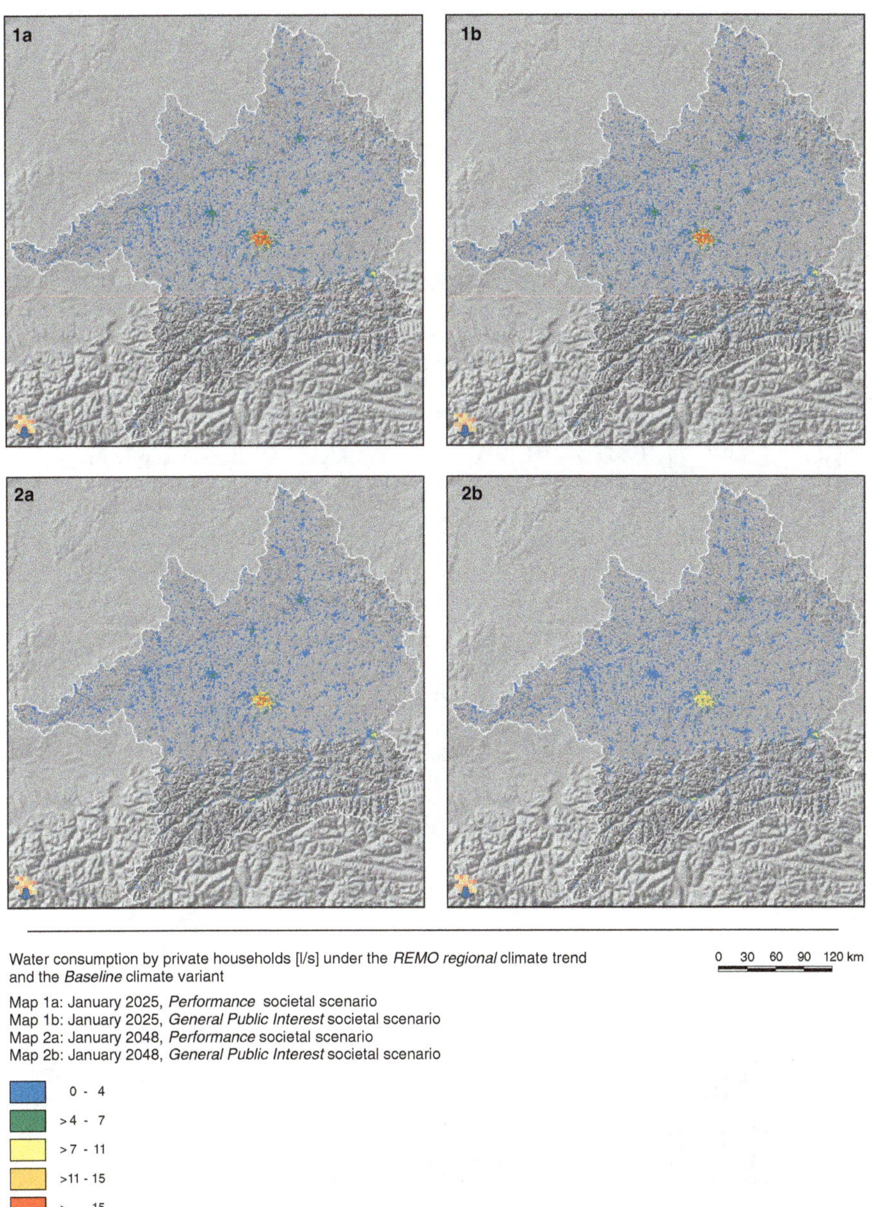

Map 65.1 Societal scenarios in *DeepHousehold* – effects of the *General Public Interest* and *Performance* scenarios on drinking water consumption by private households

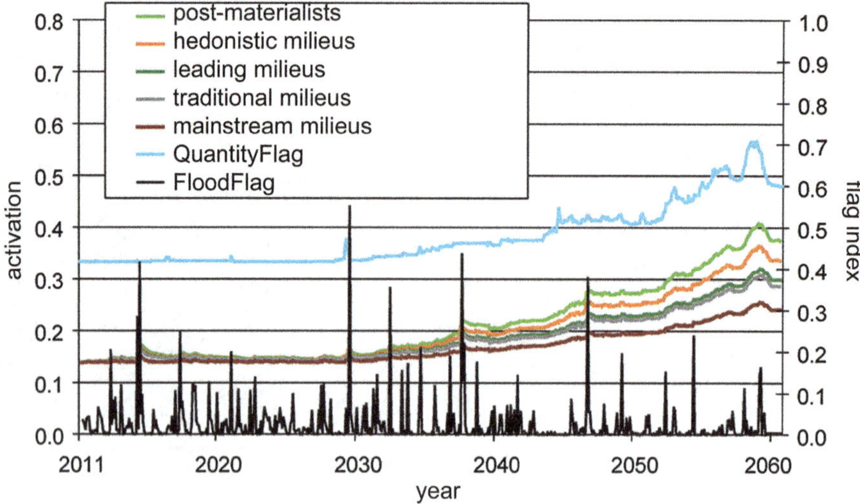

Fig. 65.3 Time courses of activation for the five actor types and flag indices (FloodFlag and QuantityFlag) in the societal scenario *Baseline* (aggregated overall proxels)

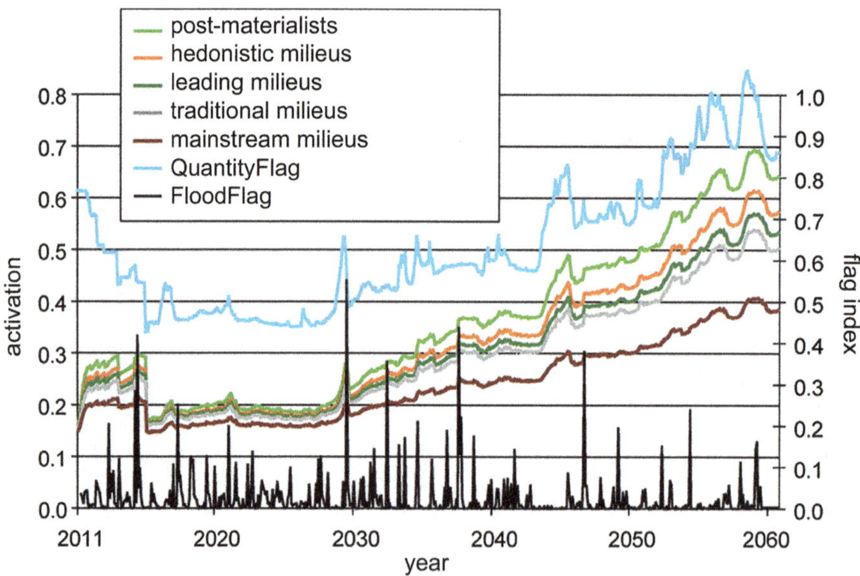

Fig. 65.4 Time courses of activation for the five actor types and flag indices (FloodFlag and QuantityFlag) in the societal scenario *General Public Interest* (aggregated overall proxels)

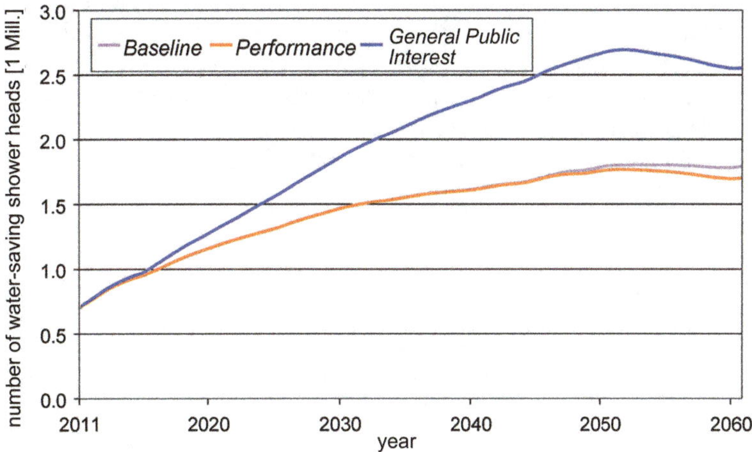

Fig. 65.5 Diffusion of water-saving showerhead innovation

It can be noted from Figs. 65.2, 65.3 and 65.4 that more *QuantityFlags* appear in the *General Public Interest* scenario, which suggests that there is more pressure on the viability dimensions of the actors and activation is correspondingly high. Activation is lowest for the *Performance* scenario; the *Baseline* scenario is between the other two societal scenarios (see Fig. 65.2).

At first glance, it appears that the *Performance* societal scenario is the "more pleasant" scenario for the households, because it has fewer flags and hence a lower activation. However, if the groundwater supplies diminish at long-term as a result of non-sustainable management, this would entail irreversible problems one day for the supply of drinking water. Accordingly, the drinking water resources should already be managed sustainably today and in the coming decades, and households should be warned of shortages. A basic understanding for this is assumed in the *General Public Interest* scenario, with a stronger orientation towards the future and a reduced tendency toward repression of complex problems; this may not be the case in the whole population, but at least in certain milieu groups. This is also accounted for in the *DeepHousehold* overall model by the increased demand for water-saving innovations (see Schwarz and Ernst 2009).

Figure 65.5 uses the example of the water-saving showerhead innovation to illustrate the differences between scenarios that result from enhanced water-saving technologies.

As a result of a generally higher environmental consciousness and lower price sensitivity that comes with innovations, there are significantly more water-saving showerheads purchased in the *General Public Interest* scenario. The differences between the *Performance* and *Baseline* scenarios only appear only quite late in the simulation run.

It is quite easy to see on Map 65.1 that water consumption in the densely populated urban areas – and particularly in the metropolitan area of Munich – is typically higher. In January of 2025 (Maps 65.1 – 1a, b), there are still no big differences between the two societal scenarios, but in January of 2048, the situation is rather different: now, the actors in the *Performance* scenario (Map 65.1 – 2a) use significantly more water than the actors in the *General Public Interest* scenario (Map 65.1 – 2b). This is the result of both greater appeals to conservation from increased warning flags and of a risen environmental social awareness in the actors.

Reference

Schwarz N, Ernst A (2009) Agent-based modeling of the diffusion of environmental innovations – an empirical approach. Technol Forecast Soc 76(4):497–511

Chapter 66
Scenarios with Economic Perspectives Under the Impact of Climate and Social Changes

Christoph Jeßberger and Markus Zimmer

Abstract Climate change and social developments affect economic parameters in different ways. To analyse the complex economic, social and environmental systems and their multidimensional interconnections, we developed the two sophisticated simulation models *Demography* and *Economy* (see Chaps. D.10 and D.11) and designed appropriate interfaces to other simulation models representing further parts of the economic, social and environmental system. In this chapter, we compare the results of different scenarios that help to quantify the impact of climate and social developments on various economic parameters. For this purpose, the differences in the trend growth rates of these parameters are analysed and illustrated in maps and figures. Exemplarily, gross regional product and industrial water use inside the Upper Danube river basin are applied. Here, key results show that climate change and demographic developments come along with a reduction in industrial water use of up to −15.75 % during the period between 2012 and 2025. Likewise, gross regional product (GRP) increases in most regions only below inflation rate, thus, resulting in a negative GRP in real terms (though per capita values are strictly positive). Additional social scenarios, e.g. one which is more capital market orientated or one with increasing environmental concerns in the society, further lever these results.

Keywords GLOWA-Danube • Industrial groundwater use • Gross regional product • Climate change • Climate scenarios • Social scenarios • Trend growth rates • Trends

C. Jeßberger (✉)
Economics and Transportation, Sustainability and Renewable Energies, Bauhaus Luftfahrt e.V, Ottobrunn, Germany
e-mail: christoph.jessberger@bauhaus-luftfahrt.net

M. Zimmer
Center for Energy, Climate and Exhaustible Resources, Ifo Institute – Leibniz Institute for Economic Research, University of Munich, Munich, Germany
e-mail: zimmer@ifo.de

W. Mauser, M. Prasch (eds.), *Regional Assessment of Global Change Impacts*, DOI 10.1007/978-3-319-16751-0_66

66.1 Introduction

The effects of future climate or social scenarios can be clarified by analysing the differences in the trends for various parameters. In this chapter, these trends are investigated for economic development for the period 2012–2025, using gross regional product (hereafter GRP; GRP = economic value of goods and services produced within a region) and industrial groundwater use as example parameters. The three existing social scenarios (*Baseline*, *Performance* and *Common Public Interest*) were used to calculate these indicator variables within the catchment area; in each case, the driver for the models was the *REMO regional* climate trend with the *Baseline* climate variant (see Chaps. 47, 48, 49 and 50). The *Baseline* social scenario describes the trend in the *Economy* that results from a continuation of the status quo and therefore serves as a reference for the comparison of the changes that arise in contrast to the two other social scenarios.

The results of the scenario calculations arise from coupled simulations of the *Demography*, *Economy*, *GroundwaterFlow*, *GroundwaterTransport*, *Household*, *Tourism* and *WaterSupply* models, where prespecified results from the *Atmosphere*, *Farming*, *Landsurface*, *Rivernetwork* and *Traffic* models and the climate scenarios (see Chaps. 47, 48, 49 and 50) serve as input data. The particularly relevant influencing variables from the various sub-models are therefore the ground- and flowing water conditions, represented by the respective flags (see Chaps. 3 and 28 and Barthel et al. 2008, 2009) for the price of water from the public water supply, the employment and capital market conditions, the climate and the regional attractiveness for tourism. In addition, other trends are predefined, such as birth or immigration rates that are exogenous to the model (see Egerer and Zimmer 2006,; Zimmer 2008).

66.2 Changes to Industrial Groundwater Use

In Map 66.1 – 1, the trend in industrial groundwater use is shown for the period 2012–2025 in the *Baseline* social scenario. This simulation run indicates the absolute magnitudes at which the industry actors respond to changing external conditions. Thus, in the map, the percent change in groundwater use by the industrial sector in January is depicted for each proxel. The changes in industrial groundwater use for January 2025 are between −15.75 and +3.33 % compared to 2012. By region, the very different change in industrial groundwater use is easy to see. The decline in industrial groundwater use in municipal agglomerates is the weakest. Hence, for example, the area in and around Munich shows a slightly positive change in industrial water use, while there is a sharp decline in the northern regions of the Upper Danube basin.

Industrial water use responds differently to changing climate conditions; for example, at Weiden/Oberpfalz and Salzburg, it is evident that the specific regions (i.e. industrial water use) are affected in a variety of ways by the changes in the supply of groundwater that result from climate change (see Fig. 66.1).

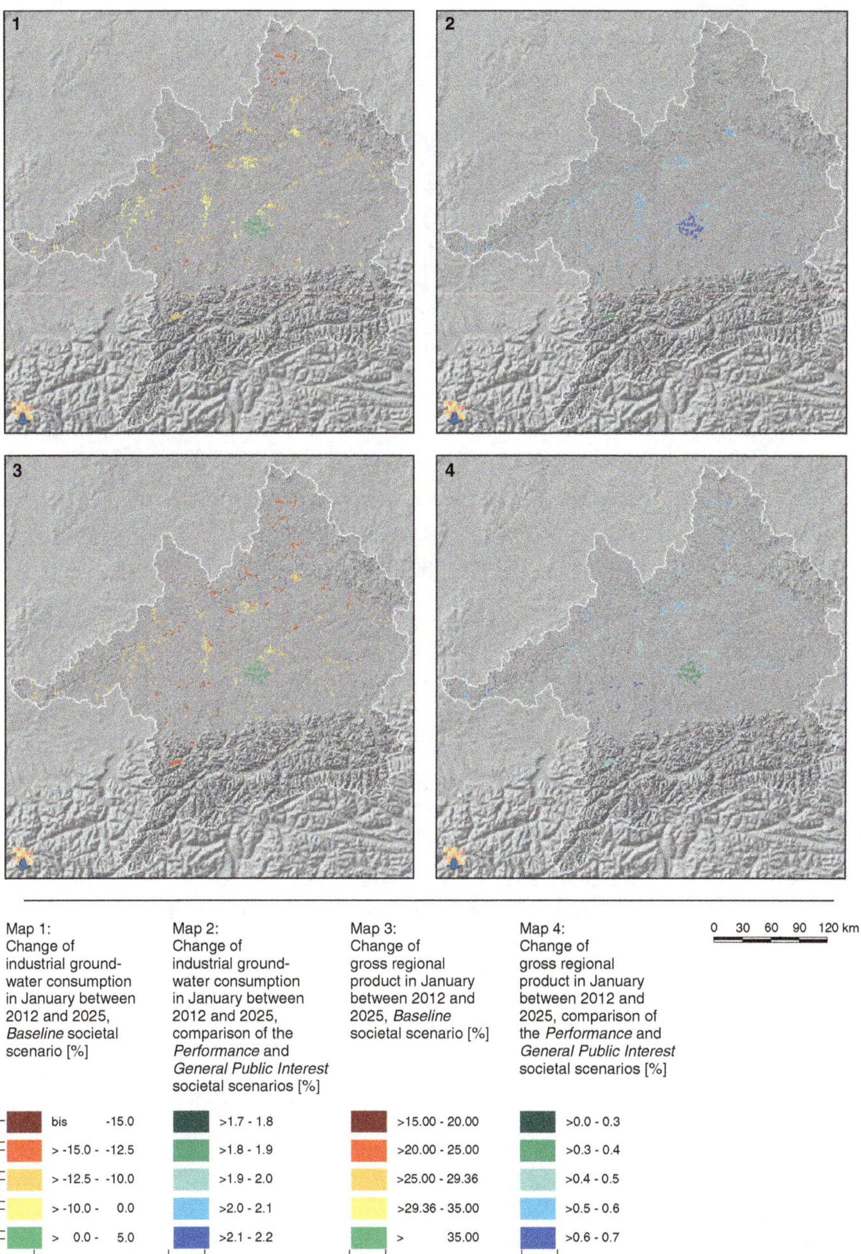

Map 66.1 Scenarios for economic perspectives under the impact of climate and social changes (*REMO regional* climate trend and *Baseline* climate variant)

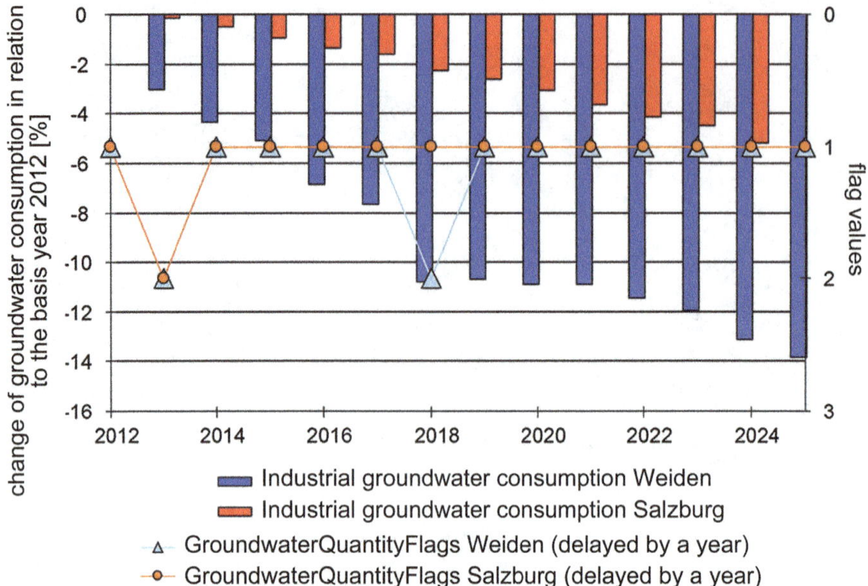

Fig. 66.1 Industrial groundwater consumption in January in comparison to the basis year 2012 and averaged delayed annual GroundwaterQuantityFlags in Weiden/Upper Palatine and Salzburg, result of a *Baseline* simulation run

The declining groundwater situation (groundwater quantity flag value 2) that has an effect in the years 2013 as well as 2018 is apparent in Weiden by a considerable drop in industrial water use. In the years in between, the supply of ground water levels is off again, such that there is only a slight reduction in industrial water use. Finally, in 2025, there is an overall reduction of 14 % in industrial groundwater use compared to 2012.

In contrast, in Salzburg there is a completely different trend (see Fig. 66.1). There is still a decline in the supply of groundwater that affects industrial groundwater use in 2013. However, the sufficient supply in all of the subsequent years means that industrial groundwater use does not decline rapidly but instead gradually reduces year by year. In this way, there is an overall reduction in industrial groundwater use in Salzburg of only 5 % in 2025 compared to 2012 (Jessberger et al. 2011).

66.3 Industrial Groundwater Use in the Different Scenarios

Map 66.1 – 2 shows the relative change in industrial groundwater use in January for every proxel where there is use by the industrial sector from 2012 to 2025. This approach involves a comparison of a simulation run using the *Performance* social scenario with a run using the *Common Public Interest* scenario. The differences over the entire period between each of the percent changes resulting from the different social scenarios were mapped to each proxel. Hence, positive values represent a higher water demand in the *Performance* scenario, and negative percentages represent a lower industrial groundwater demand in the *Performance* scenario. For the entire catchment area, the change in groundwater use from 2012 to 2025 for the *Performance* scenario is between +1.74 and +2.13 % higher than in the *Common Public Interest* scenario. In the industrial areas around Munich, Innsbruck and Salzburg, this relatively high industrial water use shows greater regional increases compared to the rest of the basin. The regional differences can in part be inferred from the more widely varying regional groundwater conditions and the associated interactions between the more and less affected regions. In addition, greater state promotion of sustainable innovations in the *Common Public Interest* scenario leads to a reinforcement of the response by industry to the declining groundwater situation (see Chap. 28).

66.4 Trends in the Gross Regional Product

Here, the results of a simulation based on the *Baseline* social scenario are used to show the degree to which the industry actors adjust their production to changing natural and socioeconomic environmental conditions. Map 66.1 – 3 depicts the percent change in the gross regional product (GRP) from 2012 to 2025 in January for each proxel. Growth rates for the GRP between +19.77 and +44.86 % are apparent. If an annual inflation rate of 2 % is assumed, then the values less than 29.36 % would be equivalent to a GRP growth that is negative in real terms. This means that only the industrial areas of the highest category (29.36–44.88 % GRP growth) may indicate growth in both nominal and real terms. The population decline and the resulting diminishing employable population are the driving factors of the real GRP decrease. The domestic migration towards Southern Germany has only a dampening effect and cannot entirely compensate for the declining population trend (see Egerer and Zimmer 2006; Zimmer 2008), and only the migration into the municipal areas around Munich is strong enough. Like industrial groundwater use, the prevailing decline in GRP in the rest of the basin is also accelerated by the worsening groundwater situation there (see Chap. 28).

In Weiden, the relationship between the predicted population trend (see Federal Statistical Office 2003) and the GRP is apparent (see Fig. 66.2). The population decline leads to a rather delayed GRP growth, such that the GRP decreases in real terms if an annual inflation rate of 2 % is assumed. As a result, in 2025, there is a real negative GRP growth of −8.39 % compared to 2012. The declining industrial groundwater use based on the diminishing supply of groundwater reinforces this trend.

Since the demographic trend in Austria shows a steady population growth (e.g. in Salzburg; see Fig. 66.2) based on higher birth rates (see Statistics Austria 2005),

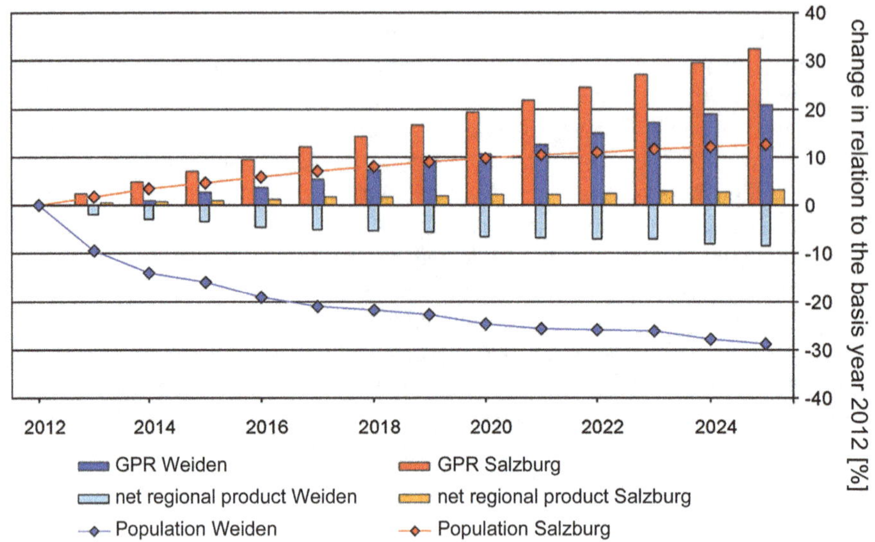

Fig. 66.2 Change of gross regional product (GRP) and population in comparison to the basis year 2012 in Weiden/Upper Palatine and Salzburg, result of a *Baseline* simulation run

there is an increase in the GRP in both nominal and real terms. The slight decrease in industrial groundwater use has little effect (Jessberger et al. 2011).

66.5 Gross Regional Product in the Different Scenarios

Map 66.1 – 4 illustrates the difference in the rates of growth of the gross regional product by proxel, between 2012 and 2025, as a comparison of a simulation using the *Performance* social scenario and *Common Public Interest* scenario. Positive values represent greater growth in the *Performance* scenario. The growth difference over the entire 13-year period is between +0.28 and +0.67 %.

The stronger GRP growth in the *Performance* scenario outweighs the one in the *Common Public Interest* scenario. One cause for this result is the enhanced response in the *Common Public Interest* scenario to a declining groundwater situation. Thus, production partially shifts to the proxels on which the availability of groundwater improves over time. Accordingly, regions with sufficient water supply benefit by comparison with an overall scarcity of the water supply caused by climate change.

References

Barthel R, Janisch S, Schwarz N, Trifkovic A, Nickel D, Schulz C, Mauser W (2008) An integrated modelling framework for simulating regional-scale actor responses to global change in the water domain. Environ Model Software 23(9):1095–1121

Barthel R, Janisch S, Nickel D, Trifkovic A, Hörhan T (2009) Using the multiactor-approach in GLOWA-Danube to simulate decisions for the water supply sector under conditions of global climate change. Water Resour Manage 24:239–275

Egerer M, Zimmer M (2006) Does global change matter? – the case of industries in the upper Danube Catchment area. Trans Ecol Environ 98:75–88

Jessberger C, Sindram M, Zimmer M (2011) Global warming induced water-cycle changes and industrial production – a scenario analysis for the Upper Danube River Basin. J Econ Stat 231(3):415–439

Statistik Austria (2005) Volkszählung 2001 – Haushalte und Familien, Wien

Statistisches Bundesamt (2003) Bevölkerung Deutschlands bis 2050 10. koordinierte Bevölkerungsvorausberechnung, Wiesbaden

Zimmer M (2008) Assessing global change from a regional perspective: an economic close-up of climate change and migration. Dissertation, Ludwig-Maximilians-Universität München (LMU Munich)

Chapter 67
Interventions in *DeepHousehold*

Andreas Ernst, Silke Kuhn, Michael Elbers, and Daniel Klemm

Abstract This section presents the modelling of policy interventions (or measures) in the *DeepHousehold* model. A measure is defined as a targeted, selective intervention to counteract or support a trend in a specified scenario. While a short-term measure is the sending of drinking water warning flags, a longer-term and more significant reduction in the consumption of drinking water can be achieved by means of a change in technology that results in more efficient use of resources. In the model, this is tested by introducing two types of measure (subsidies, informational) and two types of technical innovation that should be promoted (showerheads, toilet flushes). Model results stemming from several runs are given for a four-month measure to innovate showerheads in 2020 and a three-month measure to adapt modern toilet flushes in 2030. It becomes apparent that different actor milieu types react differentially to the measures. The results are discussed and an outlook is given.

Keywords GLOWA-Danube • Policy intervention • Subsidies • Information campaign • Innovation adoption

67.1 Introduction

In addition to the climate and societal scenarios (see Chap. 47 and Fig. 67.1), policy intervention or intervention is one component of a GLOWA-Danube scenario. In contrast to the other elements, an intervention is not an imperative component of a scenario but represents one of its possible specifications.

A. Ernst (✉) • S. Kuhn
Center for Environmental Systems Research (CESR),
University of Kassel, Kassel, Germany
e-mail: ernst@usf.uni-kassel.de

M. Elbers
Micromata GmbH, Kassel, Germany
e-mail: m.elbers@micromata.de

D. Klemm
wusoma GmbH, Munich, Germany
e-mail: mail@danielklemm.de

W. Mauser, M. Prasch (eds.), *Regional Assessment of Global Change Impacts*, DOI 10.1007/978-3-319-16751-0_67

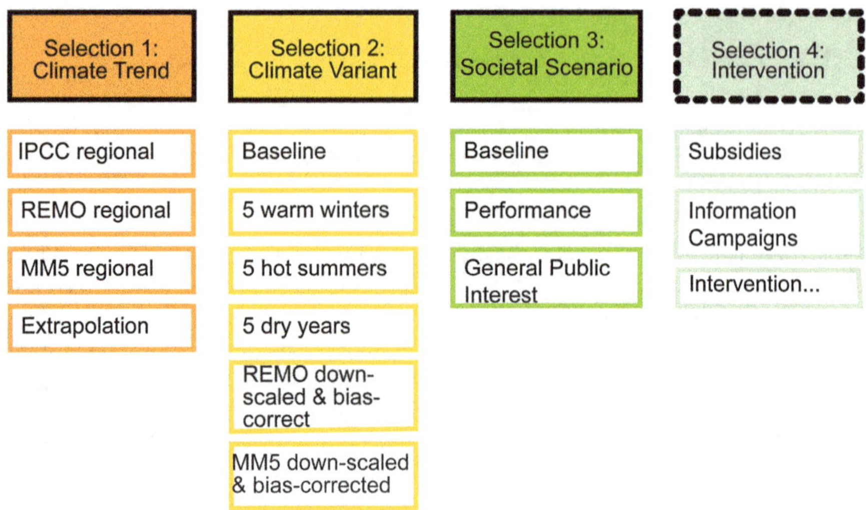

Fig. 67.1 GLOWA-Danube scenario matrix, extended by *DHH* interventions information campaigns and subsidies. The *box* "Intervention…" at the *bottom right* indicates that further interventions are conceivable

An intervention is defined as a targeted, selective intervention to counteract or support a trend in a specified scenario. In this sense, the range of interventions covers the extent of the externally specified courses of actions.

67.2 Interventions in *DeepHousehold*

DANUBIA simulation runs have indicated that the different climate scenarios lead to a shortage in the supply of groundwater and hence to drinking water deficits in some regions. One option for reducing water consumption already exists in the model: the response to drinking water warning flags which the households receive from the water suppliers (see Chaps. 27 and 28). Consequently and if necessary, households reduce their drinking water consumption (see Chap. 42). This response by the *Household* actors is specified within the model; i.e. each time the *DeepHousehold* (*DHH*) model receives a warning flag from the *WaterSupply* model, the households respond. However, this method can be viewed only as a short-term response to a shortage in the supply of drinking water, since households return to their old habits as soon as the warning flags no longer appear. A longer-term and more significant reduction in the consumption of drinking water can be achieved by means of a change in technology that results in more efficient use of resources. The diffusion of water-saving technologies therefore has an important effect on the water demand.

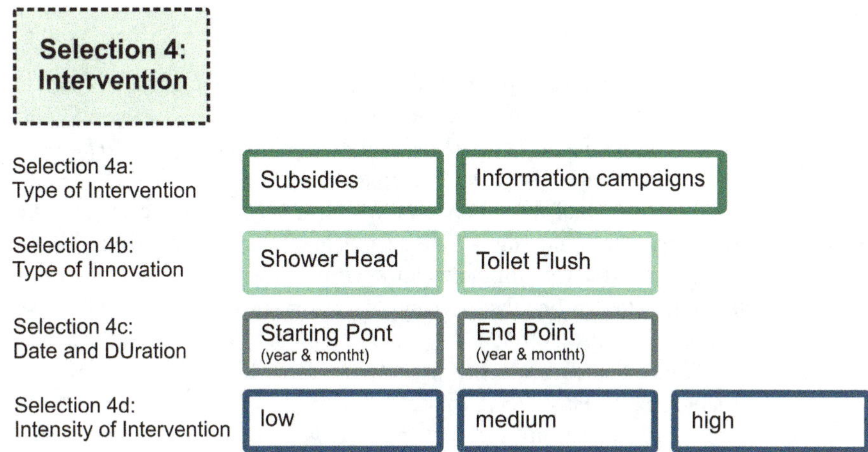

Fig. 67.2 *DHH* interventions and their specifications

For this reason, the first interventions to be introduced should be those that promote water conservation technologies within households. Encouraging the purchase of water-conserving innovations through information campaigns and financial support (subsidies) was modelled (see Fig. 67.1).

The concept of an intervention in *DHH* utilises the existing innovation model for the diffusion of water-saving innovations (see Chap. 43). In contrast to the pure innovation model, in which a device is replaced by a new device only when the old one no longer functions, interventions go beyond this kind of replacement. Households that have an old, but still, intact device think about the acquisition of a new device with lower water consumption, inspired by the intervention. In the model, this is realised by a higher evaluation rate in the actors.

There are several options for selecting and combining the interventions to provide the stakeholders with the best possible tool for supporting decision making.

In addition to the *type of intervention* (subsidies, informational), the *type of innovation* that should be promoted (showerheads, toilet flushes) is specified. Moreover, the *starting and endpoints* and hence the *duration* of the intervention (freely selectable), as well as the *intensity* of the intervention (low, medium, high), can be selected (see Fig. 67.2). The intensity determines what percentage of households will accept the campaign. In contrast to the other selection options in the GLOWA-Danube scenarios, the interventions and their specifications can be freely combined with each other. It is possible to encourage both types of innovation (showerheads and toilet flushes) and also to promote them with both types of interventions (subsidies and information campaign). Thus, the interventions can provide a key support for the stakeholders in the decision-making process, since the effectiveness and costs can be specifically estimated using systematic testing of plausible scenarios.

67.3 Results

The following results are based on the *REMO regional-Baseline* climate scenario. Two interventions were implemented for each societal scenario: a *four-month intervention to innovate showerheads in 2020* and a *three-month intervention to innovate toilet flushes in 2030* (in each case from March to June).

Figure 67.3 shows the total water consumption in m³/s for the three societal scenarios and two information campaigns. The decreasing total water consumption can be attributed both to the diffusion of the water-saving innovations and to a declining overall population. Based on the spread of water-saving showerheads (see Maps 67.1 – 1a, b), the seasonal fluctuations also diminish. The effects of enhanced spread of water-saving showerheads that are attributable to a policy intervention are indicated in the adjacent Maps 67.1 – 2a, b.

From Fig. 67.4, it can be inferred that the *Information campaign* intervention (with respect to water-saving showerheads) leads to a continuous increase in the diffusion of water-saving showerheads. The magnitude of the effect of the *Information campaign* intervention (March–June 2020) differs, depending on the societal scenario. In the *Performance* scenario, the number of water-saving showerheads increases less steeply. The differing effect of the intervention is largely attributable to the basic settings for sustainability and the market that are predominant in the respective scenarios (see also Chap. 52). Since environmental awareness in the *Baseline* and *General Public Interest* scenarios is more prominent, the *Information*

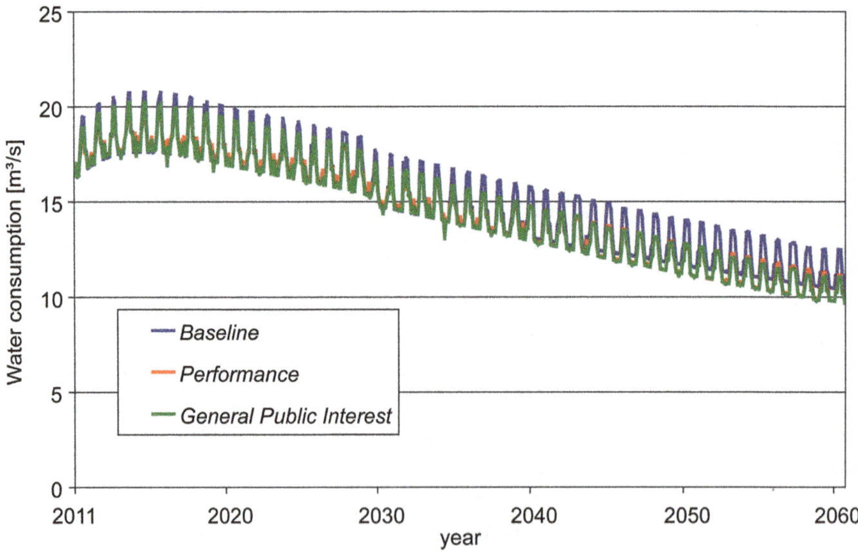

Fig. 67.3 Time course of households' total water consumption in the three societal scenarios (Scenario information in the text)

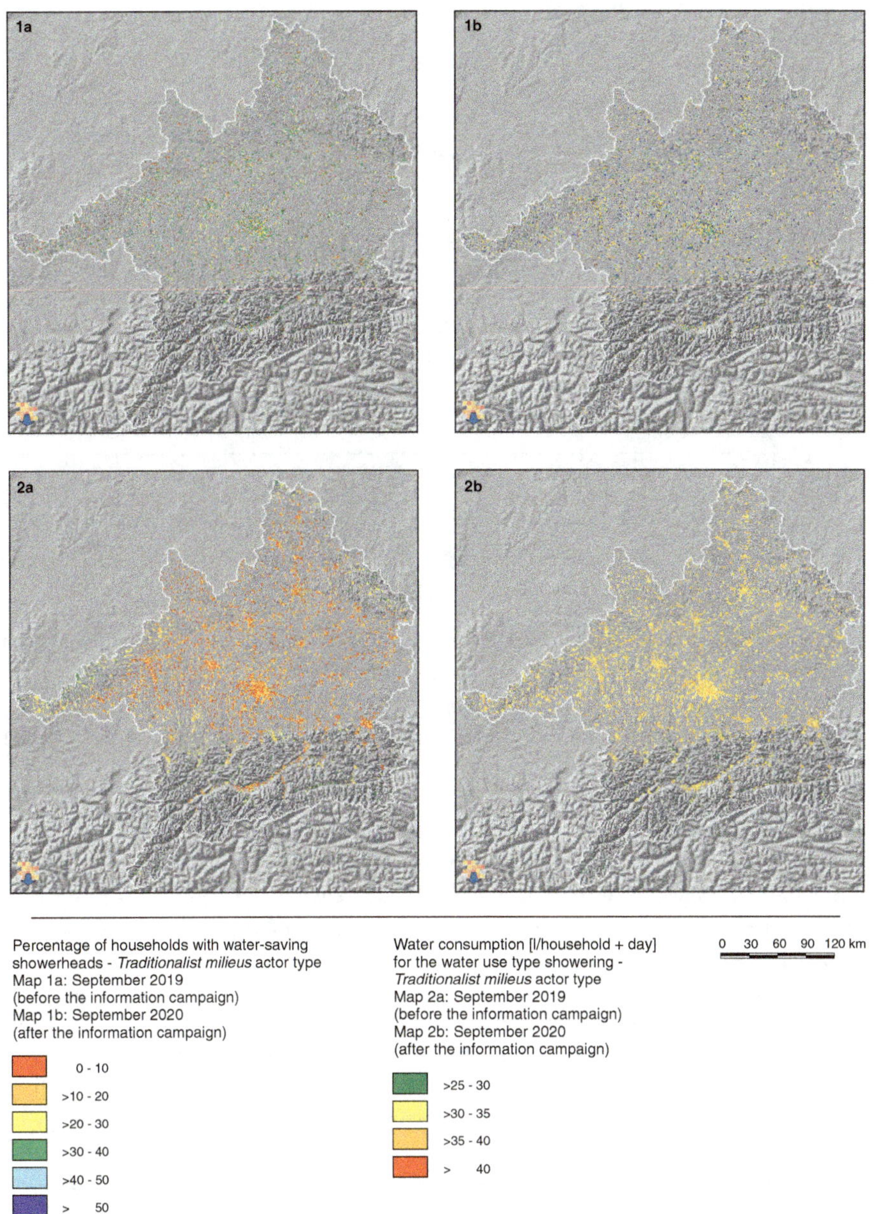

GLOWA-Danube scenario: *REMO regional* climate trend, *Baseline* climate variant, *Baseline* societal scenario, *Information campaign* intervention (March-June 2020)

Map 67.1 Interventions in *DeepHousehold*

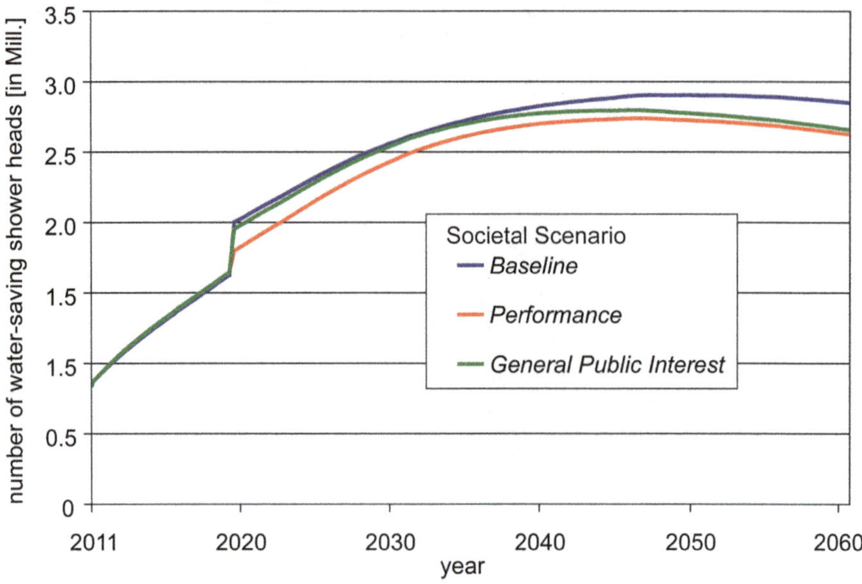

Fig. 67.4 Time course of diffusion of water-saving showerheads (absolute number in Mill.) in the three societal scenarios (Scenario information in the text)

campaign intervention, which appeals to environmental awareness, shows a larger effect. This is primarily the result of the larger diffusion of innovations and the associated market saturation.

The effects of the *Information campaign* intervention (in this case, with respect to toilet flushes) (March–June 2030) can be seen in Fig. 67.5. This figure illustrates the number of standard toilet flushes that are decreasing as a result of the intervention and being replaced by dual-flush cisterns or stop buttons.

Figure 67.6 compares the effects of the type of intervention (*Subsidies*, *Informational*) for specific actor types in the *Baseline* scenario. The interventions relate to water-saving showerheads. It is apparent that the *Hedonist milieus* actor type only opts for the water-saving showerhead in the *Subsidies* intervention. The trend in the number of water-saving showerheads even continues: triggered by a price advantage, the water-saving showerheads begin to spread among the *Hedonist milieus* actor type and influence buying decisions via the social network (see Chap. 43). In contrast, for the *Traditionalist milieus* actor type, the *Information campaign* intervention exhibits a relatively larger effect from the start. The diffusion of water-saving showerheads in the *Post-materialist* actor type is quite high at the beginning compared to the other actor types, so the influence of the two interventions is smaller.

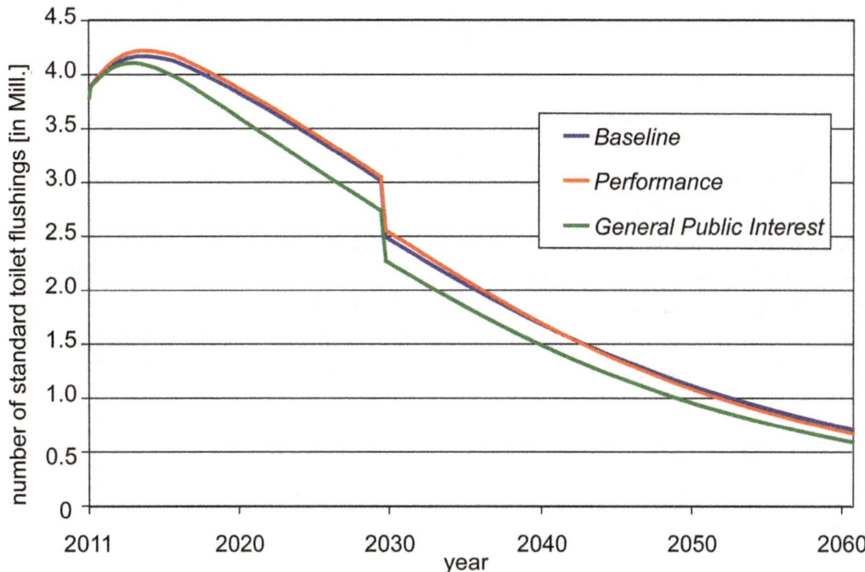

Fig. 67.5 Time course of diffusion of standard toilet flushings (absolute number in Mill.) in the three societal scenarios (Scenario information in the text)

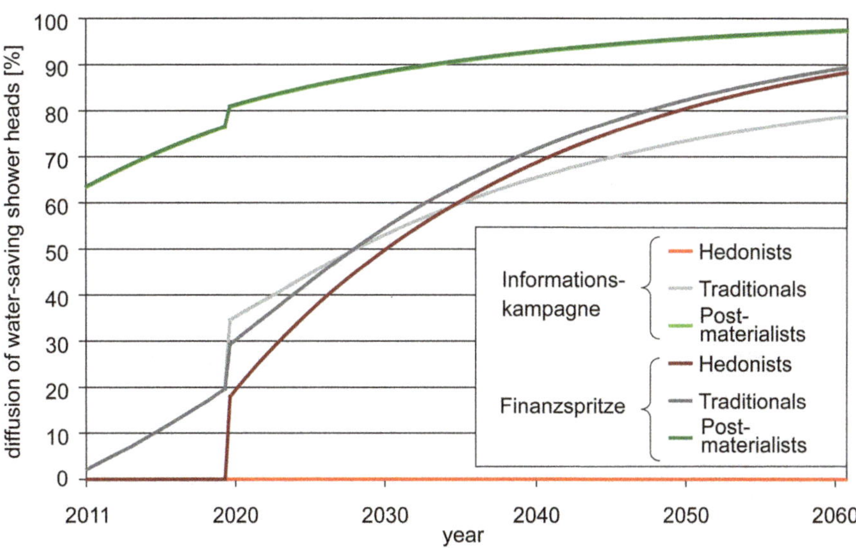

Fig. 67.6 Time course of diffusion of water-saving showerheads for three actor types and two types of interventions (Scenario information in the text)

67.4 Outlook

In addition to the implementation of extra interventions, such as legislations or prohibitions, a further refinement of the interventions is desirable. One option might be an information campaign that is geographically confined to regions with relative low groundwater supplies in order to defer other investments such as those for long-distance water supply networks, for example. Another possibility might be the targeted implementation of interventions exclusively in areas in order to reach more households with the same financial output. From this, cost estimates for specific savings targets may be potential questions that arise. In this regard, additional knowledge input from the stakeholders is necessary.

Chapter 68
Effects of Climate Change on Hydropower Generation and Reservoir Management

Franziska Koch, Andrea Reiter, and Heike Bach

Abstract Due to an increase in power demand and renewable energy, Bavaria and Tyrol plan to increase hydroelectric power generation. However, climate change may lead to a decrease in discharge and water availability and thereafter a decrease in hydroelectric power generation without further constructions considering only the current facilities. The change in the simulated annual output of the hydropower plants in the Upper Danube basin divided in the six subbasins Inn, Salzach, Isar, Lech, Iller and the remaining area of the Danube basin are shown for the two periods 2011–20135 and 2036–2060 under the GLOWA-Danube *REMO regional-Baseline* climate scenario. For the first period, a slight decrease was simulated, whereas for the second period, a significant decrease leads to a reduction in hydroelectric power generation of 9–16 % in the subbasins. Moreover, the trends in mean monthly inflow, outflow and reservoir fill volume are shown for the two periods for the example of the Gepatsch reservoir. Without changing the monthly based operation plan in the future, the change in the pattern of the inflow would have an impact on the reservoir filling volume and the pattern of the discharge at the reservoir outlet which would lead to a more balanced power generation over the year.

Keywords GLOWA-Danube • Hydropower generation • Discharge • Reservoir fill volume • Climate change impact

F. Koch (✉)
Department of Geography, Ludwig-Maximilians-Universität München (LMU Munich), Munich, Germany
e-mail: f.koch@iggf.geo.uni-muenchen.de

A. Reiter
Bavarian Research Alliance GmbH, Munich, Germany
e-mail: reiter@bayfor.org

H. Bach
Vista Geowissenschaftliche Fernerkundung GmbH, Weßling, Germany
e-mail: bach@vista-geo.de

W. Mauser, M. Prasch (eds.), *Regional Assessment of Global Change Impacts*, DOI 10.1007/978-3-319-16751-0_68

68.1 Introduction

Both the Bavarian State Government and the Austrian State of Tyrol plan to increase hydroelectric power generation in the future (E.ON Wasserkraft GmbH and Bayerische Wasserkraftwerke GmbH 2009; Amt der Tiroler Landesregierung 2008). This plan is the consequence of both predictions of increasing energy consumption and from scheduled augmentation of the proportion of renewable energies as a part of climate protection. The Bavarian State Government has set the overall goal to increase the power generated from hydroelectric sources over the coming years by 10 % compared to the reference year 2000 through modernisation, retrofitting, reactivation and construction of new and expanded facilities. In this context, several feasibility studies have been carried out both in Tyrol and Bavaria. However, to successfully assess the potential of hydropower, it is necessary to consider the further development of existing facilities and to carry these tendencies over to the planned expansions and new constructions. The forecast of discharge conditions and hence the analyses of the overall future water availability are key factors in this assessment. This chapter presents the change in the simulated annual output of the hydropower plants calculated in Chap. 23 for the periods 2011–2025 and 2036–2060 under the GLOWA-Danube *REMO regional-Baseline* climate scenario. In addition, the example of the Gepatsch reservoir (see Map 23.1 for the location) is used to illustrate the trends in inflows, outflows and reservoir fill volumes over the course of a year under this climate scenario.

68.2 Change in the Mean Simulated Annual Output in the Subbasins

Map 68.1 – 1 presents the mean annual output of the hydropower plants simulated in DANUBIA for the six subbasins of the Iller, Lech, Isar, Inn and Salzach as well as the remaining area of the Danube basin for the period 1971–2000. The Inn and Salzach subbasins originate in the high alpine glaciated headwater regions of the Upper Danube and are therefore highly characterised by glaciers and especially by snowmelt. In contrast, the other regions are not, or are only negligibly, influenced by glaciers but are largely dominated by a nival regime. Climate change also involves changes in the discharges and to glacial and snow storage (see Chaps. 53; 56 and 60); these changes in turn have a direct effect on hydropower generation.

A scenario calculation under the GLOWA-Danube climate scenario named above was carried out in order to investigate potential future changes in the annual output from hydropower plants. The adjacent Maps 68.1 – 1 and 68.1 – 2 depict the future change in the simulated mean annual output in the subbasins of the Upper Danube as a percent with respect to the reference period 1971–2000. In both future periods that are investigated, a decrease in annual output is expected. However, this

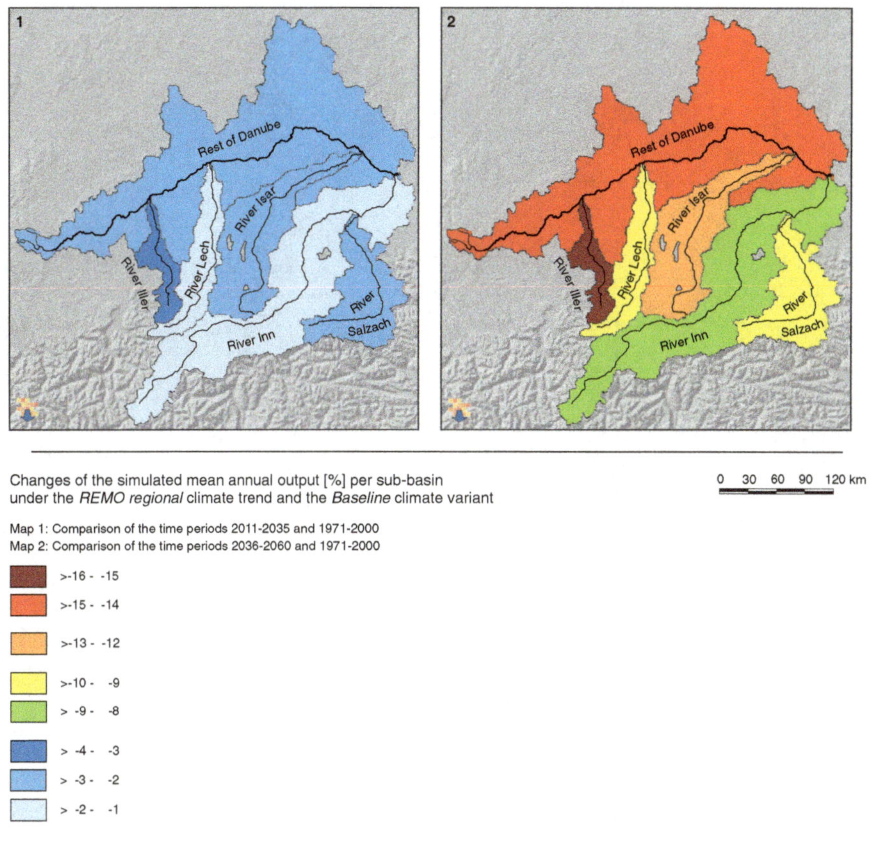

Map 68.1 Effects of climate change on hydropower generation and on reservoir management (*REMO regional* climate trend and *Baseline* climate variant)

trend appears in the first period (2011–2035) only weakly, with decreases between 1 and 4 %, while there are huge declines of 8–16 % in the second period (2036–2060). One reason for this result is the decrease in the simulated discharges over the whole Upper Danube basin (see Chap. 53).

It is quite obvious in Map 68.1 – 2 and Table 68.1 that hydropower generation for the period 2036–2060 in the subbasins that are less mountainous (the Iller and the rest of the Danube) shows the greatest declines. In contrast, the simulated annual output in the Inn, Lech and Salzach subbasins decreases less sharply. Despite the retreat of the glaciers and quantitative and temporal changes in snow storage (see Chaps. 56 and 57), the glacial and snow balance somewhat cushion the decline in hydropower generation in the future. As already explained in Chap. 54, there is even an increase of low-flow discharge in the high alpine regions, which has a positive effect on hydropower. The Isar catchment shows an intermediate position with a decrease in the simulated annual output of 12.7 %. This also applies to its regional position in the Upper Danube basin and its discharge regime, which is not completely free of alpine influences but is also not high alpine.

Table 68.1 Changes of the simulated mean annual output of the run-of-the-river and storage power plants for the sub-catchments of the Upper Danube basin under the *REMO regional-Baseline* climate scenario for the periods 2011–2035 and 2036–2060 compared to the reference period 1971–2000

Time period	Type of hydropower plant*	Danube (rest)		Iller		Lech		Isar		Inn		Salzach		Upper Danube	
		[GWh]	[%]	[GWh]	[%]	[GWh]	[%]	[GWh]	[%]	[GWh]	[%]	[GWh]	[%]	[GWh]	[%]
1971 to 2000	RPP	1909.0	–	452.8	–	1448.8	–	1281.1	–	5527.2	–	1477.4	–	12096.3	–
	SPP		–		–	154.0	–	341.0	–	3784.3	–	924.5	–	5203.9	–
	RPP+SPP	1909.0	–	452.8	–	1602.8	–	1622.1	–	9311.6	–	2401.9	–	17300.2	–
2011 to 2035	RPP	1863.3	−2.4	436.3	−3.6	1432.9	−1.1	1251.7	−2.3	5465.5	−1.1	1450.7	−1.8	11900.5	−1.6
	SPP		–		–	148.3	−3.7	335.3	−1.7	3684.9	−2.6	895.5	−3.1	5063.9	−2.7
	RPP+SPP	1863.3	−2.4	436.3	−3.6	1581.1	−1.4	1587.0	−2.2	9150.4	−1.7	2346.2	−2.3	16964.4	−1.9
2036 to 2060	RPP	1628.9	−14.8	382.9	−15.4	1318.2	−9.0	1112.4	−13.2	5054.3	−8.6	1317.5	−10.8	10814.3	−10.6
	SPP		–		–	135.0	−12.4	303.5	−11.0	3433.8	−9.3	851.3	−7.9	4723.6	−9.2
	RPP+SPP	1628.9	−14.8	382.9	−15.4	1453.2	−9.3	1415.8	−12.7	8488.1	−8.8	2168.9	−9.7	15537.9	−10.2

In addition, Table 68.1 presents the change in the simulated annual output in the subbasins for the two simulated future periods, separated for either run-of-the-river or storage power plants. However, there is no significant difference in the decrease in annual output according to the type of hydropower plant. Nevertheless, it is clear that the decline in power generation is much more notable both for all subbasins and for both types of power plants in the second period (2036–2060) compared to the first period (2011–2035).

68.3 Reservoir Management Over the Course of a Year Using the Example of the Gepatsch Reservoir

In the following, the Gepatsch reservoir in the Inn catchment is used as an example for modelling of reservoirs. It represents one of the largest hydroelectric power plants in Austria and belongs to the Kaunertal power plant of TIWAG (Tiroler Wasserkraft AG) in Prutz. The reservoir is situated at 1,767 m a.s.l. and has a storage operation area of 139 million m^3. The catchment area of the reservoir is 275 km^2, which includes both the natural basin area (107 km^2) and artificial transfer through tunnels from neighbouring valleys (TIWAG 2006). The water is conveyed via a pressure tunnel with a drop height of 895 m to the power plant in Prutz at 872 m a.s.l., where up to 54 m^3/s can flow through the turbines. The Kaunertal power plant has one of the highest output efficiencies in Austria, with a mean annual output of 661 GWh. Depending on the status of the reservoir, the bottleneck capacity lies between 325 and 392 MW (TIWAG 2006). The fill level is regulated by the operation plan such that it is primarily oriented by the snow- and glacier melt and increased in storage geared toward increased power demands in the winter months (see Chap. 29). As a result, the reservoir fill volume reaches a maximum toward the end of the summer and a minimum in early spring.

Both the future annual pattern and the future total volume of the inflow into the reservoir will change according to the simulation under the conditions of the *REMO regional-Baseline* climate scenario (see Chap. 53); hence, changes are expected as a consequence to both the discharge contribution and the reservoir fill volume. The adjacent Figs. 68.1 and 68.2 show the monthly inflow and outflow as well as the reservoir fill volumes for the Gepatsch reservoir for the reference period in the past (1971–2000) and for the two modelled future periods (2011–2035 and 2036–2060). In the period from 1971 to 2000, filling of the reservoir takes place largely in the months from April to September, with a maximum in June and July (inflow > outflow), whereas in the winter months that are generally characterised by low inflow to the reservoir, the stored water is increasingly discharged (outflow > inflow). In both of the future periods examined, this pattern changes noticeably under constant storage control. Hence, in the future, the inflow into the reservoir in the summer is significantly less than in the past. Moreover, the inflow maxima in the period 2036–2060 shift compared to 1971–2000 as a result of the

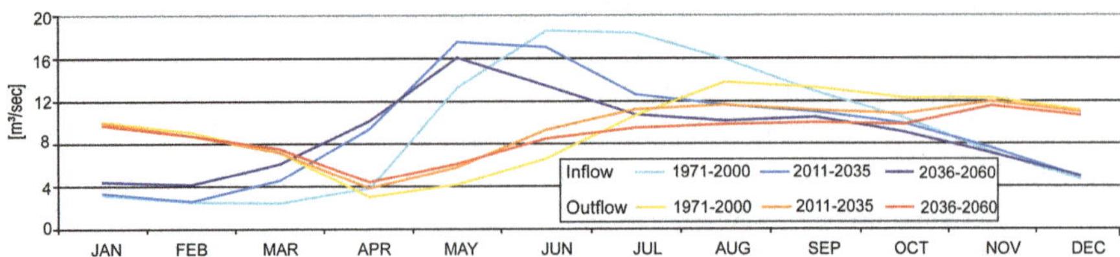

Fig. 68.1 Mean annual inflow and outflow of the Gepatsch reservoir for the reference period (1971–2000) and two simulated future periods (2011–2035 and 2036–2060). Mean inflow 1971–2000, 9.44 m^3/s; 2011–2035, 9.35 m^3/s; 2036–2060, 8.91 m^3/s

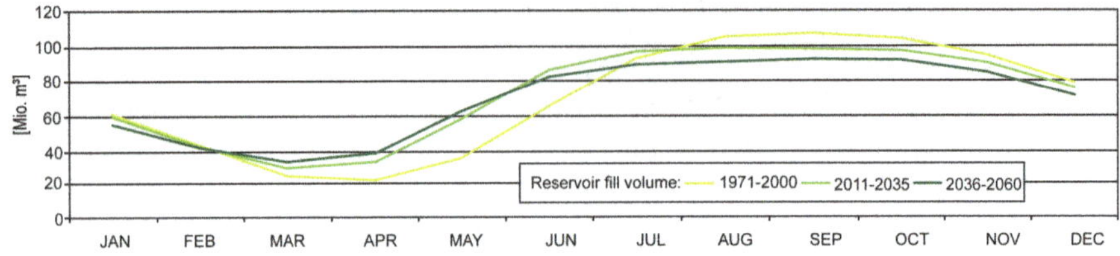

Fig. 68.2 Mean annual fill volume of the Gepatsch reservoir for the reference period (1971–2000) and two simulated future periods (2011–2035 and 2036–2060). Mean reservoir fill volume 1971–2000, 67.29 million m^3; 2011–2035, 69.60 million m^3; 2036–2060, 66.93 million m^3

simulated earlier glacier and snowmelts (see also Chaps. 56 and 60) by around two months, to May. In addition, in the climate scenario that was used, the winter months are seen with more rain in the place of snow in the future (see Chap. 57); this change results in an increased inflow into the storage reservoir in this season, primarily in the second period. Altogether over the course of the year, there will be a steady inflow with a shift of the maximum to spring. In the second period, the total volume of the inflow also decreases, since, as explained earlier in Chap. 56, the glacial storage sharply declines in the future even in the catchment of the dam, while the glacial contribution to inflow that is still increasing in the first time period likewise decreases (Koch et al. 2011).

While maintaining the operation plan, the change in the pattern of inflow has an impact on the annual pattern of the discharge at the reservoir outlet, which in turn has a direct relationship with power generation. Whereas the previously greater contributions in late summer and fall decline in the future; the minimum contributions in April to June increase in both future periods. In contrast, the discharge contributions in the winter months (November to February) remain at the same level in relative terms.

Overall, there is also a balancing-out effect of discharge that is also manifested in the reservoir fill volumes (see Fig. 68.2). If the mean annual trend in reservoir fill volume from 2036 to 2060 is compared to 1971–2000, the fluctuations are clearly reduced, and hence, the range of possible fill volumes in the operating site is no longer completely utilised.

68.4 Summary and Outlook

A sometimes significant decrease in the hydropower generation is simulated for the future under the *REMO regional-Baseline* climate scenario. In order to maintain the current annual output from the hydropower plants in the Upper Danube basin or to even increase the amount of hydropower generated within the energy sector, investment into new and expanded facilities is necessary. The aspects of reservoir management, including inflow, outflow and fill volumes for the reservoir will exhibit changed yearly patterns, which will tend to result in a more balanced power generation over the year. The changes under the climate scenario that were presented in this chapter can serve the power producers in supporting decisions regarding the development of potential adjustment strategies.

References

Amt der Tiroler Landesregierung, Abteilung Wasser, Forst und Energierecht (eds) (2008) Tiroler Energiestrategie 2020. Grundlage für die Tiroler Energiepolitik, Innsbruck

E.ON Wasserkraft GmbH, Bayerische Elektrizitätswerke GmbH (eds) (2009) Masterplan. „Ausbaupotentiale Wasserkraft in Bayern."Bericht aus Sicht der beiden großen Betreiber von Wasserkraftanlagen in Bayern. Landshut, Augsburg

Koch F, Prasch M, Bach H, Mauser W, Appel F, Weber M (2011) How will hydroelectric power generation develop under climate change scenarios? A case study in the Upper Danube Basin. Energies 4(10):1508–1541. doi:10.3390/en4101508

TIWAG – Tiroler Wasserkraft AG (ed) (2006) Das Kraftwerk Kaunertal. Informationsbroschüre, Innsbruck

Chapter 69
Estimating the Change in Groundwater Quality Resulting from Changes to Land Use and Groundwater Recharge

Roland Barthel and Fredy Alexander Peña Reyes

Abstract Global change will change the quantity and quality of groundwater resources and subsequently the prerequisites for the supply of safe drinking water. Quite often, attempts are made to simply express the status of groundwater resources by stating changes in groundwater level or concentrations. However, such results require further interpretation by experts before they can be used in decision making. This chapter presents the approach chosen in GLOWA-Danube to translate the results of model simulations of land use changes calculated by the agricultural model Farming into an index that expresses the changes of the quality status of groundwater resources in one single value. The chosen approach is based on the intrinsic groundwater vulnerability index (DRASTIC) and a predefined typical pollution load associated with different land uses. This chapter describes the methodology chosen and the main assumption underlying the approach. Results are shown for selected districts of the UDC for a variety of climatic and societal scenarios.

Keywords GLOWA-Danube • DANUBIA • Groundwater quality • DRASTIC • Vulnerability • Agriculture

69.1 Introduction

The dynamic groundwater quality index (GWQ index) presented in this chapter is calculated using the land use changes derived from the *Farming* model (see Chap. 70) and a parameter known as intrinsic groundwater vulnerability. Intrinsic vulnerability describes the (hypothetical) risk of groundwater contamination in a geographically

R. Barthel (✉)
Department of Earth Sciences, University of Gothenburg, Göteborg, Sweden
e-mail: roland.barthel@gu.se

F.A.P. Reyes
Department of Earth and Environmental Sciences, Universtity of Milano-Bicocca, Milan, Italy
e-mail: fredy.penareyes@unimib.it

W. Mauser, M. Prasch (eds.), *Regional Assessment of Global Change Impacts*, DOI 10.1007/978-3-319-16751-0_69

defined area in the event that the land surface in this area suffers from a pollutant load. Among other factors, this risk depends on the permeability of the surface layer, the groundwater level (GWL) or the groundwater recharge (GWR) in the affected area. The common approach to calculating vulnerability is the DRASTIC method (Aller et al. 1987). GWL and GWR are climate- and land use-dependent parameters which are dynamically calculated by the *Groundwater* and *Soil* models in DANUBIA.

The GWQ index presented here accounts exclusively for diffuse inputs from agriculture (fertiliser, pesticides (PSM = Pflanzenschutzmittel)) and does not consider point sources such as inherited industrial waste. Inputs from agriculture are estimated from land use. Thus, for each land use type, a probability of exposure is assigned for the export of nitrogen and pesticides. These allocations are carried out based on values provided in the literature and from expert knowledge.

The GWQ index is calculated from the combination of the exposure risk and vulnerability. In this way, a change in land use results in a change of the GWQ index. Vulnerability is also dynamic, as it accounts for changes to GWR and GWL. Since land use changes annually, the GWQ index is also calculated on a yearly basis. Short-term (daily) changes in GWR and groundwater level are not accounted for.

The relatively simple GWQ index presents an alternative to and expansion for the numerical transport models for groundwater and soil and permits a rapid estimate for trends.

69.2 Methodology

69.2.1 Calculation of Groundwater Vulnerability

The DRASTIC method is one of the most common approaches for calculating groundwater vulnerability and at the same time is one of the best suited for the regional scale, since it utilises relatively few and largely comprehensive data. The acronym DRASTIC is comprised of the letters representing the seven parameters that are used: *D*epth, *R*echarge, *A*quifer material, *S*oil texture, *T*opography, *I*mpact (of the unsaturated zone) and hydraulic *C*onductivity. Calculation of the index and allocation of the parameter usually take place at the level of grid cells.

The significance of the parameter for vulnerability is specified by the weight, usually assigned a value from 1 to 5. The parameter values are classified in order to determine the impact of the magnitude of a parameter value on vulnerability. Hence, for example, the risk of contamination is higher at lower groundwater levels. Typically, there are 10 classes used (1 = minimal risk, 10 = high risk).

Table 69.1 Weights for the calculation of the DRASTIC index

Parameter	Overall weights	PSM weights
Groundwater level	5	5
Groundwater recharge	4	4
Type of aquifer	3	3
Soil texture	2	5
Topography (gradient)	1	3
Unsaturated zone (type)	5	4
Hydraulic conductivity (aquifer)	3	2

The DRASTIC index *D* is calculated per model cell (in this case a proxel) as the sum of the products of the weights *wXi* and the classes *rXi* (rating) for the individual parameters *Xi* where *i={D,R,A,S,T,I,C}*:

$$D = \sum_{i=1}^{7} wX_i \cdot rX_i .$$

Values for weight and the classification scheme can be found in the literature. Generally, an adjustment to the individual particularities of the region under consideration is required. The different transportabilities and degradabilities of pollutants can also be accounted for by the weights and classes. In the present example, a distinction between nitrogen and PSM is possible. Table 69.1 lists the weights that are used.

Space constraints do not allow for the reproduction of the classifications of the values for the individual parameters (ratings).

69.2.2 Calculation of the Probability of Export

The (intrinsic) vulnerability initially says little about an actual contamination of the groundwater at a site. This can only be assessed if actual pollution sources are known. Calculation of the GWQ index is based on the assumption that every land use can be assigned a probability of the application and export (seepage out of the root zone) of fertilisers and pesticides. This (export) probability is related to the quantity applied and the frequency of application of products typical for the type of crop. In collaboration with the Agricultural Economics (*Farming*) subproject, the land use categories utilised in DANUBIA were assigned this kind of probability. Thus, a value of 1 means low probability, and 5 means high probability of pollutant export. The assignment is based on the knowledge from experts within the project groups and, for example, on the statements in Roßberg et al. (2002) for pesticides and the data in Schmidt and Osterburg (2005) for nitrogen. The latter evaluated the proportion of nitrogen in the overall balance that cannot be used directly. Roßberg et al. (2002) developed a treatment index for pesticides; this index is described in

Table 69.2 Export probabilities for pesticides and nitrate for selected land use classes and cultivated crops

Land use	Extensively used grassland	Deciduous forest	Coniferous forest	Natural grassland	Intensively-used rassland	Grass-clover mixture	Oat	Legumes	Maize	Corn silage	Rye	Summer barley	Summer wheat	Winter barley	Sugar beet	Winter wheat	Rapeseed	Potatoes	Special crops
Risk from PSE	1	1	1	1	1	1	2	2	2	2	3	3	3	3	3	4	4	5	5
Risk from nitrate	1	1	1	1	2	3	1	2	3	4	1	1	1	1	3	1	4	1	5

detail in the reference. The probabilities of export *pA* for a selection of land use types are given in Table 69.2.

In order to take into account the fact that a proxel can have more than one land use type, the total export probability per proxel, *pAPr*, was weighted by the proportional areas *Ai* (%) and the export probabilities for each crop type. Also accounted for is the fact that the export probabilities for pesticides *pAPSMi* and nitrogen *pANi* for *n* land uses *LNi* differ. Another different estimate of the risk from nitrogen and PSM can optionally be considered (factor *k*, here $k = 1$).

$$pA_{\mathrm{pr}} = \sum_{i=1}^{n} LN_i \cdot pAN_i + k \cdot \sum_{i=1}^{n} LN_i \cdot A_i \cdot pAPSM_i .$$

69.2.3 Calculation of the Dynamic Groundwater Quality Index (GWQ Index)

The GWQ index is the product of the DRASTIC index *D* and the export probability pA_{Pr} for each proxel.

$$\mathrm{GWQ} = D_{\mathrm{Pr}} \cdot pA_{\mathrm{Pr}} .$$

The GWQ index can have a maximum value of 230,000, with the weights and classes used for each calculation. To reach this maximum, a proxel must be completely covered with a speciality crop, and the groundwater must be virtually

Table 69.3 Nominal evaluation and classification of the dynamic GWQ index

Evaluation	*D*	*pA*	GWQ
Very good/very light	23–65	10–200	230–15,000
Good/light	66–110	201–400	15,001–45,000
Moderate	111–155	401–600	45,001–90,000
Poor/heavy	156–200	601–800	90,001–160,000
Very poor/very heavy	>200	>800	>160,000

unprotected. The highest realised values in the calculations are around 100,000 for the reference period and 140,000 for the scenarios. Since the magnitude of the calculated value does not provide direct knowledge about groundwater quality, the values are classified according to the ranking shown in Table 69.3.

69.3 Validation Using Measurement Data on Groundwater Quality

The DRASTIC method is conceptually quite a simple method that consists of a high degree of generalisation and a number of subjective specifications. The same is true for the allocation of pollutant export probabilities to land use categories and hence ultimately for the GWQ index as a whole. Therefore, validation of the calculation results using measurement data on groundwater load and concentrations of pollutant in the soil and leachate is useful, though difficult to implement. Although there are data on nitrate in groundwater for a number of sites (see Chap. 22), measurements of currently used pesticides and soil and leachate measurements are scarcer. Validation using measured values is therefore only possible within limits for the assessment of nitrogen in groundwater. The huge discrepancy between the scale of the calculations (proxels) and the observations (point) must also be considered.

69.4 Results

Map 69.1 depict the temporally pooled results for both societal scenarios (*Performance* and *Baseline*; see Chap. 52), in each case combined with the *REMO regional–Baseline* climate scenario (see Chaps. 47, 48, 49 and 50), for four sample districts. In each case, the mean of the GWQ index for the decade 2050–2059 is shown, classified according to the categories given in Table 69.3.

Each chart depicts the temporal trend in the vulnerability index (*D*), the export probability (*pA*) and the GWQ index. In each case, the mean for the reference decade (calculated from the results of reference simulation) is subtracted from the scenario results. Thus, an increase in the values shown indicates a deterioration.

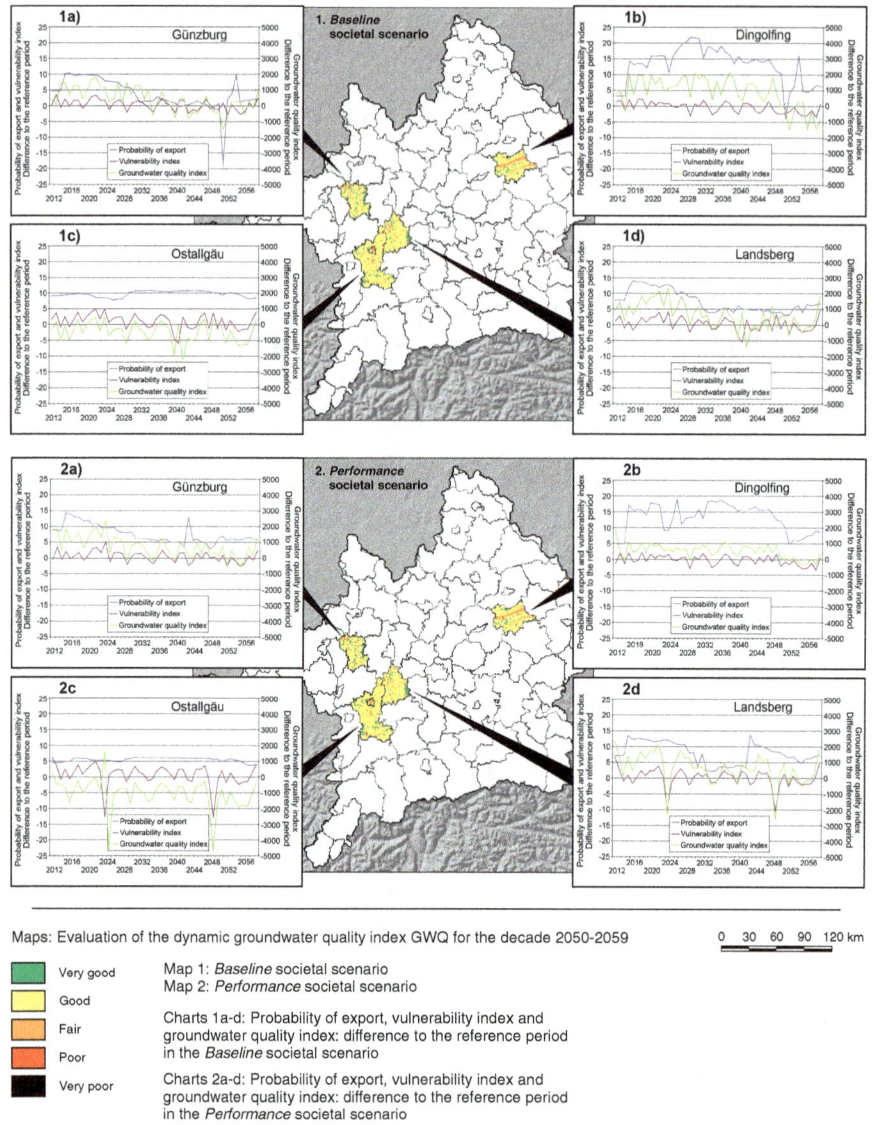

Calculated under the *REMO regional* climate trend and the *Baseline* climate variant.

Map 69.1 Development of vulnerability, the export probability and the groundwater quality resulting from changes of land use, groundwater recharge and the groundwater level (dynamically coupled model run of the DANUBIA Biological, SNT, Farming and NaturalEnvironment (PROMET) model components for the years 2011–2059 and 1995–2006 (dynamic land use and groundwater recharge); dynamically coupled model run of the DANUBIA GroundwaterFlow und WaterSupply model components for the years 2011–2059 and 1960–2006 (depth of groundwater tables); *REMO regional* climate trend, *Baseline* climate variant)

The results reveal a fundamental significant fluctuation of vulnerability for all four districts; this fluctuation results from the significant changes in the dynamics variables (GWR and groundwater level). Moreover, there is a slight tendency for a decrease in the GWQ index, i.e. there is a slight improvement of GWQ in all districts for both scenarios. The changes are quite negligible on average if the classifications are taken into account (see Table 69.1). Also, the differences in the area are limited to a few proxels. This may be explained by the fact that the changes in land use are not great and groundwater recharge only decreases slightly.

A more detailed coverage of the climate-, land use-, agriculture- and groundwater-related aspects in DANUBIA can be found in Barthel et al. (2012).

References

Aller L, Bennet T, Lehr JH, Petty RJ, Hackett G (1987) DRASTIC: a standardized system for evaluating groundwater pollution potential using hydrogeological setting. EPA/600/2-87/035. US Environmental Protection Agency, Washington, DC, p 163

Barthel R, Reichenau TG, Krimly T, Dabbert S, Schneider K, Mauser W (2012) Integrated modeling of global change impacts on agriculture and groundwater resources. Water Resour Manage 26(7):1929–1951

Roßberg D, Gutsche V, Enzian S, Wick M (2002) Neptun 2000 Erhebung von Daten zum tatsächlichen Einsatz chemischer Pflanzenschutzmittel im Ackerbau Deutschlands. In: Berichte aus der Biologischen Bundesanstalt für Land- und Forstwirtschaft, vol 98, p 27

Schmidt T, Osterburg B (2005) Aufbau des Berichtsmoduls "Landwirtschaft und Umwelt" in den Umweltökonomischen Gesamtrechnungen. FAL, Abschlußbericht 2005, Braunschweig und Wiesbaden

Chapter 70
Effects of Future Climate Changes on Yields, Land Use and Agricultural Incomes

Tatjana Krimly, Josef Apfelbeck, Marco Huigen, Stephan Dabbert, Tim G. Reichenau, Victoria I.S. Lenz-Wiedemann, Christian W. Klar, and Karl Schneider

Abstract Dynamically coupled model runs of the DANUBIA components *Biological, SNT, NaturalEnvironment* and *Farming* were performed to estimate the effects of climate change on crop yields, agricultural land use and income. Calculations are based on a GLOWA-Danube scenario including the climate trend REMO regional, the climate variant Baseline and the societal scenario Baseline. Results of the scenario calculation are compared with the reference period for four sample districts, which represent different site conditions within the drainage basin. In general, the scenario results show an increase in yields for the considered groups of crops. However, changes of individual crops within these groups differ between the districts. All districts have an increase in income that at the beginning of the scenario period is mainly caused by the increase in premium payments of the CAP compared to the reference period. The further income increase at the end of the scenario period, which is significantly higher in districts with a higher proportion of arable land, can be attributed to the increase in yields. With respect to land use, all districts show a decrease in forage crops and an increase in the cultivation of cereals. Overall, the results indicate that no negative impacts on the productivity of the agricultural land and the income situation of the farms are to be expected up to the middle of the century.

Keywords GLOWA-Danube • Climate change • Crop yields • Agricultural land use • Agricultural income • Dynamic model coupling • Scenario calculation

---

T. Krimly (✉) • J. Apfelbeck • M. Huigen • S. Dabbert
Production Theory and Resource Economics, Universität Hohenheim, Stuttgart, Germany
e-mail: T.Krimly@uni-hohenheim.de; dabbert@uni-hohenheim.de

T.G. Reichenau • V.I.S. Lenz-Wiedemann • K. Schneider
Geographisches Institut der Universität zu Köln, Cologne, Germany
e-mail: tim.reichenau@uni-koeln.de; victoria.lenz@uni-koeln.de; karl.schneider@uni-koeln.de

C.W. Klar
Forschungszentrum Jülich, Jülich, Germany
e-mail: c.klar@fz-juelich.de

W. Mauser, M. Prasch (eds.), *Regional Assessment of Global Change Impacts*, DOI 10.1007/978-3-319-16751-0_70

70.1 Introduction

Location factors related to climate have a significant impact on the cultivation suitability of crops and on their capabilities for growth, development and reproduction and hence also on the potential yields (Chmielewski 2007). The variability of the annual yields achieved is thus largely influenced by the weather conditions. In addition, crop management decisions depend to a large degree on climate and weather. Climate-induced yield changes of crops can lead to shifts in their profitability, which, from an economic viewpoint, can result in changes in agricultural land use and incomes.

The goal of the study presented in this chapter is to estimate the effects of climate change and weather variability on crop yields and, therefore, agricultural land use and the agricultural income; this goal is achieved by the dynamic coupling of the DANUBIA *Biological* (Lenz-Wiedemann et al. 2010), *SNT* (Klar et al. 2008), *NaturalEnvironment* (see Chap. 72) and *Farming* components.

70.2 Scenario Assumptions

The scenario results presented here are based on the following GLOWA-Danube scenario:

- The *REMO regional* climate trend: a temperature increase of 5.2 °C, changes in precipitation –4.9 % in winter and –31.4 % in summer (see Chaps. 47 and 48).
- The *Baseline* climate variant: the criterion is the mean annual temperature between 2011 and 2035 (see Chaps. 49 and 50).
- The *Baseline* societal scenario: *business-as-usual*, which continues with the status quo (see Chap. 52).

For the *Farming* model, *business-as-usual* means that the political situation for agriculture from the 2003 EU common agricultural policy (CAP) reform (e.g. decoupling, cross compliance) that will have reached its final stage in 2013 is maintained throughout the scenario period until 2060. This leads to significant changes compared to the reference period in premium payments from the first pillar of the CAP. Table 70.1 shows this for Bavaria by way of example. Over the course of decoupling of the premium payments from production, there are significantly higher premium payments after 2013 for permanent crops, row crops, forage crops and grassland.

70.3 Model Components Involved

The calculations were performed with the dynamically coupled DANUBIA *Biological*, *SNT*, *NaturalEnvironment* (see Chap. 72) and *Farming* components. *NaturalEnvironment* provides the necessary meteorological and hydrological site

Table 70.1 Premium payments from the first pillar of the CAP in Bavaria

	Ø Premium 1996–2005 euro/ha	Ø Premium 2013–2059 euro/ha
Permanent crops	30	340
Row crops (corn, sugar beet, potato)	189	340
Forage (corn silage, grass-clover mixture)	228	340
Winter cereals (winter wheat, winter barley, rye)	322	340
Spring cereals (summer barley, summer wheat, oats)	322	340
Other (legumes, rapeseed, fallow land)	426	377
Grassland	9	340

Barthel et al. (2012)

Table 70.2 Characteristics of the selected districts in the reference period

	Dingolfing	Günzburg	Landsberg	Ostallgäu
Altitude (m a.s.l.)	330–450	440–517	540–740	582–2,000
Annual precipitation (mm/a)	650–750	660–910	972	1,000–1,400
Annual mean air temperature (°C)	7.0–8.0	7.5	6.0–7.5	6.4–7.4
Share of arable land on UAA* (%)	89.9	61.0	52.3	9.2
Share of grassland on UAA* (%)	10.1	39.0	47.7	90.8

Amt für Ernährung, Landwirtschaft und Forsten Landau a.d. Isar, Krumbach (Schwaben), Fürstenfeldbruck, Kaufbeuren (2009)
*Utilised agricultural area

factors that influence plant growth and nitrogen fluxes and turnover. *SNT* models the plant-available nitrogen (see Chap. 38). *Biological* simulates plant growth and calculates the yields achieved (see Chap. 37). There is a tight coupling on a daily basis between the agricultural crop management activities from *DeepFarming* and the development stages reached by the crops in the plant growth model (see Chap. 40). The changes of land use and income as a result of climate-related changes in yield are calculated in *Farming*.

70.4 Results

The results of the scenario calculation (mean values for the years 2012–2021 and 2049–2058) are compared with the reference period (mean value for the years 1996–2005). Four sample districts are shown that represent differing site conditions within the drainage basin (see Table 70.2). In the district of Dingolfing, which has the highest proportion of arable land and the highest yield potential, approximately 50 % of the farms are cash-cropping farms (Bayrisches Landesamt für Statistik und Datenverarbeitung 2007). As the proportion of grassland increases, the amount of

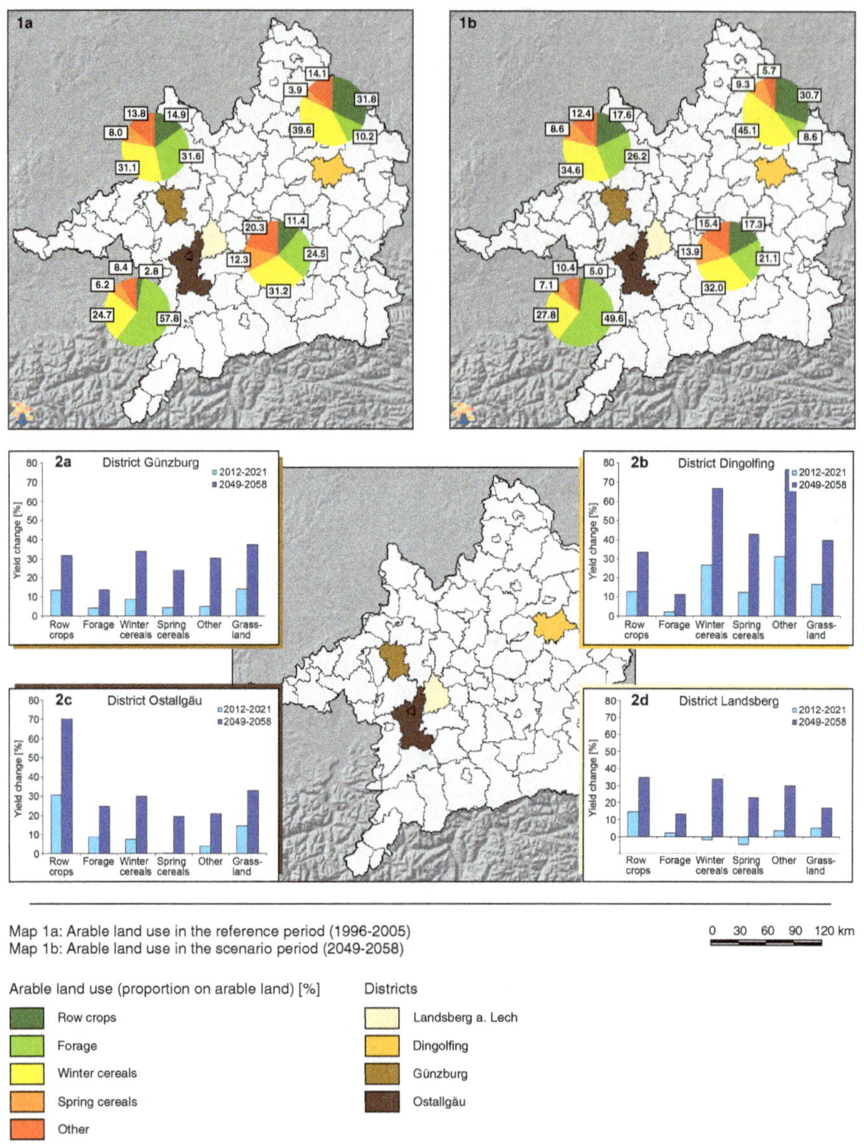

Charts 2a-d: Yield changes compared to the refence period (1996-2005)

Calculated under the *REMO regional* climate trend, the *Baseline* climate variant and the *Baseline* societal scenario.

Map 70.1 Effects of future climate changes on yields and land use (dynamically coupled model run of the DANUBIA Biological, SNT, Farming and NaturalEnvironment model components for the period 2011–2059, *REMO regional* climate trend and *Baseline* climate variant)

Table 70.3 Development of gross margins in the districts

Change in gross margin	Dingolfing (%)	Günzburg (%)	Landsberg (%)	Ostallgäu (%)
2012–2021 to reference	30	13	19	20
2049–2058 to reference	43	17	24	21

fodder-growing farms with cattle increases; in the district Ostallgäu, this proportion is approximately 95 %.

In general under the scenario conditions, there is an increase in agricultural yields compared to the reference period (see Map 70.1 – 2 and chart 2a–d). Towards the end of the scenario period (2049–2058), the increase is even higher again compared to the beginning (2012–2021). The increase is quite different among the individual crops and districts. In the Landsberg district, at the beginning of the scenario period (2012–2021), there is an overall decrease for winter and spring cereals. In Dingolfing and Ostallgäu too, there are declining yields for some crops (e.g. spring barley) within the groups listed in Table 70.1. However, these are not reflected in the group means. This is also the case for corn in Dingolfing in the period 2049–2058. No decrease in yield was simulated for Günzburg. Comparing the districts, the highest increase on average across all crops arose in Dingolfing, followed by Ostallgäu, Günzburg and Landsberg. This is the case for both periods evaluated in the scenario period. The lowest increase in yield in the second scenario period over all districts was on average 16 % for forage crops. The yields for grassland and spring cereals increase by about 30 %, and for row crops, winter cereals and other crops the increase is around 40 %. The greatest differences in the trends for yields among the districts are found for corn. In Ostallgäu, which currently has the poorest conditions for arable cropping, climate change causes a considerable increase in corn yield, whereas in Dingolfing, which today is one of the districts most favourable for arable cropping, there is a slight decrease in corn yield.

In general, in all the districts studied, the scenario leads to an improved situation in terms of income, under the assumption of constant prices (see Table 70.3). The income increases in Günzburg, Landsberg and Ostallgäu in the period 2012–2021 are mainly the result of the increase in premium payments (see Table 70.3) and in particular the premium payments for forage crops and grassland. In Dingolfing, the yield increases are significantly higher compared to the other three districts, especially for root crops and winter cereals; this leads to the positive effect on the income in this district. The further improvements of the income in the years 2049–2058 are the result of the climate-induced yield increases of the arable crops and are, therefore, significantly higher in districts with a large proportion of arable land (Dingolfing) compared to those with a higher proportion of grassland (Ostallgäu).

With respect to land use, a decrease in cultivation of forage crops for cattle is evident in all districts in the period 2049–2058 compared to the reference period (see Map 70.1 – 1a, b). This is especially apparent in Ostallgäu (−8.2 percentage points) and in Günzburg (−5.4 percentage points) and is mostly at the expense of corn silage for which yields are almost unchanged, whereas yields from grassland improve. In Landsberg, forage crop cultivation declines less sharply, since here the yield increases for grassland are significantly lower. In Dingolfing, forage crops are already only a minor crop in the reference period. The cultivation of winter and spring cereals increases in all districts. The greatest increase is in Dingolfing, where the highest yield increases take place. In contrast, cultivation of row crops increases only in the previously cooler and moister districts, while in Dingolfing this remains approximately the same. The greatest increase is +5.9 percentage points in Landsberg. Here, increased cultivation of potatoes is especially notable and yields significantly increase. In the "Other" category, which contains both cultivation of legumes and rapeseed, as well as fallow land, there is a decline in cultivated area, particularly in Dingolfing, with a decrease of −8.4 percentage points, but also in Landsberg, with a decrease of −4.9 percentage points. In Dingolfing, this can be attributed especially to the decrease in fallow land but also to the drop in rapeseed cultivation, which has a smaller yield increase compared to other crops. This is also the cause for the decreased cultivated area in Landsberg.

The scenario results indicate that the GLOWA-Danube scenario causes different effects in the different parts of the Upper Danube basin. Districts with a high proportion of arable land benefit more from yield increases, as farms with arable cropping are much more flexible than farms which mainly use grassland, because they have the option to adapt to changes in yield by changes in their crop rotations. Overall, the results of the simulation indicate that no negative impacts are expected up to the middle of the century, both with respect to productivity of agricultural land and in terms of the income situation for farms.

References

Amt für Ernährung, Landwirtschaft und Forsten Landau a.d. Isar, Krumbach (Schwaben), Fürstenfeldbruck, Kaufbeuren (2009) Daten und Fakten, Unser Dienstgebiet, Natürliche Standortverhältnisse

Barthel R, Reichenau TG, Krimly T, Dabbert S, Schneider K, Mauser W (2012) Integrated modeling of global change impacts on agriculture and groundwater resources. Water Resour Manage 26:1929–1951

Bayerisches Landesamt für Statistik und Datenverarbeitung (2007) Agrarstrukturerhebung; Betriebstypen, betriebswirtschaftliche Ausrichtung, Anzahl der landwirtschaftlichen Betriebe, München

Chmielewski F-M (2007) Folgen des Klimawandels für die Land und Forstwirtschaft. In: Endlicher W, Gerstengarbe F-W (eds) Der Klimawandel – Einblicke, Rückblicke und Ausblicke, Eigenverlag, Potsdam, pp 75–85

Klar CW, Fiener P, Neuhaus P, Lenz-Wiedemann VIS, Schneider K (2008) Modelling of soil nitrogen dynamics within the decision support system DANUBIA. Ecol Model 217:181–196

Lenz-Wiedemann VIS, Klar CW, Schneider K (2010) Development and test of a crop growth model for application within a global change decision support system. Ecol Model 221:314–329

Chapter 71
Soil Temperature under Scenario Conditions

Markus Muerth

Abstract The seasonal course of temperature in the soil layers near to the surface has a huge impact on the biological and biochemical processes of the root zone. The underlying relationship between temperature and biochemical processes results in what is commonly known as the Arrhenius equation. A hydrological aspect of the soil's energy balance is the occurrence of flooding as a result of frozen ground layers. In the event of long-lasting frosts, the flow paths in the top soil layer freeze and increased lateral discharge arises near the surface, which in turn leads to more severe flooding events.

The calculation of soil temperature in DANUBIA with the physical-based Soil Heat Transfer Module (SHTM) allows for the simulation of spatially differentiated, hourly soil temperature patterns. This pattern is dependent not only on air temperature but also on available radiation, land use and the phase transition of soil water. The presented maps illustrate the spatially differentiated, seasonal response of soil temperature to the GLOWA-Danube climate change scenarios for the Upper Danube. In general, the soil warming is slower in mountainous regions due to longer soil frosts and snow cover, while local patterns are mainly induced by the different land cover classes.

Keywords GLOWA-Danube • DANUBIA • Soil temperature • Climate change • Scenarios

71.1 Introduction

The seasonal course of temperature in the soil layers near to the surface has a huge impact on the biological and biochemical processes of the root zone. The underlying relationship between temperature and biochemical processes results in what is known as the Arrhenius equation. Moisture and temperature control the microbial decomposition of organic material in the upper soil layers and hence also regulate the release of carbon dioxide into the atmosphere (Davidson and Janssens 2006).

M. Muerth (✉)
Department of Geography, Ludwig-Maximilians-Universität München (LMU Munich), Munich, Germany
e-mail: m.muerth@lmu.de

W. Mauser, M. Prasch (eds.), *Regional Assessment of Global Change Impacts*, DOI 10.1007/978-3-319-16751-0_71

Leaching of nitrate from the soil column penetrated by roots is also dependent on the biochemical processes of nitrogen transformation in the soil. These factors are modelled in DANUBIA within *SNT* as a sub-model of the *Soil* component (see Chap. 38) and are dependent on the temperature dynamics within the soil. Specific processes of plant development are also ultimately influenced by the soil temperature. These processes include dormancy and germination, as well as root growth (Pregitzer et al. 2000), which are accounted for in part by the *Biological* model component (see Chaps. 36 and 37).

One hydrological aspect of the simulation of the soil energy balance is the occurrence of flooding (mostly in late winter) in the subbasins of the Upper Danube as a result of the frozen top layer of the ground. Using gauge measurements from some subbasins, it is evident that primarily during late winter snow melts, the amount of lateral discharge within the overall discharge (faster gauge output) is higher. In the event of long-lasting frosts, the flow paths in the top soil layer freeze, and increased lateral discharge arises near the surface, which in turn leads to more severe flooding events (Bayard et al. 2005).

The calculation of soil temperature in DANUBIA with the *Soil Heat Transfer Module* (*SHTM*) allows for the simulation of spatially differentiated temperature patterns in multiple soil layers; this pattern is dependent not only on air temperature but also on available radiation, land use and the phase transition of the soil water. Map 71.1 – 1–4 illustrates the spatially differentiated response of soil temperature to climate change in the Upper Danube.

71.2 Data Processing

The thermal properties of the different layers of each soil texture class are needed as a basis for the simulation of soil temperature in DANUBIA (see Chap. 8). Thermal capacity and thermal conductivity are parameterised based on soil texture, porosity and organic content. The actual thermal properties are calculated during runtime influenced by the current soil moisture content according to the model of de Vries (1963).

71.3 Model Documentation

SHTM is a physical model of the conductive heat transfer between the different modelled layers of the soil column (Muerth and Mauser 2012) and is a part of the *Soil* component. Transfer and storage of thermal energy is calculated hourly to simulate the daily pattern of soil temperature. The geometry of the soil layers in *SHTM* is the same as for the soil water model with an additional “virtual” soil layer below. The temperature in this layer is only affected by the annual temperature amplitude and is therefore calculated with an analytical equation that is adjusted to the past

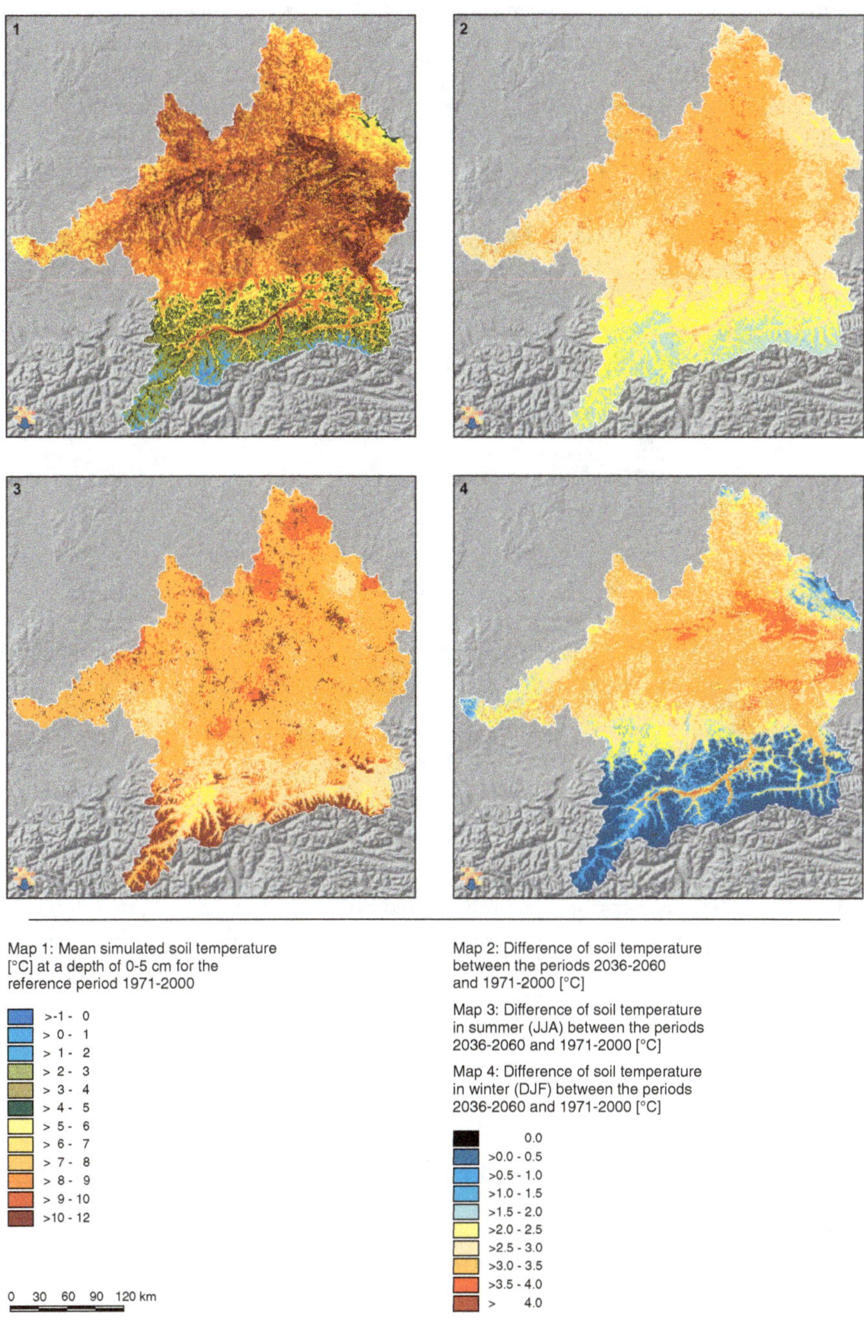

Map 71.1 Scenarios of the top soil layer temperature (REMO regional climate trend and Baseline climate variant)

annual mean air temperature during simulation. This prevents the mean temperature from drifting as a result of unknown underground heat flows in long-term simulations. If the temperature falls below freezing within a soil layer, a routine begins that calculates the proportion of frozen water for a specific temperature. Each additional withdrawal of energy from a soil layer then takes away the freed-up phase-transition energy from this proportion of frozen water before further cooling takes place.

As the driving variable for the calculation of soil temperature, heat flow to the soil surface layer is calculated in the *Surface* model component, wherein the temperature of the upper soil layer is accounted for in the calculation of the energy balance at the soil surface. The short and long wave radiation available beneath a plant population is distributed via a set of equations based on physics into the energy flows emanating from the soil surface. These include:

- The evaporation of water
- The thermal radiation
- The sensible heat transfer by convection
- The heat flow into the soil

Since evaporation is calculated using incident radiation and available soil moisture, the other energy flows can be determined by iteration of the surface temperature, such that the energy balance is closed. If there is snow cover, the energy flow in the soil is simulated with the energy balance of the *Snow* model component (see Chap. 30), where temperature of the upper soil layer is accounted for in this case as well.

71.4 Results

Map 71.1 – 1 illustrates the mean simulated soil temperature [°C] at a depth of 0–5 cm for the reference period 1971–2000. In contrast to the map of the mean air temperature (see Chap. 11), the spatial patterns are considerably more distinct as a result of the different simulated land uses. In particular, forested areas (e.g. south of Munich at the centre of the map) are characterised by a cooler mean temperature as a result of shading of the soil. Although the mean soil temperature is actually even slightly lower under tree stands compared to the mean air temperature (by −1 to 0 °C), the mean soil temperatures for other land covers are generally higher by about 1–3 °C compared to the long-term average air temperature. In high alpine regions, where the phase transition of soil water and the frequent snow cover often dampen soil temperatures in winter, the mean temperature of the top soil layer itself is up to 5 °C higher than the mean air temperature.

The average warming of the upper soil layer in the GLOWA-Danube *REMO regional–Baseline* climate scenario (see Chaps. 47, 48, 49, and 50) shown in Map 71.1 – 2 also reveals significant regional differences. The difference between the mean temperature for the years 2036–2060 (*REMO regional–Baseline*) and the reference period (1971–2000) is significantly smaller in the mountainous regions of

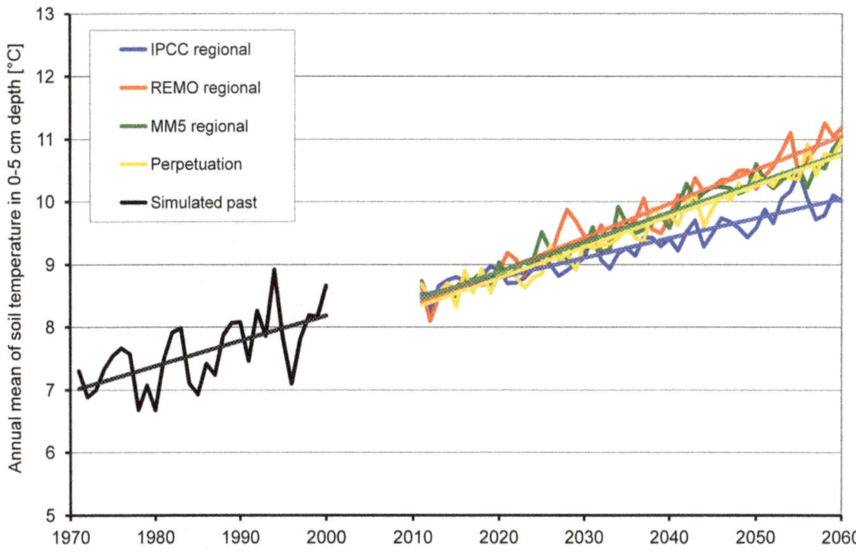

Fig. 71.1 Annual mean temperature of the top soil layer for the reference period 1971–2000 and the *IPCC regional, REMO regional, MM5 regional* and *Extrapolation* climate trends

the Alps and the Bavarian Forest compared to the lowlands. In non-alpine regions, the negligible variations in soil temperature increase are primarily the result of the heterogeneous combination of land use and soil textures in the basin.

In alpine regions, the change in soil temperature clearly depends on the climatic conditions in relation to elevation. To some extent, the smaller warming effect is already present within the air temperature data. However, mainly the freezing of soil water during winter months reduces the soil temperature increase compared to air temperature. If the trends in mean soil temperatures simulated using different climate trends are compared, the four averaged climate variants of the *REMO regional* climate trend result in the greatest warming (see Fig. 71.1).

The increase in soil temperature within the 0–5 cm depth layer based on the *Baseline* climate variant of the *REMO regional* climate trend during the summer months (June, July and August; JJA) is shown in Map 71.3 – 3; it shows a relatively homogeneous warming effect throughout in the basin. Besides settlement areas, the high altitude areas in the Central Alps exhibit the most marked warming. A comparison of the results of the averaged climate variants of the four climate trends depicted in Fig. 71.2 reveals that the *REMO regional* climate trend results in a distinct temperature increase, but that this increase is surpassed by that based on the *MM5 regional* climate trend.

The most pronounced regional differentiation in temperature increase is shown for the winter months (December, January and February; DJF), illustrated in Map 71.3 – 4. In the lowlands and Alpine Valleys, the temperature increase in the top soil layer is around 3–4 °C, whereas the simulated increase in the alpine regions

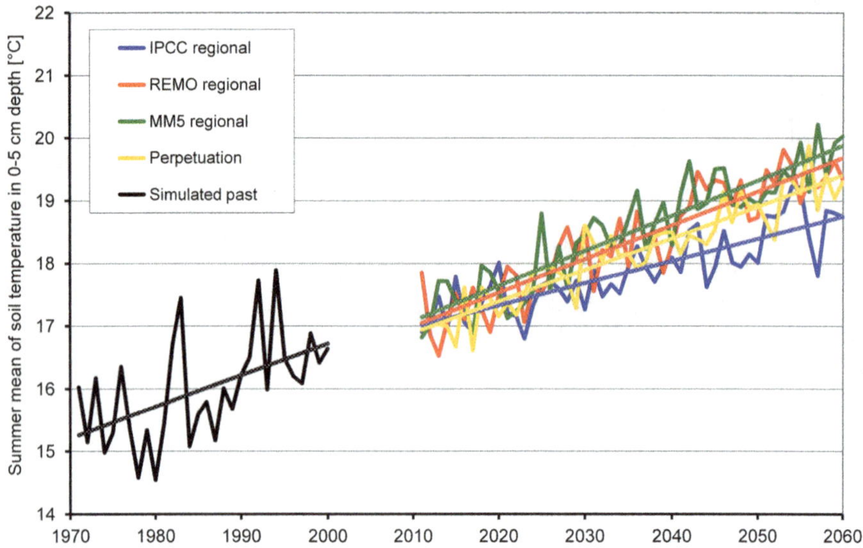

Fig. 71.2 Mean summer (JJA) temperature of the top soil layer for the reference period 1971–2000 and the *IPCC regional, REMO regional, MM5 regional* and *Extrapolation* climate trends

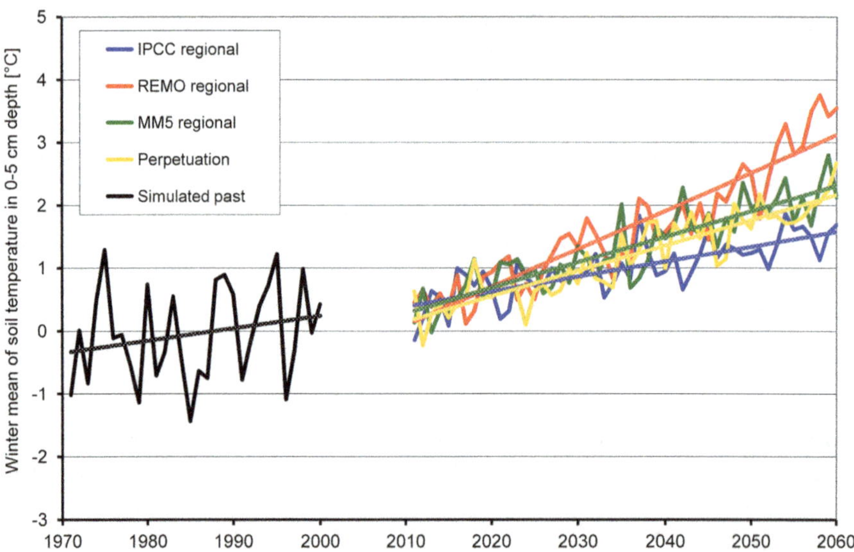

Fig. 71.3 Mean winter (DJF) temperature of the top soil layer for the reference period 1971–2000 and the *IPCC regional, REMO regional, MM5 regional* and *Extrapolation* climate trends

is smaller at 1 °C. The plausibility of this result is apparent through the evaluation of historical soil temperature trends: the summer warming of the top soil layer (see Fig. 71.2) is evidently more marked than the average annual trend (see Fig. 71.1). In contrast, the historical trend in winter soil temperatures is rather negligible and not significant (see Fig. 71.3). In the Upper Danube basin, a change in the air temperatures in winter leads to a change in the number of days with frozen ground. Only once the average air temperature in winter is well above 0 °C, as is the case in the GLOWA-Danube scenarios, does soil temperature increase as much as air temperature. For this reason, the winter soil temperatures in the lowlands show a distinct increase for the future time period, whereas in the alpine regions only the number of days with frozen ground decreases.

References

Bayard D, Stähli M, Parriaux A, Flühler H (2005) The influence of seasonally frozen soil on the snowmelt runoff at two Alpine sites in southern Switzerland. J Hydrol 309:66–84

Davidson EA, Janssens IA (2006) Temperature sensitivity of soil carbon decomposition and feedbacks to climate change. Nature 440:165–173

de Vries DA (1963) Thermal properties of soils. In: van Wjik WR (ed) Physics of plant environment. Wiley, Amsterdam, pp 210–235

Muerth M, Mauser W (2012) Rigorous evaluation of a soil heat transfer model for mesoscale climate change impact studies. Environ Model Software 35:149–162

Pregitzer KS, King JS, Burton AJ, Brown SE (2000) Responses of tree fine roots to temperature. New Phytol 147:105–115

Chapter 72
Effects of Climate Change on Nitrate Leaching

Tim G. Reichenau, Christian W. Klar, and Karl Schneider

Abstract Nitrate leaching has significant influence on groundwater quality. Climate and land-use change will affect its intensity and spatial distribution. This chapter concentrates on climate effects and Chap. 73 additionally deals with effects of land-use change.

Coupled simulations with the DANUBIA simulation system (plant growth, balances of carbon, nitrogen, water, energy) were performed for four districts in the Upper Danube catchment for the years 1996–2005 and 2011–2060. Meteorological drivers were generated using a statistical weather generator and trends from a regional climate model. Land-use and cultivation practice (timing, fertilisation) were kept constant.

Slightly increasing nitrate concentrations in the leachate were simulated during the scenario periods. District mean nitrate concentrations vary between 10 and 44 mg l^{-1}. Highest local concentrations were found for intensively used cropsites on organic soils. Maximum concentrations occur in years where reduced percolation more than compensated for the generally declining nitrate loads. This agrees with a reduction of the soil's mineral nitrogen content in the end of the growing season. However, some results for individual crops deviate from these general findings. Especially grassland shows an increase in nitrate loads due to a slightly decreased vegetation uptake, increasing the availability for leaching. This indicates the need to adjust the fertilisation scheme under climate change conditions.

The results show that climate changes alone will not lead to serious changes in nitrate leaching.

Keywords Nitrate leaching • Climate change • Plant growth • Nitrogen balance • DANUBIA • GLOWA-Danube • Agricultural area • Ecosystem model

T.G. Reichenau (✉) • K. Schneider
Institute of Geography, University of Cologne, Cologne, Germany
e-mail: tim.reichenau@uni-koeln.de; karl.schneider@uni-koeln.de

C.W. Klar
Forschungszentrum Jülich, Jülich, Germany
e-mail: c.klar@fz-juelich.de

W. Mauser, M. Prasch (eds.), *Regional Assessment of Global Change Impacts*, DOI 10.1007/978-3-319-16751-0_72

72.1 Introduction

Nitrate leaching is one of the most significant factors influencing groundwater quality. It is influenced both by natural (climate, pedologic) and by anthropogenic (cultivation) factors.

The temperature increase associated with climate change causes an intensification of the biogeochemical processes in the soil. This leads to an acceleration of the microbial turnover of nitrogen and hence also to a potentially greater nitrogen availability. Spatiotemporal changes in precipitation modify groundwater recharge, which in turn is closely associated with nitrate leaching. Humans exert a significant impact on the nitrogen balance as a consequence of choice of land-use and cultivation (e.g. fertilisation). In general, the following sequence of nitrate leaching potential exists for individual crop groups: row crops > cereals > grassland (Frede and Dabbert 1999). Furthermore, aspects of cultivation management such as quantities of fertilisers and the dates of sowing, harvest and fertilisations influence the soil nitrogen budget.

Both the consequences of climate change and the changed land-use that results from alterations in agricultural policy influence the future amount of nitrate leaching and its spatial distribution within the Upper Danube catchment.

Estimation of future nitrate leaching therefore requires a tool that can describe the mutual interactions of processes that determine the soil nitrogen balance in a dynamic and transdisciplinary manner. DANUBIA is particularly suitable for this, since it combines modelling of the natural nitrogen budget (see Chap. 38) and the management controlled by actors (see Chap. 40).

In this chapter, the influences of climate change on nitrate leaching are considered in isolation in order to separate changes caused by climate from effects due to agro-economic factors. In Chap. 73, the additional influence of agricultural policy on nitrogen fluxes is considered.

72.2 "Agriculture" Model Assembly

Depending on the given task, DANUBIA allows to combine a set of sectoral models to build the task-specific modelling system.

The models that comprise the "Agriculture" model assembly and their key interactions are shown in Fig. 72.1. An overview is provided in Chap. 70. Detailed descriptions of the plant growth and soil nitrogen models and validation information can be found in Lenz-Wiedemann et al. (2010) and in Klar et al. (2008).

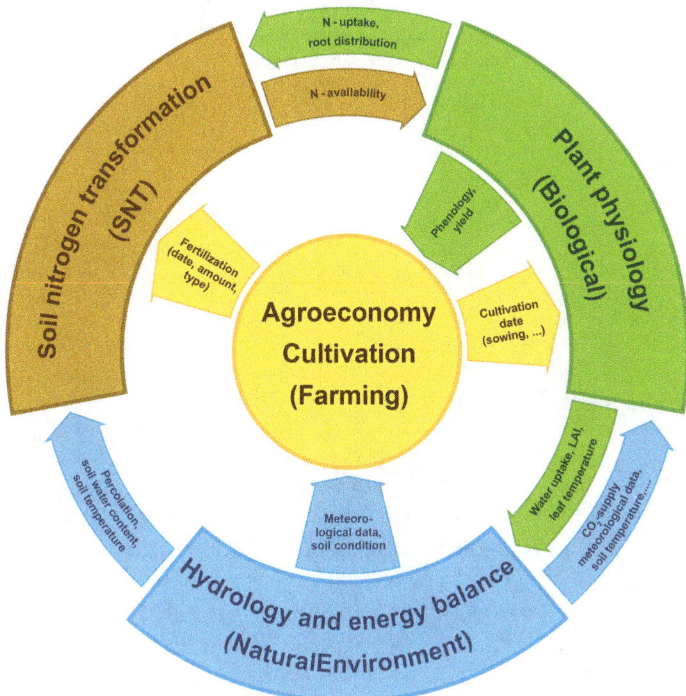

Fig. 72.1 Key interactions between the components of the "Agriculture" model assembly (Source: Reichenau et al. 2012)

72.3 Assumptions of the Scenarios

The scenario results presented in this chapter are based on a set of assumptions about the future climate as well as about the agro-political development. According to the specifications of scenario generation in GLOWA-Danube, the following climate scenario was used in the context of this study:

- *REMO regional* climate trend (see Chaps. 47 and 48)
- *Baseline* climate variant (see Chaps. 49 and 50)

The agricultural economics model ACRE which is a component of the "Agriculture" model assembly was deactivated in order to eliminate the influence of agricultural policy (see Chap. 40). Hence, changes to nitrate leaching in the scenario are exclusively the result of the effects of climate change. This scenario is referred to as the Climate-Only scenario in this chapter.

Climate change effects expressed as difference between a reference and a scenario period are presented exemplarily for four agricultural districts in Table 72.1.

Table 72.1 Change of climatic variables in the climate scenario *REMO regional–Baseline* in sample districts (2049–2058) compared to the reference period (1996–2005)

	Dingolfing	Ostallgäu	Landsberg	Günzburg
Air temperature [°C]	+2.9	+3.0	+2.8	+2.8
Precipitation [mm/a]	−81	−113	−222	+43
Precipitation summer[a] [mm]	−64	−99	−175	−3
Precipitation winter[b] [mm]	−17	−14	−47	+46
CO_2 [ppm]	+171	+171	+171	+171

[a]November 1–April 30
[b]May 1–October 30

72.4 Results

In accord with Chap. 70, the results are illustrated using the example of four districts. Because of the different regional conditions with respect to natural features (see Table 70.2) and agro-economic conditions (land-use, farm types), there are considerable differences in nitrate leaching.

Climate change effects on nitrate leaching are quantified by comparing results for the scenario period with the reference period. The reference period relates to the modelled means for 1996–2005, while the scenario data represent the means for 2049–2058.

The map presents the spatial distribution of nitrate leaching – defined as the nitrate concentration (mg l^{-1}) in the leachate from the vadose zone – for the four sample districts, using the *REMO regional–Baseline* climate scenario. The accompanying diagrams (a–d) depict the changes in nitrate concentrations in the leachate for individual crop groups as compared to the respective values for the reference period. In addition to nitrate concentrations, the change in nitrate load from the vadose soil zone is also shown.

In the *REMO regional–Baseline* climate scenario, the district mean nitrate concentration (see Fig. 72.2) varies between 10 and 44 mg l^{-1}. Higher values are few and far between. They occur in particular where intense cultivation is found on organic soils (e.g. Günzburg, see Map 72.1). There is a slight increase in nitrate load in the leachate over the entire period. In years with reduced percolation, maximum values may occur, such as those in Dingolfing (see Figs. 72.2 and 72.3). The reduced percolation leads to an increase in nitrate concentration despite the decline in the nitrate load for most districts and crop groups.

However, climate change effects upon nitrate leaching for individual crops may deviate from these general results. Winter cereals take up significantly more nitrogen as a consequence of the increase in biomass production caused by the increased temperatures and atmospheric CO_2 concentrations. The additional nitrogen demand is mostly covered by a greater mineralisation from the soil organic matter (approximately +40 %). The additional nitrogen uptake exceeds the quantity of additional nitrogen released. In the overall balance, this causes a reduction of the nitrogen

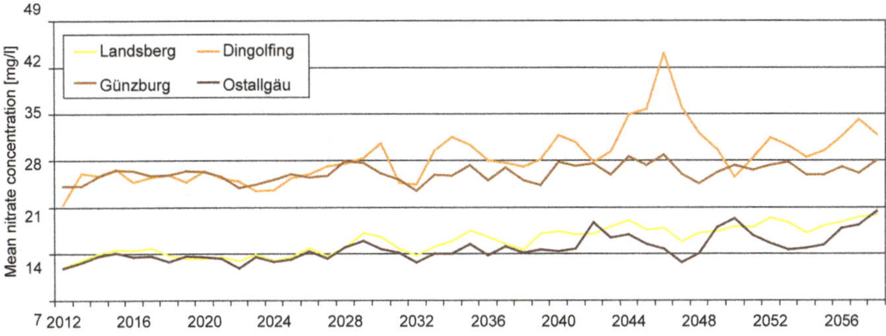

Fig. 72.2 Mean nitrate concentration in the leachate for the years 2012 until 2058 for the sample districts for the climate scenario *REMO regional–Baseline*

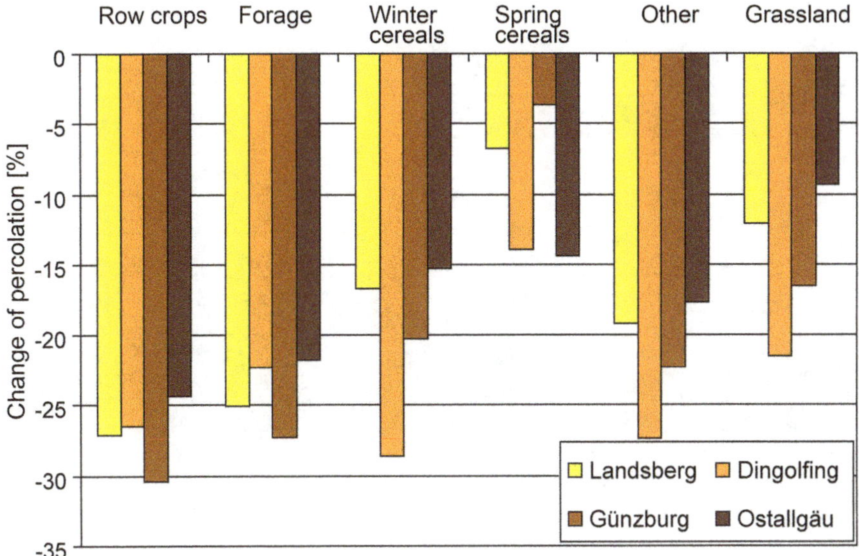

Fig. 72.3 Change of mean percolation in the scenario period (2049–2058) compared to the reference period (1996–2005)

load. Nevertheless, a decline in percolation by up to −30 % (see Fig. 72.3) causes the nitrate concentration to increase under winter cereals.

In contrast to winter cereals, grassland shows an increase in modelled nitrate load, except for Günzburg. This effect is especially notable in the grassland-dominated districts of Landsberg and Ostallgäu (see Map 72.1, Charts c and d). A slight decrease in nitrogen uptake was modelled for grassland. One likely cause for this result is that the fertilisation dates were chosen according to the current climate conditions. In the scenario, grassland develops faster, so that the period of

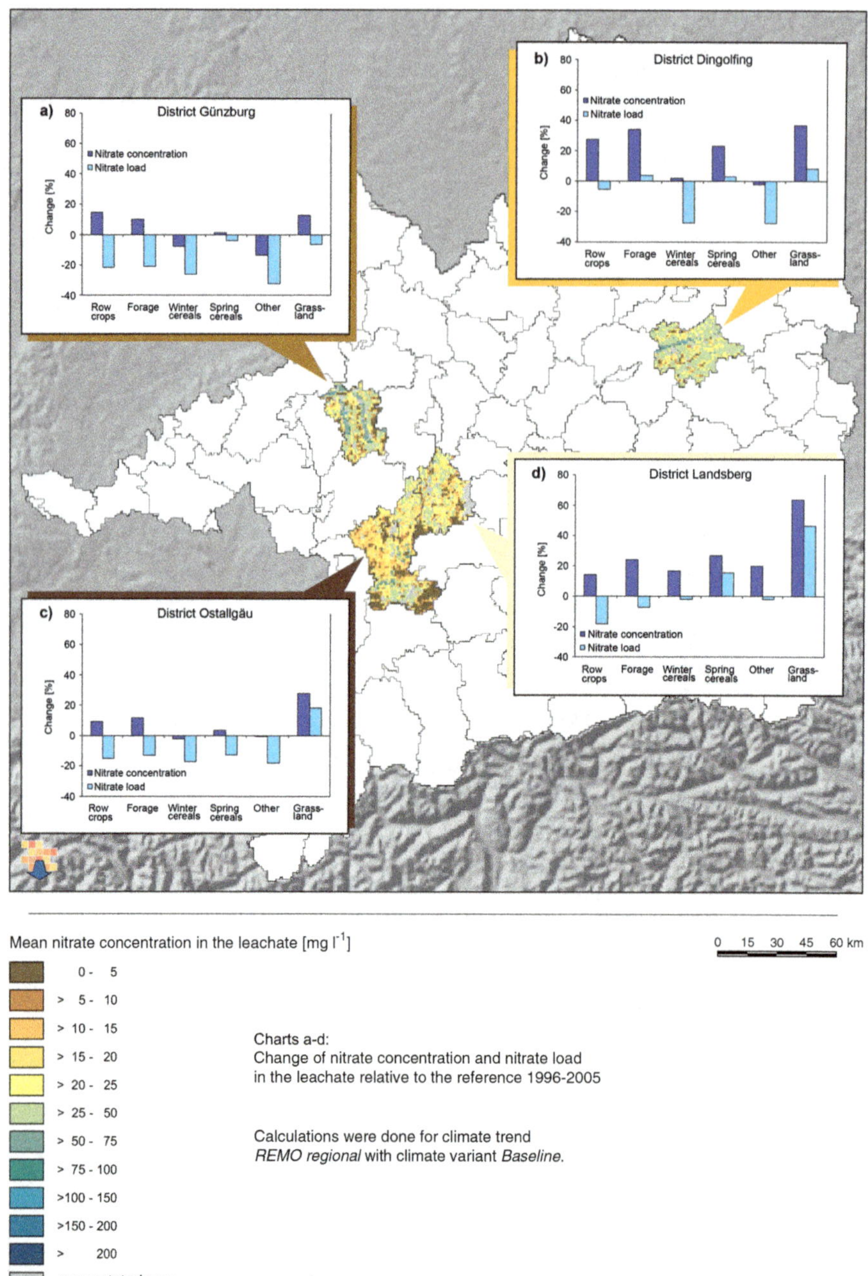

Map 72.1 Effects of future climate changes on nitrate leaching (*REMO regional* climate trend and *Baseline* climate variant)

nitrogen uptake is shorter. Thus, at the time of fertiliser application, the grass has already matured, and in this development stage, the nitrogen demand and hence nitrogen uptake are lower. In combination with the somewhat higher rates of mineralisation, this results in additional nitrate available for leaching. Moreover, the model results show a reduction of the mineral nitrogen content at the end of the growing season, which demonstrates the need to adjust the fertilisation scheme.

In general, the model results indicate that climate changes alone will not lead to serious changes in nitrate leaching. In terms of protection of water, appropriate changes in fertilisation management (e.g. adjusting fertilisation dates) can compensate for the consequences of climate change.

References

Frede H-G, Dabbert S (1999) Handbuch zum Gewässerschutz in der Landwirtschaft. Ecomed, Landsberg

Klar CW, Fiener P, Neuhaus P, Lenz-Wiedemann VIS, Schneider K (2008) Modelling of soil nitrogen dynamics within the decision support system DANUBIA. Ecol Model 217:181–196

Lenz-Wiedemann VIS, Klar CW, Schneider K (2010) Development and test of a crop growth model for application within a global change decision support system. Ecol Model 221:314–329

Reichenau TG, Krimly T, Schneider K (2012) Effects of dynamic agricultural decision making in an ecohydrological model. EGU General Assembly 2012, Vienna, Presentation

Chapter 73
Effects of Agro-Economic Decisions on Nitrate Leaching

Tim G. Reichenau, Christian W. Klar, and Karl Schneider

Abstract Nitrate leaching has significant influence on groundwater quality. Climate and land-use change will affect its future intensity and spatial distribution. This chapter concentrates on effects resulting from changes in land-use due to agro-economic decisions, and Chap. 72 deals with the effects of climate change.

Coupled simulations with the DANUBIA simulation system (plant growth, balances of carbon, nitrogen, water, energy, agro-economy) were performed for four districts in the Upper Danube catchment for the years 2011–2060. The agro-economic model which simulates land-use change based upon economic decision was run using two scenarios: (1) continuation of the status quo and (2) discontinuation of premium payments after 2013. Meteorological drivers and cultivation (timing, fertilisation) were kept constant. Results for the years 2049–2058 were compared to a model run with constant land-use.

Row crop and spring cereal area increases. The area of forage and miscellaneous crops decreases. The extent of winter cereals increases in districts dominated by arable land but decreases in grassland-dominated districts.

Despite the land-use changes, spatial patterns of nitrate leaching changed little. Critical nitrate concentrations (>75 mg l^{-1}) were calculated particularly in regions with intensive agriculture on organic soils.

The direction of changes in nitrate concentration and nitrate load are usually identical. Changes of nitrate leaching of max ±4 % occur in the grassland-dominated districts, while for the districts dominated by arable land, one shows an increase (<4 %) and the other a decrease (<9 %).

Compared to the effect of climate change, the investigated agro-economic scenarios exert only small effects on nitrate leaching.

Keywords Nitrate leaching • Land-use change • Plant growth • Nitrogen balance • DANUBIA • GLOWA-Danube • Agro-economy model • Agricultural area • Ecosystem model

T.G. Reichenau (✉) • K. Schneider
Institute of Geography, University of Cologne, Cologne, Germany
e-mail: tim.reichenau@uni-koeln.de; karl.schneider@uni-koeln.de

C.W. Klar
Forschungszentrum Jülich, Jülich, Germany
e-mail: c.klar@fz-juelich.de

W. Mauser, M. Prasch (eds.), *Regional Assessment of Global Change Impacts*, DOI 10.1007/978-3-319-16751-0_73

73.1 Introduction

Changes of the agro-political framework as well as changes in agricultural yields caused by climate change induce agro-economically motivated land-use changes. In this context the current chapter focuses on the resulting changes of nitrate leaching.

The analysis of land-use change (see Chap. 39) thus complements the study of the effects of climate change on nitrate leaching presented in Chap. 72 by investigating effects caused by agro-economic factors. On the one hand, climate-induced changes of agricultural yield exert an influence on the economic decisions by farmers about which crops to cultivate. On the other hand, these decisions are also influenced by the prevailing agro-political framework such as premium payments. The integrative effects of climate and policy thus cause region-specific changes to the cultivated areas of the various crops.

Agricultural decisions about cultivation influence the spatial and temporal patterns of nitrate leaching. The dynamic coupling of natural and socio-economic models and the interaction of natural processes with decisions by agricultural actors in DANUBIA allow to model the effects of land-use changes on nitrate leaching.

Especially in view of the implementation of the water quality targets specified in the EU Water Framework Directive (EU 2000), DANUBIA is a suitable tool for examining catchment-specific policy options for future sustainable water management.

73.2 Assumptions of the Scenarios

As in Chap. 72, the climate scenario applied consists of the *REMO regional* climate trend (see Chaps. 47 and 48) with the *Baseline* climate variant (see Chaps. 49 and 50). The scenario results presented in this chapter are also based on assumptions about the future trends in agricultural policy (see Chap. 70). Common to both agro-economic scenarios, the conditions specified in the 2003 EU Agricultural Policy Reform apply for the period from 2010 to 2014. The trend in the agricultural policy framework for the remaining period of the scenario is described by two different scenarios:

- *Baseline*: continuation of the status quo (see Table 70.1)
- *Performance*: discontinuation of all premium payments

73.3 Description of the Model

The "Agriculture" model assembly introduced in Chap. 72 was used. The distribution of agricultural areas to crop types was calculated based on agro-economic optimisation using ACRE (see Chaps. 39 and 40).

In the current study, it was assumed that the overall extent of the utilised agricultural area does not change. In addition, there is no transformation of grassland into arable land. Thus, land-use changes are possible only within the predefined arable area. The quantities of fertilisers for the individual crops are kept constant in both scenarios and are based on requirements to achieve the reference yields from 1995. The land-use changes affect the nitrogen conditions in the soil via crop rotation effects and hence also act on the nitrate leaching of subsequent crops.

73.4 Presentation of the Results

As in Chap. 70, the results relate to four sample districts and the period 2049–2058. In the current study, the results of both agro-economic scenarios are compared with the results of the scenarios without land-use change (*Climate-Only* scenario, see Chap. 72). This method allows the influence of agricultural policy on nitrate leaching to be evaluated.

Map 73.1 illustrates the spatial distribution of nitrate leaching expressed as nitrate concentration in the leachate in mg l^{-1}. The top portion of Map 73.1(1) depicts the *Baseline* scenario, while Map 73.1(2) on the bottom shows the results for the *Performance* scenario.

Despite the changed land-use, the maps indicate only little modifications in terms of spatial patterns of nitrate leaching compared to the *Climate-Only* scenario (see Map 72.1). Hardly any changes appear in the *Baseline* agro-economic scenario, whereas the *Performance* scenario shows an amplification of the pattern from the *Climate-Only* scenario, especially in the districts of Günzburg and Dingolfing which have large proportions of arable land. Critical nitrate concentrations (>75 mg l^{-1}) were calculated particularly in the regions with intensive agriculture on organic soils in the Günz and Mindel valleys.

The causes for the changes include the crop rotation effects mentioned in Sect. 73.3 as well as the varied shares of the individual crop groups as compared to the *Climate-Only* scenario. The changes in areas of the various crop groups relative to the total area of the district are shown in Fig. 73.1.

The proportional area cultivated with row crops and cereals mostly increases independent of the agro-political scenario, whereas a loss in the area of forage and other crops is projected. In both scenarios, the greatest increase of 2–4 % was calculated for spring cereals in Dingolfing. The extent of winter cereals increases in the districts dominated by arable farming (Günzburg, Dingolfing) but diminishes in the districts dominated by grassland (Landsberg, Ostallgäu).

Overall, the changes in the composition of the land-use at the level of the district are not great (see Fig. 73.1). However, this result should not hide the fact that considerable variation in the areas of individual crops was revealed. One example is the increasing extent of row crop cultivation in Landsberg in the *Performance* scenario: the increase of around 2 % as related to the district area is equivalent to a 74 % increase in the row crops area. This kind of change leads to an increase of nitrate

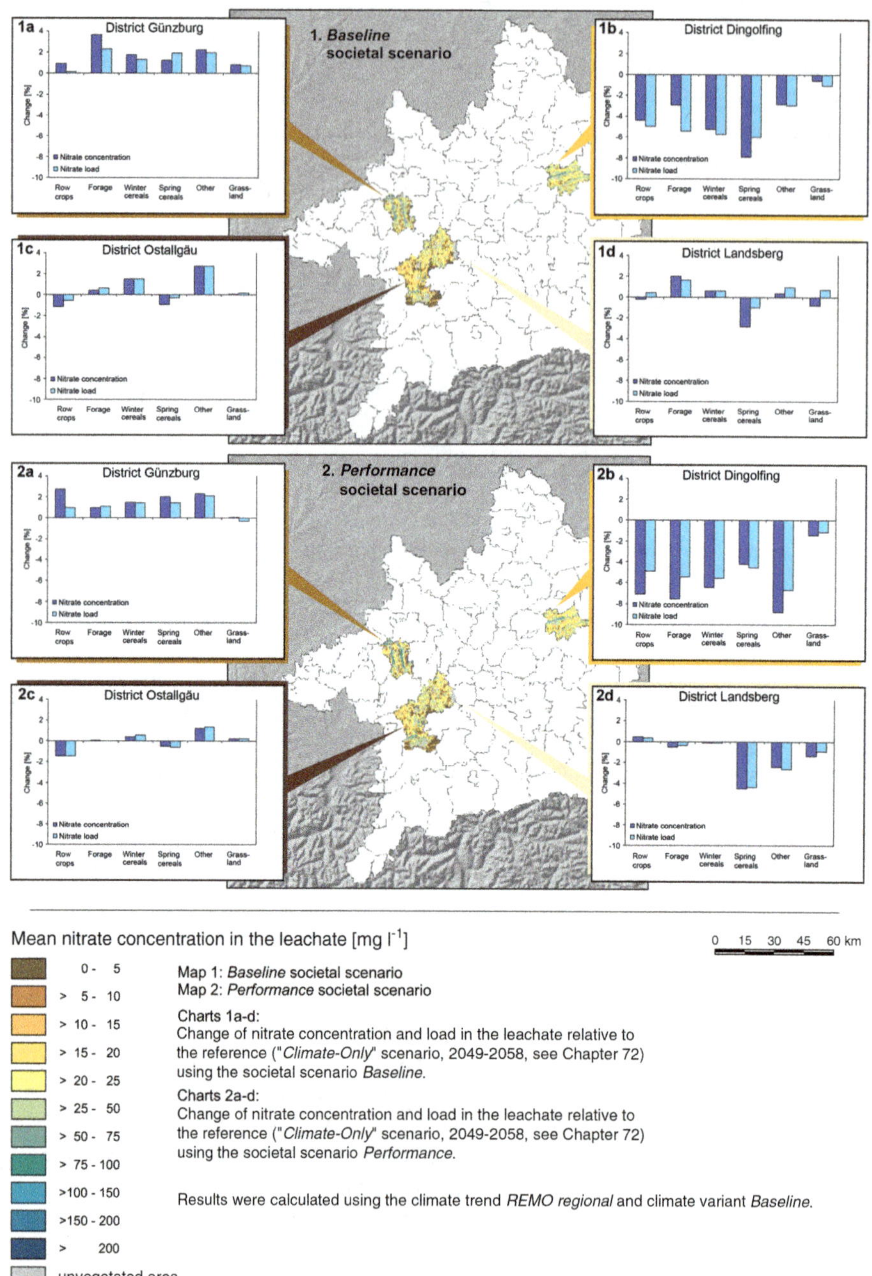

Map 73.1 Effects of agro-economic decisions on nitrate leaching (*REMO regional* climate trend and *Baseline* climate variant)

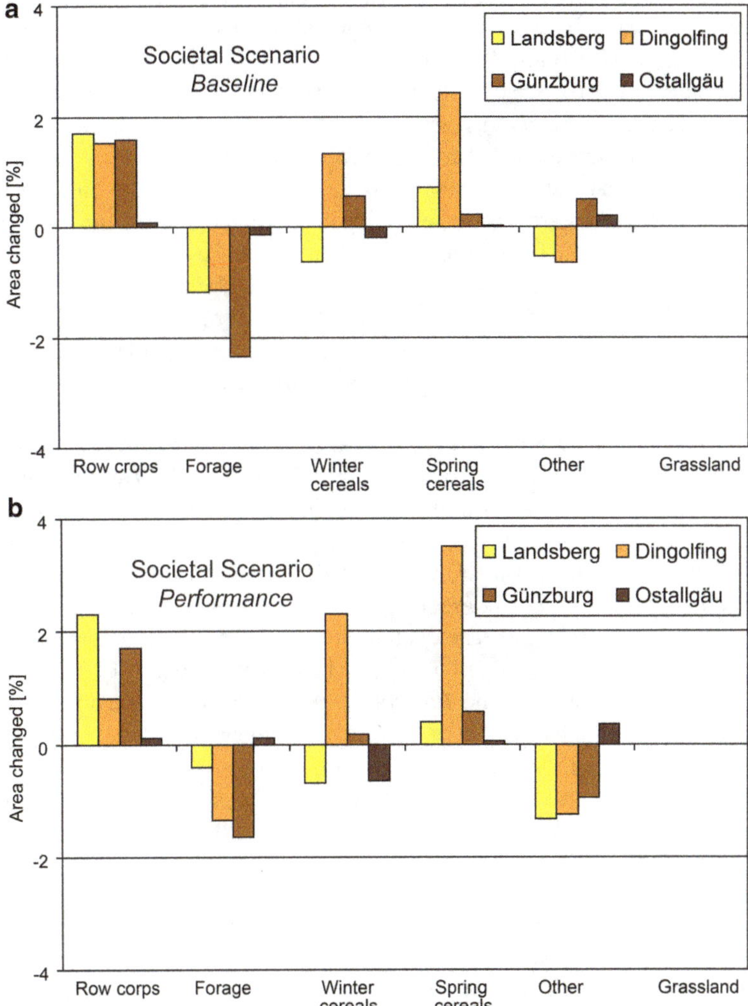

Fig. 73.1 Change of areal share of the individual crop groups in the area of each district compared to the *Climate-Only* scenario for the period 2049 until 2058 for the societal scenarios (**a**) *Baseline* and (**b**) *Performance*

leaching in areas where row crops replace crops with low nitrate leaching potential (e.g. winter cereals).

Charts 1a–d and 2a–d in Map 73.1 show the changes in nitrate concentration and nitrate load by crop group for the individual districts under the *REMO regional* climate trend and the *Baseline* climate variant. The smallest changes occur in the grassland-dominated districts of Landsberg and Ostallgäu, independent of the agro-economic scenario. For Landsberg only, the model results of the *Performance* scenario show a decline in nitrate concentrations and nitrate load by more than 4 %. In

Günzburg, an increase in nitrate concentrations and nitrate load for all crop types in both scenarios was calculated, whereas the model results for Dingolfing reveal consistently declining concentrations and loads.

Exemplarily, the patterns for Günzburg and Dingolfing are presented below. Differences between the two districts are due to the increase of the area cultivated with row crops at the expense of area with forage and other crops (e.g. rapeseed) in Günzburg. After harvesting of row crops, more nitrogen remains in the soils, in part because row crops are more heavily fertilised and more nitrogen makes its way from the harvest residuals into the soil. Due to the increase of row crop area, this results in an increased availability of nitrogen at the time of sowing of the subsequent crop. Hence, in Günzburg there is a higher nitrate load compared to the *Climate-Only* scenario. Because percolation is quite similar in all the scenarios that were examined, the results also reveal a higher nitrate concentration similar to the higher nitrogen load.

Particular in Dingolfing, there is an opposite trend. Here nitrate concentration and nitrate load decrease for all crops (by up to −9 and −7 %). The agro-economic optimisation of assigning areas to the individual crops leads to an expansion of the area of spring cereals (+141 %) at the expense of forage and other crops. In contrast to Günzburg, the nitrogen availability thereby decreases, since less nitrogen remains in the soil after harvest. As a consequence, there is a smaller nitrate load compared to the *Climate-Only* scenario. The decline in nitrate concentration is consequently explained by the similarity of percolation in the different scenarios.

In general, the results of the *Baseline* and *Performance* scenarios show only small differences in the effects of agricultural policy on nitrate leaching. The changes in land-use that result from agricultural policy and crop rotation effects result in no significant changes in nitrate leaching. The patterns of change in nitrate leaching are primarily determined by climate change; this result is evident from highly consistent results from both agro-economic scenarios.

Taken together, the analyses of the effects of the scenarios (in this chapter and in Chap. 72) suggest that neither climate change nor the agricultural policy scenarios presented here lead to a significant change in nitrate leaching. However, there may still be considerable localised risks for groundwater quality in the future as a result of nitrate leaching (see Maps 73.1(1 and 2)).

Against the background of the EU Water Framework Directive and the current nitrogen concentrations in groundwater which are somewhat too high already (LfU 2008), the slight trend to increasing nitrate leaching (see Fig. 72.2) is nonetheless an ongoing call for action in the districts that were examined.

In this respect, the "Agriculture" model assembly uses the coupled modelling of natural and socio-economic processes to provide information for assessing the change in nitrogen fluxes under changing climate and management conditions and can therefore contribute to the development of adapted agricultural management practices.

References

EU, Europäische Union (2000) RL 2000/60/EG des Europäischen Parlaments und des Rates vom 23.10.2000 zur Schaffung eines Ordnungsrahmens für Maßnahmen der Gemeinschaft im Bereich der Wasserpolitik. ABl. Nr. L372/19, pp 19–31

LfU, Bayerisches Landesamt für Umwelt (ed) (2008) Nitratbericht Bayern. Berichtsjahre 2000 bis 2004. LfU, Augsburg

Chapter 74
Climate-Related Forest Fire Risk

Judith Stagl, Monika Prasch, and Ruth Weidinger

Abstract An increase in the risk of forest fires in Central Europe is seen as a likely consequence of global warming. Therefore, timely planning of measures to adapt is necessary and requires the evaluation of specific hazards in affected regions. Knowledge about potential regional effects of climate change on the risk of forest fires is required to protect the forested regions in the Upper Danube basin. The forest fire index of Baumgartner, which is based on the forest fire statistics from Bavaria, was implemented in a forest fire module within DANUBIA. The future risk of forest fire was assessed in this study under the conditions of the *Remo regional* climate trend and the *Baseline* climate variant. The potential future development of the hydroclimatic conditions during March to September was evaluated using temperature, precipitation, potential evaporation and the climatic water balance. The model results indicate a significant increase in the forest fire risks within the Upper Danube basin as a consequence of climate warming. Under the assumbtions of the *Remo regional* climate trend and the *Baseline* climate variant, the number of days with "high" and "exceptionally high" forest fire risk more than doubled in the long-term average for the period from 2011 to 2060. Very dry years with extreme fire hazards may become more common. Such hazards were never reached in the past. For the Upper Danube area, an increase in days with significant forest fire risk may be expected especially for the summer months.

Keywords GLOWA-Danube • Upper Danube • Forest fire risk • Forest fire • Modelling • Climate change

J. Stagl (✉)
Potsdam Institute for Climate Impact Research, Potsdam, Germany
e-mail: stagl@pik-potsdam.de

M. Prasch • R. Weidinger
Department of Geography, Ludwig-Maximilians-Universität München (LMU Munich), Munich, Germany
e-mail: m.prasch@lmu.de; R.Weidinger@lmu.de

W. Mauser, M. Prasch (eds.), *Regional Assessment of Global Change Impacts*, DOI 10.1007/978-3-319-16751-0_74

74.1 Introduction

An increase in the risk of forest fires even in areas with previously low risk is seen as a likely consequence of climate warming in Central Europe (IPCC 2007). Therefore, timely planning of measures to adapt is necessary and requires the evaluation of specific hazards in affected regions. In light of the huge potential damage arising from forest fires, their precautionary prevention has great social relevance. Thus, knowledge about the regional effects of climate change on the risk of forest fires is required to protect the forested regions in the Upper Danube basin (approximately 32 % coniferous forest and 8 % deciduous and mixed forest). Extensive forested areas are mostly found in the Alpine region and the low mountain ranges (see Chap. D.3).

Generally, the formation of wildfires requires both direct sources of ignition and dispositive factors. The ignition potential of plants depends on the supply of water in and to the plant matter and is thus controlled by meteorological processes. Ignition itself is primarily caused by anthropogenic factors (mostly from arson and acts of negligence). At our latitudes, only 3–10 % of all forest fires arise as a result of lightning strikes (Badeck et al. 2003). Temperature and precipitation over the course of the year are the most important dispositive factors. Long periods without precipitation and concurrent high rates of evaporation associated with sunny weather and high temperatures lead to situations in which the litter layer and, in cases of low groundwater levels, the vegetation are very dry and hence susceptible to fire (Badeck et al. 2003). Moreover, the season and the forest structure, especially the composition of tree species and the stand coverage, are important factors. Young species up to 40 years and the generally drier pine stands of all age groups represent the forests with the greatest risks of fire (Wiese 2001). 40–60 % of all forest fires occur in the drier spring months (March to May), when the dried-out ground cover and vegetation are still not protected by fresh grass and herb growth (König 2007). The flammability of the forest ground cover significantly declines with the full development of the vegetation over the subsequent months.

For the assessment of forest fire risk and their prediction, respectively, the aim is to develop meteorological methods to assess the regional fire hazard. To respond to imminent forest fires with a preventative approach, mostly multiple day forecasts are applied. Various methods to assess the fire hazard have been used in different countries for many years (Alexander et al. 1996; Langholz and Schmidtmayer 1993; Lawson and Armitage 2008). The forest fire index of Baumgartner et al. (1967) has been used for the Upper Danube basin in this study; this method has also been the method of choice by the German Weather Service since the 1970s. This hazard key, which is based on the forest fire statistics from Bavaria for the years 1950–1959, provides the basis for the forest fire module implemented in DANUBIA; hence, this model can calculate the weather-related risk of a forest fire for each proxel. The future risk of forest fire was determined for this chapter under the conditions of the *REMO regional* climate trend and the *Baseline* climate variant (see Chaps. 47, 48,

49, 50, and 51). The development of the risk of forest fires under further GLOWA-Danube scenarios is presented in Chap. 75.

74.2 Description of the Forest Fire Module

Baumgartner et al. (1967) rely upon the fact that the susceptibility to ignition in a forest increases with the dryness of the fuel source as a consequence of the process of evaporation. The intensity of the evaporative processes in turn depends primarily on the incident solar radiation, wind speed, atmospheric humidity and the amount of precipitation. As a result, the current hazard for a forested area can be calculated from the combination of these meteorological variables for the previous days. In the forest fire module, potential evaporation is calculated according to the formula derived by Penman (1948). The forest fire risk is specified as the summing function of the difference of potential evaporation (*E*pot) and precipitation (*N*) for the last 5 days:

$$\text{Baumgartnerindex} = \sum_{i=1}^{5} E_{\text{pot}}(i) - N(i)$$

Based on their research, Baumgartner et al. (1967) determined statistically that the sum of E_{pot}-N over the last five days "sufficiently accurately" specifies one of five risk categories for the next day (see Fig. 74.1).

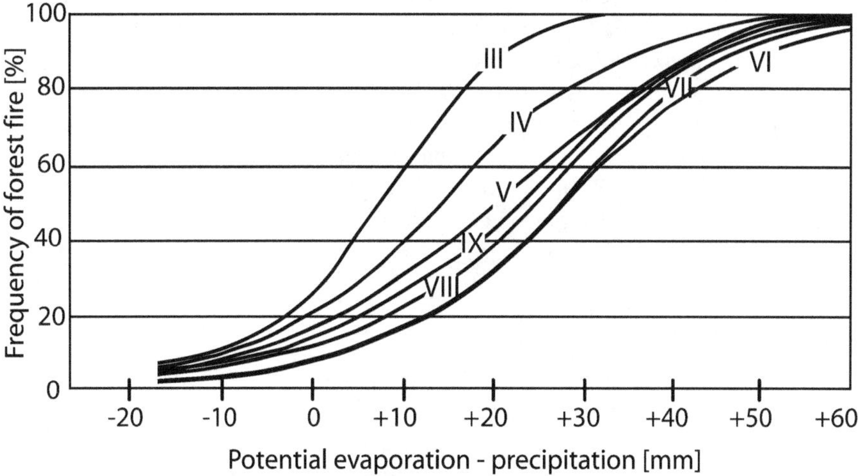

Fig. 74.1 Cumulative curve of frequency of the Baumgartner index (March–September, III to IX) of 1706 forest fires in Bavaria 1950–1959 (Baumgartner et al. 1967)

Table 74.1 Standard values (in mm) of the Baumgartner index as a measure for the forest fire risk of the following day in Bavaria (Baumgartner et al. 1967); risk category: *1* = very low, *2* = low, *3* = moderate, *4* = high, *5* = very high

Risk category	1	2	3	4	5
March	<–5	–5 to 2	3 to 8	9 to 14	>15
April	<–3	–3 to 7	8 to 15	16 to 26	>27
May	<3	3 to 15	16 to 24	25 to 34	>35
June	<12	12 to 23	24 to 31	32 to 40	>41
July	<12	12 to 23	24 to 30	31 to 39	>40
August	<8	8 to 19	20 to 27	28 to 36	>37
September	<8	6 to 17	18 to 25	26 to 34	>35

The standard values for evaluating the local forest fire risk using this method are listed in Table 74.1. A classification by month takes into account the seasonal distribution of forest fires; hence, in the early spring months, the same level of risk is reached even with a lower water deficit compared to in late summer.

In addition to precipitation sums and potential evaporation, the forest fire module calculates the meteorological risk of forest fires for each day.

The model results were compared to the results from the German Weather Office (Deutschen Wetterdienstes = DWD) for 2002 and 2003 in order to verify the forest fire module. The year 2002 was chosen as a comparatively average climatic year. In contrast, 2003 represents an exceptionally hot and dry year with particularly high values for forest fire risks. The evaluation was performed both visually and using a contingency matrix and a weighted kappa coefficient $0 \leq \kappa \leq 1$ (Grouven et al. 2007) as a measure of the agreement between the two models.

The same risk value was calculated by both models for 2002 in 76.5 % of cases for 11 climate stations (2003: 61.1 %). The DWD model calculated a risk level that was higher in 11.4 % of cases (2003: 23.3 %) and lower in 10.1 % of cases (2003: 8.6 %). The kappa coefficient for 2002 was $\kappa = 0.84$ and for 2003 was $\kappa = 0.52$. The models are thus highly consistent for 2002 and moderately so for 2003.

The deviations for 2003 show a tendency for higher risk assessment in the DWD model. Differing methods for calculating the potential evaporation underlie these deviations to some extent. In the forest fire module, the Penman method was used, whereas the DWD selected the method derived by Haude.

74.3 Results

The results of the forest fire module are presented as regional mean values and specialised as maps for the past and a future scenario. The assessment of the future forest fire risk is based on the *REMO regional–Baseline* climate scenario. Only the months from March to September were analysed, since according to Baumgartner et al. (1967), there is a real risk of forest fires only during the growing season. The conditions in spring (March to May) and summer (June to August) are evaluated separately. A second focus was set on the climatic water balance (the difference between precipitation and potential evaporation), which is considered as a measure of the potential water yield in a region over the long-term average.

The regional mean values were analysed for the periods 1960–2005 and 2011–2060 and the modelled data by area were analysed for three decades (1991–2000, 2021–2030 and 2051–2060). The years 1991–2000 represent the present conditions in the Upper Danube basin. Hence, this period served as the reference decade. The decade 2021–2030 was chosen to estimate the trend in the near future, and the decade 2051–2060 should reveal the conditions that might dominate in around 40–50 years. The effects of climate change are clearly apparent by this time. The hydroclimatic conditions and their changes within the basin are presented first, and then the forest fire risk within the Upper Danube is described and compared among the decades.

74.3.1 Hydroclimatic Change

A potential future development of the hydroclimatic situation (March to September) is illustrated using temperature, precipitation and potential evaporation as well as the climatic water balance.

Temperature Figure 74.2 shows the trend in mean temperatures within the growing season for 1960–2060. The consistent increase by 4 °C over the course of these 100 years is clearly evident. The highest values to date were recorded in the year 2003 with a mean of 12.0 °C. Such extreme years are likely to occur at greater frequencies over the next 30 years. Although the mean temperature increased by 0.38 °C per

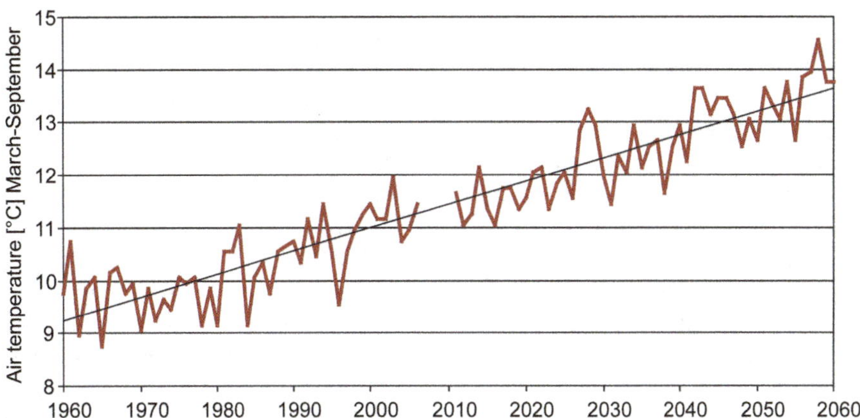

Fig. 74.2 Mean air temperature March–September for the periods 1960–2005 and 2011–2060 in the Upper Danube basin

decade in the past, the increase per decade under scenario conditions is 0.52 °C, and hence a mean daily air temperature of 14.5 °C is projected.

Precipitation Compared to the period 1991–2000, a decline in precipitation during the growing season by approximately 10 % is simulated for the future up to 2060 (see Map 74.1 – 1a–c): the means decrease from 785 to 700 mm. The driest regions are in the Nördlinger Ries district, the Upper Palatinate, in the region south of Eichstätt and in the Dungau district south of the Bavarian Forest between Regensburg and Straubing. The highest values can be seen in the high-altitude regions of the Alps, where the decrease in precipitation is the most marked (see also Chap. 53).

The precipitation sums fluctuate widely from year to year. In 2002, there was 949 mm of precipitation, but in the drier 2003, only 593 mm, which is less by a third. In the climate scenario period (2011–2060), a mean of only 520 mm precipitation sum was calculated for the growing season.

Evaporation The increasing sums for potential evaporation are closely associated with the increasing temperatures (see Figs. 74.2 and 74.3 and Map 74.1 – 2a–c). During the period 1960–2005, there was an increase in the evaporation sums of 11 mm per decade. The trend calculated for the future (2011–2060) is more than twice as high, at 23 mm per decade. The highest measured value was in 2003 (598 mm). Future annual evaporation sums of up to 656 mm are even possible (see also Chaps. 53 and 58).

Map 74.1 – 2a–c indicates that the potential evaporation is highly correlated with elevation: the lowest values are from the high altitudes in the Alps, whereas high values are found in the Hallertau, Dungau and Inn valley regions. Values above 600 mm, which are seen today only at few locations, are expected across wide areas of the lowlands. Evaporation will increase even at higher elevations. As a result, rates of evaporation will no longer differ in central uplands compared to the northern borders of the Alps.

Climatic Water Balance A significant decrease in the potential water available during the growing season is evident for the Upper Danube basin (see Fig. 74.3 and Map 74.1 – 3a–c). This result is the consequence of decreasing precipitation and increasing rates of evaporation. The measurement data from the past (1960–2005) indicate a decrease of 7.3 mm per decade. This trend is projected to strengthen in the period from 2011 to 2060: the average decrease reaches 45 mm per decade. The conditions in 2003 provide a glimpse of what future conditions could be like: that year, a negative water balance was reached for the first time (−4.2 mm). Figure 74.3 indicates that a water deficit with maximum of up to −116 mm may arise more frequently in the future, especially after the second third of the twenty-first century.

Map 74.1 – 3a–c shows the distribution of the potential water yield by forested area (see Chap. 9). The water deficits that prevail in the northern parts of the Upper Danube basin decrease towards the south. The forested regions of the low mountain ranges have values up to 600 mm in the high-altitude regions of the Bavarian Forest; these values represent a smaller water excess compared to the Alpine region, where

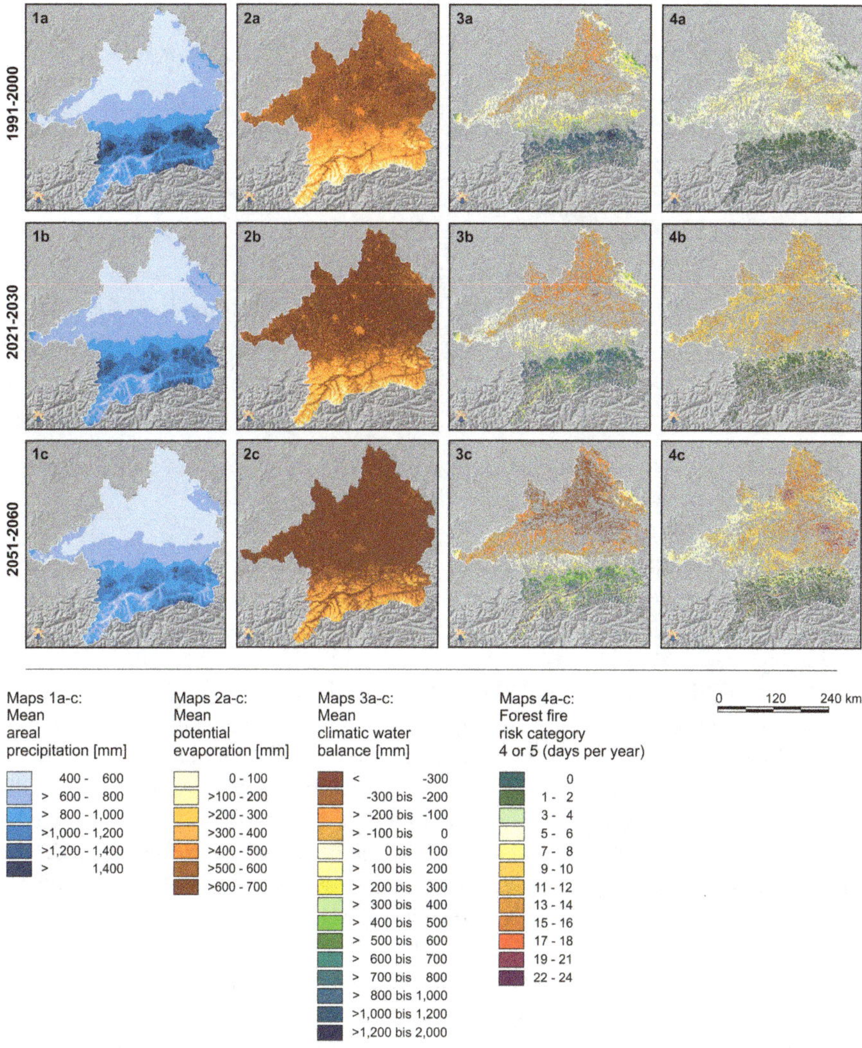

Calculated under the *REMO regional* climate trend and the *Baseline* climate variant (months: March-September).

Map 74.1 Climate-related forest fire risk (*REMO regional* climate trend and *Baseline* climate variant)

maximum values are found around the northern Alpine fringe (up to 1,827 mm). Along the central alpine dry valleys, some regions have negative values.

In the decade from 2021 to 2030, there may be significantly less water available for the forests compared to today. The potential water yield is reduced by 34 % compared to 1991–2000. Regions with the greatest water deficits include Lower Bavaria and especially the districts of Dungau and Upper and Middle Franconia. In

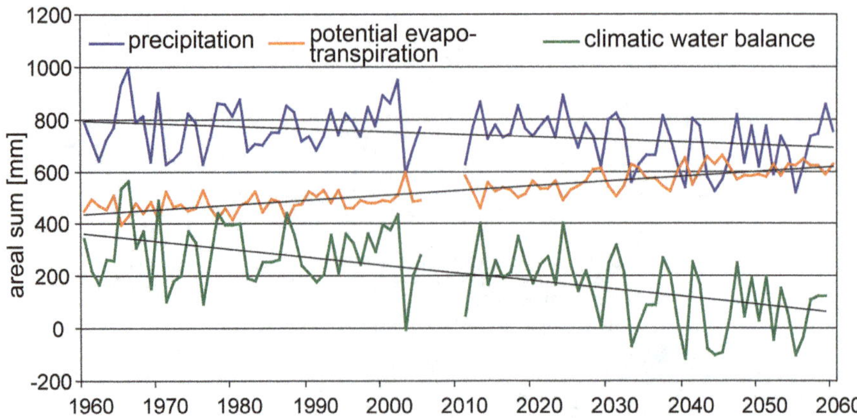

Fig. 74.3 Precipitation, potential evapotranspiration and climatic water balance in 1960–2005 and 2011–2060 in the Upper Danube basin

contrast, there is still a water excess (>100 mm) in the high-altitude regions of the low mountain ranges (the Bavarian Forest, the Swabian Jura) and in the south of Upper Bavaria, Swabia and in the Alps, although there are lower maxima in the high alpine regions.

On average during the period 2051–2060, only 45 mm of water is projected to be potentially available during the growing season. This is equivalent to a decrease of 83 % compared to 1991–2000. A marked water deficit dominates across all of Lower Bavaria, the Upper Palatinate and in Upper and Middle Franconia. In addition, lower values arise across the Alpine foothills, except in the highest altitudes of the low mountain ranges. Regions with a potential water deficit also include the river valleys of the Alps in the south-western part of the basin. The water surpluses of the Alps decline by more than 500 mm compared to 1991–2000. Even the regional maxima in the high alpine regions decrease by 36 % compared to 1991–2000.

The precipitation and evaporation sums, as well as the climatic water balance in the spring months (March to May; see Fig. 74.4), indicate only negligible changes over the 100 years considered. Increasingly drier springs are predicted for the future, although the climatic water balance shows positive values, with the exception of a few extraordinary dry years with low precipitation sums.

Figure 74.5 reveals a slight decrease in precipitation in summer for the years 1960–2005. There is a concurrent increase in evaporation sums. As a result, the water balance is increasingly negative.

A sharp decline in precipitation is calculated for the future under the assumed scenario. Along with a further increase in evaporation, the result will be a marked decline in the potential water that is available. After 2030, a positive climatic water balance is calculated only in exceptional cases. Already within 20 years under the scenario conditions, warm and dry summers like that experienced in 2003 will be the norm.

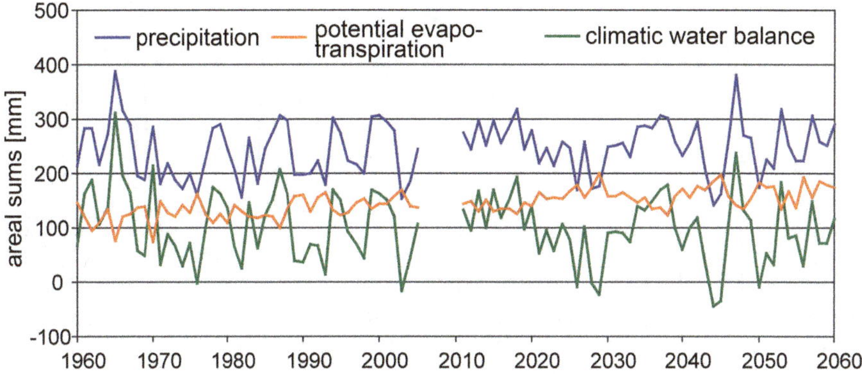

Fig. 74.4 Precipitation, potential evapotranspiration and climatic water balance in spring (March–May) 1960–2005 and 2011–2060 in the Upper Danube basin

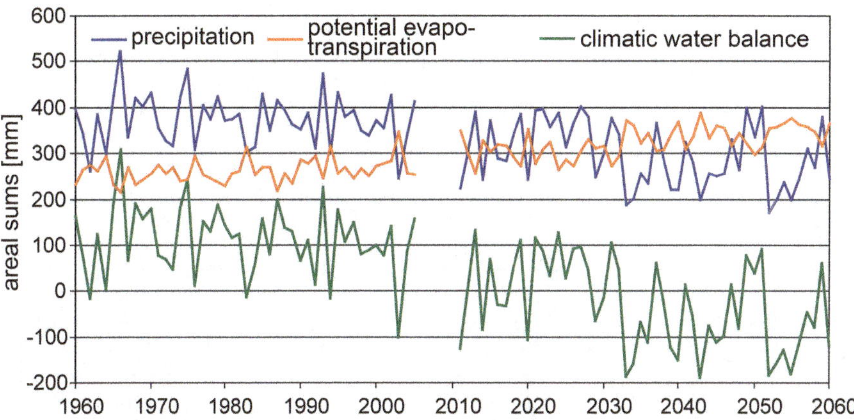

Fig. 74.5 Precipitation, potential evapotranspiration and climatic water balance in summer (May to August) 1960–2005 and 2011–2060 in the Upper Danube basin

74.3.2 Forest Fire Risk

This analysis included all days with a risk of forest fire at level four ("high risk") or five ("exceptionally high risk"). Figure 74.6 shows the regional mean values for the number of such days per year in the Upper Danube basin. There is no significant trend for the past (1960–2005), with the exception of the extraordinary year 2003; the variability was 8.4 days. However, events such as in those that occurred in 2003 are projected to be more frequent for the future. The future higher rates of evaporation and lower precipitation totals result in increasing drought conditions, which in turn are associated with higher risk of forest fires related to climatic conditions.

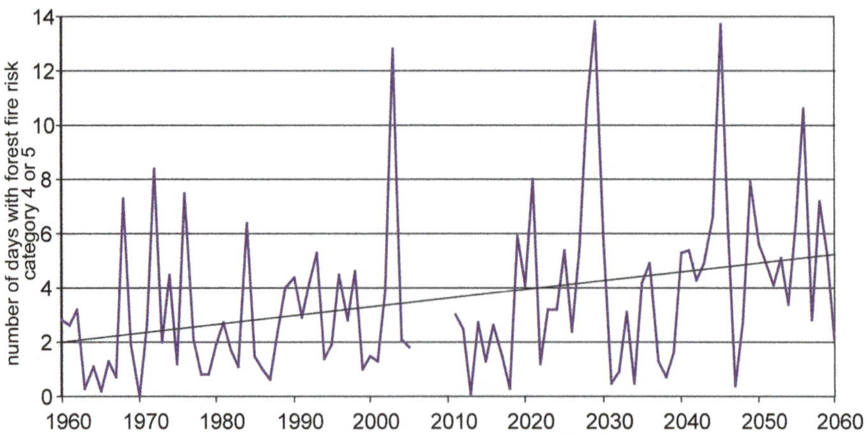

Fig. 74.6 Number of days with forest fire risk category 4 or 5, after Baumgartner et al. (1967), 1960–2005 and 2011–2060 in the Upper Danube basin

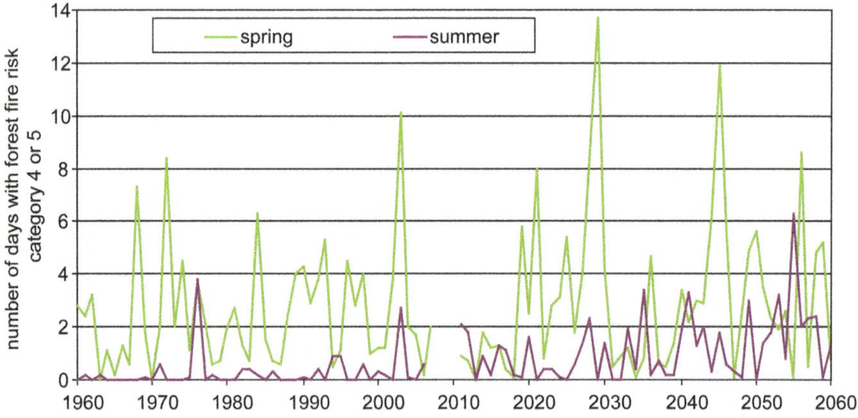

Fig. 74.7 Areal mean number of days in spring (Mar–Apr) and summer (Jun–Aug) with a forest fire risk category 4 or 5, after Baumgartner et al. (1967), 1960–2005 and 2011–2060 in the Upper Danube basin

In the scenario considered, the spring months (March to May) still have the highest forest fire risk. Figure 74.7 reveals a total number of forest fire hazard days for the future never seen in the past (up to 13.7). Figure 74.7 shows a low fire hazard for the summer months in the years 1960–2005. In the future (2011–2060), it is apparent that the forest fire risk in the scenario period generally increases faster in summer than in spring.

For the decade 1991–2000 (see Map 74.1 – 4a), there are a negligible number of days (5–8) with severe forest fire risk over the entire basin. For regions with especially high rates of evaporation, the fire hazard is classed slightly higher (9–11 days). In the low mountain ranges and the Alpine region, there is a lower forest fire risk as a result of lower evaporation rates and higher total precipitations. One exception is the central alpine dry valleys, where there is a higher forest fire risk, primarily in the Inn valley, with a maximum of 14.5 days per year.

Compared to the reference decade, the decade 2021–2030 (see Map 74.1 – 4b) shows an increase in the number of forest fire days by 84 % over the whole basin, except in the high-altitude regions of the low mountain ranges. The main reason is the gradual increase in potential evaporation. Precipitation totals remain largely unchanged in the lowlands. Mostly, the regions in Franconia and the Upper Palatinate are affected by a slight decline in precipitation sums. A negative climatic water balance is calculated for the northern portion of the drainage basin and in the Tertiary Hills. The greatest decrease in precipitation takes place in the high alpine regions, but this area remains quite moist as a result of low evaporation rates. The entire Alpine region, with the exception of the central alpine dry valleys, is also with low risk of forest fires in the future. However, primarily in the Inn valley, the risks that are already quite high today may double. The most notable increase, at 150 %, was calculated for the number of days with greater and unusually higher risk of forest fires in the Alpine foothills, especially in the Fünfseenland region of Bavaria. The highest forest fire risks (13–17 days) were calculated for the Upper Palatinate, Franconia and large areas of Lower Bavaria, with increases of 100 %.

The scenario period 2051–2060 also shows an increase in the forest fire risk by 86 % compared to the reference decade. However, there is a rather different picture for the region north of the Alps. With only a negligible increase in forest fire risk, the western part of the drainage basin (Swabia, the Swabian Jura) has little risk (<8 days) – in some areas there is even a slight decrease. The forest fire risk may double in the Upper Palatinate, Central Franconia, the Tertiary Hills and in eastern Upper Bavaria and the Bavarian Fünfseenland region (10–13 days). In some regions of Central Franconia, Lower Bavaria and the Upper Palatinate, the risk even triples (16–23 days). In the high-altitude regions of the low mountain ranges and in the Alps, the forest fire risk is projected to remain low (0–5 days) despite an ongoing slight increase. Although rates of evaporation increase and precipitation totals decrease, the water balance in these regions continues to be distinctly positive. In the central alpine dry valleys, the forest fire risk is quite high (up to 23 days; an increase of up to 150 %); this result is quite likely the consequence of the highest evaporation totals in the basin.

74.4 Conclusion

The model results indicate significant changes in the forest fire risks within the Upper Danube basin as a consequence of climate warming. Under the assumptions of the *REMO regional* climate trend and the *Baseline* climate variant, the forest fire risk more than doubles in the long-term average for the period from 2011 to 2060. Very dry years with total days with extreme fire hazards may become more common; such totals were never reached in the past. An increase in forest fire risk is expected especially for the summer months. The comparison to the situation today indicates a tendency for elevated forest fire risks associated with climate both in regions that are already quite dry today and in moist areas.

References

Alexander ME, Stocks BJ, Lawson BD (1996) The Canadian forest fire danger rating system. Initial Attack 1996(Spring):6–9

Badeck F-W, Lasch P, Hauf Y, Rock J, Suckow F, Thornicke K (2003) Steigendes klimatisches Waldbrandrisiko. AFZ/Der Wald 59(2):2–4

Baumgartner A, Klemmer L, Raschke E, Waldmann G (1967) Waldbrände in Bayern 1950–1959. Mitteilungen der Staatsverwaltung Bayern 36, München

Grouven U, Bender R, Ziegler A, Lange S (2007) Der Kappa-Koeffizient – Artikel Nr. 23 der Statistik-Serie in der DMW. In: Deutsche Medizinische Wochenschrift 2007/132, pp 65–68

IPCC – Intergovernmental panel on climate change, WMO/UNEP (ed) (2007) Klimaänderung 2007: Zusammenfassungen für politische Entscheidungsträger. Vierter Sachstandsbericht des IPCC (AR4), Bern/Wien/Berlin

König HC (2007): Waldbrandschutz. Kompendium für Forst und Feuerwehr. Edition Gefahrenabwehr; Supplement (Band 1), Berlin

Langholz H, Schmidtmayer E (1993) Meteorologische Verfahren zur Abschätzung des Waldbrandrisikos. Allgemeine Forstzeitschrift 48(8):94–396

Lawson BD, Armitage OB (2008) Weather Guide for the Canadian Forest Fire Danger Rating System. Natural Resources Canada, Canadian Forest Service, Northern Forestry Centre, Edmonton, Alberta

Penman HL (1948) Natural evaporation from open water, bare soil and grass. Proc Roy Soc Lond A 194:120–145

Wiese A (2001) Waldbrandversicherung als Ergänzung der Waldbrandprävention. In: AFZ/Der Wald, 11/2001, pp 568–569

Chapter 75
Forest Fire Risk Under Various Climate Trends

Judith Stagl

Abstract An increase in the risk of forest fires in Central Europe is seen as a likely consequence of global warming. Therefore, timely planning of adaptation measures is necessary and requires the evaluation of specific hazards in affected regions. Knowledge about potential regional effects of climate change on the risk of forest fires is required to protect the forested regions in the Upper Danube basin. The forest fire index of Baumgartner, which is based on the forest fire statistics from Bavaria, was implemented in a forest fire module within DANUBIA (see Chap. 74). This study evaluates the development of the climate-related risk of forest fires under 16 GLOWA-Danube scenarios for the period 2011–2060. Additionally the spatial distributions of the forest fire risk for four different climate trends based on the baseline climate variant were evaluated in more detail to detect long-term climate-related changes (2011–2035, 2036–2060). All climate scenario calculations indicate an increase in days with "high" and "exceptionally high" forest fire risk in the Upper Danube basin, but with different values in the extrema. On average across the region and over all 16 scenarios, 30 % more days with high forest fire risk are projected in the summer months during the period 2036–2060. Spatially similar pattern occur under the climate scenarios *IPCC regional, REMO regional* and *MM5 regional*. In contrast under the climate trend *Perpetuation*, the number of days with significant high forest fire risk is projected as almost twice as high compared to the other scenarios.

Keywords GLOWA-Danube • Upper Danube • Forest fire risk • Forest fire • Modelling • Climate change

75.1 Introduction

In order to be able to make statements about the extent of climate-related changes to the risk of forest fires for the Upper Danube basin, different climate change scenarios were analysed for the trends in forest fire risk associated with climate. The modelling undertaken in Chap. 74 based on the *REMO regional–Baseline* climate

J. Stagl (✉)
Potsdam Institute for Climate Impact Research, Potsdam, Germany
e-mail: stagl@pik-potsdam.de

W. Mauser, M. Prasch (eds.), *Regional Assessment of Global Change Impacts*, DOI 10.1007/978-3-319-16751-0_75

scenario was extended to include other GLOWA-Danube climate scenarios. The model approach for the forest fire module described in Chap. 74 remains the same.

75.2 Data Processing

The climate-related risk of forest fire in the Upper Danube was modelled and analysed with 16 GLOWA-Danube climate scenarios for the period 2011–2060. A detailed description of the scenarios can be found in Chaps. 47, 48, 49 and 50. The analysis of the outputs from these runs was at first evaluated by area means. In addition, the results were averaged for the four different variants in each climate trend to depict a mean temporal pattern. As a second step, the spatial distributions of the forest fire risks calculated for the four different climate trends based on the *Baseline* climate variant were evaluated and compared. The periods 2011–2035 and 2036–2060 are presented separately to detect the long-term climate-related changes. The mean number of days per year on which a significant "forest fire weather hazard" was classified was chosen as the unit for analysis; these classifications include a category of four ("high forest fire risk") or five ("exceptionally high forest fire risk") (see Chap. 74).

75.3 Results

The temporal trend in the forest fire risk during the growing season (March–September) is depicted in Fig. 75.1. The ensemble of 16 scenario calculations covers a range of possible trends. All the climate scenarios that were modelled indicate an increase in the climate-related forest fire risk in the Upper Danube basin but differ in the extreme values. The results for all climate trends also reveal years with only a small number of days with "forest fire weather hazard". Thus, the range in the number of days with forest fire risk significantly increases from 0–12.8 days (1960–2005) to 0–17.7 days (2011–2060).

The four statistical climate variants were averaged for each climate trend for Fig. 75.2. Over the long-term average, the *Extrapolation* climate trend shows an increase by 1.4 days per decade and hence the steepest trend. The reason for this result is primarily the very sharp decline in precipitation during the growing period that is assumed for this trend (see Chaps. 47 and 48). The forest fire risk increases by 0.34–.05 days per decade over the long-term average for the *IPCC regional, REMO regional* and *MM5 regional* climate trends, which is considerably less steep. The lowest increase in forest fire risk is calculated using the *IPCC regional* climate trend.

A detailed analysis of the four climate trends based on the *Baseline* climate variant (see Chaps. 47, 48, 49 and 50) was carried out to evaluate the spatial distribution of forest fire risk in the Upper Danube. The maps provide an idea of how the forest fire risk might change due to climate change. Map 75.1 – 1a–2d illustrate the mean

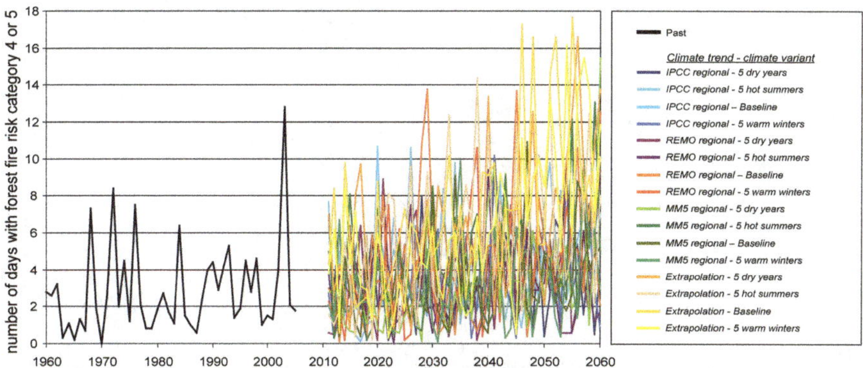

Fig. 75.1 Number of days with a forest fire risk category 4 or 5 (After Baumgartner et al. (1967), for the past 1960–2005 and for 16 statistical GLOWA-Danube climate variants 2011–2060)

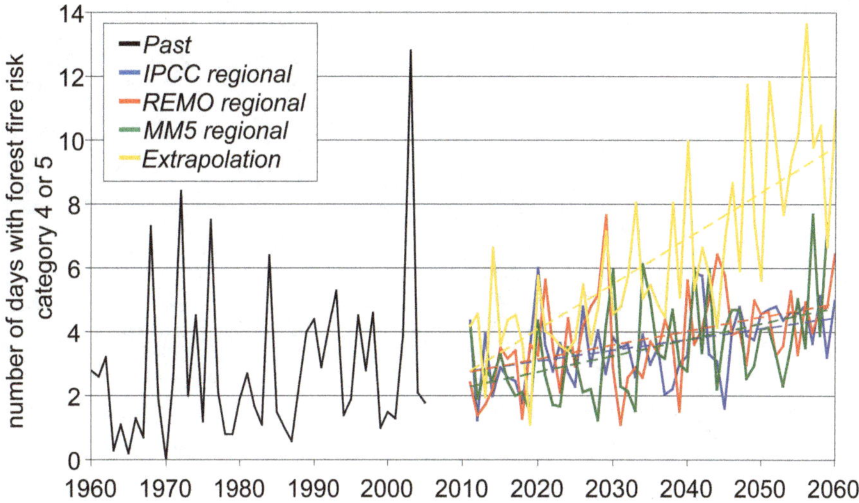

Fig. 75.2 Number of days with a forest fire risk category 4 or 5 (After Baumgartner et al. (1967) 1960–2005 and 2011–2060 averaged over the four climate variants for each climate trend (*dashed lines* represent trend lines))

number of days at forest fire risk level 4 or 5 over the whole growing season (March to September) for the periods 2011–2035 and 2036–2060. Map 74.1 – 4a from Chap. 74, which presents the historical forest fire risk (1991–2000), can serve for comparison. Map 75.1 – 3a–4d illustrate the seasonal forest fire risk in spring and summer for 2036–2060. The spring months from March to May generally have the highest forest fire risks (see Chap. 74). However, in the context of climate warming, the higher temperatures in the summer months and a decline in summer precipitation

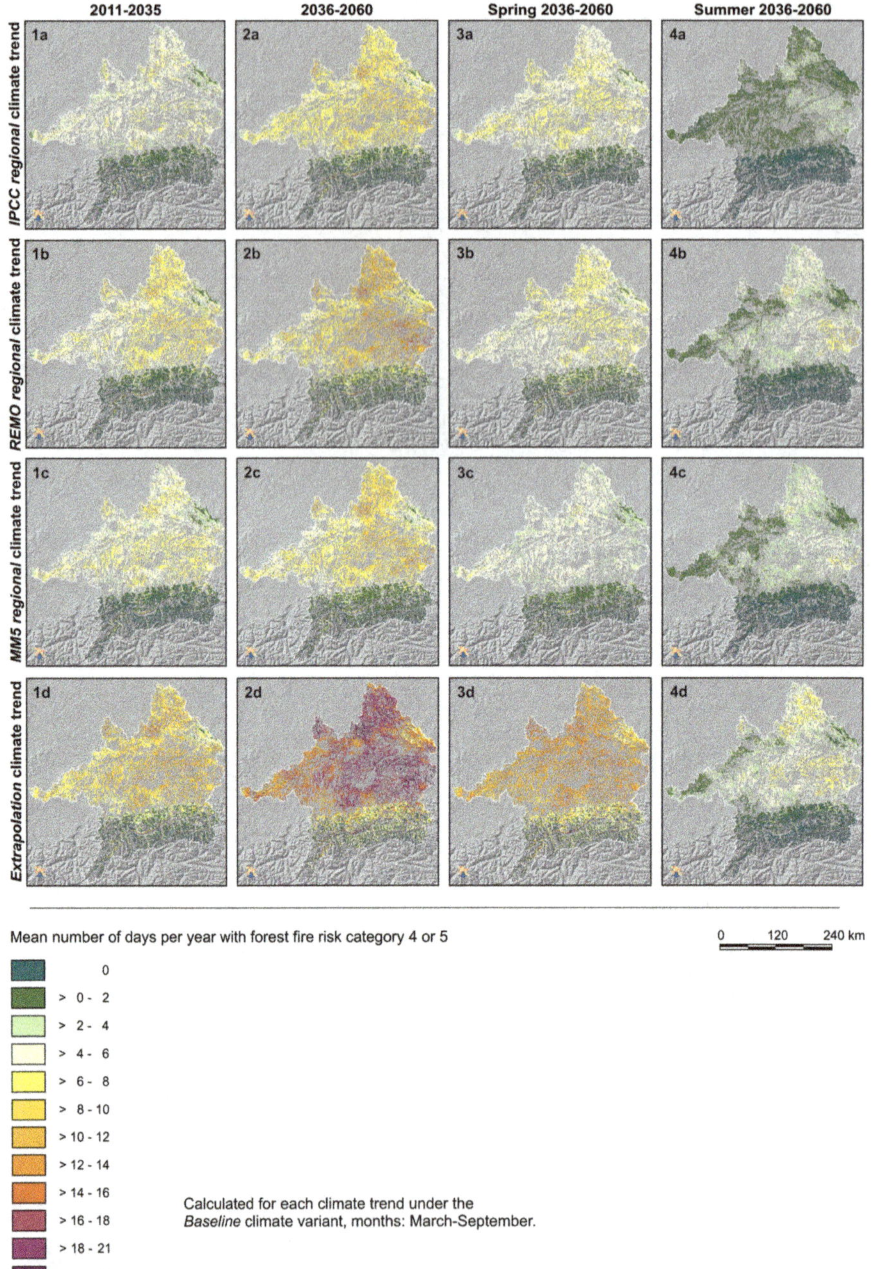

Map 75.1 Scenario analysis of the trend in climate-related forest fire risk

means that the risk of forest fire increases in the months June–August (see Fig. 74.7, Chap. 74). Map 75.1 – 4a–4d indicate the regions that are especially affected for the period 2036–2060.

The calculations under the *IPCC regional* climate trend (see Map 75.1 – 1a–4a) result in the comparatively weakest increase in forest fire risk. The mean for the years 2011–2035 suggest even a slight decline compared to the regional mean today (2011–2035, 4.2 days; 1991–2000, 4.3 days). However, in the Inn valley, maximum values (17.9 days) are reached that are more than a quarter greater than those from the past. With the *IPCC regional* climate trend, an increase by 44 % in the forest fire risk over the regional average was calculated for the period 2036–2060 with unchanged maxima (see Map 75.1 – 2a). In this example, the small number of days with forest fire risk in the summer compared to the other climate trends is notable. One reason for this result is the lower number of summery days and very hot days in the *IPCC regional* climate trend (daily high temperature ≥25 °C and ≥30 °C) (see Map 48.4). The forest fire risk under the *REMO regional* climate trend is classified around a quarter higher compared to under the *IPCC regional* climate trend, for both study periods (see Map 1b–2b). The mean number of days with a high or exceptionally high forest fire risk during the growing season is 5.3 days for the period 2011–2035. For the period 2036–2060, this number increases to 7.5 days. The *REMO regional* climate trend is based on a temperature increase of +3.7 °C and a precipitation increase of +9.1 % for the spring. While the risk of forest fire in the spring does not differ significantly from the estimate from the *IPCC regional* climate trend, the average for the months June to August is classed twice as high. The sharp decline in precipitation of −31.5 %, along with a temperature increase of +5.3 °C during the summer months under the *REMO regional* climate trend, is largely responsible for this difference.

The estimates for the forest fire risks under the *MM5 regional* scenario (see Map 75.1 – 1c–4c) lie between the values calculated under the *IPCC regional* and *REMO regional* climate trends. Map 75.1 – 3c shows only a slight increase of 7 % in the forest fire risk for the period 2011–2035 compared to the past (see Map 74.1 – 4a in Chap. 74). Based on the *MM5 regional* scenario, there is a further 33 % increase with maximum values of 18 days for the period 2036–2060. The risk levels for the spring months in the period 2036–2060 under *MM5 regional* are classified as lower compared to the other climate trends. In this scenario, a precipitation increase of +13.1 % is assumed for the spring months. A proportionally low temperature increase of +3.2 °C in the spring also contributes to the lower climate-related forest fire risk levels. Under the *MM5 regional* scenario, 38 % of days have a forest fire risk in summer for the years 2036–2060 (for comparison: under *REMO regional* 35 %, *IPCC regional* 21 % and *Extrapolation* 25 %). The *MM5 regional* climate trend is based on a precipitation decrease of −28.7 % and a temperature increase of +5.8 °C in summer over the years 1990–2100.

Map 75.1 – 1d–4d depict a classification of the level of forest fire risk under the *Extrapolation* climate trend that is twice as high as under the other climate trends, for both periods. Already within the period 2011–2035, the forest fire risk is on average 75 % higher compared to the reference decade (see Map 74.1 – 4a in Chap. 74).

The *Extrapolation* climate trend also significantly differs from the other climate variants in the period 2036–2060, because the risk is almost doubled, at 13.6 days and an increase of 81 % over the period 2011–2035. On Map 2d, values below 15 days with few exceptions are only found in the Alps, the central uplands and in the western parts of the basin. Approximately 25 % of days with forest fire risk occur in the summer months (see Map 75.1 – 4d).

75.4 Conclusion

The analysis of the forest fire risks under the *IPCC regional, REMO regional* and *MM5 regional* climate trends lead to similar results. The most differences arise under the climate trend *Extrapolation*. Under this trend, the forest fire risks are classified as almost twice as high compared to the other scenarios. This tendency is particularly striking in the distant future. However, all models indicated very similar spatial distributions for the classification of risk. The highest values in all climate trends were calculated for Franconia (especially in the Nordgau region), the Hohenloher Plain, in the eastern Tertiary Hills and in the inner alpine dry valleys. The high estimate of forest fire risk under the *Extrapolation* climate trend is primarily the result of the strong decline in precipitation in spring (−42 %) and in summer (−69 %) (see Chap. 48). In contrast, there is a slight increase in precipitation under the *IPCC regional, REMO regional* and *MM5 regional* climate trends. Under the *MM5 regional* climate trend, the precipitation increase in spring is even +13.1 % over the years 1990–2100. Thus, it is clear that in regions with the highest forest fire risks, the increase is primarily attributable to the increased risk in summer. On average across the region and over all 16 scenarios, 30 % more days with high forest fire risk are projected to occur in the summer months during the period 2036–2060.

Reference

Baumgartner A, Klemmer L, Raschke E, Waldmann G (1967) Waldbrände in Bayern 1950–1959. Mitteilungen der Staatsverwaltung Bayern 36, München

Chapter 76
Effects of Future Climate Trends on Crop Management

Tatjana Krimly, Josef Apfelbeck, Marco Huigen, and Stephan Dabbert

Abstract To assess the impact of climate change and variability in the weather on the timing of crop management activities, dynamically coupled model runs of the DANUBIA components *Biological*, *SNT*, *NaturalEnvironment* and *Farming* were performed. Calculations are based on a GLOWA-Danube scenario including the climate trend REMO regional, the climate variant Baseline and the societal scenario Baseline. Results of the scenario calculation are compared with the reference period for four sample districts, which represent different site conditions within the drainage basin. The increase in air temperature over the scenario period leads to a shortening of the growing period of winter wheat and spring barley and, therefore, also for the cultivation periods. With very similar sowing dates, on average, the harvest of winter wheat is earlier by 21 days at the end of the scenario period compared to the reference period for all districts. Furthermore, the harvest dates for winter and spring cereals move closer to each other.

Keywords GLOWA-Danube • Climate change • Crop management activities • Decision algorithms • Dynamic model coupling • Scenario calculation

76.1 Introduction

The timing of crop management, with activities including sowing/planting, fertilisation and harvest, depends to a large extent on climate and weather conditions, which in turn influence plant development. The span of time in which different crop management activities typically occur under specific climate conditions are known as fieldwork intervals (KTBL 1996).

The goal of the study presented in this Chapter is to assess the impact of climate changes and variability in the weather on the fieldwork intervals in crop growing, especially in the cultivation of cereals, and the resulting consequences for agriculture.

T. Krimly (✉) • J. Apfelbeck • M. Huigen • S. Dabbert
Production Theory and Resource Economics, Universität Hohenheim, Stuttgart, Germany
e-mail: T.Krimly@uni-hohenheim.de; josefapfelbeck@gmx.net; marco_huigen@hotmail.com; dabbert@uni-hohenheim.de

W. Mauser, M. Prasch (eds.), *Regional Assessment of Global Change Impacts*, DOI 10.1007/978-3-319-16751-0_76

76.2 Scenario Assumptions

The results presented in this Chapter are based on the following GLOWA-Danube scenario:

- *REMO regional* climate trend: a temperature increase of 5.2 °C and a change of precipitation of −4.9 % in winter and −31.4 % in summer (see Chaps. 47 and 48).
- *Baseline* climate variant: the criterion is the mean annual temperature between 2011 and 2035 (see Chaps. 49 and 50).
- *Baseline* societal scenario: the *business-as-usual* scenario, which continues with the status quo (see Chaps. 52 and 70).

76.3 Model Components Involved

The calculations were carried out using the dynamically coupled DANUBIA *Biological*, *SNT*, *NaturalEnvironment* and *Farming* components. *NaturalEnvironment* provides the necessary meteorological and hydrological site factors that influence plant growth, nitrogen flows and the processes of nitrogen turnover as well as the dates for crop management activities (see Chap. 4). *SNT* models the plant available nitrogen (see Chap. 38), and *Biological* simulates plant growth (see Chap. 37). Simulations of the coupled models run on a daily basis. *DeepFarming* combines the crop management activities with the crops' development stages simulated in the plant growth model and the weather conditions and soil saturations as measures of passability from *NaturalEnvironment* (see Chaps. 72 and 40). *DeepFarming* calculates temporal changes of the different crop management activities as a result of the climate conditions and plant growth.

The following decision algorithms and thresholds are involved in the decision-making by the actors in the *DeepFarming* model regarding the execution of the different crop management activities:

- For sowing/planting, which starts on a specific date in the year for each crop type, the soil temperature must be >3 °C and the passability of the field must be ensured.
- For each planned application of nitrogen, a crop must have reached a specified development stage. Among the cereal crops, only winter cereals receive three applications. The first application takes place at the beginning of the growing season/tillering, the second at bolting and the third at boot stage. Winter barley and rye receive one application at the beginning of the growing season/tillering and one at bolting. Spring barley receives one application at sowing and another at tillering, spring wheat at sowing and bolting. Oats are fertilised only once at sowing. Fertilising can only be executed on days with morning precipitation (between 04:00–08:00) less than 4 mm, since precipitation has a significant impact on the passability of the field. One further condition involves the soil

temperature. According to LfL (2007), application of fertiliser should not take place if the soil is saturated with water, frozen or evenly covered with more than 5 cm of snow.

- Harvest can be undertaken as soon as the crop has reached the stage necessary for this (for cereals this stage is known as full ripeness), if the morning precipitation is less than 4 mm and the field is passable. The morning precipitation is thus an important factor in the decision, since with too much precipitation, the harvest requires more than a day to dry. If these conditions are not met by a certain date, the harvest is still executed in *DeepFarming* to ensure a smooth cycle of the crop rotation in the model.

For the results presented below, it is assumed that the crop management activities are also carried out if the passability of the field is limited (i.e. soil is moist but still passable with the necessary machinery).

76.4 Results

The results of the reference situation (1996–2005) and the scenario calculation (2012–2058) are presented as time series. For each district studied and for each year, the average date (day of the year) on which the management activity was performed was calculated. The analysis included sowing, fertilisation (up to three applications) and harvest. It should be noted that the results do not take into account certain factors that limit plant growth, such as weeds, pests or extreme weather events.

Four sample districts were chosen to present the results; these four districts represent different site-specific climate conditions within the basin (see Map 76.1 – 1a–c and Table 70.2 in Chap. 70). On average, the district of Dingolfing has the highest mean annual temperature and the lowest annual precipitation over the reference period. The Ostallgäu district represents a much cooler district with more precipitation.

The different climate conditions in the districts are reflected in the patterns of crop growth. The results of the *Biological* model reveal longer growing periods for cereals from sowing to harvest in the cooler districts (Landsberg and Ostallgäu) compared to the warmer districts (Dingolfing and Günzburg). Below, the interval between sowing and harvest in the *DeepFarming* model is designated the cultivation period; this interval depends on both the development stage of each crop and the prevailing weather conditions. Considering winter wheat and spring barley as examples, the district of Dingolfing has on average the shortest cultivation period during the reference period, while the district of Ostallgäu has the longest cultivation period (see Table 76.1).

Also, over the reference period, the range of variation in the cultivation period for winter wheat is significantly higher in Ostallgäu (−23 to +26 days) compared to in Dingolfing (−16 to +16 days) (see Fig. 76.1). This is attributed to the temperature,

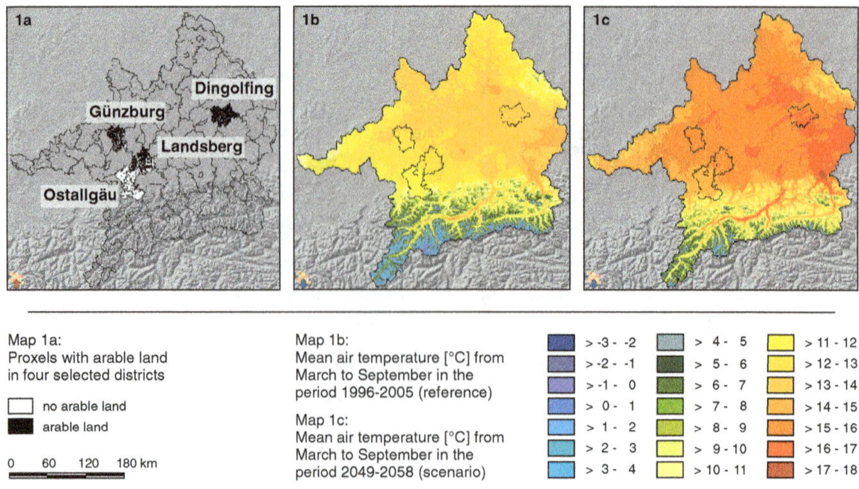

Map 76.1 Proxels with arable land and mean air temperature for the years 1996–2005 (reference) and 2049–2058 (scenario) in the selected districts

Table 76.1 Cultivation periods of winter wheat and spring barley in the selected districts in the reference period (1996–2005) and at the end of the scenario period (2049–2058)

District	Dingolfing	Günzburg	Landsberg	Ostallgäu
	Mean value 1996–2005/2049–2058 for winter wheat			
Sowing day[a]	293/293	293/293	294/293	293/293
Harvest day	214/196	220/200	223/201	233/206
Cultivation period in days	286/268	292/272	294/273	305/278
	Mean value 1996–2005/2049–2058 for spring barley			
Sowing day	71/65	70/65	70/65	73/65
Harvest day	197/189	202/185	205/188	214/194
Cultivation period in days	126/124	132/120	135/123	141/129

[a]In the preceding year of harvest

which is responsible for inhibiting or accelerating the development of plants. Hence, for example, in Ostallgäu in 1996 (the coldest year in the reference period), a mean daily temperature of only 10.3 °C was reached during the main growing season from March to September, whereas in Dingolfing, the temperature was much higher (11.8 °C).

Overall, there were no delays in the harvest and fertilisation management activities as a result of unfavourable weather conditions during the reference period in any of the districts that were analysed. There were delays in the date of sowing only for spring barley in some years, as a result of the very low soil temperatures in spring, especially in the district of Ostallgäu (see Fig. 76.2).

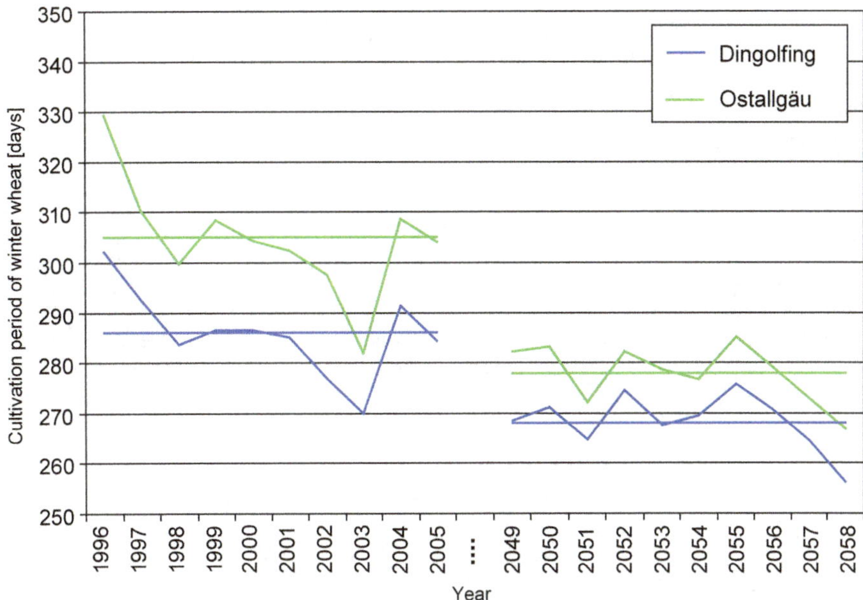

Fig. 76.1 Cultivation periods of winter wheat with the respective mean values for the reference period (1996–2005) and the end of the scenario period (2049–2058)

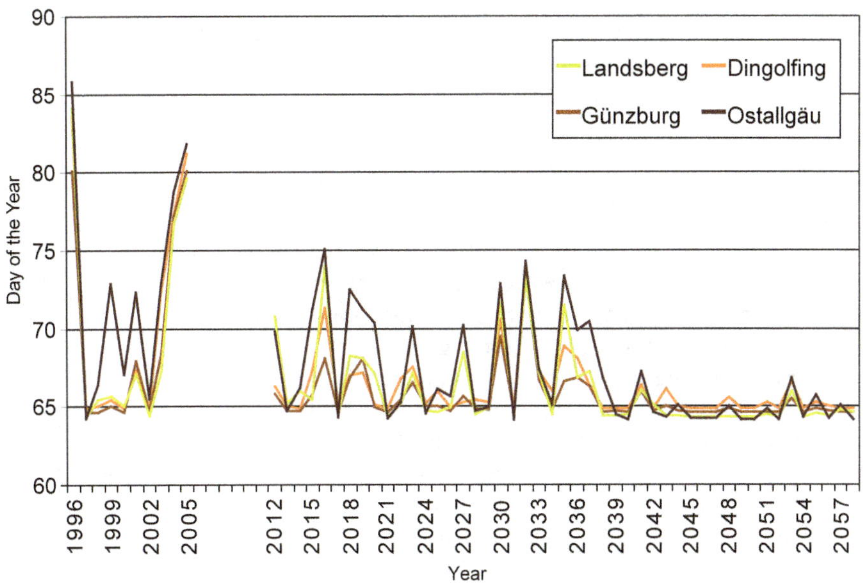

Fig. 76.2 Mean sowing date for spring barley in the selected districts for the reference period (1996–2005) and the scenario period (2012–2058)

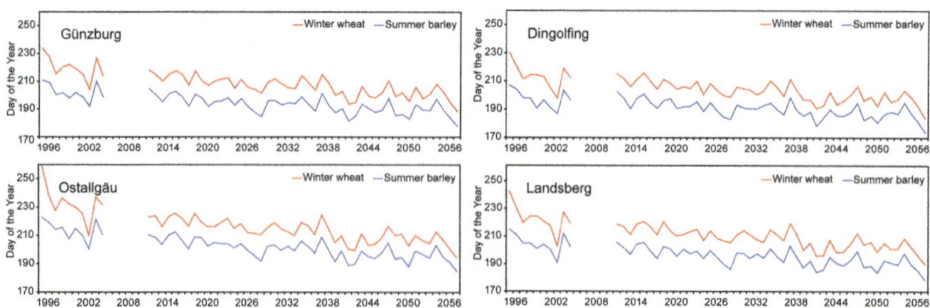

Fig. 76.3 Harvest date for spring barley and winter wheat in the reference and the scenario period in the selected districts of Günzburg (*upper left*), Dingolfing (*upper right*), Ostallgäu (*lower left*) and Landsberg (*lower right*)

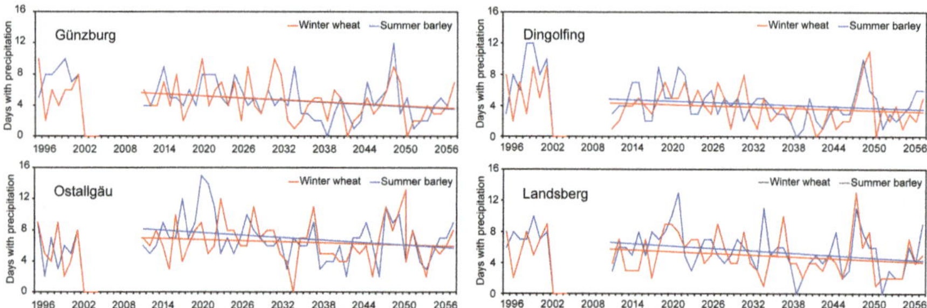

Fig. 76.4 Number of days with precipitation ±10 days around the harvest date in the reference and the scenario period in the selected districts of Günzburg (*upper left*), Dingolfing (*upper right*), Ostallgäu (*lower left*) and Landsberg (*lower right*)

In general, the scenario results from the *Biological* plant growth model indicate a significant shortening of the simulated growing period for cereals up to the end of the scenario period. This is the result of the increase in air temperature (see Map 76.1 – 1b, c). The cultivation periods are thus also shortened (see Table 76.1). Over the scenario period, the range of variation for the cultivation period decreases in all districts to −11 to +8 days (see Fig. 76.1).

With very similar sowing dates, on average, the harvest of winter wheat is therefore earlier by 21 days for all four districts over the period 2049–2058 compared to the reference period (see Fig. 76.3). The smallest change is in Dingolfing (−18 days), and the greatest change is in Ostallgäu (−27 days). This trend towards earlier harvest dates is apparent for all winter and spring cereals. At the same time, the harvest dates for winter and spring cereals move closer together (see Fig. 76.3). As shown in Fig. 76.4, the planning reliability for the harvest date is diminished. Over the scenario period, the number of rain days within the period around the harvest date (±10 days) decreases, but the variability increases.

References

KTBL (1996) Taschenbuch Landwirtschaft 1996/97. Landwirtschaftsverlag, Münster-Hiltrup

LfL (Bayerische Landesanstalt für Landwirtschaft) (2007) Leitfaden für die Düngung von Ackerund Grünland. Bayerische Landesanstalt für Landwirtschaft, Freising

Index

W. Mauser, M. Prasch (eds.), *Regional Assessment of Global Change Impacts*, DOI 10.1007/978-3-319-16751-0

CPSIA information can be obtained
at www.ICGtesting.com
Printed in the USA
LVHW080306290620
659242LV00005B/715